Introduction to the Mechanics and Physics of Solids

Introduction to the Mechanics and Physics of Solids

E W Billington

Adam Hilger Ltd
Bristol and Boston

British Library Cataloguing in Publication Data
Billington, E.W.
Introduction to the mechanics and physics of solids
1. Solid state physics
I. Title
530.4′1 QC176

ISBN 0-85274-491-9

Consultant Editor: Mr A E de Barr

Published by Adam Hilger Ltd
Techno House, Redcliffe Way, Bristol BS1 6NX, England
PO Box 230, Accord, MA 02018, USA

Typeset by Mid-County Press, London and printed in Great Britain
by J W Arrowsmith Ltd, Bristol

Contents

Preface

Because of their emphasis on basic concepts and fundamental principles, the nonlinear field theories of mechanics have an important role in modern engineering and technology. A unified discussion of these topics usually comes under the general title of nonlinear continuum mechanics. The purpose of this book is to present an alternative approach to the formulation and application of the constitutive equations describing the mechanical response of isotropic solids, no account being taken of thermodynamic restrictions, the proposed constitutive equations being purely mechanical. Furthermore, no attempt is made to take account of the subcontinuum mechanisms of deformation, that is the micromechanics of the material. Wherever possible, direct notation has been used. This has the advantage of recording physical statements in a form unencumbered by mathematical details needed only for calculations. To make the book as self-contained as possible, a discussion of tensor analysis and algebra is given in Chapter 1. Those aspects of kinematics together with the physical principles basic to any discussion of the subject are described in Chapter 2.

The development of contemporary nonlinear continuum mechanics is to a large extent centred on the concept and formulation of various deformation tensors. For an isothermal, mechanical system, stress is related to the deformation tensors through the concept of a response function, the resulting constitutive equations being referred to simply as stress relations, thus recognising that in this particular formulation the classical concept of strain is no longer relevant. However, in the context of mechanical engineering, it is often convenient, and for some systems essential, to recognise that stresses produce strains. Thus, the question arises of whether it is possible to rearrange the response function in such a way as to obtain a stress–strain relation. Such a rearrangement is possible using the Cayley–Hamilton theorem. This approach gives rise to two identically equivalent representations of the response function. When expressed in terms of the deformation tensors, the response function is referred to as the *deformation response function*, and when expresed in terms of a new measure of finite strain the response function is referred to as the *strain response function*. The formulation of the constitutive equations in terms of the two equivalent representations of the response function is considered in Chapter 4.

Plastic distortion, that is yielding, has the effect of separating the characteristic stress–strain curve into an elastic range and a plastic range. Initial yield is the upper limit of the purely elastic deformation behaviour of the material. Thus, the choice of the constitutive equation describing the purely elastic deformation behaviour determines the initial yield function. Classical plasticity theory assumes the purely elastic behaviour to be described by the classical theory of infinitesimal elasticity in the form of the generalised Hooke's law. This assumption restricts the classical theory of plasticity to those materials which satisfy von Mises' yield criterion. There is, however, a growing body of experimental evidence for the existence of materials which do not satisfy von Mises' criterion. Thus, the question arises of whether it is possible to develop a constitutive equation for a non-simple material which will satisfy a more general initial yield criterion. The development of a generalised, isotropic, yield criterion is considered in Chapter 3. Its incorporation into the constitutive equations by way of the concepts of a stress intensity function and a generalised loading function is considered in Chapter 4. The general deformation and flow properties of a class of elastic–plastic solids are considered in Chapter 5.

Some universal solutions for a class of elastic–plastic solids, set within the spatial description, are discussed in Chapter 6.

In the solution of particular problems, the analysis can on occasion be simplified by transforming the stress relation from the current configuration back to the reference configuration. This approach gives rise to the problem of determining an appropriate representation of the stress. The classical approach is by way of the first Piola–Kirchhoff stress tensor, or its transpose which is referred to as the nominal stress tensor. However, the nominal stress tensor, which involves both the current configuration of a material body and a reference configuration, is not in general a symmetric tensor. Because it is not symmetric it is difficult to incorporate into the constitutive relation between stress and strain. In this context, a more useful form is the second Piola–Kirchhoff stress tensor which is symmetric and is concerned solely with the reference configuration. However, no direct physical interpretation can be given for the second Piola–Kirchhoff stress tensor. Thus, there remains the problem of defining a measure of stress which will provide a representation that, while transformed back to the reference configuration, is still a symmetric tensor and for which a physically acceptable interpretation can be given. The formulation of a generalised referential stress tensor is discussed in Chapter 8. It is shown that in principle there are an infinite number of referential stress tensors, of which the simplest, referred to as the first referential stress tensor, is just the second Piola–Kirchhoff stress tensor. The next in sequence, referred to as the second referential stress tensor, is the only referential stress tensor to have physical significance. The interpretation of the second referential stress tensor differs from that of the nominal stress by a rigid rotation.

The predictions discussed in Chapters 6 and 8 regarding material behaviour, are compared with the results of experiment in Chapters 7 and 9.

Central to the discussion of the properties of the proposed constitutive equations is the use of Lode's stress and strain parameters, and two further parameters which have been referred to as Lode's generalised stress and strain parameters. Lode originally introduced the parameters in the context of classical plasticity theory which is set in the spatial description. Analagous parameters have been introduced in Chapter 8, in respect of the referential description. The use of these parameters greatly simplifies the analysis of various well defined simple modes of deformation.

E W Billington
PO Box 2, Hampton,
Middx TW12 2LQ
July 1985

Acknowledgments

A list of general references are given at the end of each chapter. Full historical notes and references to the literature can be found in the works of Truesdell and Touplin, and Truesdell and Noll. The author wishes to acknowledge here his indebtedness to the encyclopedia articles of Truesdell and Toupin, and Truesdell and Noll. Selected references, restricted to the particular topics considered in this book, are also given at the end of each chapter.

The author wishes to thank the following for permission to use figures in this book:

Pergamon Press for figures 7.1, 7.2, 7.3, 7.8, 7.9, 7.10, 7.11 and 7.20; G Lianis, H Ford and Pergamon Press for figures 7.5 and 7.6; the Institute of Physics for figures 7.16, 7.17, 7.18 and 7.19; Springer for figures 9.1, 9.2 and 9.3. The author is particularly indebted to Dr A Foux for permission to use his measurements of the Poynting effect to prepare figure 7.8.

Conventions and Notation

In direct notation the distinction between vectors and tensors has been made by use of a different type of letter. Capital letters in direct notation denote tensors and lower case letters in direct notation denote vectors. The only exceptions to this system are those physical quantities denoted by symbols whose usage is sufficiently general to justify their retention. The conventions regarding the type of letters used for symbols are listed below:

points: $X \quad Y \quad Z \qquad x \quad y \quad z$.
Scalars: $A \quad a$.
Vectors: $\boldsymbol{A} \quad \boldsymbol{a}$.
Tensors $\mathbf{A} \quad \mathbf{a}$.
Sets, groups, bodies and regions: $\mathscr{A}$.

Frequently used symbols are listed below; the section in which a quantity first appears is given in brackets.

$\mathbf{B}$	B^{ij}	left Cauchy–Green tensor (2.1.3)
	B	configuration of a body (2.1.1)
	$\mathscr{B}$	body (2.1.1)
$\mathbf{C}$	$C_{\alpha\beta}$	right Cauchy–Green tensor (2.1.3)
$\mathbf{D}$	D_{ij}	stretching tensor (2.1.4)
	D	deformation constant (6.5.2)
$\mathbf{E}$	$E_{\alpha\beta}$	Green–St Venant strain tensor (8.3.2)
	E_0	Young's modulus (6.3.3)
$\bar{\mathbf{E}}$	$\bar{E}^{ij}$	apparent finite strain: spatial description (4.2.2)
$\hat{\mathbf{E}}$	$\hat{E}_{\alpha\beta}$	apparent finite strain: referential description (8.3.2)
$\mathbf{F}$	$F^i{}_\alpha$	deformation gradient tensor (2.1.3)
$\mathbf{F}_t$	$F^i_{t\alpha}$	relative deformation gradient tensor (2.1.3)
$\boldsymbol{F}$	F_i	force (2.2.1)

$\mathbf{G}$	G_{ij}	function of the stress tensor $\mathbf{T}$ (4.2.2)
$\mathbf{G}'$	$G'^{i}{}_{j}$	function of the deviatoric stress tensor $\mathbf{T}'$ (4.2.1)
$\hat{\mathbf{G}}'$	$\hat{G}'_{\alpha\beta}$	function of the deviator of the second referential stress tensor $\hat{\mathbf{T}}'$ (8.3.2)
$\boldsymbol{G}$	G^i	body couple density (2.2.1)
G		tensor valued functional (2.3.1)
	G_0	classical shear modulus of rigidity (4.5)
	$\bar{G}$	generalised shear modulus: spatial description (4.2.2)
	$\hat{G}$	generalised shear modulus: referential description (8.3.2)
	$\mathscr{H}$	constitutive functional of a material (2.3.1)
$\boldsymbol{H}$	H^i	angular momentum (2.1.1)
	$H(R_t)$	rate at which thermal energy is being fed into the system (2.2.4)
	$J'_i\ (i=2,3)$	invariants of the deviatoric stress tensor $\mathbf{T}'$ (3.1.1)
	$\hat{J}'_i\ (i=2,3)$	invariants of the deviator of the second referential stress tensor $\hat{\mathbf{T}}'$ (8.3.2)
	J	Jacobean of deformation (2.1.3)
$\mathbf{K}$	$K_{\alpha\beta}$	response function: referential description (8.3.2)
$\hat{\mathbf{K}}'$	$\hat{K}'_{\alpha\beta}$	response function, $(\mathbf{K}'=2\hat{G}\hat{\mathbf{K}}')$ (8.3.2)
	$K(R_t)$	kinetic energy of the material occupying the region R_t (2.2.4)
	$K'_i\ (i=2,3)$	invariants of the deviator of the apparent finite strain $\bar{\mathbf{E}}'$ (4.2.2)
	$\hat{K}'_i\ (i=2,3)$	invariants of the deviator of the apparent finite strain $\hat{\mathbf{E}}'$ (8.3.2)
$\mathbf{L}$	L_{ij}	velocity gradient (2.1.4)
$\boldsymbol{L}$		vector-valued functional (2.3.1)
$\mathbf{M}$	M^{ij}	response function: spatial description (4.2.2)
$\bar{\mathbf{M}}'$	$\bar{M}'_{ij}$	response function, $(\mathbf{M}'=2\bar{G}\bar{\mathbf{M}}')$ (4.2.3)
$\boldsymbol{N}$	N_α	unit normal vector (2.2.2)
	P	pressure (2.3.4)
	$P(R_t)$	mechanical power (rate at which work is being done on the material) (2.2.4)
	$\mathscr{P}$	part of a body (2.1.3)
$\mathbf{Q}$		arbitrary orthogonal tensor (1.6.8)
	$R^{\rho}_{\alpha\beta\gamma\delta}$	Riemann–Christofell tensor (1.8)
$\mathbf{R}$	$R^{i}{}_{\alpha}$	local rotation tensor (2.1.3)
$\mathbf{S}$		displacement gradient tensor (4.5)
$\tilde{\mathbf{S}}$		infinitesimal displacement gradient (4.5)

	S	entropy (2.2.4)
	S_t	material surface in the current configuration (2.2.2)
$\mathbf{T}$	T^{ij}	Cauchy's stress tensor (2.2.3)
$\mathbf{T}^0$	$(T^0)^{\alpha i}$	nominal stress tensor (2.2.3)
$(\mathbf{T}^0)^{\mathrm{T}}$		first Piola–Kirchhoff stress tensor (2.2.3)
$\tilde{\mathbf{T}}$	$\tilde{T}^{\alpha\beta}$	second Piola–Kirchhoff stress tensor (2.2.3), (also called the first referential stress tensor)
$\hat{\mathbf{T}}$	$\hat{T}^{\alpha\beta}$	second referential stress tensor (8.2.1)
$\mathbf{T}_{\mathrm{E}}$		extra stress tensor (2.3.4)
$\mathbf{U}$	$U_{\alpha\beta}$	right stretch tensor (2.1.3)
	$U(R_t)$	internal energy of the material occupying the region R_t (2.2.4)
$\mathbf{V}$	V^{ij}	left stretch tensor (2.1.3)
$\mathbf{W}$	W_{ij}	spin tensor (2.1.4)
$\boldsymbol{X}$	X^{α}	body point in the reference configuration (2.1.1)
	X	typical body point (2.1.1)
	Y	yield stress in simple uniaxial tension (and compression) (3.1.3)
	X, Y, Z	Cartesian coordinates of $\boldsymbol{X}$ (2.1.1)
$\boldsymbol{a}$	a^i	acceleration (2.1.1)
$\boldsymbol{b}$	b^i	body force per unit mass (2.2.1)
$\boldsymbol{c}$	c^i	surface couple density (2.2.1)
$\mathbf{e}$	e_{ij}	Almansi–Hamel strain tensor (4.5)
	$f(\mathbf{T})$	yield function (3.1.1)
$\boldsymbol{h}$	h^i	total angular momentum density (2.2.1)
	h	rate of flow of thermal energy per unit area into the material occupying the region R_t (2.2.4)
$\boldsymbol{j}$	j^i	flux per unit area (2.2.6)
	k	yield stress in pure shear (3.1.3)
$\boldsymbol{m}$	m^i	internal angular momentum (2.2.5)
$\boldsymbol{p}$	p^i	linear momentum (2.2.1)
$\boldsymbol{q}$	q^i	heat vector (directed opposite to the flux of heat) (2.2.4)
	r	heat supply or source per unit mass and time (2.2.4)
	s	elapsed time (2.3.1)
$\boldsymbol{t}$	t^i	stress or traction vector (2.2.3)
	$\hat{t}_{\alpha}$ $(\alpha=1, 2, 3)$	principal referential stresses (8.2.1)
	t	present time (2.1.1)
$\boldsymbol{u}$	u^i	displacement vector (4.5)

$\boldsymbol{v}$	v^i	velocity (2.1.1)
	x^i, y^i, z^i	spatial coordinates (2.1.1)
	x, y, z	Cartesian coordinates of $\boldsymbol{x}$ (1.7.3)
$\boldsymbol{\Gamma}$	Γ^i	applied torque (2.2.1)
	Ω	dimensionless ratio, $(27K_3'^2/4K_2'^3)$ (4.3.2)
	$\hat{\Omega}$	dimensionless ratio, $(27\hat{K}_3'^2/4\hat{K}_2'^3)$ (8.4.2)
	Θ	polar coordinate in the referential description (6.5.2)
	$\tilde{\alpha}_i$ $(i=0, 1, 2)$	response coefficients (4.2.2)
	β_i $(\mathrm{i}=0, 1, 2)$	response coefficients (4.1.1)
	γ	shear strain (6.7)
	γ	entropy production per unit mass (2.2.4)
	δ_{ij}	Kronecker delta (1.5.1)
	δ_i $(\mathrm{i}=0, 1, 2)$	loading coefficients: spatial description (4.2.1)
	∂	boundary operator (1.10)
	ε	internal energy per unit mass (2.2.4)
	ζ	stress ratio (6.5.1)
	η	ratio of current volume to undeformed volume of a thin-walled tube (6.5.2)
	θ	absolute temperature (2.2.4)
	θ	polar coordinate in the spatial description (1.7.3)
$\boldsymbol{\theta}$		deformation function (2.1.3)
	κ	deformation parameter based on Lode's strain parameter v (4.1.2)
	λ_i $(i=1, 2, 3)$	principal stretches (2.1.3)
	μ	Lode's stress parameter: spatial description (4.1.2)
	$\bar{\mu}$	generalised form of Lode's stress parameter: spatial description (4.2.3)
	$\hat{\mu}$	Lode's stress parameter: referential description (8.3.1)
	$\tilde{\mu}$	generalised form of Lode's stress parameter: referential description (8.3.2)
	v	Lode's strain parameter: spatial description (4.2.3)
	$\bar{v}$	generalised form of Lode's strain parameter: spatial description (4.1.2)
	$\hat{v}$	Lode's strain parameter: referential description (8.3.2)
	$\tilde{v}$	generalised form of Lode's strain parameter: referential description (8.3.1)
	ξ	stress ratio (6.5.1)
	ρ	mass density: spatial description (2.2.1)

	ρ_{r}	mass density: referential description (2.2.1)
	σ_i $(i=1,2,3)$	principal stresses (2.2.3)
	τ	past time (2.3.1)
	$\tau_{(\mathrm{s})}$	principal shearing stress (2.2.3)
$\boldsymbol{\phi}$	ϕ^i	function identifying a motion (2.1.1)
	ϕ	polar coordinate (1.7.3)
	φ_i $(i=0,1,2)$	response coefficients: spatial description (4.2.3)
	$\hat{\varphi}_i$ $(i=0,1,2)$	response coefficients: referential description (8.3.2)
$\boldsymbol{\chi}$	χ^i	configuration (2.1.1)
$\boldsymbol{\chi}_t$	χ_t^i	relative deformation function (2.1.3)
	ω	dimensionless ratio, $(27J_3'^2/4J_2'^3)$ (3.1.3)
	$\hat{\omega}$	dimensionless ratio, $(27\hat{J}_3'^2/4\hat{J}_2'^3)$ (8.3.2)
	$\mathscr{O}$	orthogonal group (2.3.3)
	$\mathscr{G}$	isotropy group (2.3.3)

1

Mathematical Preliminaries

1.1 MATRICES

In this section the main properties of matrices are quoted without proof; (for proofs of the theorems the reader is referred to the books in the bibliography).

An $m \times n$ *matrix* $\boldsymbol{A}$ is a rectangular array of real or complex numbers A_{ij} arranged in m rows and n columns. Thus, using the convention that the first index denotes the row and the second index the column

$$\boldsymbol{A} = \begin{bmatrix} A_{11} & A_{12} & \dots & A_{1n} \\ A_{21} & A_{22} & \dots & A_{2n} \\ \vdots & \vdots & & \vdots \\ A_{m1} & A_{m2} & \dots & A_{mn} \end{bmatrix} \equiv [A_{ij}]. \tag{1.1.1}$$

A matrix of m rows and n columns is said to be of order m by n, or $m \times n$. The numbers A_{11}, A_{12}, etc are the *elements* of the matrix.

A matrix, every element of which is zero, is called a *zero matrix* or *null matrix*, denoted by $\mathbf{0}$.

The matrix,

$$\boldsymbol{A} = \begin{bmatrix} A_{11} \\ A_{21} \\ \vdots \\ A_{m1} \end{bmatrix} \equiv \begin{bmatrix} a_1 \\ a_2 \\ \vdots \\ a_m \end{bmatrix} \qquad (A_{i1} \equiv a_i) \tag{1.1.2}$$

is called a *column matrix* or *column vector*, of order $m \times 1$. A matrix

$$\boldsymbol{B} = [B_{11} \;\; B_{12} \;\; \dots \;\; B_{1n}] \equiv [b_1 \;\; b_2 \;\; \dots \;\; b_n] \qquad (B_{ij} \equiv b_j)$$

is a *row matrix* or *row vector* of order $1 \times n$.

Matrix addition and subtraction is defined only for matrices of the same order. They are then said to be *conformable* or *compatible* for addition and subtraction. Thus, if

$$\boldsymbol{A}=\begin{bmatrix} A_{11} & A_{12} & A_{13} \\ A_{21} & A_{22} & A_{23} \end{bmatrix} \qquad \boldsymbol{B}=\begin{bmatrix} B_{11} & B_{12} & B_{13} \\ B_{21} & B_{22} & B_{23} \end{bmatrix} \tag{1.1.3}$$

then

$$\boldsymbol{A}+\boldsymbol{B}=\begin{bmatrix} A_{11}+B_{11} & A_{12}+B_{12} & A_{13}+B_{13} \\ A_{21}+B_{21} & A_{22}+B_{22} & A_{23}+B_{23} \end{bmatrix}. \tag{1.1.4}$$

If λ is a number and $\boldsymbol{A}$ is a matrix, then $\lambda\boldsymbol{A}$ is a matrix given by

$$\lambda\boldsymbol{A}=\begin{bmatrix} \lambda A_{11} & \lambda A_{12} & \dots & \lambda A_{1n} \\ \lambda A_{21} & \lambda A_{22} & \dots & \lambda A_{2n} \\ \vdots & \vdots & & \vdots \\ \lambda A_{m1} & \lambda A_{m2} & \dots & \lambda A_{mn} \end{bmatrix} \equiv [\lambda A_{ij}]=\boldsymbol{A}\lambda. \tag{1.1.5}$$

Therefore

$$-\boldsymbol{A}=(-1)\boldsymbol{A}=[-A_{ij}]. \tag{1.1.6}$$

These definitions of addition and subtraction and multiplication by a number imply that

$$\boldsymbol{A}+\boldsymbol{B}=\boldsymbol{B}+\boldsymbol{A} \tag{1.1.7}$$

$$\boldsymbol{A}+(\boldsymbol{B}+\boldsymbol{C})=(\boldsymbol{A}+\boldsymbol{B})+\boldsymbol{C} \tag{1.1.8}$$

$$\boldsymbol{A}+\boldsymbol{0}=\boldsymbol{A} \tag{1.1.9}$$

$$\lambda(\boldsymbol{A}+\boldsymbol{B})=\lambda\boldsymbol{A}+\lambda\boldsymbol{B} \tag{1.1.10}$$

where $\boldsymbol{A}$, $\boldsymbol{B}$ and $\boldsymbol{C}$ are assumed compatible.

The pair of simultaneous equations

$$\begin{aligned} y_1 &= A_{11}x_1 + A_{12}x_2 \\ y_2 &= A_{21}x_1 + A_{22}x_2 \end{aligned} \tag{1.1.11}$$

can be represented by the matrix equation

$$\begin{bmatrix} y_1 \\ y_2 \end{bmatrix}=\begin{bmatrix} A_{11} & A_{12} \\ A_{21} & A_{22} \end{bmatrix}\begin{bmatrix} x_1 \\ x_2 \end{bmatrix} \tag{1.1.12}$$

or, alternatively

$$\boldsymbol{y}=\boldsymbol{A}\boldsymbol{x} \tag{1.1.13}$$

where $\boldsymbol{A}\boldsymbol{x}$ is the product of the (2×2) matrix

$$\boldsymbol{A}=\begin{bmatrix} A_{11} & A_{12} \\ A_{21} & A_{22} \end{bmatrix}$$

and the (2×1) column matrix $\boldsymbol{x}$. The array denoted by $\boldsymbol{A}$ is called the matrix of coefficients. Thus, from equation (1.1.11)

$$\boldsymbol{y}=\begin{bmatrix} A_{11}x_1+A_{12}x_2 \\ A_{21}x_1+A_{22}x_2 \end{bmatrix}=\boldsymbol{A}\boldsymbol{x}. \tag{1.1.14}$$

Consider now the equations

$$\begin{aligned} z_1 &= B_{11}y_1+B_{12}y_2 \\ z_2 &= B_{21}y_1+B_{22}y_2 \\ z_3 &= B_{31}y_1+B_{32}y_2 \end{aligned} \tag{1.1.15}$$

which can be represented by the matrix equation

$$\begin{bmatrix} z_1 \\ z_2 \\ z_3 \end{bmatrix} = \begin{bmatrix} B_{11} & B_{12} \\ B_{21} & B_{22} \\ B_{31} & B_{32} \end{bmatrix} \begin{bmatrix} y_1 \\ y_2 \end{bmatrix} \tag{1.1.16}$$

or

$$\boldsymbol{z}=\boldsymbol{B}\boldsymbol{y}, \tag{1.1.17}$$

where $\boldsymbol{B}\boldsymbol{y}$ is the product of the (3×2) matrix $\boldsymbol{B}$ and the (2×1) column matrix $\boldsymbol{y}$. From equation (1.1.15)

$$\boldsymbol{z}=\begin{bmatrix} B_{11}y_1+B_{12}y_2 \\ B_{21}y_1+B_{22}y_2 \\ B_{31}y_1+B_{32}y_2 \end{bmatrix}=\boldsymbol{B}\boldsymbol{y}. \tag{1.1.18}$$

Replacing the $\boldsymbol{y}$ in equation (1.1.17) with the $\boldsymbol{y}$ from equation (1.1.13) gives

$$\boldsymbol{z}=\boldsymbol{B}(\boldsymbol{A}\boldsymbol{x}). \tag{1.1.19}$$

Substituting the values of y_1, y_2 obtained from equation (1.1.11) into equation (1.1.15) gives

$$\begin{aligned} z_1 &= B_{11}(A_{11}x_1+A_{12}x_2)+B_{12}(A_{21}x_1+A_{22}x_2) \\ z_2 &= B_{21}(A_{11}x_1+A_{12}x_2)+B_{22}(A_{21}x_1+A_{22}x_2) \\ z_3 &= B_{31}(A_{11}x_1+A_{12}x_2)+B_{32}(A_{21}x_1+A_{22}x_2) \end{aligned} \tag{1.1.20}$$

which may be rearranged into the form

$$\begin{aligned} z_1 &= (B_{11}A_{11}+B_{12}A_{21})x_1+(B_{11}A_{12}+B_{12}A_{22})x_2 \\ z_2 &= (B_{21}A_{11}+B_{22}A_{21})x_1+(B_{21}A_{12}+B_{22}A_{22})x_2 \\ z_3 &= (B_{31}A_{11}+B_{32}A_{21})x_1+(B_{31}A_{12}+B_{32}A_{22})x_2 \end{aligned} \tag{1.1.21}$$

which corresponds to the form of equation (1.1.19) expressed in the form

$$z=(\boldsymbol{BA})\boldsymbol{x}. \tag{1.1.22}$$

Thus the product matrix $\boldsymbol{BA}$ is the (3×2) matrix

$$[\boldsymbol{BA}]=\begin{bmatrix} B_{11}A_{11}+B_{12}A_{21} & B_{11}A_{12}+B_{12}A_{22} \\ B_{21}A_{11}+B_{22}A_{21} & B_{21}A_{12}+B_{22}A_{22} \\ B_{31}A_{11}+B_{32}A_{21} & B_{31}A_{12}+B_{32}A_{22} \end{bmatrix}. \tag{1.1.23}$$

The (i, k) element of the product matrix $\boldsymbol{BA}$ is the product of the ith row of $\boldsymbol{B}$ and the kth column of $\boldsymbol{A}$. For this reason the *pre-multiplier* $\boldsymbol{B}$ must have the same number of columns as the number of rows in the *post-multiplier* $\boldsymbol{A}$. If this is the case, $\boldsymbol{B}$ is said to be compatible to $\boldsymbol{A}$ for multiplication. It is important to note that if $\boldsymbol{B}$ is compatible to $\boldsymbol{A}$, then $\boldsymbol{A}$ is not necessarily compatible to $\boldsymbol{B}$. Even if $\boldsymbol{BA}$ is defined it is not necessarily equal to $\boldsymbol{AB}$.

On the assumption that $\boldsymbol{A}, \boldsymbol{B}$ and $\boldsymbol{C}$ are compatible for the indicated sums and products, it is possible to show that

$$\begin{aligned} \boldsymbol{A}(\boldsymbol{B}+\boldsymbol{C})&=\boldsymbol{AB}+\boldsymbol{AC} \\ (\boldsymbol{A}+\boldsymbol{B})\boldsymbol{C}&=\boldsymbol{AC}+\boldsymbol{BC} \\ \boldsymbol{A}(\boldsymbol{BC})&=(\boldsymbol{AB})\boldsymbol{C}. \end{aligned} \tag{1.1.24}$$

However, in general, matrix multiplication is not commutative: thus

$$\boldsymbol{AB}\neq\boldsymbol{BA}. \tag{1.1.25}$$

Also it is to be noted that $\boldsymbol{AB}=\boldsymbol{BC}$ does not necessarily imply $\boldsymbol{A}=\boldsymbol{C}$, nor does $\boldsymbol{AB}=0$ imply either $\boldsymbol{A}=0$ or $\boldsymbol{B}=0$.

A *square matrix* is an $(m\times m)$ matrix, and thus has the same number of rows and columns. The elements of such a matrix for which the row and column numbers are equal constitute the *leading diagonal*. The sum of the elements of the leading diagonal of a square matrix $\boldsymbol{A}$ is called the *trace* and is written

$$\mathrm{tr}\boldsymbol{A}=A_{11}+A_{22}+A_{33}+\cdots+A_{mm}. \tag{1.1.26}$$

From equations (1.1.5) and (1.1.10) it follows that

$$\mathrm{tr}(\alpha\boldsymbol{A}+\beta\boldsymbol{B})=\mathrm{Tr}\boldsymbol{A}+\beta\,\mathrm{Tr}\boldsymbol{B}. \tag{1.1.27}$$

The square matrix $\boldsymbol{I}$ defined by

$$\boldsymbol{I}=\begin{bmatrix} 1 & 0 & \dots & 0 \\ 0 & 1 & \dots & 0 \\ \vdots & \vdots & & \vdots \\ 0 & 0 & \dots & 1 \end{bmatrix} \tag{1.1.28}$$

is the *identity matrix*.

A square matrix $\boldsymbol{A}$ whose elements satisfy $A_{ij}=0$, $i>j$, is called an *upper triangular matrix*: thus

$$[A_{ij}]=\begin{bmatrix} A_{11} & A_{12} & A_{13} & \dots & A_{1n} \\ 0 & A_{22} & A_{23} & \dots & A_{2n} \\ 0 & 0 & A_{33} & \dots & A_{3n} \\ \vdots & \vdots & \vdots & & \vdots \\ 0 & 0 & 0 & \dots & A_{nn} \end{bmatrix}. \tag{1.1.29}$$

A lower triangular matrix can be defined in a similar fashion.

If $\boldsymbol{A}$ and $\boldsymbol{B}$ are square matrices of the same order such that $\boldsymbol{AB}=\boldsymbol{BA}=\boldsymbol{I}$, then $\boldsymbol{B}$ is called the *inverse* of $\boldsymbol{A}$, thus, $\boldsymbol{B}=\boldsymbol{A}^{-1}$. Also, $\boldsymbol{A}$ is the inverse of $\boldsymbol{B}$, i.e. $\boldsymbol{A}=\boldsymbol{B}^{-1}$. If $\boldsymbol{A}$ has an inverse, it is said to be *non-singular*. For the case that $\boldsymbol{A}$ and $\boldsymbol{B}$ are square matrices of the same order with inverses $\boldsymbol{A}^{-1}$ and $\boldsymbol{B}^{-1}$, respectively, then

$$(\boldsymbol{AB})^{-1}=\boldsymbol{B}^{-1}\boldsymbol{A}^{-1} \tag{1.1.30}$$

which follows from noting that

$$(\boldsymbol{AB})\boldsymbol{B}^{-1}\boldsymbol{A}^{-1}=\boldsymbol{A}(\boldsymbol{BB}^{-1})\boldsymbol{A}^{-1}=\boldsymbol{AIA}^{-1}=\boldsymbol{AA}^{-1}=\boldsymbol{I}$$

and similarly

$$(\boldsymbol{B}^{-1}\boldsymbol{A}^{-1})\boldsymbol{AB}=\boldsymbol{I}.$$

To any matrix $\boldsymbol{A}$ there corresponds a transpose matrix, denoted by $\boldsymbol{A}^{\mathrm{T}}$, formed by interchanging the rows and columns. It is easily shown that

$$\begin{gathered}(\boldsymbol{A}^{\mathrm{T}})^{\mathrm{T}}=\boldsymbol{A}\\ (\alpha\boldsymbol{A}+\beta\boldsymbol{B})^{\mathrm{T}}=\alpha\boldsymbol{A}^{\mathrm{T}}+\beta\boldsymbol{B}^{\mathrm{T}}\\ (\boldsymbol{AB})^{\mathrm{T}}=\boldsymbol{B}^{\mathrm{T}}\boldsymbol{A}^{\mathrm{T}}.\end{gathered} \tag{1.1.31}$$

A square matrix $\boldsymbol{A}$ such that $\boldsymbol{A}^{\mathrm{T}}=\boldsymbol{A}$ is said to be *symmetric*, and if $\boldsymbol{A}^{\mathrm{T}}=-\boldsymbol{A}$, $\boldsymbol{A}$ is called a *skew-symmetric matrix*. The condition $\boldsymbol{A}^{\mathrm{T}}=-\boldsymbol{A}$ implies that the diagonal elements of a skew-symmetric matrix are all zero. The identify

$$\boldsymbol{A}=\tfrac{1}{2}(\boldsymbol{A}+\boldsymbol{A}^{\mathrm{T}})+\tfrac{1}{2}(\boldsymbol{A}-\boldsymbol{A}^{\mathrm{T}}) \tag{1.1.32}$$

demonstrates that every square matrix can be expressed as the sum of symmetric and skew-symmetric parts. This decomposition is unique.

If $\boldsymbol{A}$ is a square matrix, its *determinant* is written $\det\boldsymbol{A}$. Taking $\boldsymbol{A}$ to be the square matrix $(m\times m)$, the square matrix $(m-1)\times(m-1)$ can be constructed by removing the ith row and the jth column. The determinant of this matrix, denoted by M_{ij}, is called the *minor* of A_{ij}. The *cofactor* of A_{ij} is defined by

$$\operatorname{cof} A_{ij}=(-1)^{i+j}M_{ij}. \tag{1.1.33}$$

A determinant may be represented in terms of the elements and cofactors of any one row or column as follows

$$\det \boldsymbol{A} = \sum_{i=1}^{m} A_{ik} \operatorname{cof} A_{ik} = \sum_{j=1}^{m} A_{kj} \operatorname{cof} A_{kj} \qquad (k=1,2,\ldots,m). \tag{1.1.34}$$

For example,

$$\boldsymbol{A} = \begin{bmatrix} A_{11} & A_{12} & A_{13} \\ A_{21} & A_{22} & A_{23} \\ A_{31} & A_{32} & A_{33} \end{bmatrix} \tag{1.1.35}$$

then

$$\operatorname{cof} A_{11} = \begin{bmatrix} A_{22} & A_{23} \\ A_{32} & A_{33} \end{bmatrix} \qquad \operatorname{cof} A_{12} = \begin{bmatrix} A_{23} & A_{21} \\ A_{33} & A_{31} \end{bmatrix}$$

$$\operatorname{cof} A_{13} = \begin{bmatrix} A_{21} & A_{22} \\ A_{31} & A_{32} \end{bmatrix} \tag{1.1.36}$$

and

$$\det \boldsymbol{A} = A_{11}(A_{22}A_{33} - A_{23}A_{32}) - A_{12}(A_{21}A_{33} - A_{23}A_{31}) + A_{13}(A_{21}A_{32} - A_{22}A_{31}). \tag{1.1.37}$$

The value of a determinant is not changed by any of the following operations

(i) if the rows and columns are interchanged
(ii) an even number of interchanges of any two rows or two columns
(iii) addition of the elements of any row, (or column), all multiplied, if desired, by the same parameter α, to the respective corresponding elements of another row, (or column, respectively).

An odd number of interchanges of any two rows or two columns is equivalent to multiplication of the determinant by -1.

Multiplication of all the elements of any one row or column by a factor α is equivalent to multiplication of the determinant by α. A determinant is equal to zero if

(i) all elements of any row or column are zero
(ii) corresponding elements of any two rows or columns are equal, or proportional with the same proportionality factor.

It is easily shown that

$$\det \boldsymbol{A} = \det \boldsymbol{A}^{\mathrm{T}} \qquad \det(\boldsymbol{AB}) = \det \boldsymbol{A} \det \boldsymbol{B}. \tag{1.1.38}$$

Corresponding to any arbitrary $m \times m$ matrix A there is a unique $m \times m$

matrix $\boldsymbol{A}^{+}$, called the adjugate of $\boldsymbol{A}$, which is given by

$$\boldsymbol{A}^{+}=\begin{bmatrix} \operatorname{cof} A_{11} & \operatorname{cof} A_{12} & \ldots & \operatorname{cof} A_{1m} \\ \operatorname{cof} A_{21} & \operatorname{cof} A_{22} & \ldots & \operatorname{cof} A_{2m} \\ \vdots & & & \vdots \\ \operatorname{cof} A_{m1} & \operatorname{cof} A_{m2} & \ldots & \operatorname{cof} A_{mm} \end{bmatrix}. \tag{1.1.39}$$

It can be shown that

$$\boldsymbol{A}\boldsymbol{A}^{\mathrm{T}+}=(\det \boldsymbol{A})\boldsymbol{I}=(\det \boldsymbol{A}^{\mathrm{T}})\boldsymbol{I} \tag{1.1.40}$$

and if $\det \boldsymbol{A} \neq 0$, then equation (1.1.40) shows that the inverse $\boldsymbol{A}^{-1}$ exists and is given by

$$\boldsymbol{A}^{-1}=\frac{\boldsymbol{A}^{\mathrm{T}+}}{\det \boldsymbol{A}^{\mathrm{T}}}=\frac{\boldsymbol{A}^{+\mathrm{T}}}{\det \boldsymbol{A}} \tag{1.1.41}$$

it being noted that,

$$(\boldsymbol{A}^{+})^{+}=(\det \boldsymbol{A})^{m-2}\boldsymbol{A} \qquad \det \boldsymbol{A}^{+}=(\det \boldsymbol{A})^{m-1}. \tag{1.1.42}$$

1.2 SETS, RELATIONS AND FUNCTIONS

1.2.1 Sets and Set Algebra

The term *set* is used to denote a collection, class or family of objects in the form of numbers or symbols considered as a whole. Examples are the set of all prime numbers or the set of all matrices with determinants equal to unity. Geometrically, the objects in a set determine the *domain.* Each such object is called an *element* or *member* of the set.

The notation,

$$b \in \mathscr{B},$$

denotes that the element b is contained in the set $\mathscr{B}$, and the notation

$$b \notin \mathscr{B},$$

denotes that b is not an element of $\mathscr{B}$.

There are two common ways of specifying a set:

(i) all elements shown within braces, e.g.

$$\mathscr{B}=\{a, b, c, d\}$$

denotes that a, b, c, d are elements of the set $\mathscr{B}$;

(ii) where the elements of a set are characterised by their possession of a specific property, use is made of the notation,

$$\mathscr{C}=\{a \mid P(a)\}$$

where a denotes a typical element and $P(a)$ is the property which determines a to be in the set. For example, let $\mathscr{I}$ denotes the set of integers, and let $\mathscr{C}$ denote the set of all integers m for which coth $m<1.313$, then

$$\mathscr{C}=\{m\in\mathscr{I}\,|\,\coth m<1.313\}$$

states that $\mathscr{C}$ is the set of all integers m such that coth m is less than 1.313.

If $\mathscr{A}$ and $\mathscr{B}$ are sets, $\mathscr{B}$ is said to be *equal* to $\mathscr{A}$ if and only if both sets contain exactly the same elements. Thus, if a denotes the elements, then $\mathscr{A}=\mathscr{B}$ implies $a\in\mathscr{A}$ if and only if $a\in\mathscr{B}$.

The set $\mathscr{B}$ is said to be a *subset* of $\mathscr{A}$ if every element b of $\mathscr{B}$ is also an element of $\mathscr{A}$. Thus

$$\mathscr{B}\subseteq\mathscr{A}$$

states that if $b\in\mathscr{B}$, then $b\in\mathscr{A}$. For example, the set of integers $\mathscr{I}$ is a subset of the set of real numbers $\mathscr{R}$, i.e. $\mathscr{I}\subseteq\mathscr{R}$. The subset notation $\mathscr{B}\subseteq\mathscr{A}$ can also be written $\mathscr{A}\supseteq\mathscr{B}$. It follows that the equality of sets can be expressed in terms of the subset notation as

$$\mathscr{A}=\mathscr{B}\Leftrightarrow\mathscr{A}\subseteq\mathscr{B}\quad\text{and}\quad\mathscr{B}\subseteq\mathscr{A}. \tag{1.2.1}$$

A non-empty subset $\mathscr{B}$ of $\mathscr{A}$ is called a *proper subset* of $\mathscr{A}$ if $\mathscr{B}$ is not equal to $\mathscr{A}$. An example of a proper subset is the set of integers $\mathscr{I}$.

The *empty set*, *null set* or *void set* is the set with no elements and is denoted ϕ.

The *singleton* is the set containing a single element b and is denoted by $\{b\}$.

Sets can be combined to produce other sets. Let $\mathscr{S}$ be any set, and let $\mathscr{H}$ denote the collection of all subsets of $\mathscr{S}$. The elements of $\mathscr{H}$ are the subsets of $\mathscr{S}$. It is now required to define three operations on $\mathscr{H}$, called *union*, *intersection* and *complementation*.

Thus, if $\mathscr{A}$ and $\mathscr{B}$ are any elements of $\mathscr{H}$, then the union of the sets $\mathscr{A}$ and $\mathscr{B}$ consists of all elements m of $\mathscr{S}$ that belong to $\mathscr{A}$, or to $\mathscr{B}$, or to both. The set, $\mathscr{A}$ union $\mathscr{B}$, is denoted $\mathscr{A}\cup\mathscr{B}$ and is specified by

$$\textit{Union:}\qquad \mathscr{A}\cup\mathscr{B}\equiv\{m\in\mathscr{S}\,|\,m\in A\ \text{or}\ m\in\mathscr{B}\}. \tag{1.2.2}$$

The operation of set union is commutative

$$\mathscr{A}\cup\mathscr{B}=\mathscr{B}\cup\mathscr{A} \tag{1.2.3}$$

and associative

$$\mathscr{A}\cup(\mathscr{B}\cup\mathscr{C})=(\mathscr{A}\cup\mathscr{B})\cup\mathscr{C} \tag{1.2.4}$$

for all $\mathscr{A}$, $\mathscr{B}$, $\mathscr{C}\in\mathscr{H}$.

The set, $\mathscr{A}$ intersection $\mathscr{B}$, is denoted $\mathscr{A}\cap\mathscr{B}$ and is specified by

$$\textit{Intersection:}\qquad \mathscr{A}\cap\mathscr{B}=\{m\in\mathscr{S}\,|\,m\in\mathscr{A}\ \text{and}\ m\in\mathscr{B}\}. \tag{1.2.5}$$

The operation of set intersection is commutative

$$\mathscr{A}\cap\mathscr{B}=\mathscr{B}\cap\mathscr{A} \tag{1.2.6}$$

and associative

$$\mathscr{A}\cap(\mathscr{B}\cap\mathscr{C})=(\mathscr{A}\cap\mathscr{B})\cap\mathscr{C} \tag{1.2.7}$$

for all $\mathscr{A}$, $\mathscr{B}$, $\mathscr{C}\in\mathscr{H}$.

Two sets are said to be *disjoint* if they have no elements in common: i.e. if

$$\mathscr{A}\cap\mathscr{B}=\phi. \tag{1.2.8}$$

The operations of union and intersection are related by the following distributive laws

$$\begin{aligned}\mathscr{A}\cap(\mathscr{B}\cup\mathscr{C})&=(\mathscr{A}\cap\mathscr{B})\cup(\mathscr{A}\cap\mathscr{C})\\ \mathscr{A}\cup(\mathscr{B}\cap\mathscr{C})&=(\mathscr{A}\cup\mathscr{B})\cap(\mathscr{A}\cup\mathscr{C})\end{aligned} \tag{1.2.9}$$

for all $\mathscr{A}$, $\mathscr{B}$, $\mathscr{C}\in\mathscr{H}$.

The *complement* of $\mathscr{A}$ in $\mathscr{S}$ consists of all elements m of $\mathscr{S}$ that do not belong to $\mathscr{A}$. The complement of $\mathscr{A}$ with respect to $\mathscr{S}$ is denoted by $\mathscr{S}/\mathscr{A}$ and is specified by

$$\textit{Complement:}\qquad \mathscr{S}/\mathscr{A}\equiv\{m \mid m\in\mathscr{S} \text{ and } m\notin\mathscr{A}\}. \tag{1.2.10}$$

Complementation satisfies the following properties

$$\mathscr{S}/(\mathscr{S}/\mathscr{A})=\mathscr{A} \qquad \text{for } \mathscr{A}\subset\mathscr{S} \tag{1.2.11}$$

and

$$\mathscr{S}/\mathscr{A}=\mathscr{S}\Leftrightarrow\mathscr{S}\cap\mathscr{A}=\phi. \tag{1.2.12}$$

1.2.2 Relations and Equivalence Relations

The concept of ordering is not essential to the definition of a set. There are, however, cases of interest where it is important to order the elements of a set.

Let $\mathscr{A}$ and $\mathscr{B}$ be sets, and let $\mathscr{A}\times\mathscr{B}$ denote the set of all ordered pairs (a, b) such that $a\in\mathscr{A}$ and $b\in\mathscr{B}$. The set $\mathscr{A}\times\mathscr{B}$ is called the *cartesian product* of $\mathscr{A}$ and $\mathscr{B}$ and is specified by

$$\mathscr{A}\times\mathscr{B}=\{(a, b) \mid a\in A \text{ and } b\in\mathscr{B}\}. \tag{1.2.13}$$

Any subset $\mathscr{G}$ of $\mathscr{A}\times\mathscr{B}$ defines a *binary relation* from the set $\mathscr{A}$ to the set $\mathscr{B}$. Given any $\mathscr{G}\subseteq\mathscr{A}\times\mathscr{B}$ if $(a, b)\in\mathscr{G}$, then a is said to be *related to* b, the relation being denoted by $a\mathscr{G}b$. The *domain* of $\mathscr{G}$ is the set of all elements of $\mathscr{A}$ that are related by $\mathscr{G}$ to at least one element of $\mathscr{B}$

$$\text{dom }\mathscr{G}=\{a\in\mathscr{A} \mid a\mathscr{G}m \text{ for some } m\in\mathscr{B}\}. \tag{1.2.14}$$

The *range* of $\mathscr{G}$ is the set of all elements of $\mathscr{B}$ to which at least one element of $\mathscr{A}$ is related by $\mathscr{G}$

$$\text{range }\mathscr{G}=\{b\in\mathscr{B} \mid n\mathscr{G}b \text{ for some } n\in\mathscr{A}\}. \tag{1.2.15}$$

Let $\mathscr{S}$ be any set, and let $\mathscr{H}$ denote the collection of all subsets of $\mathscr{S}$, then the subset relation $\subseteq$ is a relation in $\mathscr{H}$. It follows that this relation has the properties

Reflexivity:	$\mathscr{A} \subseteq \mathscr{B}$ for every $\mathscr{A} \in \mathscr{H}$
Antisymmetry:	if $\mathscr{A} \subseteq \mathscr{B}$ and $\mathscr{B} \subseteq \mathscr{A}$, then $\mathscr{A} = \mathscr{B}$
Transitivity:	if $\mathscr{A} \subseteq \mathscr{B}$ and $\mathscr{B} \subseteq \mathscr{C}$, then $\mathscr{A} \subseteq \mathscr{C}$.

Any relation having these properties is called a *partial ordering*.

An *equivalence relation* on a set $\mathscr{A}$ is a binary relation $\mathscr{J}$ in $\mathscr{A}$ which is reflexive, symmetric, and transitive

Reflexive:	$a \mathscr{J} a$ for each $a \in \mathscr{A}$
Symmetric:	if $a \mathscr{J} b$, then $b \mathscr{J} a$
Transitive:	if $a \mathscr{J} b$ and $b \mathscr{J} c$, then $a \mathscr{J} c$.

As an example of an equivalence relation $\mathscr{J}$ on the set of all real numbers $\mathscr{R}$, let $a \mathscr{J} b \Leftrightarrow a = b$ for all a, b. To verify that this is an equivalence relation, note that $a = a$ for each $a \in \mathscr{J}$ (reflexivity), $a = b \Rightarrow b = a$ for each $a, b \in \mathscr{J}$ (symmetry), and $a = b$, $b = c \Rightarrow a = c$ for a, b and c in $\mathscr{J}$ (transitivity).

1.2.3 Functions and Functionals

A *function* or *mapping* f between the sets $\mathscr{X}$ and $\mathscr{Y}$ is the relation which satisfies the condition that if $(x, y_1) \in f$ and $(x, y_2) \in f$, then $y_1 = y_2$ for all $x \in \mathscr{X}$. In this context, a function is a particular kind of relation which, however specified, has the property of giving for each $x \in \mathscr{X}$ one and only one $y \in \mathscr{Y}$. Thus, a function f defines a *single-valued relation*, and can be specified by

$$y = f(x) \tag{1.2.16}$$

which indicates the element $y \in \mathscr{Y}$ that f associates with $x \in \mathscr{X}$, or, alternatively

$$f: \mathscr{X} \to \mathscr{Y}. \tag{1.2.17}$$

The element y is called the *value* of f at x. The *domain* of the function f is the set $\mathscr{X}$. The *range* of f is the set of all y for which there exists an x such that $y = f(x)$, the range of f being denoted by $f(\mathscr{X})$. When the domain of f is the set of real numbers $\mathscr{R}$, the function f is said to be a *function of a real variable*. The function f is said to be *real-valued* when $f(\mathscr{X})$ is contained in $\mathscr{R}$. Functions of a *complex-variable* and *complex-valued* functions are defined analogously. A function of a real variable need not be real-valued, nor need a function of a complex variable be complex-valued.

If $\mathscr{X}_0$ is any subset of $\mathscr{X}$, and $f: \mathscr{X} \to \mathscr{Y}$, the *image* of $\mathscr{X}_0$ under f is the set

$$f(\mathscr{X}_0) \equiv \{ y \mid y = f(x) \text{ for some } x \in \mathscr{X} \} \tag{1.2.18}$$

it being noted that

$$f(\mathscr{X}_0) \subseteq f(\mathscr{X}). \qquad (1.2.19)$$

Similarly, if $\mathscr{Y}_0$ is a subset of $\mathscr{Y}$, then the *inverse image* of $\mathscr{Y}_0$ under f is the set

$$f^{-1}(\mathscr{Y}_0) \equiv \{x \,|\, f(x) \in \mathscr{Y}_0|\} \qquad (1.2.20)$$

it being noted that

$$f^{-1}(\mathscr{Y}_0) \subseteq \mathscr{X}. \qquad (1.2.21)$$

A function $f: \mathscr{X} \to \mathscr{Y}$ is said to be *one-one* (or *injective*), if no two elements of $\mathscr{X}$ have the same f-image. That is, if for every $x_1, x_2 \in \mathscr{X}$,

$$f(x_1) = f(x_2) \Rightarrow x_1 = x_2 \qquad (1.2.22)$$

the function f is said to be one-one.

If $f(\mathscr{X})$ is a proper subset of $\mathscr{Y}$, the function f is said to be *into* $\mathscr{Y}$.

A function $f: \mathscr{X} \to \mathscr{Y}$ is said to be *onto* (or *surjective*) if the range of f is $\mathscr{Y}$. That is, if for every element $y \in \mathscr{Y}$, there exists at least one element $x \in \mathscr{X}$ such that $f(x) = y$, the function f is said to be *onto* $\mathscr{Y}$.

A function $f: \mathscr{X} \to \mathscr{Y}$ is said to form a *one-one correspondence*, or to be *bijective* if it is both one-one and onto.

If $f: \mathscr{X} \to \mathscr{Y}$ is a one-one function, then there exists a function $f^{-1}: f(\mathscr{X}) \to \mathscr{X}$ which associates with each $y \in f(\mathscr{X})$ the unique $x \in \mathscr{X}$ such that $y = f(x)$. The quantity f^{-1} is the *inverse* of f, written $x = f^{-1}(y)$. It is to be noted that inverse functions can only be defined for functions that are one-one; also note that $f^{-1}[f(\mathscr{X})] = \mathscr{X}$.

The *composition* of two functions $f: \mathscr{X} \to \mathscr{Y}$ and $g: \mathscr{Y} \to \mathscr{X}$ is a function $h: \mathscr{X} \to \mathscr{Z}$ defined by $h(x) = g[f(x)]$ for all $x \in \mathscr{X}$. The function h is written

$$h = g \circ f. \qquad (1.2.23)$$

The composition of any finite number of functions is easily defined in a manner similar to that of a pair. The operation of composition of functions is not generally commutative

$$g \circ f \neq f \circ g.$$

In particular, if $g \circ f$ is defined, $f \circ g$ may not be defined, and even if $g \circ f$ and $f \circ g$ are both defined, they may not be equal. The operation of composition is associative

$$h \circ (g \circ f) = (h \circ g) \circ f \qquad (1.2.24)$$

if each of the indicated compositions is defined.

The identity function id: $\mathscr{X} \to \mathscr{X}$ is defined as the function $\mathrm{id}(x) = x$ for all $x \in \mathscr{X}$. It follows that, if f is a one-one correspondence from $\mathscr{X} \to \mathscr{Y}$, then

$$f^{-1} \circ f = \mathrm{id}_{\mathscr{X}} \qquad f \circ f^{-1} = \mathrm{id}_{\mathscr{Y}} \qquad (1.2.25)$$

where $\mathrm{id}_{\mathscr{X}}$ and $\mathrm{id}_{\mathscr{Y}}$ denote the identity functions of $\mathscr{X}$ and $\mathscr{Y}$, respectively.

A *functional* is defined as a function whose domain $\mathscr{X}$ is a set of elements each of which is itself a function $x(t)$ over the real variable t from a to b. A functional $\mathscr{G}$ can be written in the alternative forms:

$$\mathscr{G}:x(t) \text{ over } (a,b) \rightarrow y \tag{1.2.26}$$

or

$$\mathscr{G}:\overset{b}{\underset{a}{x(t)}} \rightarrow y \tag{1.2.27}$$

$$y = \mathscr{G}\left(\overset{b}{\underset{a}{x(t)}}\right) = \overset{b}{\underset{t=a}{\mathscr{G}}}\,[x(t)]. \tag{1.2.28}$$

It is to be noted that y is not a function of t. Although the functional $\mathscr{G}$ has been shown above as being over the open interval (a, b), it could alternatively be over the closed interval $[a, b]$ or over a semi-open interval. The use of the term functional for this special type of mapping is due to Volterra (1930). An example of a functional is the integral,

$$y = \int_a^b x(t)\,\mathrm{d}t. \tag{1.2.29}$$

1.3 GROUPS

1.3.1 The Axioms for a Group

Central to the definition of a group is the concept of a *binary operation*. If $\mathscr{G}$ is a non-empty set, a binary operation on $\mathscr{G}$ is a function from $\mathscr{G} \times \mathscr{G}$ to $\mathscr{G}$. If a, $b \in \mathscr{G}$, the binary operation will be denoted by $*$ and its value by $a * b$. For any binary operation on $\mathscr{G}$, it is to be noted that $\mathscr{G}$ is *closed* with respect to $*$ in the sense that if $a, b \in \mathscr{G}$, then also, $a \times b \in \mathscr{G}$. A binary operation $*$ on $\mathscr{G}$ is associative if

$$(a * b) * c = a * (b * c) \qquad \forall\, a, b, c \in \mathscr{G} \tag{1.3.1}$$

where $\forall$ denotes *for all*.

A *semigroup* is a pair $(\mathscr{G}, *)$ consisting of a non-empty set $\mathscr{G}$ with an associative binary operation $*$.

A binary operation $*$ on $\mathscr{G}$ is commutative if

$$a * b = b * a \qquad \forall\, a, b \in \mathscr{G}. \tag{1.3.2}$$

If the binary operation of the group is commutative, the group is said to be a *commutative* or *Abelian* group.

An element $e \in \mathscr{G}$ that satisfies the condition

$$e * a = a * e = a \qquad \forall\, a \in \mathscr{G} \tag{1.3.3}$$

is called an identity element for the binary operation $*$ on the set $\mathscr{G}$. In the set of real numbers with the binary operation of multiplication, it is easy to see that the number 1 plays the role of identity element. In the set of real numbers with the binary operation of addition, o has the role of the identity element. Clearly $\mathscr{G}$ contains at most one identity element e. For if e' is another identity element in $\mathscr{G}$, then

$$e' * a = a * e' = a \qquad \forall a \in \mathscr{G}$$

also. In particular, if the choice is made that $a = e$, the $e' * e = e$. But from equation (1.3.3), it follows that $e' * e = e'$. Thus, $e' = e$. In general, $\mathscr{G}$ need not have any identity element. But if there is an identity element, and if the binary operation is regarded as multiplicative, then the identity element is often called the *unity* element; on the other hand, if the binary operation is additive, then the identity element is called the *zero* element.

In a semigroup $\mathscr{G}$ containing an identity element e with respect to the binary operation $*$, an element a^{-1} is said to be an *inverse* of the element a if

$$a * a^{-1} = a^{-1} * a = e. \tag{1.3.4}$$

In general, a need not have an inverse. But if an inverse a^{-1} of a exists, then it is unique, the proof being essentially the same as that of the uniqueness of the identity element. The identity element is its own inverse. In the set $\mathscr{R}/\{0\}$ with the binary operation of multiplication, the inverse of a number is the reciprocal of the number. In the set of real numbers with the binary operation of addition, the inverse of a number is the negative of the number.

A *group* is a pair $(\mathscr{G}, *)$ consisting of an associative binary operation $*$ and a set $\mathscr{G}$ which contains the identity element and the inverses of all elements of $\mathscr{G}$ with respect to the binary operation $*$. This definition of a group can be restated as follows. A group $\mathscr{G}$ is a set of elements with a binary operation $*$ such that

(i) $a * b \in \mathscr{G} \qquad \forall a, b \in \mathscr{G}$ (closure)
(ii) $a * (b * c) = (a * b) * c \qquad \forall a, b, c \in \mathscr{G}$ (associativity)
(iii) there exists an element $e \in \mathscr{G}$ (the identity element) such that $a * e = e * a = a \qquad \forall a \in \mathscr{G}$
(iv) for each $a \in \mathscr{G}$ there is an element $a^{-1} \in \mathscr{G}$ (the inverse of a) such that $a * a^{-1} = a^{-1} * a = e$.

A *commutative* or *Abelian group* is a group in which $a * b = b * a$ for all, $a, b \in \mathscr{G}$.

The set $\mathscr{R}/\{0\}$ with the binary operation of multiplication forms a group, and the set $\mathscr{R}$ with the binary operation of addition forms another group. The set of positive integers with the binary operation of multiplication forms a semigroup with an identity element but does not form a group because the fourth condition above is not satisfied.

When the particular $*$ to be employed is understood, the notational convention customarily employed is to denote a group simply by $\mathscr{G}$ rather than by the pair $(\mathscr{G}, *)$.

1.3.2 Properties of a Group

The basic properties of a group $\mathscr{G}$ in general are

(i) The identity element $e \in \mathscr{G}$ is unique.

(ii) The inverse element a^{-1} of any element $a \in \mathscr{G}$ is unique. If n is a positive integer, the powers of $a \in \mathscr{G}$ are defined as follows

(a) $n=1,\ a^1=a$

(b) $n>1,\ a^n=a^{n-1} * a$

(c) $a^0=e$

(d) $a^{-n}=(a^{-1})^n$.

(iii) If m, n, k are integers, positive or negative, then for $a \in \mathscr{G}$,

$$a^m * a^n = a^{m+n} \qquad (a^m)^n = a^{mn}$$

$$(a^{m+n})^k = a^{mk+nk}.$$

In particular, when $m=n=-1$:

(iv) $(a^{-1})^{-1}=a \qquad \forall a \in \mathscr{G}$.

(v) $(a * b)^{-1} = b^{-1} * a^{-1} \qquad \forall a, b \in \mathscr{G}$.

(vi) For any elements a, b in $\mathscr{G}$, the two equations $x * a = b$ and $a * y = b$ have the unique solutions $x = b * a^{-1}$ and $y = a^{-1} * b$.

The equation $x = b * a^{-1}$ satisfies the equation $x * a = b$,

$$x * a = (b * a^{-1}) * a = b * (a^{-1} * a) = b * e = b.$$

Conversely, $x * a = b$ implies

$$x = x * e = x * (a * a^{-1}) = (x * a) * a^{-1} = b * a^{-1}$$

which is unique. Similar arguments apply to the equation $a * y = b$.

(vii) For any three elements a, b, c in $\mathscr{G}$, either $a * c = b * c$ or $c * a = c * b$ implies $a = b$.

(viii) For any two elements a, b in the group $\mathscr{G}$, either $a * b = b$ or $b * a = b$ implies that a is the identity element.

A non-empty subset $\mathscr{G}'$ of $\mathscr{G}$ is a subgroup of $\mathscr{G}$ if $\mathscr{G}'$ is a group with respect to the binary operation of $\mathscr{G}$, i.e. $\mathscr{G}'$ is a subgroup of $\mathscr{G}$ if and only if

(a) $e \in \mathscr{G}'$

(b) $a \in \mathscr{G}' \Rightarrow a^{-1} \in \mathscr{G}'$

(c) $a, b \in \mathscr{G}' \Rightarrow a * b \in \mathscr{G}'$.

(ix) Let $\mathscr{G}'$ be a non-empty subset of $\mathscr{G}$. Then $\mathscr{G}'$ is a subgroup if

$$a, b \in \mathscr{G}' \Rightarrow a * b^{-1} \in \mathscr{G}';$$

(a) since $\mathscr{G}'$ is non-empty, it contains an element a, hence $a * a^{-1} = e \in \mathscr{G}'$;

(*b*) if $b \in \mathscr{G}'$, then $e * b^{-1} = b^{-1} \in \mathscr{G}'$;
(*c*) if $a, b \in \mathscr{G}'$ then $a * (b^{-1})^{-1} = a * b \in \mathscr{G}'$.

If $\mathscr{G}$ is a group, then $\mathscr{G}$ itself is a subgroup of $\mathscr{G}$, and the group consisting only of the element e is also a subgroup of $\mathscr{G}$. A subgroup of $\mathscr{G}$ other than $\mathscr{G}$ itself and the group e is called a *proper* subgroup of $\mathscr{G}$.

(x) The intersecticn of any two subgroups of a group remains a subgroup.

1.3.3 Group Homomorphisms

A *group homomorphism* relates to groups having the same type of formation or structure. Specifically, if $(\mathscr{G}, *)$ and $(\mathscr{H}, \circ)$ are two groups with the binary operations $*$ and $\circ$, respectively, a homomorphism of the group $(\mathscr{G}, *)$ with the group $(\mathscr{H}, \circ)$ is a function $f: \mathscr{G} \to \mathscr{H}$ such that

$$f(a * b) = f(a) \circ f(b) \qquad \forall\, a, b \in \mathscr{G}. \tag{1.3.5}$$

If a homomorphism exists between two groups, the groups are said to be *homomorphic*.

In accordance with the definition of a bijective function given in §1.2.3, a homomorphism $f: \mathscr{G} \to \mathscr{H}$ is an isomorphism if f is both one-to-one and onto. If an isomorphism exists between two groups, the groups are said to be *isomorphic*.

A homomorphism $f: \mathscr{G} \to \mathscr{G}$ is called an *endomorphism*.

An isomorphism $f: \mathscr{G} \to \mathscr{G}$ is called an *automorphism*.

Using the definition of a homomorphism in the form of equation (1.3.5) on the identity $a = a * e$, it is found that

$$f(a) = f(a * e) = f(a) \circ f(e).$$

From property (viii) of a group it follows that $f(e)$ is the identity element e_0 of $\mathscr{H}$. Now, using this result and the definition of a homomorphism in the form of equation (1.3.5) applied to the equation $e = a * a^{-1}$, it follows that

$$e_0 = f(e) = f(a * a^{-1}) = f(a) \circ f(a^{-1}).$$

Hence, it follows that, if $f: \mathscr{G} \to \mathscr{H}$ is a homomorphism, then $f(e)$ coincides with the identity element e_0 of $\mathscr{H}$ and

$$f(a^{-1}) = f(a)^{-1}$$

is the inverse of $f(a)$ in accord with property (ii) of a group. Thus, a homomorphism maps identities into identities and inverses into inverses.

If $\mathscr{G}'$ is a subgroup of $\mathscr{G}$, and since the homomorphism $f: \mathscr{G} \to \mathscr{H}$ maps identities into identities, it follows that since $e \in \mathscr{G}'$ the identity element e_0 of $\mathscr{H}$ is contained in $f(\mathscr{G}')$. Also, since a homomorphism maps inverses into inverses, it follows that for any $a \in \mathscr{G}'$, $f(a)^{-1} \in f(\mathscr{G}')$ and similarly for any

$$a, b \in \mathscr{G}' \qquad f(a) \circ f(b) \in f(\mathscr{G}')$$

since

$$f(a) \circ f(b) = f(a * b) \in f(\mathscr{G}').$$

Hence, it follows that $f(\mathscr{G}')$ is a subgroup of $\mathscr{H}$. Thus a homomorphism takes a subgroup into a subgroup. It also follows that $f(\mathscr{G})$ is itself a subgroup of $\mathscr{H}$.

The converse result can be stated as follows. If $f: \mathscr{G} \rightarrow \mathscr{H}$ is a homomorphism and if $\mathscr{H}'$ is a subgroup of $\mathscr{H}$, then the pre-image $f^{-1}(\mathscr{H}')$ is a subgroup of $\mathscr{G}$.

The *kernel* of a homomorphism $f: \mathscr{G} \rightarrow \mathscr{H}$ is the subgroup $f^{-1}(e_0)$ of $\mathscr{G}$. Thus, the kernel of f is the set of elements of $\mathscr{G}$ that are mapped by f to the identity element e_0 of $\mathscr{H}$. The notation $K(f)$ will be used to denote the kernel of f.

If $f(a) = f(b)$, then $f(a) \circ f(b)^{-1} = e_0$, and hence $f(a * b^{-1}) = e_0$ and it follows that $a * b^{-1} \in K(f)$. Thus, if $K(f) = \{e\}$ then $a * b^{-1} = e$ or $a = b$. Conversely, assuming f is one-to-one and since, from above, $K(f)$ if a subgroup of $\mathscr{G}$, it must contain e. If $K(f)$ were to contain any other element a such that $f(a) = e_0$, a contradiction would arise since f is one-to-one; therefore $K(f) = \{e\}$. Thus, a homomorphism $f: \mathscr{G} \rightarrow \mathscr{H}$ is one-to-one if and only if $K(f) = \{e\}$.

Since an isomorphism is one-to-one and onto, it has an inverse which is a function from $\mathscr{H}$ onto $\mathscr{G}$.

If f is one-to-one and onto, it follows from §1.2.3 that f^{-1} is also one-to-one and onto. Let a_0 be the element of $\mathscr{H}$ such that $a = f^{-1}(a_0)$ for any $a \in \mathscr{G}$; then

$$a * b = f^{-1}(a_0) * f^{-1}(b_0).$$

But $a * b$ is the inverse image of the element $a_0 \circ b_0 \in \mathscr{H}$ because

$$f(a * b) = f(a) \circ f(b) = a_0 \circ b_0$$

since f is a homomorphism. Therefore

$$f^{-1}(a_0) * f^{-1}(b_0) = f^{-1}(a_0 \circ b_0)$$

which shows that f^{-1} satisfies the definition given in equation (1.3.5) of a homomorphism. Thus, if $f: \mathscr{G} \rightarrow \mathscr{H}$ is an ismorphism, then $f^{-1}: \mathscr{H} \rightarrow \mathscr{G}$ is an isomorphism.

A homomorphism $f: \mathscr{G} \rightarrow \mathscr{H}$ is an isomorphism if it is onto and if its kernel contains only the identity element of $\mathscr{G}$.

1.3.4 Rings and Fields

A *ring* $\mathscr{R}$ is a set of elements with two binary operations $+$, usually called addition and $\cdot$ usually called multiplication, such that

(i) $\mathscr{R}$ is a commutative group under addition with identity $\circ$,

(ii) $a \cdot b \in \mathscr{R} \qquad \forall a, b \in \mathscr{R}$;

(iii) $a \cdot (b \cdot c) = (a \cdot b) \cdot c \qquad \forall a, b, c \in \mathscr{R}$;

(iv) $a \cdot (b+c) = a \cdot b + a \cdot c$ and $(b+c) \cdot a = b \cdot a + c \cdot a$ $\quad \forall a, b, c \in \mathscr{R}$ (distributivity).

A commutative ring is a ring in which multiplication is commutative. A ring with an element 1 such that $1 \cdot a = a \cdot 1$ $\forall a \in \mathscr{R}$ is a ring with unity. The ring $\mathscr{R}$ with operations + and · may be denoted by $(\mathscr{R}, +, \cdot)$.

A field $\mathscr{F}$ is a set of elements with two binary operations + (usually called addition) and · (usually called multiplication) such that

(i) $\mathscr{F}$ is a commutative group under addition with identity $\circ$;

(ii) the elements of $\mathscr{F}$, other than $\circ$, form a commutative group under multiplication,

(iii) $a \cdot (b+c) = a \cdot b + a \cdot c \quad \forall a, b, c \in \mathscr{F}$.

The field $\mathscr{F}$ with operations + and · may be denoted by $(\mathscr{F}, +, \cdot)$.

1.4 SCALARS AND SCALAR FIELDS

Certain physical quantities are described completely when their magnitude is stated in appropriate units. Such quantities are called *scalar* quantities. Typical examples of scalar quantities are temperature, mass, volume and time. Speed is a further example of a scalar quantity but velocity, which depends on direction, is not. A scalar quantity is independent of the units in which it is specified, and it is also independent of the coordinate system used to specify its location in space. It is thus invariant with respect to coordinate transformations. In terms of tensors, a scalar is a tensor of order zero.

A *scalar field* is defined as a region of space, with each point of which there is associated a *scalar point function*. Suppose a scalar ϕ is defined at every point P within a domain D by the relation $\phi = \phi(P)$. If, for every point P, there is a unique value $\phi(P)$, then $\phi = \phi(P)$ defines a scalar function of position called a scalar field. The temperature throughout a given material is an example of a scalar field.

1.5 VECTORS

1.5.1 Vector Spaces: I, *n*-dimensional

In general, first contact with the concept of a vector is a geometrical one in the form of a directed line segment. In this context, the reader is assumed to be familiar with the elementary treatment of vector theory, including the *scalar (or inner) product* $\boldsymbol{a} \cdot \boldsymbol{b}$ and the *vector (or outer) product* $\boldsymbol{a} \wedge \boldsymbol{b}$ of two arbitrary vectors $\boldsymbol{a}$ and $\boldsymbol{b}$, based on the geometrical representation of vectors. This type of vector is a special example of the more general aspect of a vector presented in this section. The concept of a vector, as advanced in this section is purely algebraic. Thus, for present purposes a vector is to be defined as a member of a set that satisfies certain algebraic rules. These algebraic rules can be taken as an axiomatic definition of the general concept of a *vector space*. However, before preceeding to state these rules, the concept of a vector space is considered in general terms.

A *vector space* (often called a *linear space*) $\mathscr{V}$, consists of the following:

(i) a set $\mathscr{V}$ of elements, called *vectors*;

(ii) a field $\mathscr{F}$ of scalars;

(iii) a function called *vector addition*: $\mathscr{V} \times \mathscr{V} \to \mathscr{V}$, ensures that any two vectors $\boldsymbol{a}$ and $\boldsymbol{b}$ of $\mathscr{V}$ determine a unique vector $(\boldsymbol{a}+\boldsymbol{b})$ as the *sum*, the difference of $\boldsymbol{a}$ and $\boldsymbol{b}$, written as $(\boldsymbol{a}-\boldsymbol{b})$, being defined by

$$\boldsymbol{a}-\boldsymbol{b}=\boldsymbol{a}+(-\boldsymbol{b}); \tag{1.5.1}$$

(iv) a function called *scalar multiplication*: $\mathscr{F} \times \mathscr{V} \to \mathscr{V}$, ensures that any vector $\boldsymbol{a}$ from $\mathscr{V}$ and any scalar λ from $\mathscr{F}$ determine the element $\lambda\boldsymbol{a}$ in $\mathscr{V}$.

Examples of scalar fields are real numbers, rational numbers and complex numbers.

The axioms governing multiplication by scalars λ, $\mu \in \mathscr{F}$ and the laws defining the addition and subtraction of vectors can be stated in terms of arbitrary vectors $\boldsymbol{a}$, $\boldsymbol{b}$ and $\boldsymbol{c}$ as follows:

(i) There exists a binary operation in $\mathscr{V}$ called addition and denoted by + such that:

$$(a)\quad \boldsymbol{a}+\boldsymbol{b}=\boldsymbol{c} \qquad \forall\, \boldsymbol{a}, \boldsymbol{b}, \boldsymbol{c} \in \mathscr{V} \tag{1.5.2}$$

$$(b)\quad \boldsymbol{a}+\boldsymbol{b}=\boldsymbol{b}+\boldsymbol{a} \qquad \forall\, \boldsymbol{a}, \boldsymbol{b} \in \mathscr{V} \tag{1.5.3}$$

$$(c)\quad \boldsymbol{a}+(\boldsymbol{b}+\boldsymbol{c})=(\boldsymbol{a}+\boldsymbol{b})+\boldsymbol{c} \qquad \forall\, \boldsymbol{a}, \boldsymbol{b}, \boldsymbol{c} \in \mathscr{V} \tag{1.5.4}$$

(*d*) There exists an element $\boldsymbol{0} \in \mathscr{V}$ such that

$$\boldsymbol{a}+\boldsymbol{0}=\boldsymbol{a} \qquad \forall \boldsymbol{a} \in \mathscr{V} \tag{1.5.5}$$

(*e*) There exists an element $-\boldsymbol{a} \in \mathscr{V}$ such that

$$\boldsymbol{a}+(-\boldsymbol{a})=\boldsymbol{0} \qquad \forall\, \boldsymbol{a} \in \mathscr{V}. \tag{1.5.6}$$

(ii) There exists an operation called *scalar multiplication* in which every scalar $\lambda \in \mathscr{F}$ can be combined with every element $\boldsymbol{a} \in \mathscr{V}$ to give an element $\lambda\boldsymbol{a} \in \mathscr{V}$ such that

$$(a)\quad 1\boldsymbol{a}=\boldsymbol{a} \qquad \forall\, \boldsymbol{a} \in \mathscr{V}, \tag{1.5.7}$$

$$(b)\quad \lambda(\mu\boldsymbol{a})=\mu(\lambda\boldsymbol{a})=(\mu\lambda)\boldsymbol{a} \qquad \forall \boldsymbol{a} \in \mathscr{V} \qquad \lambda, \mu \in \mathscr{F}, \tag{1.5.8}$$

$$(c)\quad (\lambda+\mu)\boldsymbol{a}=\lambda\boldsymbol{a}+\mu\boldsymbol{a} \qquad \forall \boldsymbol{a} \in \mathscr{V} \qquad \lambda, \mu \in \mathscr{F}, \tag{1.5.9}$$

$$(d)\quad \lambda(\boldsymbol{a}+\boldsymbol{b})=\lambda\boldsymbol{a}+\lambda\boldsymbol{b} \qquad \forall\, \boldsymbol{a}, \boldsymbol{b} \in \mathscr{V} \qquad \lambda \in \mathscr{F}. \tag{1.5.10}$$

Equation (1.5.2) formulates the axiom of vector addition, the resultant vector $\boldsymbol{c}$ being defined as the vector sum of $\boldsymbol{a}$ and $\boldsymbol{b}$ and having the properties defined by the axioms formulated by way of equations (1.5.3) and (1.5.4). Thus, vectors can be added in any order.

The axiom formulated by equation (1.5.5) defines the unique vector called the *zero vector*.

Equation (1.5.6) formulates the axiom which defines the negative of a vector.

The axiom formulated by equation (1.5.7) expresses the fact that multiplication by unity does not change the arbitrary vector $\boldsymbol{a}$.

Equation (1.5.8) formulates an axiom which asserts that multiplication of vectors by finite scalars (in $\mathscr{F}$) is *associative* and *commutative*.

The axioms formulated by equations (1.5.9) and (1.5.10) define the relationship between scalar multiplication of vectors by a scalar (in $\mathscr{F}$) and the addition of vectors.

From equations (1.5.2) to (1.5.10) it is evident that the laws which govern the addition and subtraction of vectors and the multiplication of vectors by scalars (in $\mathscr{F}$) are *associative*, *commutative* and *distributive*.

A finite set of m ($m \geqslant 1$) non-zero vectors $\{\boldsymbol{v}_1, \boldsymbol{v}_2, \ldots, \boldsymbol{v}_m\}$ in a vector space $\mathscr{V}$ is said to form a *linearly independent* system of order m if and only if, for all scalars λ_i in $\mathscr{F}$

$$\lambda_1 \boldsymbol{v}_1 + \lambda_2 \boldsymbol{v}_2 + \cdots + \lambda_m \boldsymbol{v}_m = \boldsymbol{0} \tag{1.5.11}$$

implies

$$\lambda_1 = \lambda_2 = \cdots \lambda_m = 0. \tag{1.5.12}$$

If, however, there exists a set of scalars $\{\lambda_1, \lambda_2, \ldots, \lambda_m\}$, not all zero, such that $\sum_{i=1}^{m} \lambda_i \boldsymbol{v}_i = \boldsymbol{0}$, then the given system of vectors is said to be *linearly dependent*. Every non-empty subset of a linearly independent set is linearly independent.

If the order of the linearly independent system is bounded, it is possible to determine an integer n such that there exist linearly independent systems of order n but not of order $n+1$. Such a linearly independent set in a vector space is said to be *maximal*. A vector space that contains a finite, (maximal), linearly independent set is said to be *finite dimensional*.

The *basis* of a vector space $\mathscr{V}$ is any maximal, linearly independent system of vectors. In general, any indexed set $e = \{\boldsymbol{e}_1, \boldsymbol{e}_2, \ldots, \boldsymbol{e}_n\}$ of n linearly independent vectors spans an n-dimensional vector space $\mathscr{V}$ and hence is called a *basis* of $\mathscr{V}$.

Let $\boldsymbol{v}$ be any vector in $\mathscr{V}$. The system of $n+1$ vectors $\{\boldsymbol{v}, \boldsymbol{e}_1, \boldsymbol{e}_2, \ldots, \boldsymbol{e}_n\}$ is necessarily linearly dependent, and so there must exist $n+1$ numbers $a, \lambda_1, \lambda_2, \ldots, \lambda_n$ (in $\mathscr{F}$), such that

$$a\boldsymbol{v} + \lambda_1 \boldsymbol{e}_1 + \lambda_2 \boldsymbol{e}_2 + \cdots + \lambda_n \boldsymbol{e}_n = \boldsymbol{0}. \tag{1.5.13}$$

For the set $e = \{\boldsymbol{e}_1, \boldsymbol{e}_2, \ldots, \boldsymbol{e}_n\}$ to be linearly independent, a must be different from zero and hence the above relation can be solved for $\boldsymbol{v}$, and thus there exist numbers $v^1, v^2, \ldots, v^n$ (in $\mathscr{F}$) such that

$$\boldsymbol{v} = v^1 \boldsymbol{e}_1 + v^2 \boldsymbol{e}_2 + \cdots + v^n \boldsymbol{e}_n. \tag{1.5.14}$$

Thus, the vector $\boldsymbol{v}$ is expressible as a linear combination of the $\boldsymbol{e}_i$. It is to be noted that this combination is unique, i.e. the set of n scalars $\{v^1, v^2, \ldots, v^n\}$ are unique. To pursue this uniqueness condition, a lack of uniqueness is assumed by assigning to $\boldsymbol{v}$ the two representations

$$\boldsymbol{v} = a^1 e_1 + a^2 \boldsymbol{e}_2 + \cdots + a^n \boldsymbol{e}_n. \tag{1.5.15}$$

$$\boldsymbol{v} = a^1 \boldsymbol{e}_1 + a^2 \boldsymbol{e}^2 + \cdots + a^n \boldsymbol{e}_n. \tag{1.5.16}$$

Taking the difference of equations (1.5.15) and (1.5.16) gives

$$(v^1 - a^1)\boldsymbol{e}_1 + (v^2 - a^2)\boldsymbol{e}_2 + \cdots + (v^n - a^n)\boldsymbol{e}_n = 0$$

and the linear independence of the basis requires that

$$a^i = v^i \qquad (i = 1, 2, \ldots, n).$$

The numbers $v^1, v^2, \ldots, v^n$ are called the *components* of $\boldsymbol{v}$ with respect to the basis $e = \{\boldsymbol{e}_1, \boldsymbol{e}_2, \ldots, \boldsymbol{e}_n\}$.

It can be shown that if a vector space $\mathscr{V}$ has a basis consisting of a finite number of elements then any other basis for $\mathscr{V}$ will have the same number of elements. This number is called the dimension of $\mathscr{V}$, written dim $\mathscr{V}$. For an n-dimensional vector space,

$$\dim \mathscr{V} = n. \tag{1.5.17}$$

Thus far the concept of a vector space has been defined without reference to the concept of *length* or *magnitude*. The concept of length, or magnitude is defined through the concept of the *scalar*, or *inner product* $\boldsymbol{a}\cdot\boldsymbol{b}$ of two arbitrary vectors $\boldsymbol{a}$ and $\boldsymbol{b}$. An *inner product space* is simply a vector space equipped with an inner product. The properties of the inner product can be stated in terms of arbitrary vectors $\boldsymbol{a}$, $\boldsymbol{b}$ and $\boldsymbol{c}$ as follows:

(iii) There exists an operation called *inner product* by which any ordered pair of vectors $\boldsymbol{a}$ and $\boldsymbol{b}$ in $\mathscr{V}$ determines an element of $\mathscr{F}$ denoted by $\boldsymbol{a}\cdot\boldsymbol{b}$ such that:

$$(a)\ \boldsymbol{b}\cdot\boldsymbol{a} = \boldsymbol{a}\cdot\boldsymbol{b} \qquad \forall\, \boldsymbol{a}, \boldsymbol{b} \in \mathscr{V} \tag{1.5.18}$$

$$(b)\ (\lambda\boldsymbol{a} + \mu\boldsymbol{b})\cdot\boldsymbol{c} = \lambda(\boldsymbol{a}\cdot\boldsymbol{c}) + \mu(\boldsymbol{b}\cdot\boldsymbol{c}) \qquad \forall\, \boldsymbol{a}, \boldsymbol{b}, \boldsymbol{c} \in \mathscr{V} \qquad \lambda, \mu \in \mathscr{F} \tag{1.5.19}$$

$$(c)\ \boldsymbol{a}\cdot\boldsymbol{a} \geqslant 0 \qquad \forall\, \boldsymbol{a} \in \mathscr{V} \text{ with } \boldsymbol{a}\cdot\boldsymbol{a} = 0 \Leftrightarrow \boldsymbol{a} = \boldsymbol{0}. \tag{1.5.20}$$

Equation (1.5.18) formulates an axiom which asserts that the scalar product is commutative. Axioms which assert that the scalar product is associative and distributive are formulated by equation (1.5.19). The axiom formulated by equation (1.5.20) asserts that for all non-zero $\boldsymbol{a}$ the scalar product is positive definite.

The *magnitude* of a vector $\boldsymbol{a}$ written $|\boldsymbol{a}|$, is an example of a mathematical concept known as a *norm*, and thus in this context an inner product space is a *normed space*. A *norm* on $\mathscr{V}$ is an operation that assigns to each non-zero vector $\boldsymbol{a} \in \mathscr{V}$ a positive real number,

$$|\boldsymbol{a}| = (\boldsymbol{a} \cdot \boldsymbol{a})^{1/2} \tag{1.5.21}$$

and which satisfies the following axioms

$$\begin{aligned} &\text{(i) } |\boldsymbol{a}| \geqslant 0 \text{ and } |\boldsymbol{a}| = 0 \Leftrightarrow \boldsymbol{a} = \boldsymbol{0} \\ &\text{(ii) } |\lambda \boldsymbol{a}| = |\lambda||\boldsymbol{a}| \\ &\text{(iii) } |\boldsymbol{a}| + |\boldsymbol{b}| \geqslant |\boldsymbol{a} + \boldsymbol{b}| \end{aligned} \tag{1.5.22}$$

for all $\boldsymbol{a}, \boldsymbol{b} \in \mathscr{V}$ and all $\lambda \in \mathscr{F}$.

A fundamental inequality exists between the norms of two vectors and the magnitude of their scalar product: thus

$$|\boldsymbol{a} \cdot \boldsymbol{b}| \leqslant |\boldsymbol{a}||\boldsymbol{b}| \tag{1.5.23}$$

which is known as the *Schwarz inequality*. The Schwarz inequality enables us to define the angle θ between two vectors $\boldsymbol{a}$ and $\boldsymbol{b}$ in a *real* inner product space by means of

$$\cos \theta = \frac{\boldsymbol{a} \cdot \boldsymbol{b}}{|\boldsymbol{a}||\boldsymbol{b}|} \tag{1.5.24}$$

where θ is an angle between 0 and π.

A vector with a unit norm is termed a *unit vector*. Two vectors $\boldsymbol{a}$ and $\boldsymbol{b}$ are said to be *orthogonal* if $\boldsymbol{a} \cdot \boldsymbol{b} = 0$.

A system of vectors in an inner product space $\mathscr{V}$ is said to be *orthogonal* and *normalised*, or simply *orthonormal* if all the vectors in the system are normalised to unity and mutually orthogonal. It is to be noted that the existence of an orthonormal basis is a fundamental property of a finite-dimensional vector space equipped with a scalar product. An orthonormal set $\hat{e} = \{\hat{\boldsymbol{e}}_1, \hat{\boldsymbol{e}}_2, \ldots, \hat{\boldsymbol{e}}_n\}$ satisfies the conditions

$$\hat{\boldsymbol{e}}_i \cdot \hat{\boldsymbol{e}}_j = \delta^{ij} \tag{1.5.25}$$

where

$$\delta_{ij} = \begin{cases} 1 & \text{if } i = j \\ 0 & \text{if } i \neq j \end{cases} \tag{1.5.26}$$

is the *Kronecker delta*.

A non-empty subset $\mathscr{W}$ of a vector space $\mathscr{V}$ is a subspace if

$$\text{(i) } \boldsymbol{a}, \boldsymbol{w} \in \mathscr{W} \Rightarrow \boldsymbol{a} + \boldsymbol{w} \in \mathscr{W} \qquad \forall \boldsymbol{a}, \boldsymbol{w} \in \mathscr{W} \tag{1.5.27}$$

$$\text{(ii) } \boldsymbol{a} \in \mathscr{W} \Rightarrow \lambda \boldsymbol{a} \in \mathscr{W} \qquad \forall \lambda \in \mathscr{F}. \tag{1.5.28}$$

The conditions formulated by equations (1.5.27) and (1.5.28) can be replaced by the equivalent condition:

$$\boldsymbol{a}, \boldsymbol{w} \in \mathscr{W} \Rightarrow \lambda \boldsymbol{a} + \mu \boldsymbol{w} \in \mathscr{W} \qquad \forall \lambda, \mu \in \mathscr{F}. \tag{1.5.29}$$

The vector spaces $\{\boldsymbol{0}\}$ and $\mathscr{V}$ itself are *trivial subspaces* of the vector space $\mathscr{V}$. If $\mathscr{W}$ is not a trivial subspace, it is said to be a *proper subspace* of $\mathscr{V}$.

1.5.2 Vector Spaces: II, Euclidean

It is a fundamental fact of nature that the space we inhabit is three-dimensional. For this reason many branches of applied mathematics and theoretical physics are concerned with physical quantities defined in a three-dimensional space. Examples are Newtonian mechanics, many aspects of modern continuum mechanics, non-relativistic quantum mechanics, and several aspects of solid state physics, to mention but a few. The geometrical aspects, basic to any understanding of the physical world have their origins in Euclidean geometry, that is three-dimensional geometry, and hence the term Euclidean space.

The theory of scalar- and vector-valued functions defined on subsets of a three-dimensional Euclidean space forms an important basic part of the mathematical foundations central to the formulation of *modern continuum mechanics.*

The concepts underlying the definition of a Euclidean vector space are just those given above for an n-dimensional vector space for the particular case $n=3$, together with one significant addition. The additional concept is that of the *vector* or *outer product* $\boldsymbol{a} \wedge \boldsymbol{b}$ of two arbitrary vectors $\boldsymbol{a}$ and $\boldsymbol{b}$.

To emphasise the importance of a three-dimensional vector space $\mathscr{V}$ over the field $\mathscr{R}$ of real numbers, that is a *real* Euclidean vector space, the definition of a finite-dimensional inner product space with $n=3$ is restated in detail for the field of real numbers $\mathscr{R}$, together with the laws defining the vector product.

The axioms governing a real Euclidean vector space $\mathscr{V}$ can be stated in terms of scalars $\lambda, \mu \in \mathscr{R}$ and arbitrary vectors $\boldsymbol{a}, \boldsymbol{b}, \boldsymbol{c} \in \mathscr{V}$ as follows.

(i) There exists a binary operation in $\mathscr{V}$ called addition and denoted by $+$ such that:

$$(a)\ \boldsymbol{a}+\boldsymbol{b}=\boldsymbol{c} \qquad \forall\, \boldsymbol{a}, \boldsymbol{b}, \boldsymbol{c} \in \mathscr{V} \tag{1.5.30}$$

$$(b)\ \boldsymbol{a}+\boldsymbol{b}=\boldsymbol{b}+\boldsymbol{a} \qquad \forall\, \boldsymbol{a}, \boldsymbol{b} \in \mathscr{V} \tag{1.5.31}$$

$$(c)\ \boldsymbol{a}+(\boldsymbol{b}+\boldsymbol{c})=(\boldsymbol{a}+\boldsymbol{b})+\boldsymbol{c} \qquad \forall\, \boldsymbol{a}, \boldsymbol{b}, \boldsymbol{c} \in \mathscr{V} \tag{1.5.32}$$

(d) there exists an element $\boldsymbol{0} \in \mathscr{V}$ such that

$$\boldsymbol{a}+\boldsymbol{0}=\boldsymbol{a} \qquad \forall\, \boldsymbol{a} \in \mathscr{V} \tag{1.5.33}$$

(e) there exists an element $-\boldsymbol{a} \in \mathscr{V}$ such that

$$\boldsymbol{a}+(-\boldsymbol{a})=\boldsymbol{0} \qquad \forall\, \boldsymbol{a} \in \mathscr{V}. \tag{1.5.34}$$

(ii) There exists an operation called *scalar multiplication* in which every scalar $\lambda \in \mathscr{R}$ can be combined with every element $\boldsymbol{a} \in \mathscr{V}$ to give an element $\lambda\boldsymbol{a} \in \mathscr{V}$ such that:

$$(a)\ 1\boldsymbol{a}=\boldsymbol{a} \qquad \forall\, \boldsymbol{a} \in \mathscr{V} \tag{1.5.35}$$

$$(b)\ \lambda(\mu\boldsymbol{a})=\mu(\lambda\boldsymbol{a})=(\mu\lambda)\boldsymbol{a} \qquad \forall\, \boldsymbol{a} \in \mathscr{V} \qquad \lambda, \mu \in \mathscr{R} \tag{1.5.36}$$

$$(c)\ (\lambda+\mu)\boldsymbol{a}=\lambda\boldsymbol{a}+\mu\boldsymbol{a} \qquad \forall\, \boldsymbol{a} \in \mathscr{V} \qquad \lambda, \mu \in \mathscr{R} \tag{1.5.37}$$

$$(d)\ \lambda(\boldsymbol{a}+\boldsymbol{b})=\lambda\boldsymbol{a}+\lambda\boldsymbol{b} \qquad \forall\, \boldsymbol{a}, \boldsymbol{b} \in \mathscr{V} \qquad \lambda \in \mathscr{R}. \tag{1.5.38}$$

(iii) There exists an operation called *inner product* by which any ordered pair of vectors $\boldsymbol{a}$ and $\boldsymbol{b}$ in $\mathscr{V}$ determines an element of $\mathscr{R}$ denoted by $\boldsymbol{a}\cdot\boldsymbol{b}$ such that

$$(a)\ \boldsymbol{b}\cdot\boldsymbol{a}=\boldsymbol{a}\cdot\boldsymbol{b} \qquad \forall\, \boldsymbol{a}, \boldsymbol{b} \in \mathscr{V} \tag{1.5.39}$$

$$(b)\ (\lambda\boldsymbol{a}+\mu\boldsymbol{b})\cdot\boldsymbol{c}=\lambda(\boldsymbol{a}\cdot\boldsymbol{c})+\mu(\boldsymbol{b}\cdot\boldsymbol{c}) \qquad \forall \boldsymbol{a}, \boldsymbol{b}, \boldsymbol{c} \in \mathscr{V} \qquad \lambda, \mu \in \mathscr{R} \tag{1.5.40}$$

$$(c)\ \boldsymbol{a}\cdot\boldsymbol{a} \geqslant 0 \qquad \forall\, \boldsymbol{a} \in \mathscr{V} \text{ with } \boldsymbol{a}\cdot\boldsymbol{a}=0 \Leftrightarrow \boldsymbol{a}=\boldsymbol{0}. \tag{1.5.41}$$

(iv) There exists an operation called *outer product* by which any ordered pair of vectors $\boldsymbol{a}$ and $\boldsymbol{b}$ in $\mathscr{V}$ determines an element in $\mathscr{V}$ denoted by $\boldsymbol{a} \wedge \boldsymbol{b}$ such that:

$$(a)\ \boldsymbol{b} \wedge \boldsymbol{a}=-\boldsymbol{a} \wedge \boldsymbol{b} \qquad \forall\, \boldsymbol{a}, \boldsymbol{b} \in \mathscr{V} \tag{1.5.42}$$

$$(b)\ (\lambda\boldsymbol{a}+\mu\boldsymbol{b}) \wedge \boldsymbol{c}=\lambda(\boldsymbol{a} \wedge \boldsymbol{c})+\mu(\boldsymbol{b} \wedge \boldsymbol{c}) \qquad \forall\, \boldsymbol{a}, \boldsymbol{b}, \boldsymbol{c} \in \mathscr{V} \qquad \lambda, \mu \in \mathscr{R} \tag{1.5.43}$$

$$(c)\ \boldsymbol{a}\cdot(\boldsymbol{a} \wedge \boldsymbol{b})=0 \qquad \forall\, \boldsymbol{a}, \boldsymbol{b} \in \mathscr{V} \tag{1.5.44}$$

$$(d)\ (\boldsymbol{a} \wedge \boldsymbol{b})\cdot(\boldsymbol{a} \wedge \boldsymbol{b})=(\boldsymbol{a}\cdot\boldsymbol{a})(\boldsymbol{b}\cdot\boldsymbol{b})-(\boldsymbol{a}\cdot\boldsymbol{b})^2 \qquad \forall\, \boldsymbol{a}, \boldsymbol{b} \in \mathscr{V}. \tag{1.5.45}$$

With regard to the vector product, it can be shown, as follows, that $\boldsymbol{a} \wedge \boldsymbol{b}=\boldsymbol{0}$ if and only if $\boldsymbol{a}$ and $\boldsymbol{b}$ are linearly dependent. If $\boldsymbol{a}$ and $\boldsymbol{b}$ are linearly dependent, either $\boldsymbol{a}=\boldsymbol{0}$ or there is a scalar λ (in $\mathscr{R}$) such that $\boldsymbol{b}=\lambda\boldsymbol{a}$. For $\boldsymbol{a}=\boldsymbol{0}$ it follows from the axiom formulated by equation (1.5.43) that $\boldsymbol{a} \wedge \boldsymbol{b}=\boldsymbol{0}$, and substituting $\lambda\boldsymbol{a}$ for $\boldsymbol{b}$ in equation (1.5.42) and applying the axiom formulated by equation (1.5.43) (with $\mu=0$) gives $2(\boldsymbol{a} \wedge \boldsymbol{b})=\boldsymbol{0}$ and hence must have $\boldsymbol{a} \wedge \boldsymbol{b}=\boldsymbol{0}$. Alternatively, with $\boldsymbol{a} \wedge \boldsymbol{b}=\boldsymbol{0}$, the axioms formulated by equations (1.5.41) and (1.5.45) together with equation (1.5.21) give $\boldsymbol{a}\cdot\boldsymbol{b}=\pm|\boldsymbol{a}||\boldsymbol{b}|$. Taking first the plus sign, use of the axioms formulated by equations (1.5.39) and (1.5.40) gives

$$(|\boldsymbol{b}|\boldsymbol{a}-|\boldsymbol{a}|\boldsymbol{b})\cdot(|\boldsymbol{b}|\boldsymbol{a}-|\boldsymbol{a}|\boldsymbol{b})=2|\boldsymbol{a}|^2|\boldsymbol{b}|^2-2|\boldsymbol{a}||\boldsymbol{b}|\boldsymbol{a}\cdot\boldsymbol{b}=0$$

and as a consequence of the axiom formulated by equation (1.5.41), $|\boldsymbol{b}|\boldsymbol{a}=|\boldsymbol{a}|\boldsymbol{b}$. When the minus sign holds, a similar argument gives $|\boldsymbol{b}|\boldsymbol{a}=-|\boldsymbol{a}|\boldsymbol{b}$. For both plus and minus, it follows that either $\boldsymbol{a}=\boldsymbol{0}$ or $\boldsymbol{a}$ is a scalar multiple of $\boldsymbol{b}$. Thus, $\boldsymbol{a}$ and $\boldsymbol{b}$ are linearly dependent.

The *scalar triple product* of three vectors $\boldsymbol{a}, \boldsymbol{b}, \boldsymbol{c}$, often referred to simply as the triple product, is denoted by $[\boldsymbol{a}, \boldsymbol{b}, \boldsymbol{c}]$ and defined by the relation

$$[\boldsymbol{a}, \boldsymbol{b}, \boldsymbol{c}]=\boldsymbol{a}\cdot(\boldsymbol{b} \wedge \boldsymbol{c}). \tag{1.5.46}$$

Having regard to the axioms formulated by equations (1.5.39), (1.5.40) and (1.5.42) the sign of a triple product is reversed when the second and third members of the product are exchanged. From the axioms formulated by equations (1.5.40), (1.5.43) and (1.5.44)

$$\begin{aligned}\boldsymbol{0}&=(\boldsymbol{a}+\boldsymbol{b})\cdot\{(\boldsymbol{a}+\boldsymbol{b}) \wedge \boldsymbol{c}\}\\ &=\boldsymbol{a}\cdot(\boldsymbol{a} \wedge \boldsymbol{c})+\boldsymbol{a}\cdot(\boldsymbol{b} \wedge \boldsymbol{c})+\boldsymbol{b}\cdot(\boldsymbol{a} \wedge \boldsymbol{c})+\boldsymbol{b}\cdot(\boldsymbol{b} \wedge \boldsymbol{c})\\ &=[\boldsymbol{a}, \boldsymbol{b}, \boldsymbol{c}]+[\boldsymbol{b}, \boldsymbol{a}, \boldsymbol{c}]\end{aligned} \tag{1.5.47}$$

from which it is evident that sign reversal also results from interchanging the first and second members of a triple product. Repeated application of these properties leads to the identities:

$$\begin{aligned}[\boldsymbol{a}, \boldsymbol{b}, \boldsymbol{c}]&=[\boldsymbol{b}, \boldsymbol{c}, \boldsymbol{a}]=[\boldsymbol{c}, \boldsymbol{a}, \boldsymbol{b}]\\ &=-[\boldsymbol{a}, \boldsymbol{c}, \boldsymbol{b}]=-[\boldsymbol{b}, \boldsymbol{a}, \boldsymbol{c}]=-[\boldsymbol{c}, \boldsymbol{b}, \boldsymbol{a}] \qquad \forall\, \boldsymbol{a}, \boldsymbol{b}, \boldsymbol{c}\in\mathscr{V}.\end{aligned} \tag{1.5.48}$$

Replacing $\boldsymbol{c}$ by $\boldsymbol{c} \wedge \boldsymbol{d}$ in the axiom formulated by equation (1.5.40) gives

$$[\lambda\boldsymbol{a}+\mu\boldsymbol{b}, \boldsymbol{c}, \boldsymbol{d}]=\lambda[\boldsymbol{a}, \boldsymbol{c}, \boldsymbol{d}]+\mu[\boldsymbol{b}, \boldsymbol{c}, \boldsymbol{d}] \qquad \forall \boldsymbol{a}, \boldsymbol{b}, \boldsymbol{c}, \boldsymbol{d}\in\mathscr{V} \quad \lambda, \mu\in\mathscr{R}. \tag{1.5.49}$$

The axiom formulated by equation (1.5.40) together with the identities of equation (1.5.48), imply that a triple product having the zero vector as one of its members vanishes. If $\boldsymbol{a}, \boldsymbol{b}, \boldsymbol{c}$ are linearly dependent there exist scalars l, m, n, not all zero, such that $l\boldsymbol{a}+m\boldsymbol{b}+n\boldsymbol{c}=\boldsymbol{0}$. Hence the triple products,

$$[l\boldsymbol{a}+m\boldsymbol{b}+n\boldsymbol{c}, \boldsymbol{b}, \boldsymbol{c}] \qquad [\boldsymbol{a}, l\boldsymbol{a}+m\boldsymbol{b}+n\boldsymbol{c}, \boldsymbol{c}]$$
$$[\boldsymbol{a}, \boldsymbol{b}, l\boldsymbol{a}+m\boldsymbol{b}+n\boldsymbol{c}]$$

are all zero and, using equations (1.5.48) and (1.5.49), they reduce in turn to $l[\boldsymbol{a}, \boldsymbol{b}, \boldsymbol{c}], m[\boldsymbol{a}, \boldsymbol{b}, \boldsymbol{c}], n[\boldsymbol{a}, \boldsymbol{b}, \boldsymbol{c}]$. Since at least one of the scalars l, m, n is non-zero, $[\boldsymbol{a}, \boldsymbol{b}, \boldsymbol{c}]=0$. Now take $[\boldsymbol{a}, \boldsymbol{b}, \boldsymbol{c}]=0$ and assume that $\boldsymbol{a}, \boldsymbol{b}, \boldsymbol{c}$ are linearly independent vectors. It has been shown above, in relation to the vector product that if $\boldsymbol{b}$ and $\boldsymbol{c}$ are linearly independent vectors then $\boldsymbol{b} \wedge \boldsymbol{c}\neq\boldsymbol{0}$, and it then follows from equations (1.5.46) and (1.5.48) that $\boldsymbol{a}, \boldsymbol{b}$ and $\boldsymbol{c}$ are each orthogonal to $\boldsymbol{b} \wedge \boldsymbol{c}$. Since $\boldsymbol{a}, \boldsymbol{b}, \boldsymbol{c}$ form a basis of $\mathscr{V}$ it follows that every vector is orthogonal to $\boldsymbol{b} \wedge \boldsymbol{c}$. This conclusion is clearly false, since $\boldsymbol{b} \wedge \boldsymbol{c}$ is not orthogonal to itself, thus implying that $\boldsymbol{a}, \boldsymbol{b}, \boldsymbol{c}$ must be linearly dependent in direct contrast to the initial assumption regarding $\boldsymbol{a}, \boldsymbol{b}, \boldsymbol{c}$.

1.5.3 Component Forms

In this section the discussion is restricted to a real Euclidean vector space $\mathscr{V}$.

The basis of a vector space is any linearly independent system of vectors of maximum order. Thus, in the Euclidean vector space $\mathscr{V}$ there is a set of three vectors $e=\{\boldsymbol{e}_1, \boldsymbol{e}_2, \boldsymbol{e}_3\}$. Corresponding to an arbitrary vector $\boldsymbol{a}$ (in $\mathscr{V}$) there is an ordered triplet of scalars, (a^1, a^2, a^3) (in $\mathscr{R}$) such that

$$\boldsymbol{a}=a^1\boldsymbol{e}_1+a^2\boldsymbol{e}_2+a^3\boldsymbol{e}_3, \tag{1.5.50}$$

where the a^i $(i=1,2,3)$ are called the *components of* $\boldsymbol{a}$ relative to the basis e. The relation of equation (1.5.50) can be expressed by way of the *summation convention* in the indicial form

$$\boldsymbol{a}=a^p\boldsymbol{e}_p \tag{1.5.51}$$

summation over the values of $p=1,2,3$ being understood, unless the condition *no sum* is specifically stated. In general, for any vector relation expressed in indicial form, the summation convention is such that if a suffix appears twice, then it is understood to be summed over the range 1, 2, 3 of that suffix. The axioms formulated by way of equations (1.5.39) and (1.5.40) allow the scalar product of an arbitrary vector $\boldsymbol{a}$ with a vector $\boldsymbol{e}_i$ to be expressed by way of equation (1.5.51) in the form

$$\boldsymbol{a}\cdot\boldsymbol{e}_j=a^p\boldsymbol{e}_p\cdot\boldsymbol{e}_j=a^pg_{pj}=a_j \tag{1.5.52}$$

where

$$\boldsymbol{e}_i\cdot\boldsymbol{e}_j=g_{ij} \qquad (g_{ij}=g_{ji}) \tag{1.5.53}$$

it being noted that the g_{ij} are the elements of a square matrix, (see §1.1). In equation (1.5.52) the repeated index p is a *dummy index* and thus can be replaced by any other symbol that has not already been used.

Associated with any arbitrary basis is another basis, called a *dual* basis, that can be derived from it. The concept of a dual basis can be formulated as follows. For a basis $e=\{\boldsymbol{e}_1, \boldsymbol{e}_2, \boldsymbol{e}_3\}$, there can be obtained, using equations (1.5.46) and (1.5.51), the relation

$$\boldsymbol{a}\cdot(\boldsymbol{e}_1\wedge\boldsymbol{e}_2)=a^3\boldsymbol{e}_3\cdot(\boldsymbol{e}_1\wedge\boldsymbol{e}_2)=a^3[\boldsymbol{e}_1,\boldsymbol{e}_2,\boldsymbol{e}_3]$$

since $\boldsymbol{e}_1\wedge\boldsymbol{e}_2$ is orthogonal to both $\boldsymbol{e}_1$ and $\boldsymbol{e}_2$. Solving for a^3 gives

$$a^3=\boldsymbol{a}\cdot\frac{(\boldsymbol{e}_1\wedge\boldsymbol{e}_2)}{[\boldsymbol{e}_1,\boldsymbol{e}_2,\boldsymbol{e}_3]} \tag{1.5.54}$$

and similarly, there can be obtained

$$a^1=\boldsymbol{a}\cdot\frac{\boldsymbol{e}_2\wedge\boldsymbol{e}_3}{[\boldsymbol{e}_1,\boldsymbol{e}_2,\boldsymbol{e}_3]} \qquad a^2=\boldsymbol{a}\cdot\frac{\boldsymbol{e}_3\wedge\boldsymbol{e}_1}{[\boldsymbol{e}_1,\boldsymbol{e}_2,\boldsymbol{e}_3]}. \tag{1.5.55}$$

Setting

$$\boldsymbol{e}^1=\frac{\boldsymbol{e}_2\wedge\boldsymbol{e}_3}{[\boldsymbol{e}_1,\boldsymbol{e}_2,\boldsymbol{e}_3]} \qquad \boldsymbol{e}^2=\frac{\boldsymbol{e}_3\wedge\boldsymbol{e}_1}{[\boldsymbol{e}_1,\boldsymbol{e}_2,\boldsymbol{e}_3]} \qquad \boldsymbol{e}^3=\frac{\boldsymbol{e}_1\wedge\boldsymbol{e}_2}{[\boldsymbol{e}_1,\boldsymbol{e}_2,\boldsymbol{e}_3]} \tag{1.5.56}$$

in equations (1.5.54) and (1.5.55) gives

$$a^1 = \boldsymbol{a}\cdot\boldsymbol{e}^1 \qquad a^2 = \boldsymbol{a}\cdot\boldsymbol{e}^2 \qquad a^3 = \boldsymbol{a}\cdot\boldsymbol{e}^3 \tag{1.5.57}$$

it being noted that the use of equation (1.5.46) gives

$$\boldsymbol{e}^1\cdot\boldsymbol{e}_1 = \boldsymbol{e}^2\cdot\boldsymbol{e}_2 = \boldsymbol{e}^3\cdot\boldsymbol{e}_3 = 1 \tag{1.5.58}$$

and hence, in general

$$\boldsymbol{e}^i\cdot\boldsymbol{e}_j = \delta^i{}_j \equiv \delta_{ij}. \tag{1.5.59}$$

The set of vectors $e^* = \{\boldsymbol{e}^1, \boldsymbol{e}^2, \boldsymbol{e}^3\}$ constitute the *dual* or *reciprocal* basis. Consider the linear relation

$$u_1\boldsymbol{e}^1 + u_2\boldsymbol{e}^2 + u_3\boldsymbol{e}^3 = \boldsymbol{0}.$$

By taking the scalar product of the above relation with $\boldsymbol{e}_i$ it follows from equation (1.5.59) that $u_i = 0$, $(i = 1, 2, 3)$, and hence the dual basis is linearly independent, and it is therefore possible to express a vector in terms of the dual basis:

$$\boldsymbol{a} = a_p\boldsymbol{e}^p. \tag{1.5.60}$$

The axioms formulated by way of equations (1.5.39) and (1.5.40) allow the scalar product of an arbitrary vector $\boldsymbol{a}$ with a vector $\boldsymbol{e}^i$ to be expressed by way of equation (1.5.60) in the form

$$\boldsymbol{a}\cdot\boldsymbol{e}^j = a_p\boldsymbol{e}^p\cdot\boldsymbol{e}^j = a_p g^{pj} = a^j \tag{1.5.61}$$

where

$$\boldsymbol{e}^i\cdot\boldsymbol{e}^j = g^{ij} \qquad (g^{ij} = g^{ji}) \tag{1.5.62}$$

it being noted that the g^{ij} are the elements of a square matrix.

It is evident from equations (1.5.51) and (1.5.60) that the term vector refers to an element

$$\boldsymbol{a} = a^p\boldsymbol{e}_p = a_p\boldsymbol{e}^p \tag{1.5.63}$$

the components being given by

$$a^i = \boldsymbol{a}\cdot\boldsymbol{e}^i \qquad a_i = \boldsymbol{a}\cdot\boldsymbol{e}_i \tag{1.5.64}$$

it being noted that there is really only one basis since the dual basis was derived from the given basis. However, it is evident that there exist two ways of expressing the same vector, for a given basis, in component form. The components a^i are called the *contravariant* components of the vector $\boldsymbol{a}$ and the components a_i are called the *covariant* components of $\boldsymbol{a}$. The following relations are easily proved:

$$\begin{gathered} \boldsymbol{e}_i = g_{ip}\boldsymbol{e}^p \qquad \boldsymbol{e}^i = g^{ip}\boldsymbol{e}_p \\ g^{ip}g_{pj} = g^i{}_j = \delta^i{}_j = \delta_{ij} \end{gathered} \tag{1.5.65}$$

$$a_i = g_{ip} a^p \qquad a^i = g^{ip} a_p. \tag{1.5.66}$$

The axioms formulated by way of equations (1.5.39) and (1.5.40), together with equations (1.5.63), give the scalar product of two arbitrary vectors $\boldsymbol{a}$ and $\boldsymbol{b}$ in the form

$$\begin{aligned} \boldsymbol{a} \cdot \boldsymbol{b} &= (a^p \boldsymbol{e}_p) \cdot (b_q \boldsymbol{e}^q) = a^p b_q \, (\boldsymbol{e}_p \cdot \boldsymbol{e}^q) \\ &= a^p b_q \delta_{pq} = a^p b_p. \end{aligned} \tag{1.5.67}$$

The form of equation (1.5.67) for the scalar product of two vectors in terms of their components confirms that there is one and only one scalar product on $\mathscr{V}$ satisfying the axioms formulated by equations (1.5.39), (1.5.40) and (1.5.41). For $\boldsymbol{a} = \boldsymbol{b}$, equations (1.5.21) and (1.5.67) give

$$|\boldsymbol{a}| = (a_p a^p)^{1/2}. \tag{1.5.68}$$

Substituting for the a_i in equation (1.5.60) from equation (1.5.64) gives

$$\boldsymbol{a} = (\boldsymbol{a} \cdot \boldsymbol{e}_p) \boldsymbol{e}^p. \tag{1.5.69}$$

Set $\boldsymbol{a} = \boldsymbol{e}_2 \wedge \boldsymbol{e}_3$ in equation (1.5.69) to give

$$\boldsymbol{e}_2 \wedge \boldsymbol{e}_3 = \{(\boldsymbol{e}_2 \wedge \boldsymbol{e}_3) \cdot \boldsymbol{e}_p\} \boldsymbol{e}^p = [\boldsymbol{e}_2, \boldsymbol{e}_3, \boldsymbol{e}_p] \boldsymbol{e}^p = [\boldsymbol{e}_1, \boldsymbol{e}_2, \boldsymbol{e}_3] \boldsymbol{e}^1$$

use being made of equations (1.5.46) and (1.5.48). In this way the following relations can be obtained:

$$\begin{aligned} \boldsymbol{e}_2 \wedge \boldsymbol{e}_3 &= [\boldsymbol{e}_1, \boldsymbol{e}_2, \boldsymbol{e}_3] \boldsymbol{e}^1 \\ \boldsymbol{e}_3 \wedge \boldsymbol{e}_1 &= [\boldsymbol{e}_1, \boldsymbol{e}_2, \boldsymbol{e}_3] \boldsymbol{e}^2 \\ \boldsymbol{e}_1 \wedge \boldsymbol{e}_2 &= [\boldsymbol{e}_1, \boldsymbol{e}_2, \boldsymbol{e}_3] \boldsymbol{e}^3. \end{aligned} \tag{1.5.70}$$

From the relations of equation (1.5.70) the following can be obtained:

$$\begin{aligned} (\boldsymbol{e}_2 \wedge \boldsymbol{e}_3) \cdot (\boldsymbol{e}_2 \wedge \boldsymbol{e}_3) &= [\boldsymbol{e}_1, \boldsymbol{e}_2, \boldsymbol{e}_3]^2 g^{11} \\ (\boldsymbol{e}_2 \wedge \boldsymbol{e}_3) \cdot (\boldsymbol{e}_3 \wedge \boldsymbol{e}_1) &= [\boldsymbol{e}_1, \boldsymbol{e}_2, \boldsymbol{e}_3]^2 g^{12} \\ (\boldsymbol{e}_2 \wedge \boldsymbol{e}_3) \cdot (\boldsymbol{e}_1 \wedge \boldsymbol{e}_2) &= [\boldsymbol{e}_1, \boldsymbol{e}_2, \boldsymbol{e}_3]^2 g^{13}. \end{aligned} \tag{1.5.71}$$

Use of the vector identity

$$(\boldsymbol{a} \wedge \boldsymbol{b}) \cdot (\boldsymbol{c} \wedge \boldsymbol{d}) = (\boldsymbol{a} \cdot \boldsymbol{c})(\boldsymbol{b} \cdot \boldsymbol{d}) - (\boldsymbol{a} \cdot \boldsymbol{d})(\boldsymbol{b} \cdot \boldsymbol{c}) \qquad 1.5.72)$$

together with equation (1.5.53) enables the following relations to be obtained:

$$\begin{aligned} (\boldsymbol{e}_2 \wedge \boldsymbol{e}_3) \cdot (\boldsymbol{e}_2 \wedge \boldsymbol{e}_3) &= \operatorname{cof} g_{11} \\ (\boldsymbol{e}_2 \wedge \boldsymbol{e}_3) \cdot (\boldsymbol{e}_3 \wedge \boldsymbol{e}_1) &= \operatorname{cof} g_{12} \\ (\boldsymbol{e}_2 \wedge \boldsymbol{e}_3) \cdot (\boldsymbol{e}_1 \wedge \boldsymbol{e}_2) &= \operatorname{cof} g_{13}. \end{aligned} \tag{1.5.73}$$

Since

$$g^{ij} = \frac{\text{cof } g_{ij}}{g} \tag{1.5.74}$$

where

$$g \equiv \det[g_{ij}] \tag{1.5.75}$$

it follows from equations (1.5.71) and (1.5.72) that

$$[\boldsymbol{e}_1, \boldsymbol{e}_2, \boldsymbol{e}_3]^2 = g. \tag{1.5.76}$$

Substituting for $[\boldsymbol{e}_1, \boldsymbol{e}_2, \boldsymbol{e}_3]$ in equation (1.5.70) from equation (1.5.76) gives

$$\begin{gathered} \boldsymbol{e}_2 \wedge \boldsymbol{e}_3 = \pm\sqrt{g}\,\boldsymbol{e}^1 \qquad \boldsymbol{e}_3 \wedge \boldsymbol{e}_1 = \pm\sqrt{g}\,\boldsymbol{e}^2 \\ \boldsymbol{e}_1 \wedge \boldsymbol{e}_2 = \pm\sqrt{g}\,\boldsymbol{e}^3. \end{gathered} \tag{1.5.77}$$

The relations of equation (1.5.77) can be expressed in the form of the single relation,

$$\boldsymbol{e}_i \wedge \boldsymbol{e}_j = \pm\varepsilon_{ijp}\boldsymbol{e}^p \tag{1.5.78}$$

where the covariant components of the *general permutation symbol* are given by

$$\varepsilon_{ijk} \equiv (\boldsymbol{e}_i \wedge \boldsymbol{e}_j)\cdot\boldsymbol{e}_k = \sqrt{g}\,e_{ijk} \tag{1.5.79}$$

and where the *permutation symbol*, or *alternator*

$$e_{ijk} = \begin{cases} 0, & \text{when any two indices are equal} \\ +1, & \text{when } i, j, k \text{ are a cyclic permutation of } 1, 2, 3 \\ -1, & \text{when } i, j, k \text{ are a non-cyclic permutation of } 1, 2, 3. \end{cases} \tag{1.5.80}$$

Analogous to equations (1.5.78) and (1.5.79)

$$\boldsymbol{e}^i \wedge \boldsymbol{e}^j = \pm\varepsilon^{ijp}\boldsymbol{e}_p \tag{1.5.81}$$

$$\varepsilon^{ijk} \equiv (\boldsymbol{e}^i \wedge \boldsymbol{e}^j)\cdot\boldsymbol{e}^k = \frac{1}{\sqrt{g}}\,e^{ijk}. \tag{1.5.82}$$

where the ε^{ijk} are the contravariant components of the general permutation symbol.

The vector product $\boldsymbol{a} \wedge \boldsymbol{b}$ of the arbitrary vectors $\boldsymbol{a}$ and $\boldsymbol{b}$ can be expressed in component form, using equations (1.5.66), (1.5.78) and (1.5.81): thus

$$\begin{aligned} \boldsymbol{a} \wedge \boldsymbol{b} &= (a^p\boldsymbol{e}_p) \wedge (b^q\boldsymbol{e}_q) = a^p b^q(\boldsymbol{e}_p \wedge \boldsymbol{e}_q) \\ &= \pm\varepsilon_{pqr}a^p b^q\boldsymbol{e}^r \end{aligned} \tag{1.5.83}$$

and

$$\boldsymbol{a} \wedge \boldsymbol{b} = (a_p \boldsymbol{e}^p) \wedge (b_q \boldsymbol{e}^q) = a_p b_q (\boldsymbol{e}^p \wedge \boldsymbol{e}^q)$$
$$= \pm \varepsilon^{pqr} a_p b_q \boldsymbol{e}_r. \tag{1.5.84}$$

Similarly for the triple product:

$$[\boldsymbol{a}, \boldsymbol{b}, \boldsymbol{c}] = \pm \varepsilon_{pqr} a^p b^q c^r \tag{1.5.85}$$

and

$$[\boldsymbol{a}, \boldsymbol{b}, \boldsymbol{c}] = \pm \varepsilon^{pqr} a_p b_q c_r. \tag{1.5.86}$$

The existence of an orthonormal basis is a fundamental property of a finite-dimensional vector space characterised by a scalar product. If such a basis has base vectors $\hat{\boldsymbol{e}}_1, \hat{\boldsymbol{e}}_2, \ldots, \hat{\boldsymbol{e}}_n$, then

$$\hat{\boldsymbol{e}}_i \cdot \hat{\boldsymbol{e}}_j = \delta_{ij}. \tag{1.5.87}$$

For an orthonormal basis $\hat{e} = \{\hat{\boldsymbol{e}}_1, \hat{\boldsymbol{e}}_2, \ldots, \hat{\boldsymbol{e}}_n\}$, it is clear that the basis is identical with its dual and that the contravariant and covariant components of a vector are also identical, and hence equation (1.5.63) reduces to the single relation

$$\boldsymbol{a} = a_p \hat{\boldsymbol{e}}_p \tag{1.5.88}$$

which in turn gives

$$\boldsymbol{a} \cdot \hat{\boldsymbol{e}}_i = a_p \hat{\boldsymbol{e}}_p \cdot \hat{\boldsymbol{e}}_i = a_p \delta_{pi} = a_i \tag{1.5.89}$$

and

$$\boldsymbol{a} = a_p \hat{\boldsymbol{e}}_p = (\boldsymbol{a} \cdot \hat{\boldsymbol{e}}_p) \hat{\boldsymbol{e}}_p \tag{1.5.90}$$

which is to be compared with equation (1.5.69).

For a real Euclidean vector space $\mathscr{V}$, the orthonormal basis is $\hat{e} = \{\hat{\boldsymbol{e}}_1, \hat{\boldsymbol{e}}_2, \hat{\boldsymbol{e}}_3\}$. The relations equivalent to equations (1.5.78), (1.5.81) and equations (1.5.83) to (1.5.86) can be expressed in the form

$$\hat{\boldsymbol{e}}_i \wedge \hat{\boldsymbol{e}}_j = \pm e_{ijp} \hat{\boldsymbol{e}}_p \tag{1.5.91}$$

$$\boldsymbol{a} \wedge \boldsymbol{b} = \pm e_{pqr} a_p b_q \hat{\boldsymbol{e}}_r \tag{1.5.92}$$

$$[\boldsymbol{a}, \boldsymbol{b}, \boldsymbol{c}] = \pm e_{pqr} a_p b_q c_r \tag{1.5.93}$$

it being noted that

$$\boldsymbol{a} \cdot \boldsymbol{b} = a_p b_p \tag{1.5.94}$$

where $\boldsymbol{a}, \boldsymbol{b}, \boldsymbol{c}$ are arbitrary vectors having components a_i, b_i, c_i relative to $\hat{e}$.

It is seen from equations (1.5.83), (1.5.84) and (1.5.92) that there are two vector products on $\mathscr{V}$ obeying the axioms formulated by equations (1.5.42) to (1.5.45).

There now arises the question of how to resolve the apparent ambiguity of sign in equations (1.5.77), (1.5.78), (1.5.81), (1.5.83), (1.5.84), (1.5.85), (1.5.86) and equations (1.5.91), (1.5.92) and (1.5.93). It is evident that what is required is some procedure for partitioning the collection of all ordered bases of $\mathscr{V}$ into two classes. Noting that $\mathscr{V}$ is said to be a Euclidean vector space if it supports both a scalar product and a vector product, it would appear that the two classes of $\mathscr{V}$ can be distinguished and defined by way of the scalar triple product. That this is so can be seen as follows. If the scalar triple products of two ordered bases of $\mathscr{V}$ have the same sign, then the bases are said to be *similar*. Similarity in this sense is thus seen to be an equivalence relation which achieves the required partitioning of the collection of all ordered bases of $\mathscr{V}$ into two classes. One class contains the members with positive scalar triple products and the other those with negative scalar triple products. The partitioning of $\mathscr{V}$ into two classes by way of a rule identifying the equivalence class to which each of its ordered bases belongs is called an orientation of $\mathscr{V}$, the existence of two classes of $\mathscr{V}$ implying that there are two such orientations, denoted by $\mathscr{V}^+$ and $\mathscr{V}^-$. An ordered basis of $\mathscr{V}$ is said to be *positive* in $\mathscr{V}^+$ if its scalar triple product is positive and *positive* in $\mathscr{V}^-$ if its scalar triple product is negative. In this way, the ambiguity of sign, wherever it appears in equations (1.5.77) to (1.5.93), is removed by taking the upper or lower sign according as to whether the basis is positive in $\mathscr{V}^+$ or $\mathscr{V}^-$.

1.5.4 Change of Basis

Formulae for changing from one basis to another basis in an inner product space can be developed in terms of both the base vectors and the associated components. Let $\{\boldsymbol{e}_1, \boldsymbol{e}_2, \ldots, \boldsymbol{e}_n\}$ be one set of linearly independent vectors and take as a second basis, the set of linearly independent vectors $\{\bar{\boldsymbol{e}}_1, \bar{\boldsymbol{e}}_2, \ldots, \bar{\boldsymbol{e}}_n\}$. Since both sets of linearly independent vectors are bases of $\mathscr{V}$, it follows that

$$\boldsymbol{e}_i = \bar{G}^p{}_i \bar{\boldsymbol{e}}_p \qquad i = 1, 2, \ldots, n \tag{1.5.95}$$

and

$$\bar{\boldsymbol{e}}_j = G^p{}_j \boldsymbol{e}_p \qquad j = 1, 2, \ldots, n \tag{1.5.96}$$

where $G^p{}_j$ and $\bar{G}^p{}_i$ are both sets of n^2 scalars.

Substituting for $\bar{\boldsymbol{e}}_p$ in equation (1.5.95) from equation (1.5.96) and for $\boldsymbol{e}_p$ in equation (1.5.96) from equation (1.5.95) gives

$$\boldsymbol{e}_i = \bar{G}^p{}_i G^q{}_p \boldsymbol{e}_q \qquad \bar{\boldsymbol{e}}_j = G^p{}_j \bar{G}^q{}_p \bar{\boldsymbol{e}}_q. \tag{1.5.97}$$

Comparing these relations with those of equation (1.5.65), it follows that

$$\bar{G}^p{}_i G^j{}_p = G^p{}_i \bar{G}^j{}_p = \delta_i{}^j. \tag{1.5.98}$$

These equations state that the matrix $[G^i{}_j]$ is the inverse of the matrix $[\bar{G}^j{}_i]$ and, reciprocally, the matrix $[\bar{G}^i{}_j]$ is the inverse of the matrix $[G^j{}_i]$. The dual, or reciprocal bases are related by,

$$e^i = \bar{H}^i{}_p \bar{e}^p \qquad i = 1, 2, \ldots, n \tag{1.5.99}$$

and

$$\bar{e}^j = H^j{}_p e^p \qquad j = 1, 2, \ldots, n. \tag{1.5.100}$$

The covariant and contravariant components of $\boldsymbol{a} \in \mathscr{V}$ relative to the two pairs of bases are given by

$$\boldsymbol{a} = a_p e^p = a^p e_p = \bar{a}_p \bar{e}^p = \bar{a}^p \bar{e}_p \tag{1.5.101}$$

and their components by

$$\begin{aligned} a^i &= G^i{}_p \bar{a}^p \qquad & \bar{a}^p &= \bar{G}^p{}_i a^i \\ a_i &= H_i{}^q \bar{a}_q \qquad & \bar{a}_q &= \bar{H}_q{}^i a_i. \end{aligned} \tag{1.5.102}$$

1.6 TENSORS

1.6.1 Linear Transformations

The concept of a function or transformation from one set into another can be extended to vector spaces. A transformation **A** from a linear vector space $\mathscr{V}$ into another linear vector space $\mathscr{W}$, both vector spaces being defined on the same field of scalars $\mathscr{F}$, is a correspondence which assigns to each element $\boldsymbol{v}$ in $\mathscr{V}$ a unique element $\boldsymbol{w}$ in $\mathscr{W}$. The terms, *transformation*, *mapping*, *operator* and *tensor* are used interchangeably.

If $\mathscr{V}$ and $\mathscr{W}$ are vector spaces, a *linear transformation*, or tensor, is a function $\mathbf{A}: \mathscr{V} \to \mathscr{W}$ such that

$$\mathbf{A}(\boldsymbol{u} + \boldsymbol{v}) = \mathbf{A}(\boldsymbol{u}) + \mathbf{A}(\boldsymbol{v}) \tag{1.6.1}$$

$$\mathbf{A}(\lambda \boldsymbol{v}) = \lambda \mathbf{A}(\boldsymbol{v}) \tag{1.6.2}$$

for all $\boldsymbol{u}, \boldsymbol{v} \in \mathscr{V}$ and $\lambda \in \mathscr{F}$. The condition of equation (1.6.1) asserts that **A** is a *homomorphism* on $\mathscr{V}$ with respect to the operation of addition and thus the theorems of §1.3.3 can be applied here. Equation (1.6.2) shows that **A**, in addition to being a homomorphism, is also *homogeneous* with respect to the operation of scalar multiplication. It is to be noted that the same + symbol is used in equation (1.6.1) to denote addition in $\mathscr{V}$ on the left-hand side in contrast to the right-hand side where it denotes addition in $\mathscr{W}$. To emphasise linearity, the brackets are often omitted:

$$\boldsymbol{w} = \mathbf{A}(\boldsymbol{v}) = \mathbf{A}\boldsymbol{v}. \tag{1.6.3}$$

The conditions stated in terms of equations (1.6.1) and (1.6.2) can be combined into one; a transformation **A** is linear if and only if

$$\mathbf{A}(\lambda \boldsymbol{u} + \mu \boldsymbol{v}) = \lambda \mathbf{A}(\boldsymbol{u}) + \mu \mathbf{A}(\boldsymbol{v}) \tag{1.6.4}$$

for all $\boldsymbol{u}, \boldsymbol{v} \in \mathscr{V}$ and $\lambda, \mu \in \mathscr{F}$.

For a linear transformation $\mathbf{A}: \mathscr{V} \to \mathscr{W}$, it follows by way of §1.3.3 that

$$\mathbf{A}\boldsymbol{0} = \boldsymbol{0} \qquad \text{and} \qquad \mathbf{A}(-\boldsymbol{v}) = -\mathbf{A}\boldsymbol{v}. \tag{1.6.5}$$

If $\boldsymbol{v} \in \mathscr{V}$, the value $\mathbf{A}(\boldsymbol{v})$ is called the *image* of $\boldsymbol{v}$ under $\mathbf{A}$; if $\boldsymbol{w} \in \mathscr{W}$ and $\boldsymbol{w} = \mathbf{A}(\boldsymbol{v})$, the argument $\boldsymbol{v}$ is called a *pre-image* of $\boldsymbol{w}$ under $\mathbf{A}$.

The *kernel*, also known as the *null space* of a linear transformation $\mathbf{A}: \mathscr{V} \to \mathscr{W}$, denoted $K(\mathbf{A})$, is the set of all *pre-images* of the null vector under $\mathbf{A}$:

$$K(\mathbf{A}) = \{\boldsymbol{v} \mid \mathbf{A}\boldsymbol{v} = \boldsymbol{0}\} \tag{1.6.6}$$

it being noted that the set $K(\mathbf{A})$ is a subspace of $\mathscr{V}$ and hence it follows that

$$\dim K(\mathbf{A}) \leqslant \dim \mathscr{V}. \tag{1.6.7}$$

A linear transformation $\mathbf{A}: \mathscr{V} \to \mathscr{W}$ is *one-to-one* if and only if $K(\mathbf{A}) = \{\boldsymbol{0}\}$. If $K(\mathbf{A})$ contains only $\boldsymbol{0}$, then $\mathbf{A}$ is said to be *non-singular*; otherwise it is *singular*.

For a linear transformation $\mathbf{A}: \mathscr{V} \to \mathscr{W}$, the range of $\mathbf{A}$ is denoted by

$$R(\mathbf{A}) = \{\mathbf{A}\boldsymbol{v} \mid \boldsymbol{v} \in \mathscr{V}\} \tag{1.6.8}$$

it being noted that since $R(\mathbf{A})$ is a subspace of $\mathscr{W}$,

$$\dim R(\mathbf{A}) \leqslant \dim \mathscr{W}. \tag{1.6.9}$$

A linear transformation $\mathbf{A}: \mathscr{V} \to \mathscr{W}$ is said to be onto if $R(\mathbf{A}) = \mathscr{W}$, i.e. for every vector $\boldsymbol{w} \in \mathscr{W}$ there exists $\boldsymbol{v} \in \mathscr{V}$ such that $\mathbf{A}\boldsymbol{v} = \boldsymbol{w}$. The $\dim R(\mathbf{A}) = \dim \mathscr{W}$ if and only if $\mathbf{A}$ is *onto*. The *rank* of a linear transformation is defined as the dimension of $R(\mathbf{A})$, i.e. $\dim R(\mathbf{A})$. If $\{\boldsymbol{e}_1, \boldsymbol{e}_2, \ldots, \boldsymbol{e}_n\}$ is a basis for $\mathscr{V}$, where $n = \dim \mathscr{V}$, then

$$\boldsymbol{v} = v^j \boldsymbol{e}_j \qquad (j = 1, 2, \ldots, n)$$

and hence any vector $\mathbf{A}\boldsymbol{v} \in R(\mathbf{A})$ can be written

$$\mathbf{A}\boldsymbol{v} = v^j \mathbf{A}\boldsymbol{e}_j \qquad (j = 1, 2, \ldots, n).$$

Hence the vectors $\{\mathbf{A}\boldsymbol{e}_1, \mathbf{A}\boldsymbol{e}_2, \ldots, \mathbf{A}\boldsymbol{e}_n\}$ generate $R(\mathbf{A})$, which in turn implies

$$\dim R(\mathbf{A}) \leqslant \dim \mathscr{V}. \tag{1.6.10}$$

If $\mathbf{A}: \mathscr{V} \to \mathscr{W}$ is a linear transformation, then

$$\dim K(\mathbf{A}) + \dim R(\mathbf{A}) = \dim \mathscr{V}. \tag{1.6.11}$$

For two given spaces $\mathscr{V}$ and $\mathscr{W}$, the set of all linear transformations, that is, the tensors from $\mathscr{V}$ to $\mathscr{W}$, is denoted by $\mathscr{L}(\mathscr{V}; \mathscr{W})$.

If $\mathbf{A}$ and $\mathbf{B}$ are linear transformations $\mathscr{V} \to \mathscr{W}$, then their sum $\mathbf{A} + \mathbf{B}$, defined by

$$(\mathbf{A} + \mathbf{B})\boldsymbol{v} = \mathbf{A}\boldsymbol{v} + \mathbf{B}\boldsymbol{v} \qquad \forall\, \boldsymbol{v} \in \mathscr{V} \tag{1.6.12}$$

is a linear transformation satisfying the rule for vector addition, that is the axioms formulated by equations (1.5.2) to (1.5.6). Similarly, if $\lambda \in \mathscr{F}$, then $\lambda\mathbf{A}$ is a linear transformation $\mathscr{V} \to \mathscr{W}$ defined by

$$(\lambda \mathbf{A})\boldsymbol{v} = \lambda(\mathbf{A}\boldsymbol{v}) \tag{1.6.13}$$

for all $\boldsymbol{v} \in \mathscr{V}$.

The zero element in $\mathscr{L}(\mathscr{V}; \mathscr{W})$ is the linear transformation $\mathbf{O}$ defined by

$$\mathbf{O}\boldsymbol{v} = \boldsymbol{0} \tag{1.6.14}$$

for all $\boldsymbol{v} \in \mathscr{V}$.

The *negative* of $\mathbf{A} \in \mathscr{L}(\mathscr{V}; \mathscr{W})$ is a linear transformation $-\mathbf{A} \in \mathscr{L}(\mathscr{V}; \mathscr{W})$ defined by

$$-\mathbf{A} = -1\mathbf{A}. \tag{1.6.15}$$

Using equation (1.6.15), the difference $\mathbf{A} - \mathbf{B}$ is defined by the relation

$$\mathbf{A} - \mathbf{B} = \mathbf{A} + (-\mathbf{B}) \tag{1.6.16}$$

where $\mathbf{A}$ and $\mathbf{B}$ are linear transformations.

The conditions underlying equations (1.6.12) to (1.6.16) make $\mathscr{L}(\mathscr{V}; \mathscr{W})$ a vector space, it being noted that,

$$\dim \mathscr{L}(\mathscr{V}; \mathscr{W}) = \dim \mathscr{V} \dim \mathscr{W}.$$

If $\mathscr{U}$, $\mathscr{V}$ and $\mathscr{W}$ are vector spaces, and if $\mathbf{A}: \mathscr{V} \to \mathscr{U}$ and $\mathbf{B}: \mathscr{U} \to \mathscr{W}$ are linear transformations then $\mathbf{BA}: \mathscr{V} \to \mathscr{W}$ is a linear transformation defined by

$$\mathbf{BA}\boldsymbol{v} = \mathbf{B}(\mathbf{A}\boldsymbol{v}) \tag{1.6.17}$$

for all $\boldsymbol{v} \in \mathscr{V}$. As in the case of the composition of two functions, (§1.2.3), the product $\mathbf{AB}$ may not be defined. The properties of the product operation can be stated as follows:

$$\mathbf{C}(\mathbf{BA}) = (\mathbf{CB})\mathbf{A} \tag{1.6.18}$$

$$(\lambda\mathbf{A} + \mu\mathbf{B})\mathbf{C} = \lambda\mathbf{AC} + \mu\mathbf{BC} \tag{1.6.19}$$

$$\mathbf{C}(\lambda\mathbf{A} + \mu\mathbf{B}) = \lambda\mathbf{CA} + \mu\mathbf{CB} \tag{1.6.20}$$

for all $\lambda, \mu \in \mathscr{F}$.

A linear map $\mathbf{A}: \mathscr{V} \to \mathscr{W}$ is called an *isomorphism* if it is one-to-one and surjective, (see §§1.2.3 and 1.3.3). If such a map exists, the two vector spaces are said to be isomorphic. The two vector spaces $\mathscr{V}$ and $\mathscr{W}$ are isomorphic if and only if they have the same dimension. Thus, using the definitions given after equations (1.6.7) and (1.6.9), the conditions for two vector spaces to be isomorphic can be stated as follows. If $\mathscr{V}$ and $\mathscr{W}$ are isomorphic, then a linear map $\mathbf{A}: \mathscr{V} \to \mathscr{W}$ is an isomorphism if and only if the following conditions are satisfied

(i) $K(\mathbf{A}) = \{\boldsymbol{0}\}$ (1.6.21)
(ii) $R(\mathbf{A}) = \mathscr{W}$ (1.6.22)
(iii) $\mathbf{A}$ maps a basis of $\mathscr{V}$ onto a basis of $\mathscr{W}$.

Since an isomorphism $\mathbf{A}\colon \mathscr{V} \to \mathscr{W}$ establishes a one-to-one correspondence between the elements of $\mathscr{V}$ and $\mathscr{W}$, it follows that there must exist a unique inverse function $\mathbf{B}\colon \mathscr{W} \to \mathscr{V}$ with the property, that if

$$\boldsymbol{w} = \mathbf{A}\boldsymbol{v} \tag{1.6.23}$$

then

$$\boldsymbol{v} = \mathbf{B}(\boldsymbol{w}) \tag{1.6.24}$$

for all $\boldsymbol{v} \in \mathscr{V}$ and $\boldsymbol{w} \in \mathscr{W}$. For vectors $v_1, v_2 \in \mathscr{V}$, $w_1, w_2 \in \mathscr{W}$ and scalars λ, $\mu \in \mathscr{F}$, equations (1.6.23) and (1.6.24), together with the properties of the linear transformation $\mathbf{A}$, give

$$\begin{aligned} v\mathbf{B}(\lambda \boldsymbol{w}_1 + \mu \boldsymbol{w}_2) &= \mathbf{B}[(\lambda \mathbf{A}\boldsymbol{v}_1 + \mu \mathbf{A}\boldsymbol{v}_2)] \\ &= \mathbf{B}[\mathbf{A}(\lambda \boldsymbol{v}_1 + \mu \boldsymbol{v}_2)] \\ &= \lambda \boldsymbol{v}_1 + \mu \boldsymbol{v}_2 = \lambda \mathbf{B}(\boldsymbol{w}_1) + \mu \mathbf{B}(\boldsymbol{w}_2) \end{aligned}$$

and hence $\mathbf{B}$ is a linear transformation, which will be written $\mathbf{A}^{-1}$. The linear transformation $\mathbf{A}^{-1}$ is also an isomorphism whose inverse is $\mathbf{A}$, such that

$$(\mathbf{A}^{-1})^{-1} = \mathbf{A}. \tag{1.6.25}$$

If $\mathbf{A}\colon \mathscr{V} \to \mathscr{U}$ and $\mathbf{B}\colon \mathscr{U} \to \mathscr{W}$ are isomorphisms, then $\mathbf{BA}\colon \mathscr{V} \to \mathscr{W}$ is an isomorphism whose inverse satisfies the condition

$$(\mathbf{BA})^{-1} = \mathbf{A}^{-1}\mathbf{B}^{-1}. \tag{1.6.26}$$

The *identity* linear transformation $\mathbf{I}\colon \mathscr{V} \to \mathscr{V}$ is defined by

$$\mathbf{I}\boldsymbol{v} = \boldsymbol{v} \tag{1.6.27}$$

for all $\boldsymbol{v} \in \mathscr{V}$. It follows from equation (1.6.24) and the condition $\dim \mathscr{V} = \dim \mathscr{W}$ that if $\mathbf{A}$ is an isomorphism, then

$$\mathbf{A}\mathbf{A}^{-1} = \mathbf{I}_{\mathscr{W}} \qquad \text{and} \qquad \mathbf{A}^{-1}\mathbf{A} = \mathbf{I}_{\mathscr{V}} \tag{1.6.28}$$

where the $\mathbf{I}_{\mathscr{W}}$ and $\mathbf{I}_{\mathscr{V}}$ are the identity linear transformations on the appropriate vector space.

Isomorphisms are often referred to as *invertible* or *non-singular* linear transformations.

The set $\mathscr{L}(\mathscr{V};\mathscr{V})$ corresponds to the vector space of linear transformations $\mathscr{V} \to \mathscr{V}$. An element of $\mathscr{L}(\mathscr{V};\mathscr{V})$ is called an *endomorphism* of $\mathscr{V}$. A *one-one* endomorphism is called an *automorphism*. An example of an automorphism is the identity linear transformation $\mathbf{I}\colon \mathscr{V} \to \mathscr{V}$ defined by equation (1.6.27). If $\mathbf{A} \in \mathscr{L}(\mathscr{V};\mathscr{V})$, then it follows that

$$\mathbf{AI} = \mathbf{IA} = \mathbf{A}. \tag{1.6.29}$$

Let $\mathscr{V}$ be a real vector space of dimension n. Consider the space of linear functions $\mathscr{L}(\mathscr{V};\mathscr{R})$ from $\mathscr{V}$ into the field of real numbers $\mathscr{R}$. Since dim $\mathscr{L}(\mathscr{V};\mathscr{R})=\dim\mathscr{V}=n$, it follows that $\mathscr{V}$ and $\mathscr{L}(\mathscr{V};\mathscr{R})$ are isomorphic. The space of linear functions $\mathscr{L}(\mathscr{V};\mathscr{R})$ is called the *dual space* of $\mathscr{V}$, and will generally be denoted by $\mathscr{V}^*$. To distinguish the elements of $\mathscr{V}$ from those of $\mathscr{V}^*$, the latter elements are called *covectors*. Since $\mathscr{V}^*$ is an n-dimensional vector space by itself, any result valid for a vector space in general can be applied to $\mathscr{V}^*$ as well as to $\mathscr{V}$.

In general, if $\mathscr{V}_{1,2},\ldots,\mathscr{V}_m$ is a collection of vector spaces, then an *m-linear function* is a function

$$\mathbf{A}: \mathscr{V}_1\times\mathscr{V}_2\times\cdots\times\mathscr{V}_m\to\mathscr{R} \tag{1.6.30}$$

which is linear in each of its variables while the other variables are held constant. If the vector spaces $\mathscr{V}_1,\mathscr{V}_2,\ldots,\mathscr{V}_m$ are the vector space $\mathscr{V}$, or its dual space $\mathscr{V}^*$, then $\mathbf{A}$ is called a *tensor* on $\mathscr{V}$. A tensor of *order* (p,q) on $\mathscr{V}$, where p and q are positive integers, is a $(p+q)$-linear function,

$$\mathbf{A}:\underbrace{\mathscr{V}^*\times\cdots\times\mathscr{V}^*}_{p\text{ times}}\times\underbrace{\mathscr{V}\times\cdots\times\mathscr{V}}_{q\text{ times}}\to\mathscr{R} \tag{1.6.31}$$

The particular case, $p=q=0$, corresponds to a tensor of order (0, 0), and is said to define a *zero-order tensor* which is itself a scalar in $\mathscr{R}$. A tensor of order $(p,0)$ is a pure *contravariant* tensor of order p and a tensor of order $(0,q)$ is a pure *covariant* tensor of order q. In this context, a vector $\boldsymbol{v}\in\mathscr{V}$ is a pure contravariant tensor of order one, that is a *first-order tensor*. A tensor which is neither a pure contravariant tensor nor a pure covariant tensor is said to be a *mixed tensor* of order (p,q), it being noted that p is the contravariant order and q is the covariant order.

The main subject of this book, that is the deformation and flow of solids, is primarily concerned with second-order tensors, (though not entirely, in so far as brief reference is made to third- and fourth-order tensors).

1.6.2 Endomorphisms of a Euclidean Vector Space

Let $\mathscr{V}$ be a three-dimensional vector space over the field $\mathscr{R}$ of real numbers. A linear transformation $\mathbf{A}$ of the Euclidean vector space $\mathscr{V}$ into itself is called a *second-order tensor*. The set of all such tensors on $\mathscr{V}$ is denoted by $\mathscr{L}(\mathscr{V};\mathscr{V})$. A second-order tensor is defined as that linear transformation $\mathbf{A}:\mathscr{V}\to\mathscr{V}$ which assigns to an arbitrary vector $\boldsymbol{a}$ a vector $\mathbf{A}\boldsymbol{a}$, such that

$$\mathbf{A}(\lambda\boldsymbol{a}+\mu\boldsymbol{b})=\lambda(\mathbf{A}\boldsymbol{a})+\mu(\mathbf{A}\boldsymbol{b}) \qquad \forall\,\mathbf{A}\in\mathscr{L} \qquad \boldsymbol{a},\boldsymbol{b}\in\mathscr{V} \qquad \lambda,\mu\in\mathscr{R}. \tag{1.6.32}$$

If the actions of two tensors on an arbitrary vector are identical, the two tensors are equal. The rules for the addition, scalar multiplication and multiplication, that is composition of tensors follow from equations (1.6.12), (1.6.13) and (1.6.17):

$$(\mathbf{A}+\mathbf{B})\boldsymbol{a}=\mathbf{A}\boldsymbol{a}+\mathbf{B}\boldsymbol{a} \qquad \forall\,\mathbf{A},\mathbf{B}\in\mathscr{L} \qquad \boldsymbol{a}\in\mathscr{V} \tag{1.6.33}$$

$$(\lambda \mathbf{A})\boldsymbol{a} = \lambda(\mathbf{A}\boldsymbol{a}) \qquad \forall\, \mathbf{A} \in \mathscr{L} \qquad \boldsymbol{a} \in \mathscr{V} \qquad \lambda \in \mathscr{R} \tag{1.6.34}$$

$$(\mathbf{A}\mathbf{B})\boldsymbol{a} = \mathbf{A}(\mathbf{B}\boldsymbol{a}) \qquad \forall\, \mathbf{A}, \mathbf{B} \in \mathscr{L} \qquad \boldsymbol{a} \in \mathscr{V}. \tag{1.6.35}$$

The *zero tensor* $\mathbf{0}$ (see equation (1.6.14)) assigns to $\boldsymbol{a}$ the zero vector, and the *identity tensor* $\mathbf{I}$ (see equation (1.6.27)) assigns to $\boldsymbol{a}$ the vector $\boldsymbol{a}$ itself, thus

$$\mathbf{0}\boldsymbol{a} = \boldsymbol{0} \qquad \mathbf{I}\boldsymbol{a} = \boldsymbol{a} \qquad \forall\, \boldsymbol{a} \in \mathscr{V}. \tag{1.6.36}$$

With equations (1.6.18), (1.6.19), (1.6.20) and equations (1.6.32) to (1.6.36) as a basis, the following properties can be established:

(i) $\mathbf{A} + \mathbf{B} = \mathbf{B} + \mathbf{A} \qquad \forall\, \mathbf{A}, \mathbf{B} \in \mathscr{L}$ (1.6.37)

(ii) $\lambda(\mathbf{A}\mathbf{B}) = (\lambda\mathbf{A})\mathbf{B} = \mathbf{A}(\lambda\mathbf{B}) \qquad \forall\, \mathbf{A}, \mathbf{B} \in \mathscr{L} \qquad \lambda \in \mathscr{R}$ (1.6.38)

(iii) $\left.\begin{aligned} \mathbf{A}(\mathbf{B} + \mathbf{C}) &= \mathbf{A}\mathbf{B} + \mathbf{A}\mathbf{C} \\ (\mathbf{A} + \mathbf{B})\mathbf{C} &= \mathbf{A}\mathbf{C} + \mathbf{B}\mathbf{C} \end{aligned}\right\} \qquad \forall\, \mathbf{A}, \mathbf{B}, \mathbf{C} \in \mathscr{L}$ (1.6.39)

(iv) $\mathbf{A}(\mathbf{B}\mathbf{C}) = (\mathbf{A}\mathbf{B})\mathbf{C} \qquad \forall\, \mathbf{A}, \mathbf{B}, \mathbf{C} \in \mathscr{L}$ (1.6.40)

(v) $\mathbf{A}\mathbf{0} = \mathbf{0}\mathbf{A} = \mathbf{0} \qquad \mathbf{A}\mathbf{I} = \mathbf{I}\mathbf{A} = \mathbf{A} \qquad \forall \mathbf{A} \in \mathscr{L}.$ (1.6.41)

Given a linear transformation $\mathbf{A}: \mathscr{V} \to \mathscr{V}$, a function $\mathbf{A}^{\mathrm{T}}: \mathscr{V} \to \mathscr{V}$ is called the *transpose* of $\mathbf{A}$ if

$$\boldsymbol{a} \cdot (\mathbf{A}\boldsymbol{b}) = (\mathbf{A}^{\mathrm{T}}\boldsymbol{a}) \cdot \boldsymbol{b} \qquad \forall\, \boldsymbol{a}, \boldsymbol{b} \in \mathscr{V}. \tag{1.6.42}$$

It follows from equations (1.6.42), (1.6.33), (1.6.34), (1.6.35) and the axioms formulated by equations (1.5.39) and (1.5.40), that

$$(\mathbf{A}^{\mathrm{T}})^{\mathrm{T}} = \mathbf{A} \qquad \forall\, \mathbf{A} \in \mathscr{L} \tag{1.6.43}$$

$$(\lambda\mathbf{A} + \mu\mathbf{B})^{\mathrm{T}} = \lambda\mathbf{A}^{\mathrm{T}} + \mu\mathbf{B}^{\mathrm{T}} \qquad \forall\, \mathbf{A}, \mathbf{B} \in \mathscr{L} \qquad \lambda, \mu \in \mathscr{R} \tag{1.6.44}$$

$$(\mathbf{A}\mathbf{B})^{\mathrm{T}} = \mathbf{B}^{\mathrm{T}}\mathbf{A}^{\mathrm{T}} \qquad \forall\, \mathbf{A}, \mathbf{B} \in \mathscr{L}. \tag{1.6.45}$$

If $\mathbf{A}^{\mathrm{T}} = \mathbf{A}$, the tensor $\mathbf{A}$ is said to be *symmetric*, and if $\mathbf{A}^{\mathrm{T}} = -\mathbf{A}$, the tensor $\mathbf{A}$ is said to be *skew-symmetric*. An arbitrary tensor can be expressed as the sum of symmetric and skew-symmetric parts, as can be seen from the identity

$$\mathbf{A} = \tfrac{1}{2}(\mathbf{A} + \mathbf{A}^{\mathrm{T}}) + \tfrac{1}{2}(\mathbf{A} - \mathbf{A}^{\mathrm{T}}) \tag{1.6.46}$$

this decomposition being unique. It is of interest to compare equations (1.1.32) and (1.6.46).

1.6.3 The Tensor Product

To an ordered pair of vectors $(\boldsymbol{u}, \boldsymbol{v})$ there corresponds a tensor, denoted by $\boldsymbol{u} \otimes \boldsymbol{v}$ and which is called the tensor product of $\boldsymbol{u}$ and $\boldsymbol{v}$. The tensor product is defined through its action on an arbitrary vector $\boldsymbol{a}$; thus

$$(\boldsymbol{u} \otimes \boldsymbol{v})\boldsymbol{a} = (\boldsymbol{a} \cdot \boldsymbol{v})\boldsymbol{u} \qquad \forall\, \boldsymbol{a} \in \mathscr{V}. \tag{1.6.47}$$

Using the axioms formulated by equations (1.5.39) and (1.5.40), together with equations (1.6.33) and (1.6.42), the following properties of the tensor product can be deduced from equation (1.6.47)

(i) $(\lambda \boldsymbol{u}+\mu \boldsymbol{v}) \otimes \boldsymbol{w}=\lambda(\boldsymbol{u} \otimes \boldsymbol{w})+\mu(\boldsymbol{v} \otimes \boldsymbol{w}) \quad \forall \boldsymbol{u}, \boldsymbol{v}, \boldsymbol{w} \in \mathscr{V} \quad \lambda, \mu \in \mathscr{R}$ (1.6.48)

(ii) $\boldsymbol{u} \otimes (\lambda \boldsymbol{v}+\mu \boldsymbol{w})=\lambda(\boldsymbol{u} \otimes \boldsymbol{v})+\mu(\boldsymbol{u} \otimes \boldsymbol{w}) \quad \forall \boldsymbol{u}, \boldsymbol{v}, \boldsymbol{w} \in \mathscr{V} \quad \lambda, \mu \in \mathscr{R}$ (1.6.49)

(iii) $(\boldsymbol{u} \otimes \boldsymbol{v})^{\mathrm{T}}=\boldsymbol{v} \otimes \boldsymbol{u} \quad \forall \boldsymbol{u}, \boldsymbol{v} \in \mathscr{V}.$ (1.6.50)

Replacing $\boldsymbol{u}$ by $\boldsymbol{e}^p$ and $\boldsymbol{v}$ by $\boldsymbol{e}_p$ in equation (1.6.47) gives

$$(\boldsymbol{e}^p \otimes \boldsymbol{e}_p)\boldsymbol{a}=(\boldsymbol{a}\cdot\boldsymbol{e}_p)\boldsymbol{e}^p=\boldsymbol{a}=\mathbf{I}\boldsymbol{a}$$

and hence

$$\boldsymbol{e}^p \otimes \boldsymbol{e}_p=\boldsymbol{e}_p \otimes \boldsymbol{e}^p=\mathbf{I} \tag{1.6.51}$$

where use has been made of equations (1.6.36) and (1.5.69).

Set $\mathbf{A} \equiv (\boldsymbol{u} \otimes \boldsymbol{v})(\boldsymbol{w} \otimes \boldsymbol{x})$ where $\boldsymbol{u}$, $\boldsymbol{v}$, $\boldsymbol{w}$ and $\boldsymbol{x}$ are arbitrary vectors. Using equations (1.6.32), (1.6.35) and (1.6.47) there can be obtained the relation

$$\begin{aligned}\mathbf{A}\boldsymbol{a}&=(\boldsymbol{u} \otimes \boldsymbol{v})\{(\boldsymbol{w} \otimes \boldsymbol{x})\boldsymbol{a}\}=(\boldsymbol{u} \otimes \boldsymbol{v})\{(\boldsymbol{a}\cdot\boldsymbol{x})\boldsymbol{w}\}\\&=(\boldsymbol{a}\cdot\boldsymbol{x})\{(\boldsymbol{u} \otimes \boldsymbol{v})\boldsymbol{w}\}=(\boldsymbol{a}\cdot\boldsymbol{x})(\boldsymbol{w}\cdot\boldsymbol{v})\boldsymbol{u}\\&=(\boldsymbol{w}\cdot\boldsymbol{v})\{(\boldsymbol{a}\cdot\boldsymbol{x})\boldsymbol{u}\}=(\boldsymbol{w}\cdot\boldsymbol{v})(\boldsymbol{u} \otimes \boldsymbol{x})\boldsymbol{a}\end{aligned} \tag{1.6.52}$$

where $\boldsymbol{a}$ is also an arbitrary vector. Similarly, there can be established the relations:

$$\begin{aligned}\{\mathbf{A}(\boldsymbol{u} \otimes \boldsymbol{v})\}\boldsymbol{a}&=\mathbf{A}\{(\boldsymbol{u} \otimes \boldsymbol{v})\boldsymbol{a}\}=\mathbf{A}\{(\boldsymbol{a}\cdot\boldsymbol{v})\boldsymbol{u}\}\\&=(\boldsymbol{a}\cdot\boldsymbol{v})\mathbf{A}\boldsymbol{u}=(\mathbf{A}\boldsymbol{u} \otimes \boldsymbol{v})\boldsymbol{a}\end{aligned} \tag{1.6.53}$$

$$\begin{aligned}\{(\boldsymbol{u} \otimes \boldsymbol{v})\mathbf{A}\}\boldsymbol{a}&=(\boldsymbol{u} \otimes \boldsymbol{v})(\mathbf{A}\boldsymbol{a})=\{(\mathbf{A}\boldsymbol{a})\cdot\boldsymbol{v}\}\boldsymbol{u}\\&=\{(\boldsymbol{v}\cdot\mathbf{A}\boldsymbol{a})\}\boldsymbol{u}=\{(\mathbf{A}^{\mathrm{T}}\boldsymbol{v})\cdot\boldsymbol{a}\}\boldsymbol{u}\\&=\{\boldsymbol{a}\cdot(\mathbf{A}^{\mathrm{T}}\boldsymbol{v})\}\boldsymbol{u}=\{\boldsymbol{u} \otimes (\mathbf{A}^{\mathrm{T}}\boldsymbol{v})\}\boldsymbol{a}.\end{aligned} \tag{1.6.54}$$

From equations (1.6.52), (1.6.53) and (1.6.54) the following properties can be obtained

(iv) $(\boldsymbol{u} \otimes \boldsymbol{v})(\boldsymbol{w} \otimes \boldsymbol{x})=(\boldsymbol{w}\cdot\boldsymbol{v})(\boldsymbol{u} \otimes \boldsymbol{x})$ (1.6.55)

(v) $\mathbf{A}(\boldsymbol{u} \otimes \boldsymbol{v})=(\mathbf{A}\boldsymbol{u}) \otimes \boldsymbol{v}$ (1.6.56)

(vi) $(\boldsymbol{u} \otimes \boldsymbol{v})\mathbf{A}=\boldsymbol{u} \otimes (\mathbf{A}^{\mathrm{T}}\boldsymbol{v}).$ (1.6.57)

Let the arbitrary tensor $\mathbf{A}$ be referred to the basis $e=\{\boldsymbol{e}_1, \boldsymbol{e}_2, \boldsymbol{e}_3\}$, and set the vector

$$\mathbf{A}\boldsymbol{e}^j=A^{pj}\boldsymbol{e}_p \tag{1.6.58}$$

so that

$$A^{ij}=\boldsymbol{e}^i\cdot(\mathbf{A}\boldsymbol{e}^j). \tag{1.6.59}$$

Taking $\boldsymbol{a}=a_r\boldsymbol{e}^r$ to be an arbitrary vector, it follows that:

$$\begin{aligned}(\mathbf{A}-A^{pq}\boldsymbol{e}_p\otimes\boldsymbol{e}_q)\boldsymbol{a}&=(\mathbf{A}-A^{pq}\boldsymbol{e}_p\otimes\boldsymbol{e}_q)a_r\boldsymbol{e}^r\\&=a_r\{\mathbf{A}\boldsymbol{e}^r-A^{pq}(\boldsymbol{e}^r\cdot\boldsymbol{e}_q)\boldsymbol{e}_p\}\\&=a_r(A^{pr}-A^{pq}\delta_{rq})\boldsymbol{e}_p=\boldsymbol{0}.\end{aligned}$$

Hence $\mathbf{A}$ admits the representation,

$$\mathbf{A}=A^{pq}\boldsymbol{e}_p\otimes\boldsymbol{e}_q.$$

In an analogous manner, it can be shown that

$$\mathbf{A}=A^p{}_q\boldsymbol{e}_p\otimes\boldsymbol{e}^q=A_{pq}\boldsymbol{e}^p\otimes\boldsymbol{e}^q=A^{pq}\boldsymbol{e}_p\otimes\boldsymbol{e}_q \qquad (1.6.60)$$

the various sets of components being given by

$$A^i{}_j=\boldsymbol{e}^i\cdot(\mathbf{A}\boldsymbol{e}_j) \qquad A^{ij}=\boldsymbol{e}^i\cdot(\mathbf{A}\boldsymbol{e}^j)$$

$$A_{ij}=\boldsymbol{e}_i\cdot(\mathbf{A}\boldsymbol{e}_j). \qquad (1.6.61)$$

The *product* $\mathbf{AB}$ of two tensors is the tensor

$$\mathbf{AB}=\mathbf{A}\circ\mathbf{B}.$$

If $\mathbf{B}$ has components $B^i{}_j$ relative to e, then

$$\begin{aligned}\mathbf{AB}&=(A^i{}_p\boldsymbol{e}_i\otimes\boldsymbol{e}^p)(B^q{}_j\boldsymbol{e}_q\otimes\boldsymbol{e}^j)\\&=A^i{}_pB^q{}_j(\boldsymbol{e}_q\cdot\boldsymbol{e}^p)(\boldsymbol{e}_i\otimes\boldsymbol{e}^j)\\&=A^i{}_pB^q{}_j\delta_q{}^p\boldsymbol{e}_i\otimes\boldsymbol{e}^j\\&=A^i{}_pB^p{}_j\boldsymbol{e}_i\otimes\boldsymbol{e}^j\end{aligned}$$

from which it is evident that $\mathbf{AB}$ has components $A^i{}_pB^p{}_j$ relative to e, use having been made of equations (1.5.59), (1.6.55) and (1.6.60). Generally

$$\mathbf{AB}\neq\mathbf{BA}$$

but if

$$\mathbf{AB}=\mathbf{BA}$$

then the tensors $\mathbf{A}$ and $\mathbf{B}$ are said to *commute*. If $\mathbf{AB}=\mathbf{0}$, neither $\mathbf{A}$ nor $\mathbf{B}$ need be $\mathbf{0}$.

The powers of an arbitrary tensor $\mathbf{A}$ are defined as follows,

$$\mathbf{A}^0=\mathbf{I} \qquad \mathbf{A}^1=\mathbf{A} \qquad \mathbf{A}^2=\mathbf{AA} \quad \text{etc.}$$

The space of all tensors has a natural *inner product*,

$$\mathbf{A}\cdot\mathbf{B}=\mathrm{tr}(\mathbf{A}^{\mathrm{T}}\mathbf{B})=\mathbf{B}\cdot\mathbf{A}$$

which is referred to as the trace of $\mathbf{A}^{\mathrm{T}}\mathbf{B}$.

Two algebras are said to be isomorphic if there exists an isomorphism that is also a homomorphism with respect to the multiplication operations of the algebras. In this context, the algebra $\mathscr{L}(\mathscr{V};\mathscr{V})$ is isomorphic to the matrix algebra $\mathscr{R}^{n^2}$ (see §1.1). Specifically, relative to any basis $\{e_i\}$ of $\mathscr{V}$, a linear transformation $\mathbf{A}$ corresponds to its component matrix $[A^i{}_j]$. It is important to note that tr $\mathbf{T}=T^p{}_p \neq T^{pp} \neq T_{pp}$, and similarly

$$\det \mathbf{T} = \det T^i{}_j \neq \det T_{ij} \neq \det T^{ij}.$$

1.6.4 Tensor Algebra

Let $\mathbf{A}$ be an arbitrary tensor, and consider the scalar

$$\Phi = \{[\mathbf{A}\boldsymbol{a}, \boldsymbol{b}, \boldsymbol{c}] + [\boldsymbol{a}, \mathbf{A}\boldsymbol{b}, \boldsymbol{c}] + [\boldsymbol{a}, \boldsymbol{b}, \mathbf{A}\boldsymbol{c}]\} \tag{1.6.62}$$

where $\boldsymbol{a}, \boldsymbol{b}, \boldsymbol{c}$ are arbitrary vectors. Using equations (1.5.48) and (1.5.49), Φ can be expressed in the form

$$\Phi = a^p b^q c^r \{[\mathbf{A}\boldsymbol{e}_p, \boldsymbol{e}_q, \boldsymbol{e}_r] + [\boldsymbol{e}_p, \mathbf{A}\boldsymbol{e}_q, \boldsymbol{e}_r] + [\boldsymbol{e}_p, \boldsymbol{e}_q, \mathbf{A}\boldsymbol{e}_r]\}$$

where the a^i, b^i, c^i are the contravariant components of $\boldsymbol{a}, \boldsymbol{b}, \boldsymbol{c}$ relative to the basis $e=\{\boldsymbol{e}_1, \boldsymbol{e}_2, \boldsymbol{e}_3\}$. Using equations (1.5.79) and (1.5.85), Φ can be expressed in the form

$$\begin{aligned}\Phi &= \pm a^p b^q c^r \varepsilon_{pqr} \frac{\{[\mathbf{A}\boldsymbol{e}_1, \boldsymbol{e}_2, \boldsymbol{e}_3] + [\boldsymbol{e}_1, \mathbf{A}\boldsymbol{e}_2, \boldsymbol{e}_3] + [\boldsymbol{e}_1, \boldsymbol{e}_2, \mathbf{A}\boldsymbol{e}_3]\}}{[\boldsymbol{e}_1, \boldsymbol{e}_2, \boldsymbol{e}_3]} \\ &= \pm [\boldsymbol{a}, \boldsymbol{b}, \boldsymbol{c}] \frac{\{[\mathbf{A}\boldsymbol{e}_1, \boldsymbol{e}_2, \boldsymbol{e}_3] + [\boldsymbol{e}_1, \mathbf{A}\boldsymbol{e}_2, \boldsymbol{e}_3] + [\boldsymbol{e}_1, \boldsymbol{e}_2, \mathbf{A}\boldsymbol{e}_3]\}}{[\boldsymbol{e}_1, \boldsymbol{e}_2, \boldsymbol{e}_3]}\end{aligned} \tag{1.6.63}$$

If e is positive in $\mathscr{V}^+$, the upper sign applies in equation (1.6.63) and $[\boldsymbol{e}_1, \boldsymbol{e}_2, \boldsymbol{e}_3] = +\sqrt{g}$, where if e is positive in $\mathscr{V}^-$, the lower sign must be taken in equation (1.6.63) and $[\boldsymbol{e}_1, \boldsymbol{e}_2, \boldsymbol{e}_3] = -\sqrt{g}$. Thus,

$$\Phi = \frac{\{[\mathbf{A}\boldsymbol{e}_1, \boldsymbol{e}_2, \boldsymbol{e}_3] + [\boldsymbol{e}_1, \mathbf{A}\boldsymbol{e}_2, \boldsymbol{e}_3] + [\boldsymbol{e}_1, \boldsymbol{e}_2, \mathbf{A}\boldsymbol{e}_3]\}}{[\boldsymbol{e}_1, \boldsymbol{e}_2, \boldsymbol{e}_3]} [\boldsymbol{a}, \boldsymbol{b}, \boldsymbol{c}] \tag{1.6.64}$$

regardless of orientation. From equation (1.6.60) the following relation can be obtained

$$\mathbf{A}\boldsymbol{e}_i = A^p{}_q(\boldsymbol{e}_p \otimes \boldsymbol{e}^q)\boldsymbol{e}_i = A^p{}_q[(\boldsymbol{e}_i \cdot \boldsymbol{e}^q)\boldsymbol{e}_p] = A^p{}_q \delta_i{}^q \boldsymbol{e}_p$$

and hence

$$\mathbf{A}\boldsymbol{e}_i = A^p{}_i \boldsymbol{e}_p. \tag{1.6.65}$$

Substituting for $\mathbf{A}\boldsymbol{e}_i$ in equation (1.6.64) from equation (1.6.65), and using the axiom formulated by equation (1.5.44) gives

$$[\mathbf{A}\boldsymbol{a}, \boldsymbol{b}, \boldsymbol{c}] + [\boldsymbol{a}, \mathbf{A}\boldsymbol{b}, \boldsymbol{c}] + [\boldsymbol{a}, \boldsymbol{b}, \mathbf{A}\boldsymbol{c}] = I_{\mathbf{A}}[\boldsymbol{a}, \boldsymbol{b}, \boldsymbol{c}] \qquad \forall\, \boldsymbol{a}, \boldsymbol{b}, \boldsymbol{c} \in \mathscr{V} \tag{1.6.66}$$

where

$$I_{\mathrm{A}} = A^1{}_1 + A^2{}_2 + A^3{}_3. \tag{1.6.67}$$

When expressed in terms of the covariant components a_i, b_i, c_i of $\boldsymbol{a}, \boldsymbol{b}, \boldsymbol{c}$ relative to the dual basis, the resulting expression for Φ is just equation (1.6.66), and hence I_{A} is a scalar invariant of $\mathbf{A}$.

Using the same approach as that above, it can be shown that

$$[\boldsymbol{a}, \mathbf{A}\boldsymbol{b}, \mathbf{A}\boldsymbol{c}] + [\mathbf{A}\boldsymbol{a}, \boldsymbol{b}, \mathbf{A}\boldsymbol{c}] + [\mathbf{A}\boldsymbol{a}, \mathbf{A}\boldsymbol{b}, \boldsymbol{c}] = II_{\mathrm{A}}[\boldsymbol{a}, \boldsymbol{b}, \boldsymbol{c}] \qquad \forall\, \boldsymbol{a}, \boldsymbol{b}, \boldsymbol{c} \in \mathscr{V} \tag{1.6.68}$$

$$[\mathbf{A}\boldsymbol{a}, \mathbf{A}\boldsymbol{b}, \mathbf{A}\boldsymbol{c}] = III_{\mathrm{A}}[\boldsymbol{a}, \boldsymbol{b}, \boldsymbol{c}] \qquad \forall \boldsymbol{a}, \boldsymbol{b}, \boldsymbol{c} \in \mathscr{V} \tag{1.6.69}$$

where

$$II_{\mathrm{A}} = \tfrac{1}{2}(A^p{}_p A^q{}_q - A^p{}_q A^q{}_p) \tag{1.6.70}$$

$$III_{\mathrm{A}} = \tfrac{1}{6}(A^p{}_p A^q{}_q A^r{}_r - 3A^p{}_q A^q{}_p A^r{}_r + 2A^p{}_q A^q{}_r A^r{}_p) \tag{1.6.71}$$

are scalar invariants of $\mathbf{A}$. The scalars I_{A}, II_{A} and III_{A} are the *principal invariants* of $\mathbf{A}$. The use of the term invariant recognises that the principal invariants must be independent of the choice of basis. This condition is satisfied by the above derivation since the choice of basis in equation (1.6.62) is arbitrary. These invariants can be expressed in the form

$$I_{\mathrm{A}} = \operatorname{tr} \mathbf{A} \qquad II_{\mathrm{A}} = \tfrac{1}{2}[(\operatorname{tr} \mathbf{A})^2 - \operatorname{tr} \mathbf{A}^2] \qquad III_{\mathrm{A}} = \det \mathbf{A} \tag{1.6.72}$$

it being noted that I_{A} is referred to as the *trace* of $\mathbf{A}$.

Use of equations (1.5.48), (1.5.49), (1.6.33) and (1.6.34) gives

$$\operatorname{tr}(\lambda \mathbf{A} + \mu \mathbf{B}) = \lambda \operatorname{tr} \mathbf{A} + \mu \operatorname{tr} \mathbf{B} \qquad \forall \mathbf{A}, \mathbf{B} \in \mathscr{L} \qquad \lambda, \mu \in \mathscr{R}. \tag{1.6.73}$$

From equation (1.6.67)

$$\operatorname{tr} \mathbf{A}^{\mathrm{T}} = \operatorname{tr} \mathbf{A} \quad \forall \mathbf{A} \in \mathscr{L} \qquad \operatorname{tr}(\mathbf{AB}) = \operatorname{tr}(\mathbf{BA}) \quad \forall \mathbf{A}, \mathbf{B} \in \mathscr{L}. \tag{1.6.74}$$

Replacing $\mathbf{A}$ by $\boldsymbol{u} \otimes \boldsymbol{v}$ in equation (1.6.66) gives

$$\begin{aligned}[\boldsymbol{a}, \boldsymbol{b}, \boldsymbol{c}] \operatorname{tr}(\boldsymbol{u} \otimes \boldsymbol{v}) &= [(\boldsymbol{a} \cdot \boldsymbol{v})\boldsymbol{u}, \boldsymbol{b}, \boldsymbol{c}] + [\boldsymbol{a}, (\boldsymbol{b} \cdot \boldsymbol{v})\boldsymbol{u}, \boldsymbol{c}] + [\boldsymbol{a}, \boldsymbol{b}, (\boldsymbol{c} \cdot \boldsymbol{v})\boldsymbol{u}] \\ &= (\boldsymbol{a} \cdot \boldsymbol{v})[\boldsymbol{u}, \boldsymbol{b}, \boldsymbol{c}] + (\boldsymbol{b} \cdot \boldsymbol{v})[\boldsymbol{a}, \boldsymbol{u}, \boldsymbol{c}] + (\boldsymbol{c} \cdot \boldsymbol{v})[\boldsymbol{a}, \boldsymbol{b}, \boldsymbol{u}].\end{aligned} \tag{1.6.75}$$

Regarding $\{\boldsymbol{a}, \boldsymbol{b}, \boldsymbol{c}\}$ as an arbitrary basis of $\mathscr{V}$, it follows that there are scalars λ, μ, ν such that

$$\boldsymbol{u} = \lambda \boldsymbol{a} + \mu \boldsymbol{b} + \nu \boldsymbol{c}. \tag{1.6.76}$$

Substituting for $\boldsymbol{u}$ from equation (1.6.76) into equation (1.6.75) gives

$$[\boldsymbol{a},\boldsymbol{b},\boldsymbol{c}]\,\mathrm{tr}(\boldsymbol{u}\otimes\boldsymbol{v})=\{\lambda(\boldsymbol{a}\cdot\boldsymbol{v})+\mu(\boldsymbol{b}\cdot\boldsymbol{v})+\nu(\boldsymbol{c}\cdot\boldsymbol{v})\}[\boldsymbol{a},\boldsymbol{b},\boldsymbol{c}]$$
$$=\{(\lambda\boldsymbol{a}+\mu\boldsymbol{b}+\nu\boldsymbol{c})\cdot\boldsymbol{v}\}[\boldsymbol{a},\boldsymbol{b},\boldsymbol{c}]$$

and since $[\boldsymbol{a},\boldsymbol{b},\boldsymbol{c}]\neq 0$ it follows that

$$\mathrm{tr}(\boldsymbol{u}\otimes\boldsymbol{v})=\boldsymbol{u}\cdot\boldsymbol{v}. \tag{1.6.77}$$

From equation (1.6.69) it follows that

$$\det(\lambda\mathbf{A})=\lambda^3\det\mathbf{A}\qquad\forall\,\mathbf{A}\in\mathscr{L}\qquad\lambda\in\mathscr{R} \tag{1.6.78}$$

$$\det(\mathbf{AB})=\det\mathbf{A}\det\mathbf{B}\qquad\forall\,\mathbf{A},\mathbf{B}\in\mathscr{L} \tag{1.6.79}$$

$$\det\mathbf{I}=1 \tag{1.6.80}$$

It follows from equation (1.6.71) that

$$\det\mathbf{A}^{\mathrm{T}}=\det\mathbf{A}\qquad\forall\mathbf{A}\in\mathscr{L}. \tag{1.6.81}$$

Replacing $\mathbf{A}$ by $\boldsymbol{u}\otimes\boldsymbol{v}$ in equation (1.6.69) gives

$$[\boldsymbol{a},\boldsymbol{b},\boldsymbol{c}]\det(\boldsymbol{u}\otimes\boldsymbol{v})=[(\boldsymbol{a}\cdot\boldsymbol{v})\boldsymbol{u},(\boldsymbol{b}\cdot\boldsymbol{v})\boldsymbol{u},(\boldsymbol{c}\cdot\boldsymbol{v})\boldsymbol{u}]=0$$

and since $[\boldsymbol{a},\boldsymbol{b},\boldsymbol{c}]\neq 0$, it follows that

$$\det(\boldsymbol{u}\otimes\boldsymbol{v})=0. \tag{1.6.82}$$

As a consequence of equation (1.6.28), it follows for a linear transformation $\mathbf{A}:\mathscr{V}\rightarrow\mathscr{V}$, that

$$\mathbf{AA}^{-1}=\mathbf{A}^{-1}\mathbf{A}=\mathbf{I} \tag{1.6.83}$$

and hence, using equations (1.6.75) and (1.6.76)

$$\det\mathbf{A}^{-1}=(\det\mathbf{A})^{-1}. \tag{1.6.84}$$

If $\mathbf{B}$ is also an invertible tensor, then equation (1.6.83) gives

$$(\mathbf{AB})^{-1}=\mathbf{B}^{-1}\mathbf{A}^{-1} \tag{1.6.85}$$

where use has been made of equation (1.6.29).

For $\mathbf{A}$ to be *invertible*, i.e. for $\mathbf{A}^{-1}$ to exist, we must have $\det\mathbf{A}\neq 0$. The invertible tensors constitute a group under multiplication.

A tensor $\mathbf{A}$ such that $\det\mathbf{A}=\pm 1$ is unimodular. The unimodular tensors constitute a subgroup of the group of invertible tensors. This group is called the *unimodular group* $\mathscr{U}$.

For an arbitrary, invertible tensor $\mathbf{A}$, and arbitrary vectors $\boldsymbol{a},\boldsymbol{b},\boldsymbol{c}$, it can be shown, using equations (1.1.39) and (1.1.40), that

$$\mathbf{A}^{+}(\boldsymbol{a}\wedge\boldsymbol{b})=(\mathbf{A}\boldsymbol{a})\wedge(\mathbf{A}\boldsymbol{b})\qquad\forall\boldsymbol{a},\boldsymbol{b}\in\mathscr{V} \tag{1.6.86}$$

and that

$$\{\mathbf{A}^{\mathrm{T}+}(\boldsymbol{a} \wedge \boldsymbol{b})\}\cdot\boldsymbol{c} = [\mathbf{A}^{\mathrm{T}}\boldsymbol{a}, \mathbf{A}^{\mathrm{T}}\boldsymbol{b}, \mathbf{A}^{\mathrm{T}}(\mathbf{A}^{-1})^{\mathrm{T}}\boldsymbol{c}]$$
$$= (\det \mathbf{A}^{\mathrm{T}})(\boldsymbol{a} \wedge \boldsymbol{b})\cdot\{(\mathbf{A}^{-1})^{\mathrm{T}}\boldsymbol{c}\}$$
$$= (\det \mathbf{A}^{\mathrm{T}})\{\mathbf{A}^{-1}(\boldsymbol{a} \wedge \boldsymbol{b})\}\cdot\boldsymbol{c}$$

which implies the relation

$$\mathbf{A}^{-1} = (\det \mathbf{A}^{\mathrm{T}})^{-1}\mathbf{A}^{\mathrm{T}+} \tag{1.6.87}$$

where $\mathbf{A}^{+}$ is the adjugate of $\mathbf{A}$ (see equation (1.1.41)).

For an invertible tensor $\mathbf{A} = \mathbf{A}(\tau)$, there can be obtained from equation (1.6.69),

$$[\boldsymbol{a}, \boldsymbol{b}, \boldsymbol{c}]\frac{\mathrm{d}}{\mathrm{d}\tau}(\det \mathbf{A}) = [\mathbf{BA}\boldsymbol{a}, \mathbf{A}\boldsymbol{b}, \mathbf{A}\boldsymbol{c}] + [\mathbf{A}\boldsymbol{a}, \mathbf{BA}\boldsymbol{b}, \mathbf{A}\boldsymbol{c}] + [\mathbf{A}\boldsymbol{a}, \mathbf{A}\boldsymbol{b}, \mathbf{BA}\boldsymbol{c}]$$
$$= (\operatorname{tr}\mathbf{B})[\mathbf{A}\boldsymbol{a}, \mathbf{A}\boldsymbol{b}, \mathbf{A}\boldsymbol{c}]$$
$$= \operatorname{tr}\mathbf{B}\det\mathbf{A}[\boldsymbol{a}, \boldsymbol{b}, \boldsymbol{c}]$$

where $\mathbf{B} = (\mathrm{d}\mathbf{A}/\mathrm{d}\tau)\mathbf{A}^{-1}$, and where use has been made of equation (1.6.67). Hence, since $\boldsymbol{a}, \boldsymbol{b}, \boldsymbol{c}$ are arbitrary vectors,

$$\frac{\mathrm{d}}{\mathrm{d}\tau}(\det \mathbf{A}) = (\det \mathbf{A})\operatorname{tr}\left(\frac{\mathrm{d}\mathbf{A}}{\mathrm{d}\tau}\mathbf{A}^{-1}\right). \tag{1.6.88}$$

1.6.5 Proper Vectors and Proper Numbers of Tensors

Closely related to the study of the *spectral* properties of an endomorphism $\mathbf{A} \in \mathscr{L}(\mathscr{V}; \mathscr{V})$ is the problem of identifying a direct sum decomposition of $\mathbf{A}$.

A scalar λ is said to be a *proper number* of $\mathbf{A} \in \mathscr{L}(\mathscr{V}; \mathscr{V})$ if there exists a non-zero vector $\boldsymbol{u} \in \mathscr{V}$ such that

$$\mathbf{A}\boldsymbol{u} = \lambda\boldsymbol{u}; \tag{1.6.89}$$

in this context, the vector $\boldsymbol{u}$ in equation (1.6.89) is called the *proper vector* of $\mathbf{A}$ associated with λ. Reciprocally, a vector $\boldsymbol{u}$ is a proper vector of $\mathbf{A}$ if there is a real number λ such that equation (1.6.89) holds, and in this context λ is said to be a proper number of $\mathbf{A}$ associated with $\boldsymbol{u}$. The set of all proper numbers of $\mathbf{A}$ is the spectrum of $\mathbf{A}$, denoted by $\sigma(\mathbf{A})$. For any $\lambda \in \sigma(\mathbf{A})$ the set

$$\mathscr{V}(\lambda) = \{\boldsymbol{u} \in \mathscr{V} \mid \mathbf{A}\boldsymbol{u} = \lambda\boldsymbol{u}\}$$

is a subspace of $\mathscr{V}$, called the *proper space* associated with λ. The *geometric multiplicity* of λ is the dimension of $\mathscr{V}(\lambda)$.

Equation (1.6.89) can be expressed in the form

$$(\mathbf{A} - \lambda\mathbf{I})\boldsymbol{u} = \mathbf{0}. \tag{1.6.90}$$

Let $\{\boldsymbol{l}, \boldsymbol{m}, \boldsymbol{n}\}$ be an arbitrary basis of $\mathscr{V}$, and suppose there are scalars of α, β, γ, not all zero, such that $\boldsymbol{u}=\alpha\boldsymbol{l}+\beta\boldsymbol{m}+\gamma\boldsymbol{n}\neq\boldsymbol{0}$, then, by way of equations (1.6.32) and (1.6.33), it follows that

$$\begin{aligned}(\mathbf{A}-\lambda\mathbf{I})\boldsymbol{u} &= (\mathbf{A}-\lambda\mathbf{I})(\alpha\boldsymbol{l}+\beta\boldsymbol{m}+\gamma\boldsymbol{n})\\ &= \alpha(\mathbf{A}\boldsymbol{l}-\lambda\boldsymbol{l})+\beta(\mathbf{A}\boldsymbol{m}-\lambda\boldsymbol{m})+\gamma(\mathbf{A}\boldsymbol{n}-\lambda\boldsymbol{n})=\boldsymbol{0}\end{aligned}$$

from which it is inferred that the three vectors $(\mathbf{A}\boldsymbol{l}-\lambda\boldsymbol{l})$, $(\mathbf{A}\boldsymbol{m}-\lambda\boldsymbol{m})$, $(\mathbf{A}\boldsymbol{n}-\lambda\boldsymbol{n})$, are linearly dependent, which is just the condition (proved at the end of §1.5.2) for

$$[(\mathbf{A}-\lambda\mathbf{I})\boldsymbol{l}, (\mathbf{A}-\lambda\mathbf{I})\boldsymbol{m}, (\mathbf{A}-\lambda\mathbf{I})\boldsymbol{n}]=0.$$

Thus, by replacing $\{\boldsymbol{a}, \boldsymbol{b}, \boldsymbol{c}\}$ by $\{\boldsymbol{l}, \boldsymbol{m}, \boldsymbol{n}\}$ in equation (1.6.69) it has been concluded that the scalar λ is a proper number of $\mathbf{A}$ if and only if it is a real root of the equation

$$\det(\mathbf{A}-\lambda\mathbf{I})=0. \tag{1.6.91}$$

Equation (1.6.91) is known as the *characteristic* equation of $\mathbf{A}$ and from above it can also be expressed in the form

$$[\mathbf{A}\boldsymbol{a}-\lambda\boldsymbol{a}, \mathbf{A}\boldsymbol{b}-\lambda\boldsymbol{b}, \mathbf{A}\boldsymbol{c}-\lambda\boldsymbol{c}]=0 \tag{1.6.92}$$

where $\boldsymbol{a}, \boldsymbol{b}, \boldsymbol{c}$ are arbitrary vectors.

Use of equations (1.5.48), (1.5.49), (1.6.66), (1.6.68) and (1.6.69) enables equation (1.6.92) to be expanded in such a way as to allow the arbitrary factor $[\boldsymbol{a}, \boldsymbol{b}, \boldsymbol{c}]$ to be removed, thus giving the equation

$$\lambda^3-I_{\mathbf{A}}\lambda^2+II_{\mathbf{A}}\lambda-III_{\mathbf{A}}=0 \tag{1.6.93}$$

as an alternative form of the characteristic equation. Since the principal invariants $I_{\mathbf{A}}, II_{\mathbf{A}}, III_{\mathbf{A}}$ are real, equation (1.6.93) implies that $\mathbf{A}$ has either three proper numbers or only one.

Let $\boldsymbol{u}$ be a proper vector of $\mathbf{A}$ associated with λ. With equation (1.6.89) as basis, let it be supposed that

$$\mathbf{A}^a\boldsymbol{u}=\lambda^a\boldsymbol{u} \tag{1.6.94}$$

holds for $a=1,2,\ldots,n$. Equation (1.6.94) holds for $a=1$, since then it is just equation (1.6.89). For any other value of a, it follows that

$$\begin{aligned}\mathbf{A}^{n+1}\boldsymbol{u} &= \mathbf{A}(\mathbf{A}^n\boldsymbol{u})=\mathbf{A}(\lambda^n\boldsymbol{u})\\ &= \lambda^n\mathbf{A}\boldsymbol{u}=\lambda^{n+1}u.\end{aligned}$$

Hence, by induction, it can be concluded that equation (1.6.94) holds for all positive integers a. Applying this conclusion to the characteristic polynomial

$$f(\lambda)=\lambda^3-I_{\mathbf{A}}\lambda^2+\mathrm{II}_{\mathbf{A}}\lambda-III_{\mathbf{A}}=0$$

it follows that

$$\mathbf{f}(\mathbf{A})\boldsymbol{u}=f(\lambda)\boldsymbol{u} \tag{1.6.95}$$

where the tensor-valued function is

$$\mathbf{f}(\mathbf{A})=\mathbf{A}^3-I_{\mathbf{A}}\mathbf{A}^2+II_{\mathbf{A}}\mathbf{A}-III_{\mathbf{A}}\mathbf{I}.$$

It follows from equation (1.6.95) that $f(\lambda)$ is a proper number of $\mathbf{f}(\mathbf{A})$ and $\boldsymbol{u}$ an associated proper vector. Hence, if $\mathbf{A}$ has three proper numbers the tensor $\mathbf{f}$ has three proper numbers each equal to zero. Thus, for arbitrary $\mathbf{A}$, it follows that $\mathbf{f}(\mathbf{A})=\mathbf{0}$, a conclusion in accord with the Cayley–Hamilton theorem which states that a tensor satisfies its own characteristic equation:

$$\mathbf{A}^3-I_{\mathbf{A}}\mathbf{A}^2+II_{\mathbf{A}}\mathbf{A}-III_{\mathbf{A}}\mathbf{I}=\mathbf{0} \qquad \forall\,\mathbf{A}\in\mathscr{L}. \tag{1.6.96}$$

1.6.6 Symmetric Tensors

The existence of a basis for a vector space $\mathscr{V}$ in which the matrix of a given tensor $\mathbf{S}\in\mathscr{L}(\mathscr{V};\mathscr{V})$ is diagonal is of particular interest in mathematical physics. It can be shown that the matrices of symmetric tensors, $\mathbf{S}(=\mathbf{S}^{\mathrm{T}})$, can be expressed in a diagonal form. Because many of the tensors which occur in mathematical physics are symmetric, particular emphasis will be given to their basic properties.

Suppose a symmetric tensor $\mathbf{S}$ has three distinct proper numbers $\lambda_1,\lambda_2,\lambda_3$ and three proper vectors $\boldsymbol{u}_1,\boldsymbol{u}_2,\boldsymbol{u}_3$ associated with $\lambda_1,\lambda_2,\lambda_3$, respectively. Use of equation (1.6.89) gives

$$\mathbf{S}\boldsymbol{u}_a=\lambda_a\boldsymbol{u}_a \text{ (no sum)} \tag{1.6.97}$$

and hence

$$\lambda_1\boldsymbol{u}_1\cdot\boldsymbol{u}_2=\boldsymbol{u}_2\cdot\mathbf{S}\boldsymbol{u}_1 \tag{1.6.98}$$

$$\lambda_2\boldsymbol{u}_2\cdot\boldsymbol{u}_1=\boldsymbol{u}_1\cdot\mathbf{S}\boldsymbol{u}_2. \tag{1.6.99}$$

Noting that $\mathbf{S}=\mathbf{S}^{\mathrm{T}}$, the difference of equations (1.6.98) and (1.6.99) gives

$$(\lambda_1-\lambda_2)\boldsymbol{u}_1\cdot\boldsymbol{u}_2=0$$

where use has been made of equations (1.5.39) and (1.6.42). But $\lambda_1\neq\lambda_2$ hence $\boldsymbol{u}_1\cdot\boldsymbol{u}_2=0$ and hence the proper vectors $\boldsymbol{u}_1$ and $\boldsymbol{u}_2$ are orthogonal. In the same way, it can be shown that $\boldsymbol{u}_3$ is orthogonal to both $\boldsymbol{u}_1$ and $\boldsymbol{u}_2$. Hence it is concluded that if the three proper numbers $\lambda_1,\lambda_2,\lambda_3$ are distinct, then the three associated proper vectors form a unique orthonormal set $\{\boldsymbol{u}_1,\boldsymbol{u}_2,\boldsymbol{u}_3\}$. Such a basis is called *principal*.

A symmetric tensor $\mathbf{S}$ can be expressed in the form

$$\mathbf{S}=\mathbf{S}\mathbf{I}=\mathbf{S}(\boldsymbol{u}_a\otimes\boldsymbol{u}_a)=(\mathbf{S}\boldsymbol{u}_a)\otimes\boldsymbol{u}_a=\sum_{a=1}^{3}\lambda_a(\boldsymbol{u}_a\otimes\boldsymbol{u}_a)$$

where use has been made of equations (1.6.41), (1.6.51), (1.6.56), (1.6.97) and (1.6.48). The expression

$$\mathbf{S}=\sum_{a=1}^{3} \lambda_a(\boldsymbol{u}_a \otimes \boldsymbol{u}_a) \tag{1.6.100}$$

is referred to as the *spectral representation* of a symmetric tensor.

The matrix of components of **S** relative to the principal basis is diagonal, thus

$$[\mathbf{S}]=\begin{bmatrix} \lambda_1 & 0 & 0 \\ 0 & \lambda_2 & 0 \\ 0 & 0 & \lambda_3 \end{bmatrix} \tag{1.6.101}$$

and hence the principal invariants of **S** have the simple form

$$I_{\mathrm{S}}=\operatorname{tr}\mathbf{S}=\lambda_1+\lambda_2+\lambda_3 \tag{1.6.102}$$

$$II_{\mathrm{S}}=\lambda_1\lambda_2+\lambda_2\lambda_3+\lambda_3\lambda_1 \tag{1.6.103}$$

$$III_{\mathrm{S}}=\det\mathbf{S}=\lambda_1\lambda_2\lambda_3 \tag{1.6.104}$$

where use has been made of equations (1.6.72) and (1.6.93).

A tensor **A** is defined to be,

$$\left.\begin{matrix}\text{positive semi-definite}\\ \text{negative semi-definite}\end{matrix}\right\} \text{ if } \boldsymbol{a}\cdot(\mathbf{A}\boldsymbol{a}) \left.\begin{matrix}\geqslant 0\\ \leqslant 0\end{matrix}\right\} \quad \forall\, \boldsymbol{a}\in\mathscr{V} \tag{1.6.105}$$

and

$$\left.\begin{matrix}\text{positive definite}\\ \text{negative definite}\end{matrix}\right\} \text{ if } \boldsymbol{a}\cdot(\mathbf{A}\boldsymbol{a}) \left.\begin{matrix}> 0\\ < 0\end{matrix}\right\} \quad \forall\, \text{non-zero } \boldsymbol{a}\in\mathscr{V}. \tag{1.6.106}$$

Let **S** be a positive semi-definite symmetric tensor with proper numbers λ_a and associated unit proper vectors $\boldsymbol{u}_a$ $(a=1,2,3)$. From equation (1.6.105),

$$\begin{aligned}\boldsymbol{a}\cdot(\mathbf{S}\boldsymbol{a})&=\lambda_a\{(\boldsymbol{u}_a\otimes\boldsymbol{u}_a)\boldsymbol{a}\}\cdot\boldsymbol{a}\\ &=\lambda_a(\boldsymbol{a}\cdot\boldsymbol{u}_a)^2\geqslant 0 \qquad \forall\,\boldsymbol{a}\in\mathscr{V}\end{aligned}$$

and hence, $\lambda_1\geqslant 0, \lambda_2\geqslant 0, \lambda_3\geqslant 0$. Since **S** has positive proper numbers, it follows from equation (1.6.104) that $\det\mathbf{S}>0$ implying that **S** is invertible: the spectral representation of the inverse is

$$\mathbf{S}^{-1}=\sum_{a=1}^{3}\lambda_a^{-1}(\boldsymbol{u}_a\otimes\boldsymbol{u}_a). \tag{1.6.107}$$

If **S** is a positive semi-definite symmetric tensor, then there is a unique positive semi-definite tensor **T** such that $\mathbf{T}^2=\mathbf{S}$. The tensor **T** is denoted by $\mathbf{S}^{1/2}$. In the context of the discussion relating to equations (1.6.94) to (1.6.96), the proper numbers of $\mathbf{S}^{1/2}$ are the positive semi-definite square roots of those of $\mathbf{S}$. Thus, $\mathbf{S}^{1/2}$ has the spectral representation,

$$\mathbf{S}^{1/2}=\sum_{a=1}^{3}\lambda_a^{1/2}(\boldsymbol{u}_a\otimes\boldsymbol{u}_a).$$

1.6.7 Skew-symmetric Tensors

Let $\mathbf{W}$ be a skew-symmetric tensor, and let $\boldsymbol{a}, \boldsymbol{b}$ be arbitrary vectors. Using the condition $\mathbf{W}^{\mathrm{T}}=-\mathbf{W}$ with equation (1.6.42) gives

$$\boldsymbol{a}\cdot(\mathbf{W}\boldsymbol{b})=(\mathbf{W}^{\mathrm{T}}\boldsymbol{a})\cdot\boldsymbol{b}=-(\mathbf{W}\boldsymbol{a})\cdot\boldsymbol{b}. \tag{1.6.108}$$

Setting $\boldsymbol{a}=\boldsymbol{b}$ in equation (1.6.108) gives

$$\boldsymbol{a}\cdot(\mathbf{W}\boldsymbol{a})=0. \tag{1.6.109}$$

From equation (1.6.61)

$$W^i{}_j=\boldsymbol{e}^i\cdot(\mathbf{W}\boldsymbol{e}_j) \qquad W^{ij}=\boldsymbol{e}^i\cdot(\mathbf{W}\boldsymbol{e}^j) \qquad W_{ij}=\boldsymbol{e}_i\cdot(\mathbf{W}\boldsymbol{e}_j). \tag{1.6.110}$$

Applying the condition of equation (1.6.109) to equation (1.6.110) gives

$$\begin{gathered} W^p{}_p=W^{pp}=W_{pp}=0 \\ W^i{}_j=-W_j{}^i \qquad W^{ij}=-W^{ji} \qquad W_{ij}=-W_{ji} \end{gathered} \tag{1.6.111}$$

from which it is evident that a skew-symmetric tensor has only three independent components. This conclusion raises the question of whether a skew-symmetric tensor can, in some respects, be regarded as a special form of vector.

For a skew-symmetric tensor $\mathbf{W}$, equation (1.6.89) gives

$$\mathbf{W}\boldsymbol{u}=\mu\lambda\boldsymbol{u}$$

and applying the condition of equation (1.6.109) it follows that

$$\boldsymbol{u}\cdot(\mathbf{W}\boldsymbol{u})=\lambda(\boldsymbol{u}\cdot\boldsymbol{u})=0 \tag{1.6.112}$$

and hence $\mathbf{W}$ has either a single proper number which is zero or three proper numbers each equal to zero. Equation (1.6.112) also shows that there must exist a unit vector $\boldsymbol{u}$ for which $\mathbf{W}\boldsymbol{u}=\boldsymbol{0}$.

Replacing the basis $\{\hat{\boldsymbol{e}}_i\}$ in equation (1.5.91) by the basis $\{\boldsymbol{f}, \boldsymbol{g}, \boldsymbol{h}\}$ and taking the upper sign gives

$$\begin{aligned} \boldsymbol{f}&=\boldsymbol{g}\wedge\boldsymbol{h} \qquad & \boldsymbol{g}&=\boldsymbol{h}\wedge\boldsymbol{f} \\ \boldsymbol{h}&=\boldsymbol{f}\wedge\boldsymbol{g} \qquad & [\boldsymbol{f},\boldsymbol{g},\boldsymbol{h}]&=1. \end{aligned} \tag{1.6.113}$$

Using equations (1.6.60) and (1.6.61), expressed in terms of the basis $\{\boldsymbol{f}, \boldsymbol{g}, \boldsymbol{h}\}$ of $\mathscr{V}$, gives

$$\mathbf{W}=\omega[\boldsymbol{h}\otimes\boldsymbol{g}-\boldsymbol{g}\otimes\boldsymbol{h}] \tag{1.6.114}$$

where

$$\omega = \boldsymbol{h} \cdot (\mathbf{W}\boldsymbol{g}).$$

Taking $\boldsymbol{a}$ to be an arbitrary vector, and setting

$$\boldsymbol{w} = \omega \boldsymbol{f}, \tag{1.6.115}$$

it follows that

$$\begin{aligned}\mathbf{W}\boldsymbol{a} - \boldsymbol{w} \wedge \boldsymbol{a} &= \omega\{(\boldsymbol{h} \otimes \boldsymbol{g})\boldsymbol{a} - (\boldsymbol{g} \otimes \boldsymbol{h})\boldsymbol{a} - \boldsymbol{f} \wedge [(\boldsymbol{a}\cdot\boldsymbol{f})\boldsymbol{f} + (\boldsymbol{a}\cdot\boldsymbol{g})\boldsymbol{g} + (\boldsymbol{a}\cdot\boldsymbol{h})\boldsymbol{h}]\} \\ &= \omega\{(\boldsymbol{a}\cdot\boldsymbol{g})(\boldsymbol{h} - \boldsymbol{f} \wedge \boldsymbol{g}) - (\boldsymbol{a}\cdot\boldsymbol{h})(\boldsymbol{g} + \boldsymbol{f} \wedge \boldsymbol{h})\} = \boldsymbol{0}.\end{aligned} \tag{1.6.116}$$

The left-hand side of equation (1.6.116) gives

$$\mathbf{W}\boldsymbol{a} = \boldsymbol{w} \wedge \boldsymbol{a} \qquad \forall \boldsymbol{a} \in \mathscr{V}, \tag{1.6.117}$$

which shows that associated with a skew-symmetric tensor $\mathbf{W}$ there is a vector $\boldsymbol{w}$ which is called the *axial vector* of $\mathbf{W}$.

Replacing the $\boldsymbol{a}, \boldsymbol{b}, \boldsymbol{c}$ in equations (1.6.66), (1.6.68) and (1.6.69) by the unit vectors $\boldsymbol{f}, \boldsymbol{g}, \boldsymbol{h}$ introduced in equation (1.6.113) (and noting in relation to equation (1.6.112) that there must exist a unit vector $\boldsymbol{f}$ for which $\mathbf{W}\boldsymbol{f} = \boldsymbol{0}$), gives, for a skew-symmetric tensor $\mathbf{W}$, the relations

$$I_{\mathrm{W}} = [\boldsymbol{f}, \mathbf{W}\boldsymbol{g}, \boldsymbol{h}] + [\boldsymbol{f}, \boldsymbol{g}, \mathbf{W}\boldsymbol{h}] \tag{1.6.118}$$

$$II_{\mathrm{W}} = [\boldsymbol{f}, \mathbf{W}\boldsymbol{g}, \mathbf{W}\boldsymbol{h}] \tag{1.6.119}$$

$$III_{\mathrm{W}} = 0. \tag{1.6.120}$$

From equations (1.6.113) and (1.6.117), with $\boldsymbol{w} = \omega \boldsymbol{f}$, the following relations can be obtained

$$\mathbf{W}\boldsymbol{g} = \omega \boldsymbol{f} \wedge \boldsymbol{g} = \omega \boldsymbol{h} \qquad \mathbf{W}\boldsymbol{h} = \omega \boldsymbol{f} \wedge \boldsymbol{h} = -\omega \boldsymbol{g}$$

which when entered into equations (1.6.118) and (1.6.119) give

$$I_{\mathrm{W}} = 0 \qquad II_{\mathrm{W}} = [\boldsymbol{f}, \omega\boldsymbol{h}, -\omega\boldsymbol{g}] = \omega^2[\boldsymbol{f}, \boldsymbol{h}, -\boldsymbol{g}] = \omega^2 = |\boldsymbol{w}|^2.$$

Hence, in terms of the axial vector $\boldsymbol{w}$ of $\mathbf{W}$, the principal invariants of the skew-symmetric tensor $\mathbf{W}$ take the form

$$I_{\mathrm{W}} = 0 \qquad II_{\mathrm{W}} = |\boldsymbol{w}|^2 \qquad III_{\mathrm{W}} = \det \mathbf{W} = 0. \tag{1.6.121}$$

Thus substituting for the principal invariants of $\mathbf{W}$ in equation (1.6.93) gives the characteristic equation of $\mathbf{W}$ in the form

$$\lambda^3 - |\boldsymbol{w}|^2 \lambda = 0 \tag{1.6.122}$$

which has a single real root, equal to zero.

Various relations involving skew-symmetric tensors are now derived for later use, particularly in §2.2.5.

Application of equations (1.6.44) and (1.6.50) shows that the tensor

$\boldsymbol{v} \otimes \boldsymbol{u} - \boldsymbol{u} \otimes \boldsymbol{v}$, for any of the vectors $\boldsymbol{u}, \boldsymbol{v} \in \mathscr{V}$, is a skew-symmetric second-order tensor. Noting that

$$(\boldsymbol{v} \otimes \boldsymbol{u} - \boldsymbol{u} \otimes \boldsymbol{v})(\boldsymbol{u} \wedge \boldsymbol{v}) = \{(\boldsymbol{u} \wedge \boldsymbol{v}) \cdot \boldsymbol{u}\}\boldsymbol{v} - \{(\boldsymbol{u} \wedge \boldsymbol{v}) \cdot \boldsymbol{v}\}\boldsymbol{u} = \boldsymbol{0}$$

is of the form $\mathbf{W}\boldsymbol{u} = \boldsymbol{0}$, which was discussed in relation to equation (1.6.112) it follows by analogy with equation (1.6.115) that the axial vector of $\boldsymbol{v} \otimes \boldsymbol{u} - \boldsymbol{u} \otimes \boldsymbol{v}$ is a scalar multiple of $\boldsymbol{u} \wedge \boldsymbol{v}$: thus

$$(\boldsymbol{v} \otimes \boldsymbol{u} - \boldsymbol{u} \otimes \boldsymbol{v})\boldsymbol{a} = (\boldsymbol{a} \cdot \boldsymbol{u})\boldsymbol{v} - (\boldsymbol{a} \cdot \boldsymbol{v})\boldsymbol{u} = \lambda(\boldsymbol{u} \wedge \boldsymbol{v}) \wedge \boldsymbol{a} \qquad \forall \boldsymbol{a} \in \mathscr{V}. \tag{1.6.123}$$

Setting $\boldsymbol{a} = \boldsymbol{u}$ in equation (1.6.123) and forming the scalar product of the resulting vector with $\boldsymbol{v}$ gives

$$(\boldsymbol{u} \cdot \boldsymbol{u})(\boldsymbol{v} \cdot \boldsymbol{v}) - (\boldsymbol{u} \cdot \boldsymbol{v})^2 = \lambda\{(\boldsymbol{u} \wedge \boldsymbol{v}) \wedge \boldsymbol{u}\} \cdot \boldsymbol{v} = \lambda(\boldsymbol{u} \wedge \boldsymbol{v}) \cdot (\boldsymbol{u} \wedge \boldsymbol{v}) \tag{1.6.124}$$

and by comparison with the axiom formulated by equation (1.5.45) it is evident that $\lambda = 1$. With $\lambda = 1$, equation (1.6.123) gives, the required identities in the form

$$\boldsymbol{a} \wedge (\boldsymbol{b} \wedge \boldsymbol{c}) = (\boldsymbol{b} \otimes \boldsymbol{c} - \boldsymbol{c} \otimes \boldsymbol{b})\boldsymbol{a} = (\boldsymbol{a} \cdot \boldsymbol{c})\boldsymbol{b} - (\boldsymbol{a} \cdot \boldsymbol{b})\boldsymbol{c} \qquad \forall \boldsymbol{a}, \boldsymbol{b}, \boldsymbol{c} \in \mathscr{V}. \tag{1.6.125}$$

The identity $\{(\boldsymbol{u} \wedge \boldsymbol{v})\boldsymbol{u}\} \cdot \boldsymbol{v} = (\boldsymbol{u} \wedge \boldsymbol{v}) \cdot (\boldsymbol{u} \wedge \boldsymbol{v})$ appearing in equation (1.6.124) can be established by first deriving the expansion

$$(\boldsymbol{a} \wedge \boldsymbol{b}) \wedge \boldsymbol{c} = (\boldsymbol{c} \cdot \boldsymbol{a})\boldsymbol{b} - (\boldsymbol{b} \cdot \boldsymbol{c})\boldsymbol{a} \tag{1.6.126}$$

by way of the component forms of $\boldsymbol{a}, \boldsymbol{b}, \boldsymbol{c}$.

1.6.8 Orthogonal Tensors

A mapping $\mathbf{Q} \in \mathscr{L}(\mathscr{V}; \mathscr{V})$ of an inner-product space onto itself is orthogonal if it preserves the inner product: thus, if $\boldsymbol{a}$ and $\boldsymbol{b}$ are arbitrary vectors

$$(\mathbf{Q}\boldsymbol{a}) \cdot (\mathbf{Q}\boldsymbol{b}) = \boldsymbol{a} \cdot \boldsymbol{b} \qquad \forall \boldsymbol{a}, \boldsymbol{b} \in \mathscr{V}. \tag{1.6.127}$$

Use of equation (1.6.42) gives

$$(\mathbf{Q}\boldsymbol{a}) \cdot (\mathbf{Q}\boldsymbol{b}) = \{\mathbf{Q}^{\mathrm{T}}(\mathbf{Q}\boldsymbol{a})\} \cdot \boldsymbol{b} = \{(\mathbf{Q}^{\mathrm{T}}\mathbf{Q})\boldsymbol{a}\} \cdot \boldsymbol{b}$$

and hence a necessary and sufficient condition for $\mathbf{Q}$ to be orthogonal is

$$\mathbf{Q}^{\mathrm{T}}\mathbf{Q} = \mathbf{I}. \tag{1.6.128}$$

Equations (1.6.79), (1.6.80) and (1.6.81) give

$$\det(\mathbf{Q}^{\mathrm{T}}\mathbf{Q}) = \det \mathbf{Q}^{\mathrm{T}} \det \mathbf{Q} = (\det \mathbf{Q})^2 = 1 \tag{1.6.129}$$

and hence $\mathbf{Q}$ is invertible. Thus, from equations (1.6.83) and (1.6.128)

$$\mathbf{Q}^{-1} = \mathbf{Q}^{\mathrm{T}} \qquad \mathbf{Q}\mathbf{Q}^{\mathrm{T}} = \mathbf{I}. \tag{1.6.130}$$

Thus

$$\det \mathbf{Q} = \pm 1.$$

If det $\mathbf{Q}=1$, the orthogonal tensor $\mathbf{Q}$ is said to be a *proper orthogonal tensor*, and if det $\mathbf{Q}=-1$, then $\mathbf{Q}$ is said to be an *improper orthogonal tensor.*

The orthogonal tensors constitute a proper subgroup of $\mathscr{U}$ called the full *orthogonal group*, $\mathscr{O}$.

From equation (1.6.128)

$$\mathbf{Q}^{\mathrm{T}}(\mathbf{Q}-\mathbf{I})=-(\mathbf{Q}-\mathbf{I})^{\mathrm{T}}$$

and

$$\det[\mathbf{Q}^{\mathrm{T}}(\mathbf{Q}-\mathbf{I})]=\det[-(\mathbf{Q}-\mathbf{I})^{\mathrm{T}}]=\det \mathbf{Q}^{\mathrm{T}} \det(\mathbf{Q}-\mathbf{I})$$

can be rearranged into the form

$$\{(\det \mathbf{Q})+1\} \det(\mathbf{Q}-\mathbf{I})=0 \qquad (1.6.131)$$

where use has been made of equations (1.6.78), (1.6.79) and (1.6.81). The condition det $\mathbf{Q}=1$ applied to equation (1.6.131) gives $\det(\mathbf{Q}-\mathbf{I})=0$ which is just equation (1.6.91) with $\lambda=1$, a condition taken to imply that $\mathbf{Q}$ has a proper number equal to 1. It will be assumed for the remainder of this section that det $\mathbf{Q}=1$. Since $\mathbf{Q}$ has a proper number equal to 1 there must exist a unit vector $\boldsymbol{u}$ such that

$$\mathbf{Q}\boldsymbol{u}=\boldsymbol{u}=\mathbf{Q}^{\mathrm{T}}\boldsymbol{u}. \qquad (1.6.132)$$

With the unit vectors $\boldsymbol{u}, \boldsymbol{v}, \boldsymbol{w}$ related by equation (1.6.113), it follows from equation (1.6.132) that

$$\begin{gathered}\boldsymbol{v}\cdot(\mathbf{Q}\boldsymbol{u})=\boldsymbol{w}\cdot(\mathbf{Q}\boldsymbol{u})=\boldsymbol{u}\cdot(\mathbf{Q}\boldsymbol{v})=\boldsymbol{u}\cdot(\mathbf{Q}\boldsymbol{w})=0\\ \boldsymbol{u}\cdot(\mathbf{Q}\boldsymbol{u})=1\end{gathered} \qquad (1.6.133)$$

and from equation (1.6.127) that

$$(\mathbf{Q}\boldsymbol{v})\cdot(\mathbf{Q}\boldsymbol{w})=0 \qquad |\mathbf{Q}\boldsymbol{v}|=|\mathbf{Q}\boldsymbol{w}|=1 \qquad (1.6.134)$$

from which it follows that $\mathbf{Q}\boldsymbol{v}$ and $\mathbf{Q}\boldsymbol{w}$ are a pair of unit vectors, orthogonal to $\boldsymbol{u}$. From equation (1.6.69)

$$\det \mathbf{Q}=[\mathbf{Q}\boldsymbol{u}, \mathbf{Q}\boldsymbol{v}, \mathbf{Q}\boldsymbol{w}]=[\boldsymbol{u}, \mathbf{Q}\boldsymbol{v}, \mathbf{Q}\boldsymbol{w}]=1 \qquad (1.6.135)$$

where use has been made of equation (1.6.132). Thus, since both $\boldsymbol{v}, \boldsymbol{w}$ and $\mathbf{Q}\boldsymbol{v}$, $\mathbf{Q}\boldsymbol{w}$ are orthogonal pairs of unit vectors all orthogonal to $\boldsymbol{u}$, they satisfy relations of the form $\mathbf{Q}\boldsymbol{v}=\lambda\boldsymbol{v}+\mu\boldsymbol{w}$, $\mathbf{Q}\boldsymbol{w}=\zeta\boldsymbol{v}+\xi\boldsymbol{w}$ which can be entered into the relations of equations (1.6.134) and (1.6.135) to give four equations relating the λ, μ, ζ and ξ, which in turn give $\lambda=\xi$ and $\mu=-\zeta$. The four equations relating the λ, μ, ζ and ξ reduce for $\lambda=\xi$, $\mu=-\zeta$ to a form which can be satisfied by the substitution $\lambda=\xi=\cos\theta$, $\mu=-\zeta=\sin\theta(-\pi<\theta\leqslant\pi)$, which in turn gives

$$\begin{gathered}-\boldsymbol{v}\cdot(\mathbf{Q}\boldsymbol{w})=\boldsymbol{w}\cdot(\mathbf{Q}\boldsymbol{v})=\sin\theta\\ \boldsymbol{v}\cdot(\mathbf{Q}\boldsymbol{v})=\boldsymbol{w}\cdot(\mathbf{Q}\boldsymbol{w})=\cos\theta.\end{gathered}$$

Thus,

$$\mathbf{Q}=\boldsymbol{u}\otimes\boldsymbol{u}+(\boldsymbol{v}\otimes\boldsymbol{v}+\boldsymbol{w}\otimes\boldsymbol{w})\cos\theta-(\boldsymbol{v}\otimes\boldsymbol{w}-\boldsymbol{w}\otimes\boldsymbol{v})\sin\theta. \tag{1.6.136}$$

Equations (1.6.66), (1.6.68) and (1.6.69) can be expressed in the form

$$I_{\mathrm{Q}}=1+[\boldsymbol{u},\mathbf{Q}\boldsymbol{v},\boldsymbol{w}]+[\boldsymbol{u},\boldsymbol{v},\mathbf{Q}\boldsymbol{w}] \tag{1.6.137}$$

$$II_{\mathrm{Q}}=[\boldsymbol{u},\mathbf{Q}\boldsymbol{v},\mathbf{Q}\boldsymbol{w}]+[\boldsymbol{u},\boldsymbol{v},\mathbf{Q}\boldsymbol{w}]+[\boldsymbol{u},\mathbf{Q}\boldsymbol{v},\boldsymbol{w}] \tag{1.6.138}$$

$$III_{\mathrm{Q}}=[\boldsymbol{u},\mathbf{Q}\boldsymbol{v},\mathbf{Q}\boldsymbol{w}]=\det\mathbf{Q}=1. \tag{1.6.139}$$

It has been shown above that $\mathbf{Q}\boldsymbol{v}$ and $\mathbf{Q}\boldsymbol{w}$ can be expressed in the form

$$\mathbf{Q}\boldsymbol{v}=\cos\theta\boldsymbol{v}+\sin\theta\boldsymbol{w} \qquad \mathbf{Q}\boldsymbol{w}=-\sin\theta\boldsymbol{v}+\cos\theta\boldsymbol{w}$$

and hence, entering these forms for $\mathbf{Q}\boldsymbol{v}$ and $\mathbf{Q}\boldsymbol{w}$ into equations (1.6.137) and (1.6.138) gives

$$I_{\mathrm{Q}}=II_{\mathrm{Q}}=1+2\cos\ \theta. \tag{1.6.140}$$

Substituting for the principal invariants of $\mathbf{Q}$ in equation (1.6.93) from equations (1.6.139) and (1.6.140) gives

$$\lambda^3-(1+2\cos\theta)\lambda(\lambda-1)-1=(\lambda-1)(\lambda^2-2\lambda\cos\theta+1)=0.$$

If $\theta\neq 0,\pi, |\cos\theta|<1$ and the quadratic factor has a negative discriminant, with this as basis, $\mathbf{Q}$ has a single proper number, equal to 1.

1.6.9 Polar Decompositions

If $\mathbf{A}$ is an arbitrary invertible tensor, then there are unique positive-definite symmetric tensors $\mathbf{U}$ and $\mathbf{V}$ and a unique orthogonal tensor $\mathbf{Q}$ such that

$$\mathbf{A}=\mathbf{QU} \qquad \text{and} \qquad \mathbf{A}=\mathbf{VQ} \tag{1.6.141}$$

these two unique multiplicative decompositions being known as the *polar decomposition theorem.*

From equation (1.6.45),

$$(\mathbf{A}^{\mathrm{T}}\mathbf{A})^{\mathrm{T}}=\mathbf{A}^{\mathrm{T}}(\mathbf{A}^{\mathrm{T}})^{\mathrm{T}}=\mathbf{A}^{\mathrm{T}}\mathbf{A} \qquad (\mathbf{A}\mathbf{A}^{\mathrm{T}})^{\mathrm{T}}=(\mathbf{A}^{\mathrm{T}})^{\mathrm{T}}\mathbf{A}^{\mathrm{T}}=\mathbf{A}\mathbf{A}^{\mathrm{T}} \tag{1.6.142}$$

and use of equation (1.6.42) gives

$$\begin{aligned}\boldsymbol{a}\cdot\{(\mathbf{A}^{\mathrm{T}}\mathbf{A})\boldsymbol{a}\}&=\boldsymbol{a}\cdot\{\mathbf{A}^{\mathrm{T}}(\mathbf{A}\boldsymbol{a})\}=(\mathbf{A}\boldsymbol{a})\cdot(\mathbf{A}\boldsymbol{a})\\ \boldsymbol{a}\cdot\{(\mathbf{A}\mathbf{A}^{\mathrm{T}})\boldsymbol{a}\}&=\boldsymbol{a}\cdot\{\mathbf{A}(\mathbf{A}^{\mathrm{T}}\boldsymbol{a})\}=(\mathbf{A}^{\mathrm{T}}\boldsymbol{a})\cdot(\mathbf{A}^{\mathrm{T}}\boldsymbol{a})\end{aligned} \tag{1.6.143}$$

where $\boldsymbol{a}$ is an arbitrary non-zero vector. In the context of the axiom formulated by equation (1.5.41) and equation (1.6.106) it is evident from equations (1.6.142) and (1.6.143) that $\mathbf{AA}^{\mathrm{T}}$ and $\mathbf{A}^{\mathrm{T}}\mathbf{A}$ are positive definite symmetric tensors. Having regard to the discussion at the end of §1.6.6, it is evident that $\mathbf{A}^{\mathrm{T}}\mathbf{A}$ and $\mathbf{AA}^{\mathrm{T}}$ have unique positive definite symmetric square roots, $\mathbf{U}$ and $\mathbf{V}$

respectively. Setting $\mathbf{Q}=\mathbf{A}\mathbf{U}^{-1}$ it follows that

$$\mathbf{Q}^{\mathrm{T}}\mathbf{Q}=(\mathbf{A}\mathbf{U}^{-1})^{\mathrm{T}}(\mathbf{A}\mathbf{U}^{-1})=\mathbf{U}^{-1}\mathbf{A}^{\mathrm{T}}\mathbf{A}\mathbf{U}^{-1}=\mathbf{U}^{-1}\mathbf{U}^{2}\mathbf{U}^{-1}=\mathbf{I}$$

and similarly, setting $\mathbf{R}=\mathbf{V}^{-1}\mathbf{A}$ gives

$$\mathbf{R}\mathbf{R}^{\mathrm{T}}=(\mathbf{V}^{-1}\mathbf{A})(\mathbf{V}^{-1}\mathbf{A}^{\mathrm{T}})=\mathbf{V}^{-1}\mathbf{A}\mathbf{A}^{\mathrm{T}}\mathbf{V}^{-1}=\mathbf{V}^{-1}\mathbf{V}^{2}\mathbf{V}^{-1}=\mathbf{I}. \quad (1.6.144)$$

Hence, the tensors $\mathbf{Q}=\mathbf{A}\mathbf{U}^{-1}$ and $\mathbf{R}=\mathbf{V}^{-1}\mathbf{A}$ are orthogonal: thus

$$\mathbf{A}=\mathbf{Q}\mathbf{U} \qquad \text{and} \qquad \mathbf{A}=\mathbf{V}\mathbf{R}. \quad (1.6.145)$$

To prove the uniqueness, assume

$$\mathbf{Q}\mathbf{U}=\hat{\mathbf{Q}}\hat{\mathbf{U}} \quad (1.6.146)$$

where $\hat{\mathbf{Q}}$ is orthogonal and $\hat{\mathbf{U}}$ positive definite and symmetric. From equations (1.6.128) and (1.6.145)

$$\mathbf{U}^{2}=\mathbf{U}\mathbf{Q}^{\mathrm{T}}\mathbf{Q}\mathbf{U}=(\mathbf{Q}\mathbf{U})^{\mathrm{T}}\mathbf{Q}\mathbf{U}=(\hat{\mathbf{Q}}\hat{\mathbf{U}})^{\mathrm{T}}\hat{\mathbf{Q}}\hat{\mathbf{U}}=\hat{\mathbf{U}}^{2}.$$

Since the positive-definite square root of $\mathbf{U}^2$ is unique, it follows that $\mathbf{U}=\hat{\mathbf{U}}$, and hence from equation (1.6.144)

$$\hat{\mathbf{Q}}=\mathbf{A}\hat{\mathbf{U}}^{-1}=\mathbf{A}\mathbf{U}^{-1}=\mathbf{Q}.$$

A similar argument establishes the uniqueness of the decomposition of equation (1.6.145)

Equation (1.6.145) can be expressed in the form

$$\begin{aligned}\mathbf{A}=\mathbf{Q}\mathbf{U}=\mathbf{Q}(\mathbf{Q}^{\mathrm{T}}\mathbf{V}\mathbf{Q})&=\mathbf{Q}\{(\mathbf{V}^{1/2}\mathbf{Q})^{\mathrm{T}}(\mathbf{V}^{1/2}\mathbf{Q})\}\\&=(\mathbf{R}\mathbf{R}^{\mathrm{T}})\mathbf{V}\mathbf{R}=\mathbf{R}(\mathbf{R}^{\mathrm{T}}\mathbf{V}\mathbf{R})\\&=\mathbf{R}\{(\mathbf{V}^{1/2}\mathbf{R})^{\mathrm{T}}(\mathbf{V}^{1/2}\mathbf{R})\}\end{aligned}$$

from which it is concluded that $\mathbf{Q}=\mathbf{R}$, use having been made of equation (1.6.144) and the fact that $\mathbf{V}$ possesses a positive definite square root. It also follows that

$$\mathbf{U}=\mathbf{R}^{\mathrm{T}}\mathbf{V}\mathbf{R} \quad (1.6.147)$$

Let λ_a, μ_a be the proper numbers of $\mathbf{U}$ and $\mathbf{V}$, respectively, such that

$$\mathbf{U}\boldsymbol{u}_a=\lambda_a\boldsymbol{u}_a \qquad \mathbf{V}\boldsymbol{v}_a=\mu_a\boldsymbol{v}_a$$

where $\boldsymbol{u}_a$ and $\boldsymbol{v}_a$ are the associated proper vectors of $\mathbf{U}$ and $\mathbf{V}$ respectively. Use of equation (1.6.141) gives

$$(\mathbf{Q}\mathbf{U})\boldsymbol{u}_a=\mathbf{Q}(\mathbf{U}\boldsymbol{u}_a)=(\mathbf{V}\mathbf{Q})\boldsymbol{u}_a=\mathbf{V}(\mathbf{Q}\boldsymbol{u}_a)=\lambda_a(\mathbf{Q}\boldsymbol{u}_a)$$

from which it follows that λ_a is a proper number of $\mathbf{V}$ and $\mathbf{Q}\boldsymbol{u}_a$ is an associated proper vector. Hence λ_a and $\boldsymbol{v}_a$ must take the following values

$$\lambda_a=\mu_a \qquad \boldsymbol{v}_a=\mathbf{Q}\boldsymbol{u}_a \quad (1.6.148)$$

it being noted that

$$\mathbf{Q} = \mathbf{QI} = \mathbf{Q}(\boldsymbol{u}_a \otimes \boldsymbol{u}_a) = \boldsymbol{v}_a \otimes \boldsymbol{u}_a \tag{1.6.149}$$

where use has been made of equation (1.6.148).

1.6.10 Higher Order Tensors

Tensors of higher order have been considered briefly at the end of §1.6.1. The introduction of the tensor product affords another method of defining tensors.

Let the symbol $\boldsymbol{u} \otimes \boldsymbol{v} \otimes \boldsymbol{w}$ denote that linear mapping of a given vector space into second order tensors such that

$$(\boldsymbol{u} \otimes \boldsymbol{v} \otimes \boldsymbol{w})\boldsymbol{a} = (\boldsymbol{a} \cdot \boldsymbol{w})(\boldsymbol{u} \otimes \boldsymbol{v}) \qquad \forall\, \boldsymbol{a} \in \mathscr{V} \tag{1.6.150}$$

where $\boldsymbol{a}$ is an arbitrary vector. The products $\boldsymbol{e}_p \otimes \boldsymbol{e}_q \otimes \boldsymbol{e}_r$ of elements of a basis for the given vector space form a basis for the set of such transformations which give rise to equation (1.6.150). Thus, if $\mathbb{P}$ is any linear mapping of the given vector space into tensors, it may be expressed in the form

$$\mathbb{P} = P^{pqr}\boldsymbol{e}_p \otimes \boldsymbol{e}_p \otimes \boldsymbol{e}_r \tag{1.6.151}$$

and from the discussion given at the end of §1.6.1, it is evident that the tensors so defined are of *third-order*.

As for third order tensors, let the symbol $\boldsymbol{u} \otimes \boldsymbol{v} \otimes \boldsymbol{w} \otimes \boldsymbol{x}$ denote that linear mapping of a given vector space into third order tensors such that

$$(\boldsymbol{u} \otimes \boldsymbol{v} \otimes \boldsymbol{w} \otimes \boldsymbol{x})\boldsymbol{a} = (\boldsymbol{a} \cdot \boldsymbol{x})(\boldsymbol{u} \otimes \boldsymbol{v} \otimes \boldsymbol{w}) \qquad \forall\, \boldsymbol{a} \in \mathscr{V} \tag{1.6.152}$$

where $\boldsymbol{a}$ is an arbitrary vector. If H is any linear mapping of the given vector space into tensors, it may be expressed in the form

$$\mathsf{H} = H^{pqrs}\boldsymbol{e}_p \otimes \boldsymbol{e}_q \otimes \boldsymbol{e}_r \otimes \boldsymbol{e}_s \tag{1.6.153}$$

and from the discussion given at the end of §1.6.1, it is evident that the tensors so defined are of *fourth-order*. For the result of operating with a fourth-order tensor H upon a second-order tensor $\mathbf{D}$, use is made of the special notation $\mathsf{H}[\mathbf{D}]$. The components of $\mathsf{H}[\mathbf{D}]$ are $H^i{}_j{}^p{}_q D_p{}^q$.

The properties of third- and fourth-order tensors are similar to those of second-order tensors.

1.7 GEOMETRY

1.7.1 Euclidean Point Space

Although reference has been made, if only by implication, to general manifolds, the only specific geometry with which this book is concerned is that of Euclidean space.

The term *Euclidean point space* or *Euclidean manifold* applies to a set $\mathscr{E}$, with elements called *points*.

The relation between a Euclidean point space $\mathscr{E}$ and a Euclidean vector space $\mathscr{V}$ is defined as follows. To each ordered pair of points (x, y) in $\mathscr{E}$ there corresponds a unique vector in $\mathscr{V}$, denoted by $\overrightarrow{xy}$, with the properties:

(i) $\overrightarrow{yx} = -\overrightarrow{xy} \qquad \forall x, y \in \mathscr{E}$

(ii) $\overrightarrow{xy} = \overrightarrow{xz} + \overrightarrow{zy} \qquad \forall x, y, z \in \mathscr{E}$ (1.7.1)

(iii) given a point o, chosen arbitrarily from $\mathscr{E}$ there corresponds to each vector $\boldsymbol{x} \in \mathscr{V}$ a unique point $x \in \mathscr{E}$ such that $\boldsymbol{x} \equiv \overrightarrow{ox}$.

With regard to the third axiom, the point o is referred to, in the context of the geometrical representation of a vector, as the *origin* and $\boldsymbol{x}$ as the *position* of the point x relative to o. In the present context, the inner product space $\mathscr{V}$ is called the *translation space* of $\mathscr{E}$.

The concepts of distance and angle in $\mathscr{E}$ can be formulated by way of the definition of the scalar product on the supporting vector space $\mathscr{V}$. The difference between the arbitrary points x and y can be formulated in terms of the difference between the position $\boldsymbol{x}$ and the position $\boldsymbol{y}$ which defines the *displacement*

$$\boldsymbol{u} = \boldsymbol{y} - \boldsymbol{x} (\equiv \overrightarrow{xy}) \tag{1.7.2}$$

it being said that $\boldsymbol{u}$ translates $\boldsymbol{x}$ into $\boldsymbol{y}$; similarly, $-\boldsymbol{u}$ translates $\boldsymbol{y}$ into $\boldsymbol{x}$.

The *distance* d between the arbitrary points x and y is the magnitude of the vector $\boldsymbol{u}$ that translates $\boldsymbol{x}$ into $\boldsymbol{y}$; that is

$$d(\boldsymbol{y}, \boldsymbol{x}) = |\boldsymbol{y} - \boldsymbol{x}| \equiv |\overrightarrow{xy}|. \tag{1.7.3}$$

It follows that

$$d(\boldsymbol{y}, \boldsymbol{x}) = d(\boldsymbol{x}, \boldsymbol{y}) \tag{1.7.4}$$

$$d(\boldsymbol{y}, \boldsymbol{x}) \leqslant d(\boldsymbol{y}, \boldsymbol{z}) + d(\boldsymbol{z}, \boldsymbol{x}) \tag{1.7.5}$$

$$d(\boldsymbol{y}, \boldsymbol{x}) \geqslant 0 \qquad \text{and} \qquad d(\boldsymbol{y}, \boldsymbol{x}) = 0 \Leftrightarrow \boldsymbol{y} = \boldsymbol{x}. \tag{1.7.6}$$

The properties (1.7.4), (1.7.5) and (1.7.6) establish that $\mathscr{E}$ is a *metric space*.

The angle θ subtended by x and y at a third arbitrary point z is defined by

$$\cos\theta = \left(\frac{(\overrightarrow{zx}) \cdot (\overrightarrow{zy})}{|\overrightarrow{zx}||\overrightarrow{zy}|}\right) \qquad (0 \leqslant \theta \leqslant \pi) \tag{1.7.7}$$

which is to be compared with equation (1.5.24).

A *neighbourhood* of a given point is the set of all points whose distances from the given point are less than some positive real number. With regard to a neighbourhood considered relative to a subset P of points:

(i) an *interior point* possesses a neighbourhood which belongs entirely to P,
(ii) an *exterior point* has a neighbourhood containing no point of P,
(iii) a *boundary point* has a neighbourhood containing at least one point of P and at least one point not belonging to P, it being noted that a boundary point does not of necessity belong to P.

The totalities of interior and boundary points are called the *interior* and the *boundary* of P, respectively. The boundary of P is generally denoted by ∂P. A point set P is said to be:

(i) *open* if it coincides with its interior,
(ii) *connected* if every pair of its members can be joined by a path lying wholly in P,
(iii) *bounded* if one of its members has a neighbourhood containing every point of P.

A *domain* D of $\mathscr{E}$ is a connected open point set and a *region* R of $\mathscr{E}$ is a connected point set with non-empty interior which contains the whole of its boundary.

1.7.2 Coordinate Systems

Let $e=\{\boldsymbol{e}_1, \boldsymbol{e}_2, \ldots, \boldsymbol{e}_n\}$ be a basis for an n-dimensional vector space $\mathscr{V}$. The vector $\boldsymbol{u}(\boldsymbol{x}) \in \mathscr{V}$ can be expressed in terms of its components:

$$\boldsymbol{u}(\boldsymbol{x}) = u^p(\boldsymbol{x})\boldsymbol{e}_p \qquad \forall\, \boldsymbol{x} \in \mathscr{E}. \tag{1.7.8}$$

It is often convenient to select a different basis at each $\boldsymbol{x}$, so that,

$$\boldsymbol{e}_p = \boldsymbol{e}_p(\boldsymbol{x}) \tag{1.7.9}$$

and hence, in general, both the components and the basis may be functions of $\boldsymbol{x}$. This consideration gives rise to the subject of curvilinear coordinate analysis.

In the context of §1.5.1, the real n-space $\mathscr{R}_n$, is the n-dimensional vector space whose elements are ordered lists of n real numbers, $(x^1, x^2, \ldots, x^n)$. A *coordinate system* on an open set of a real n-dimensional space is a one-to-one mapping of that set into $\mathscr{R}_n$, and which has an invertible gradient and a continuous second gradient. If $\boldsymbol{x}$ is such a mapping, then

$$\bar{\boldsymbol{x}}(\boldsymbol{x}) = \{\bar{x}^1(\boldsymbol{x}), \bar{x}^2(\boldsymbol{x}), \ldots, \bar{x}^n(\boldsymbol{x})\} \tag{1.7.10}$$

where $\bar{x}^k$ is a scalar field having the same degree of smoothness as that assumed for $\bar{\boldsymbol{x}}$. The number $\bar{x}^k(\boldsymbol{x})$ is the kth coordinate of $\boldsymbol{x}$ in the coordinate system $\bar{\boldsymbol{x}}$.

If $\hat{\boldsymbol{x}}$ denotes the inverse of $\bar{\boldsymbol{x}}$, then

$$\bar{x}^k[\hat{x}(x^1, x^2, \ldots, x^n)] = x^k \qquad k = 1, 2, \ldots, n \tag{1.7.11}$$

for all $(x^1, x^2, \ldots, x^n)$ that lie in the range of $\bar{x}$. The vector

$$e_k(x) \equiv \frac{\partial}{\partial \bar{x}^k} \hat{x}(x^1, x^2, \ldots, x^n)|_{x^j = \bar{x}^j(x)} \tag{1.7.12}$$

is tangent to the kth *coordinate curve* at x, that curve being the set of points near x for which every coordinate but x^k has the same value as it does at x. The curves traced out by x for which only one of the x^k varies are called *coordinate lines*. Since the coordinate lines are, in general, curved rather than straight, the variables x^k are called the *curvilinear coordinates* of the point x.

The sets of vectors $\{e_1(x), e_2(x), \ldots, e_n(x)\}$ and $\{e^1(x), e^2(e), \ldots, e^n(x)\}$ are reciprocal bases of the translation space of $\mathscr{E}$. The basis $\{e_1(x), e_2(x), \ldots, e_n(x)\}$ is called the *natural basis* of the coordinate system $\bar{x}$ at x, and $\{e^1(x), e^2(x), \ldots, e^n(x)\}$ is the corresponding *reciprocal natural basis.* As already noted, in relation to equations (1.7.8) and (1.7.9), as the point x varies over the domain of $\bar{x}$, fields of natural bases and their reciprocals are obtained. In general, these bases are not orthonormal. When the coordinate surfaces are mutually perpendicular, the system is curvilinear or orthogonal, and when the surfaces are orthogonal planes, the system is rectangular. If the surfaces do not intersect at right angles, the system is said to be affine or non-orthogonal.

Euclidean manifolds possess certain special coordinate systems of particular interest. Three such systems are the rectangular Cartesian coordinate system, the cylindrical polar coordinate system and the spherical polar coordinate system. These three systems are considered in the next section.

1.7.3 The Fundamental Metric

In the context of equation (1.7.12), the square of the infinitesimal distance

$$\mathrm{d}x = e_p \, \mathrm{d}x^p \tag{1.7.13}$$

between two points can be written in terms of the natural basis as

$$\begin{aligned} \mathrm{d}s^2 = \mathrm{d}x \cdot \mathrm{d}x &= e_p \cdot e_q \, \mathrm{d}x^p \, \mathrm{d}x^q \\ &= g_{pq} \, \mathrm{d}x^p \, \mathrm{d}x^q \end{aligned} \tag{1.7.14}$$

where

$$g_{ij} = e_i \cdot e_j \tag{1.7.15}$$

is an important quantity characterising the geometrical properties of space and is called the *fundamental metric tensor*. The fundamental metric tensor has already been encountered in §1.5.3 in the form of equation (1.5.53).

The contravariant metric components g^{ij} are defined by the relation,

$$g^{ij} = e^i \cdot e^j. \tag{1.7.16}$$

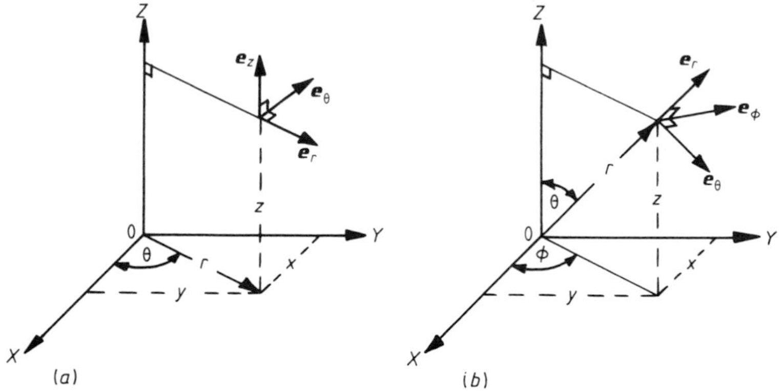

Figure 1.1 Some simple coordinate systems (*a*) Cartesian and cylindrical polar coordinates (*b*) spherical polar coordinates.

The metric tensor of the three special coordinate systems, referred to above, are now given.

Rectangular Cartesian coordinates (x, y, z). The natural basis field is orthonormal only if it is a constant field, in which case the coordinates are called rectangular Cartesian. The values of the Cartesian coordinate fields may be interpreted as the distances measured parallel to a particular set of three mutually orthogonal lines (see figure 1.1(*a*)). From figure 1.1(*a*) it follows that

$$ds^2 = dx^2 + dy^2 + dz^2 \tag{1.7.17}$$

and hence

$$[g_{ij}] = \begin{bmatrix} 1 & 0 & 0 \\ 0 & 1 & 0 \\ 0 & 0 & 1 \end{bmatrix}. \tag{1.7.18}$$

Cylindrical polar coordinates (r, θ, z). These are shown in figure 1.1(*a*), from which it follows that

$$ds^2 = dr^2 + (r\, d\theta)^2 + dz^2 \tag{1.7.19}$$

and hence

$$[g_{ij}] = \begin{bmatrix} 1 & 0 & 0 \\ 0 & r^2 & 0 \\ 0 & 0 & 1 \end{bmatrix} \qquad [g^{ij}] = \begin{bmatrix} 1 & 0 & 0 \\ 0 & r^{-2} & 0 \\ 0 & 0 & 1 \end{bmatrix}. \tag{1.7.20}$$

It folows from equation (1.7.12) that

$$e_r=\frac{\partial x}{\partial r}=e^r \qquad e_\theta=\frac{\partial x}{\partial \theta}=r^2 e^\theta \qquad e_2=\frac{\partial x}{\partial z}=e^z \tag{1.7.21}$$

where use has been made of equations (1.7.20), (1.5.59) and (1.5.65).

Spherical polar coordinates (r, θ, ϕ). These are shown in figure 1.1(b), from which it follows that

$$ds^2=dr^2+(r\,d\theta)^2+(r\sin\theta\,d\phi)^2 \tag{1.7.22}$$

and hence

$$[g_{ij}]=\begin{bmatrix} 1 & 0 & 0 \\ 0 & r^2 & 0 \\ 0 & 0 & r^2\sin^2\theta \end{bmatrix} \qquad [g^{ij}]=\begin{bmatrix} 1 & 0 & 0 \\ 0 & r^{-2} & 0 \\ 0 & 0 & (r\sin\theta)^{-2} \end{bmatrix}. \tag{1.7.23}$$

It follows from equation (1.7.12) that

$$e_r=\frac{\partial x}{\partial r}=e^r \qquad e_\theta=\frac{\partial x}{\partial \theta}=r^2 e^\theta \qquad e_\phi=\frac{\partial x}{\partial \phi}=r^2\sin^2\theta e^\phi \tag{1.7.24}$$

where use has been made of equations (1.7.23), (1.5.59) and (1.5.65).

For the arbitrary vector $a\in\mathscr{V}$ and the arbitrary tensor $\mathbf{A}\in\mathscr{L}(\mathscr{V};\mathscr{V})$, the various sets of components are related as follows

$$a_i=g_{ip}a^p \qquad a^i=g^{ip}a_p \tag{1.7.25}$$

$$A^{ij}=g^{ip}A_p{}^j=g^{pj}A^i{}_p=g^{ip}g^{jq}A_{pq}. \tag{1.7.26}$$

Consider two coordinate systems $\bar{x}$ and $\tilde{x}$, for which the coordinates of x with respect to the two systems are functionally related

$$\tilde{x}^i(x)=f^i(\bar{x}^1,\bar{x}^2,\ldots,\bar{x}^n)$$

$$\bar{x}^j(x)=g^j(\tilde{x}^1,\tilde{x}^2,\ldots,\tilde{x}^n).$$

Let $\tilde{e}=\{\tilde{e}_1(x),\tilde{e}_2(x),\ldots,\tilde{e}_n(x)\}$ be the natural basis of the coordinate system $\tilde{x}$ at x. Using equation (1.7.12), it follows that

$$\begin{aligned}\tilde{e}_i&=\frac{\partial}{\partial\tilde{x}^i}\hat{x}[g^1(\tilde{x}^1,\ldots,\tilde{x}^n),\ldots,g^n(\tilde{x}^1,\ldots,\tilde{x}^n)]\\
&=\left(\frac{\partial}{\partial\bar{x}^p}\hat{x}(\bar{x}^1,\ldots,\bar{x}^n)\right)\frac{\partial}{\partial\tilde{x}^i}g^p(\tilde{x}^1,\ldots,\tilde{x}^n)\\
&=\left(\frac{\partial}{\partial\tilde{x}^i}g^p\right)\bar{e}_p=\frac{\partial\bar{x}^p}{\partial\tilde{x}^i}\bar{e}_p.\end{aligned} \tag{1.7.27}$$

Setting

$$H^i{}_j = \frac{\partial \bar{x}^i}{\partial \tilde{x}^j} \qquad G^p{}_q = \frac{\partial \tilde{x}^p}{\partial \bar{x}^q}$$

it follows from equations (1.5.102), for an arbitrary vector $\boldsymbol{a}$ that

$$\tilde{a}_i = \frac{\partial \bar{x}^p}{\partial \tilde{x}^i} \bar{a}_p \qquad \tilde{a}^i = \frac{\partial \tilde{x}^i}{\partial \bar{x}^p} \bar{a}^p \tag{1.7.28}$$

and for a tensor $\mathbf{A}$, it follows that

$$\begin{aligned} \tilde{A}_{ij} &= \bar{A}_{pq} \frac{\partial \bar{x}^p}{\partial \tilde{x}^i} \frac{\partial \bar{x}^q}{\partial \tilde{x}^j} \\ \tilde{A}^i{}_j &= \bar{A}^p{}_q \frac{\partial \tilde{x}^i}{\partial \bar{x}^p} \frac{\partial \bar{x}^q}{\partial \tilde{x}^j} \\ \tilde{A}_i{}^j &= \bar{A}_p{}^q \frac{\partial \bar{x}^p}{\partial \tilde{x}^i} \frac{\partial \tilde{x}^j}{\partial \bar{x}^q}. \end{aligned} \tag{1.7.29}$$

For a rectangular Cartesian set of coordinates x_i and the orthonormal base vectors $\hat{\boldsymbol{e}}_1, \hat{\boldsymbol{e}}_2, \hat{\boldsymbol{e}}_3$,

$$\bar{\boldsymbol{e}}_i = \frac{\partial x_p}{\partial \bar{x}^i} \hat{\boldsymbol{e}}_p. \tag{1.7.30}$$

Hence,

$$\begin{aligned} \bar{g}_{ij} &= \frac{\partial x_p}{\partial \bar{x}^i} \frac{\partial x_q}{\partial \bar{x}^j} \hat{\boldsymbol{e}}_p \cdot \hat{\boldsymbol{e}}_q \\ &= \frac{\partial x_p}{\partial \bar{x}^i} \frac{\partial x_q}{\partial \bar{x}^j} \delta_{pq} = \frac{\partial x_p}{\partial \bar{x}^i} \frac{\partial x_p}{\partial \bar{x}^j} \end{aligned} \tag{1.7.31}$$

where the rectangular Cartesian coordinates $x_i = x^i$ are presumed given as functions of the general coordinates $\bar{x}^j$

$$x_i = x^i = f^i(\bar{x}^1, \bar{x}^2, \ldots, \bar{x}^n) \qquad i = 1, 2, \ldots, n. \tag{1.7.32}$$

1.7.4 Physical Components Relative to an Orthogonal Coordinate System

For some coordinate systems, the coordinates may not have the same dimensions, as for example in spherical polar coordinates where r has the dimension of length while θ and ϕ are dimensionless. If the coordinates do not have the same dimensions, then the components of a vector may not have the same dimensions, which is unacceptable in physical problems. If the vectors are resolved relative to unit vectors in the direction of the base vectors, then all the components will have the same dimensions; these are called the *physical components* of a vector $\boldsymbol{a}$ and will be written $a\langle i \rangle$.

From equation (1.5.21), the magnitude of the base vector $\boldsymbol{e}_i$ is given by

$$|\boldsymbol{e}_p| = (\boldsymbol{e}_p \cdot \boldsymbol{e}_p)^{1/2} = \sqrt{g_{pp}} \qquad \text{(no sum)} \tag{1.7.33}$$

and hence, in general, the base vectors are not unit vectors.

When the coordinate surfaces are mutually orthogonal, the coordinate curves are normal to the coordinate surfaces, and hence $\boldsymbol{e}^i$ is parallel to $\boldsymbol{e}_i$, and thus it follows from equation (1.5.65) that

$$\boldsymbol{e}^p = \boldsymbol{e}_p / g_{pp} \qquad (p \text{ not summed}). \tag{1.7.34}$$

Equations (1.5.65) and equation (1.7.30) give

$$g^{pp} = 1/g_{pp}. \tag{1.7.35}$$

Clearly, the contravariant base vectors are also not unit vectors.

From equation (1.7.33) the unit vector $\boldsymbol{\varepsilon}_i$ in the direction of $\boldsymbol{e}_i$ is given by

$$\boldsymbol{\varepsilon}_i = \frac{\boldsymbol{e}_i}{|\boldsymbol{e}_i|} = \frac{\boldsymbol{e}_i}{\sqrt{g_{ii}}} \qquad (i \text{ not summed}). \tag{1.7.36}$$

Thus

$$\boldsymbol{a} = a\langle i \rangle \boldsymbol{\varepsilon}_i = a^i \boldsymbol{e}_i$$

and hence, use of equation (1.7.36) gives

$$a\langle i \rangle = \sqrt{g_{ii}}\, a^i = \sqrt{g_{ii}}\, g^{ip} a_p. \tag{1.7.37}$$

From equation (1.6.61)

$$A^i{}_j = \boldsymbol{e}^i \cdot (\mathbf{A} \boldsymbol{e}_j)$$

which can be rearranged, using equations (1.7.34), (1.7.35) and (1.7.36), to give

$$A\langle ij \rangle = \left(\frac{g_{ii}}{g_{jj}}\right)^{1/2} A^i{}_j \tag{1.7.38}$$

where the $A\langle ij \rangle$ are the physical components of the arbitrary tensor $\mathbf{A}$.

It is to be noted that the physical components do not transform in a tensorial manner.

1.7.5 Christoffel Components

Since in a general curvilinear coordinate system the base vectors depend on position, it follows that the gradients of the natural basis of a coordinate system exist and are themselves tensors. Consider the partial derivative of the ith vector with respect to the jth coordinate. This vector can be expressed in the form

$$\frac{\partial \boldsymbol{e}_i}{\partial x^j} = \left(\frac{\partial \boldsymbol{e}_i}{\partial x^j} \cdot \boldsymbol{e}^p\right) \boldsymbol{e}_p = \Gamma_{ij}{}^p \boldsymbol{e}_p \tag{1.7.39}$$

where the coefficients

$$\Gamma_{ij}{}^{p} \equiv \frac{\partial \boldsymbol{e}_i}{\partial x^j} \cdot \boldsymbol{e}^p = \boldsymbol{e}^p \cdot (\Gamma_{,i} \boldsymbol{e}_j) \tag{1.7.40}$$

are known as the Christoffel symbols of the second kind for a given coordinate system. Since,

$$\boldsymbol{e}_i = \partial \boldsymbol{x} / \partial x^i$$

it follows that,

$$\frac{\partial \boldsymbol{e}_i}{\partial x^j} = \frac{\partial^2 \boldsymbol{x}}{\partial x^j \, \partial x^i} = \frac{\partial \boldsymbol{e}_j}{\partial x^i}$$

and hence

$$\Gamma_{ij}{}^{p} = \Gamma_{ji}{}^{p}. \tag{1.7.41}$$

From equation (1.5.53), the following can be obtained

$$\frac{\partial g_{ij}}{\partial x^q} = \frac{\partial \boldsymbol{e}_i}{\partial x^q} \cdot \boldsymbol{e}_j + \boldsymbol{e}_i \cdot \frac{\partial \boldsymbol{e}_j}{\partial x^q}.$$

Using equation (1.7.39)

$$\frac{\partial g_{ij}}{\partial x^k} = \Gamma_{ik}{}^{p} g_{pj} + \Gamma_{jk}{}^{p} g_{pi}. \tag{1.7.42}$$

Similarly

$$\frac{\partial g_{jk}}{\partial x^i} = \Gamma_{ji}{}^{p} g_{pk} + \Gamma_{ki}{}^{p} g_{pj} \tag{1.7.43}$$

and

$$\frac{\partial g_{ki}}{\partial x^j} = \Gamma_{kj}{}^{p} g_{pi} + \Gamma_{ij}{}^{p} g_{pk}. \tag{1.7.44}$$

Taking the difference of equations (1.7.42) and (1.7.43) and the difference of equations (1.7.42) and (1.7.44) gives the Christoffel symbol of the first kind

$$A_{ijk} = g_{kp} \Gamma_{ij}{}^{p} = \frac{1}{2}\left(\frac{\partial g_{jk}}{\partial x^i} + \frac{\partial g_{ik}}{\partial x^j} - \frac{\partial g_{ij}}{\partial x^k} \right) \tag{1.7.45}$$

where use has been made of equation (1.7.41). Multiplying through equation (1.7.45) by g^{kq} gives,

$$\Gamma_{ij}{}^{q} = \tfrac{1}{2} g^{kq} \left(\frac{\partial g_{jk}}{\partial x^i} + \frac{\partial g_{ik}}{\partial x^j} - \frac{\partial g_{ij}}{\partial x^k} \right) \tag{1.7.46}$$

and hence the Christoffel components can be calculated from the metric components g^{kq} and g_{pj} of the coordinate system.

The Christoffel symbols of some simple coordinate systems are now stated for future reference:

Rectangular Cartesian coordinates: all symbols are zero.
Cylindrical polar coordinates: the only non-zero symbols are:

$$\Gamma_{\theta\theta}{}^{r} = -r \qquad \Gamma_{r\theta}{}^{\theta} = 1/r. \tag{1.7.47}$$

Spherical polar coordinates: the only non-zero symbols are

$$\begin{aligned} &\Gamma_{\theta\theta}{}^{r} = -r \qquad \Gamma_{\phi\phi}{}^{r} = -r\sin^2\theta \qquad \Gamma_{\phi\phi}{}^{\theta} = -\sin\theta\cos\theta \\ &\Gamma_{r\theta}{}^{\theta} = 1/r \qquad \Gamma_{r\phi}{}^{\phi} = 1/r \qquad \Gamma_{\theta\phi}{}^{\phi} = \cot\theta. \end{aligned} \tag{1.7.48}$$

It is of interest to note that the Christoffel components of a coordinate system vanish identically if and only if the natural basis field is constant, i.e. for a rectangular Cartesian coordinate system.

1.7.6 Principal Axes

A positive definite symmetric tensor **S** admits the spectral representation

$$\mathbf{S} = \sum_{a=1}^{3} \lambda_a(\hat{\boldsymbol{e}}_a \otimes \hat{\boldsymbol{e}}_a) \tag{1.7.49}$$

where the λ_i $(i=1,2,3)$ are the positive proper numbers and the associated proper vectors $\hat{\boldsymbol{e}}_i$ form an orthonormal basis $\hat{e} = \{\hat{\boldsymbol{e}}_1, \hat{\boldsymbol{e}}_2, \hat{\boldsymbol{e}}_3\}$ of $\mathscr{V}$.

If $\boldsymbol{x}$ is the position of an arbitrary point x relative to an origin o, then

$$\begin{aligned} \mathbf{S}\boldsymbol{x} &= \sum_{a=1}^{3} \lambda_a(\hat{\boldsymbol{e}}_a \otimes \hat{\boldsymbol{e}}_a)\boldsymbol{x} = \sum_{a=1}^{3} \lambda_a(\boldsymbol{x}\cdot\hat{\boldsymbol{e}}_a)\hat{\boldsymbol{e}}_a \\ &= \sum_{a=1}^{3} \lambda_a x_a \hat{\boldsymbol{e}}_a \end{aligned} \tag{1.7.50}$$

where use has been made of equations (1.6.47) and (1.5.89) and where the x_i are the coordinates of x in the rectangular Cartesian coordinate system $(o, \hat{e})$. From equation (1.7.50) it is evident that the action of **S** on $\boldsymbol{x}$ is to translate x into the point y having position $\mathbf{S}\boldsymbol{x}$ relative to the origin o and whose coordinates in the system $(o, \hat{e})$ are the $\lambda_i x_i$. It is also evident from equation (1.7.50) that the distance of every point of $\mathscr{E}$ from the coordinate plane $x_1 = 0$ is changed by the positive factor λ_1, being increased if $\lambda_1 > 1$ and reduced if $(0 < \lambda_1 < 1)$. Similarly, the distances from the planes $x_2 = 0$ and $x_3 = 0$ through o are changed by the positive factors λ_2 and λ_3, respectively. Thus, the tensor **S** gives rise to a transformation of $\mathscr{E}$ which takes the form of proportional *extensions*, or *stretches*, of amounts λ_i in the mutually orthogonal directions defined by the unit proper vectors $\hat{\boldsymbol{e}}_i$, these distances being known as the *principal axes* of **S**.

1.7.7 Rotations

A proper orthogonal tensor $\mathbf{Q}$ on $\mathscr{V}$ can be expressed in the form of equation (1.6.136)

$$\mathbf{Q}=\boldsymbol{u}\otimes\boldsymbol{u}+(\boldsymbol{v}\otimes\boldsymbol{v}+\boldsymbol{w}\otimes\boldsymbol{w})\cos\theta-(\boldsymbol{v}\otimes\boldsymbol{w}-\boldsymbol{w}\otimes\boldsymbol{v})\sin\theta \tag{1.7.51}$$

where $-\pi<\theta\leqslant\pi$ and where the $\boldsymbol{u}$, $\boldsymbol{v}$, $\boldsymbol{w}$ form an orthonormal basis $u=\{\boldsymbol{u}, \boldsymbol{v}, \boldsymbol{w}\}$ of $\mathscr{V}$.

If $\boldsymbol{x}$ is the position of an arbitrary point x relative to an origin o, then

$$\begin{aligned}\mathbf{Q}\boldsymbol{x}&=\{\boldsymbol{u}\otimes\boldsymbol{u}+(\boldsymbol{v}\otimes\boldsymbol{v}+\boldsymbol{w}\otimes\boldsymbol{w})\cos\theta-(\boldsymbol{v}\otimes\boldsymbol{w}-\boldsymbol{w}\otimes\boldsymbol{v})\sin\theta\}\boldsymbol{x}\\&=u\boldsymbol{u}+(v\cos\theta-w\sin\theta)\boldsymbol{v}+(v\sin\theta+w\cos\theta)\boldsymbol{w}\end{aligned} \tag{1.7.52}$$

where

$$u=\boldsymbol{u}\cdot\boldsymbol{x} \qquad v=\boldsymbol{v}\cdot\boldsymbol{x} \qquad w=\boldsymbol{w}\cdot\boldsymbol{x}. \tag{1.7.53}$$

In terms of a rectangular Cartesian coordinate system with origin o and base vectors $\boldsymbol{u}$, $\boldsymbol{v}$, $\boldsymbol{w}$, the arbitrary point x and the point y with position $\mathbf{Q}\boldsymbol{x}$ relative to o are seen from equations (1.7.52) and (1.7.53) to have coordinates $(u, \operatorname{Re} z, \operatorname{Im} z)$ and $(u, \operatorname{Re} z\exp(j\theta), \operatorname{Im} z\exp(j\theta))$, respectively, where

$$z=v+jw \qquad j=\sqrt{-1}. \tag{1.7.54}$$

Thus both the arbitrary point x and the point y with position $\mathbf{Q}\boldsymbol{x}$ are at the same distance u from the plane through o containing $\boldsymbol{v}$ and $\boldsymbol{w}$. Similarly, both the arbitrary point x and the point y with position $\mathbf{Q}\boldsymbol{x}$ are at the same distance $(v^2+w^2)^{1/2}$ from an axis ou through o and coincident with the direction defined by $\boldsymbol{u}$. Thus, x is translated into y along the circumference of a circle of radius $(v^2+w^2)^{1/2}$. The action of $\mathbf{Q}$ on $\boldsymbol{x}$ is therefore a rotation of $\mathscr{E}$ of amount θ about the axis through o in the direction defined by $\boldsymbol{u}$.

If the arbitrary vectors $\boldsymbol{u}$ and $\boldsymbol{v}$ are rotated under an orthogonal transformation $\mathbf{Q}$ into new vectors $\bar{\boldsymbol{u}}$, $\bar{\boldsymbol{v}}$, respectively, so that

$$\bar{\boldsymbol{u}}=\boldsymbol{Q}\boldsymbol{u} \qquad \bar{\boldsymbol{v}}=\mathbf{Q}\boldsymbol{v} \tag{1.7.55}$$

then the scalar product

$$\bar{\boldsymbol{u}}\cdot\bar{\boldsymbol{v}}=(\mathbf{Q}\boldsymbol{u})\cdot(\mathbf{Q}\boldsymbol{v})=\boldsymbol{u}\cdot\mathbf{Q}^{\mathrm{T}}\mathbf{Q}\boldsymbol{v}=\boldsymbol{u}\cdot\boldsymbol{v}$$

is invariant, having used equations (1.6.42) and (1.6.128). Thus the magnitude of vectors is unchanged and also the angle between them. Hence, it is again evident that an orthogonal transformation merely rotates a vector and an orthonormal basis remains orthonormal.

1.8 COVARIANT DIFFERENTIATION

At a neighbouring point let an arbitrary vector $\boldsymbol{a}$ become $\boldsymbol{a}+\mathrm{d}\boldsymbol{a}$; with this as basis the following can be obtained from equation (1.5.51),

$$\mathrm{d}\boldsymbol{a}=\mathrm{d}a^p\boldsymbol{e}_p+a^p\,\mathrm{d}\boldsymbol{e}_p. \tag{1.8.1}$$

By the ordinary rules of calculus,

$$\mathrm{d}a^i=\frac{\partial a^i}{\partial x^q}\,\mathrm{d}x^q \tag{1.8.2}$$

and by equation (1.7.39),

$$\mathrm{d}\boldsymbol{e}_i=\frac{\partial \boldsymbol{e}_i}{\partial x^q}\,\mathrm{d}x^q=\Gamma_{iq}{}^r\boldsymbol{e}_r\,\mathrm{d}x^q. \tag{1.8.3}$$

Entering the relations of equations (1.8.2) and (1.8.3) into equation (1.8.1) and interchanging the roles of r and p in the last term as they are repeated indices gives

$$\mathrm{d}\boldsymbol{a}=\left(\frac{\partial a^p}{\partial x^q}+a^r\Gamma_{rq}{}^p\right)\boldsymbol{e}_p\,\mathrm{d}x^q. \tag{1.8.4}$$

Under a coordinate transformation $\mathrm{d}\boldsymbol{a}$ is invariant, $\boldsymbol{e}_i$ transforms as a covariant vector, and $\mathrm{d}x^j$ as a contravariant vector. Thus, the term in the brackets of equation (1.8.4) transforms as a mixed tensor. This term is called the covariant derivative of the contravariant vector $a^p\boldsymbol{e}_p$ and will be denoted by $a^i{}_{,j}$. Thus,

$$a^i{}_{,j}=\frac{\partial a^i}{\partial x^j}+a^p\Gamma^i_{pj}. \tag{1.8.5}$$

Similarly, it can be shown that the covariant derivative of a covariant vector is given by

$$a_{i,j}=\frac{\partial a_i}{\partial x^j}-a_p\Gamma^p_{ij}. \tag{1.8.6}$$

For an arbitrary second-order tensor $\mathbf{A}$, the covariant derivatives have the form

$$\begin{aligned}
A_{ij,k}&=\frac{\partial A_{ij}}{\partial x^k}-\Gamma_{ik}{}^pA_{pj}-\Gamma_{jk}{}^pA_{ip}\\
A^i{}_{j,k}&=\frac{\partial A^i{}_j}{\partial x^k}+\Gamma_{pk}{}^iA^p{}_j-\Gamma_{jk}{}^pA^i{}_p\\
A^{ij}{}_{,k}&=\frac{\partial A^{ij}}{\partial x^k}+\Gamma_{pk}{}^iA^{pj}+\Gamma_{pk}{}^jA^{ip}.
\end{aligned} \tag{1.8.7}$$

Applying covariant differentiation to g^{ij} and $\delta^i{}_j$ gives

$$g_{ij,k} = g^{ij}{}_{,k} = \delta^i{}_{j,k} = 0 \tag{1.8.8}$$

which is known as Ricci's theorem. Hence, in the process of covariant differentiation, g_{ij}, g^{ij} and δ_{ij} are not affected: thus,

$$\begin{aligned} a^i{}_{,k} &= (g^{ip} a_p)_{,k} = g^{ip} a_{p,k} \\ a_{i,k} &= (g_{ip} a^p)_{,k} = g_{ip} a^p{}_{,k}. \end{aligned} \tag{1.8.9}$$

Similarly, all the covariant derivatives of ε_{ijk} and ε^{ijk} vanish.

The second covariant derivatives can be found by repeating the above process, and so on for higher-order derivatives.

According to a theorem of calculus, the order of mixed partial differentiation is not important, an observation which prompts the question of under what conditions do the second-order covariant partial derivatives of a tensor commute. The following can be obtained from equation (1.8.6)

$$a_{i,jk} = \frac{\partial a_{i,j}}{\partial x^k} - \Gamma_{ik}{}^p a_{p,j} - \Gamma_{jk}{}^p a_{i,p}$$

or

$$a_{i,jk} = \frac{\partial^2 a_i}{\partial x^j \, \partial x^k} - \frac{\partial \Gamma_{ij}{}^p}{\partial x^k} a_p - \Gamma_{ij}{}^p \frac{\partial a_p}{\partial x^k} - \Gamma_{ik}{}^p \frac{\partial a_p}{\partial x^j}$$

$$+ \Gamma_{ik}{}^p \Gamma_{pj}{}^r a_r - \Gamma_{jk}{}^p \frac{\partial a_i}{\partial x^p} + \Gamma_{jk}{}^p \Gamma_{ip}{}^r a_r. \tag{1.8.10}$$

Interchanging the indices j and k and subtracting the result from equation (1.8.10) gives

$$a_{i,jk} - a_{i,kj} = R^p_{ijk} a_p \tag{1.8.11}$$

where

$$R^p_{ijk} = \frac{\partial \Gamma_{ik}{}^p}{\partial x^j} - \frac{\partial \Gamma_{ij}{}^p}{\partial x^k} + \Gamma_{ik}{}^r \Gamma_{rj}{}^p - \Gamma_{ij}{}^r \Gamma_{rk}{}^p \tag{1.8.12}$$

is called the Riemann–Christoffel tensor. It is thus evident that covariant derivatives of any vector are equal if and only if the Riemann–Christoffel tensor vanishes identically.

By lowering the contravariant index, a covariant fourth-order tensor called the *curvature tensor* is obtained,

$$R_{lijk} = g_{lp} R^p{}_{ijk} \tag{1.8.13}$$

where R_{ijkl} satisfies the relations

$$R_{ijkl} = -R_{jikl} = -R_{ijlk} = R_{klij} \tag{1.8.14}$$

which can be deduced from the definition

$$R_{lijk}=\frac{1}{2}\left(\frac{\partial^2 g_{il}}{\partial x^j\,\partial x^k}+\frac{\partial^2 g_{jk}}{\partial x^i\,\partial x^l}-\frac{\partial^2 g_{ik}}{\partial x^j\,\partial x^l}-\frac{\partial^2 g_{jl}}{\partial x^i\,\partial x^k}\right)$$

$$+g^{pr}(A_{jkp}A_{ilr}-A_{jlp}A_{ikr}) \qquad (1.8.15)$$

and where the Christoffel symbols A_{ijk} of the first kind are defined by equation (1.7.45).

Both tensors $R^p{}_{ijk}$ and R_{lijk} possess certain symmetry properties. It can be shown that in an n-dimensional space there are $n^2(n^2-1)/12$ non-vanishing components of R_{lijk}. In three dimensions, the non-vanishing components are

$$R_{1212}, R_{1313}, R_{2323}, R_{1213}, R_{2123}, R_{3132} \qquad (1.8.16)$$

and in two dimensions, the only non-vanishing component is R_{1212}.

By taking the covariant derivatives of $R^p{}_{ijk}$ and using equations (1.8.7) and (1.8.12) it can be shown that

$$\begin{aligned} R^p{}_{ijk,r}+R^p{}_{ikr,j}+R^p{}_{irj,k}&=0\\ R_{lijk,r}+R_{likr,j}+R_{lirj,k}&=0 \end{aligned} \qquad (1.8.17)$$

which are known as Bianchi's identities.

The Riemann–Christoffel tensor is a measure of the curvature of the space. Thus, when the Riemann–Christoffel tensor vanishes in a space, that space is called *flat*. In Euclidean space the Riemann–Christoffel tensor vanishes identically. Therefore the Euclidean space is a flat space.

1.9 VECTOR AND TENSOR FIELDS

It is now possible to restate the definition, given in §1.4 for a scalar field. If, for every point of a domain D, there corresponds a scalar, the function as defined from D to $\mathscr{R}$ is called a scalar field on D. When an origin o is selected the points of D stand in bijective correspondence with their positions relative to o: for all such positions, the set $\{\boldsymbol{x}|x\in D\}$ will be denoted by D_o. With this notation, a scalar field $\phi: D\rightarrow\mathscr{R}$ gives rise to a scalar-valued function $\phi_o: D_o\rightarrow\mathscr{R}$ defined by,

$$\phi_o(\boldsymbol{x})=\phi(x) \qquad \forall\,\boldsymbol{x}\in D_o. \qquad (1.9.1)$$

Although the function ϕ_o depends upon the choice of origin of immediate interest, this is of no physical significance since, having regard to equation (1.7.1), the effect of a change of origin is merely to increment all positions by a fixed vector. Hence, no distinction will be made between ϕ and ϕ_o, and $\phi(\boldsymbol{x})$ will be understood to denote the value of ϕ at the representative point x with position $\boldsymbol{x}$ relative to an arbitrarily chosen origin o.

In a similar manner, functions from D to $\mathscr{V}$ which assign a vector to each point of D are called *vector fields*. With the same notation, a vector field $\boldsymbol{u}$: $D \rightarrow \mathscr{V}$ gives rise to a vector-valued function $\boldsymbol{u}_o$: $D_o \rightarrow \mathscr{V}$ defined by

$$\boldsymbol{u}_o(\boldsymbol{x}) = \boldsymbol{u}(x) \qquad \forall\, \boldsymbol{x} \in D_o. \tag{1.9.2}$$

As for scalars, although the function $\boldsymbol{u}_o$ depends upon the choice of origin, it follows from equation (1.7.1) that there is again no need to make any distinction between $\boldsymbol{u}$ and $\boldsymbol{u}_o$ and as such $\boldsymbol{u}(\boldsymbol{x})$ will be taken to denote the value of $\boldsymbol{u}$ at the representative point x with position $\boldsymbol{x}$ relative to an arbitrarily chosen origin o.

A scalar field ϕ, defined on a domain D, is said to be *continuous* if

$$\lim_{\varepsilon \to 0} |\phi(\boldsymbol{x} + \varepsilon \boldsymbol{a}) - \phi(\boldsymbol{x})| = 0 \qquad \forall\, \boldsymbol{x} \in D_o \qquad \boldsymbol{a} \in \mathscr{V}. \tag{1.9.3}$$

It will be differentiable if there exists a vector field $\boldsymbol{q}$ such that

$$\lim_{\varepsilon \to 0} \left| \boldsymbol{q}(\boldsymbol{x}) \cdot \boldsymbol{a} - \varepsilon^{-1}\{\phi(\boldsymbol{x} + \varepsilon \boldsymbol{a}) - \phi(\boldsymbol{x})\} \right| = 0 \qquad \forall \boldsymbol{x} \in D_o \qquad \boldsymbol{a} \in \mathscr{V}. \tag{1.9.4}$$

There is at most one vector field with the property specified by equation (1.9.4). That this is so can be seen by considering two vector fields $\boldsymbol{q}$ and $\boldsymbol{q}^*$ in relation to the inequalities

$$\begin{aligned} 0 \leqslant |\{\boldsymbol{q}(\boldsymbol{x}) - \boldsymbol{q}^*(\boldsymbol{x})\} \cdot \boldsymbol{a}| \leqslant & \left| \boldsymbol{q}(\boldsymbol{x}) \cdot \boldsymbol{a} - \varepsilon^{-1}\{\phi(\boldsymbol{x} + \varepsilon \boldsymbol{a}) - \phi(\boldsymbol{x})\} \right| \\ & + \left| \boldsymbol{q}^*(\boldsymbol{x}) \cdot \boldsymbol{a} - \varepsilon^{-1}\{\phi(\boldsymbol{x} + \varepsilon \boldsymbol{a}) - \phi(\boldsymbol{x})\} \right|. \end{aligned} \tag{1.9.5}$$

As $\varepsilon \rightarrow 0$, it follows from these inequalities and the arbitrariness of $\boldsymbol{a}$, that $\boldsymbol{q}^* = \boldsymbol{q}$. When ϕ is differentiable, the unique vector field $\boldsymbol{q}$ is called the *gradient* of ϕ and is written in the form

$$\boldsymbol{q} = \operatorname{grad} \phi. \tag{1.9.6}$$

The properties of continuity and differentiability are attributed to a vector field $\boldsymbol{u}$ defined on D if it applies to the scalar field $\boldsymbol{u} \cdot \boldsymbol{a}$ for all choices of the vector $\boldsymbol{a}$. It is to be noted that here $\boldsymbol{a}$ is an arbitrary vector in $\mathscr{V}$ and not a vector field on D. If grad ϕ exists and is continuous, ϕ is said to be *continuously* differentiable, and this property extends to $\boldsymbol{u}$ if $\operatorname{grad}(\boldsymbol{u} \cdot \boldsymbol{a})$ exists and is continuous for all $\boldsymbol{a} \in \mathscr{V}$.

Since ϕ is differentiable it follows from equation (1.9.4) that on replacing $\boldsymbol{a}$ by the base vectors $\boldsymbol{e}^i$ in turn, the partial derivatives $\partial\phi/\partial x^i$ exist in D. In this way, it follows that

$$\boldsymbol{q} \cdot \boldsymbol{e}_p = q_p = \partial\phi/\partial x^p$$

and hence

$$\operatorname{grad} \phi = \frac{\partial \phi}{\partial x^p} \boldsymbol{e}^p. \tag{1.9.7}$$

Given that the vector field $\boldsymbol{u}$ is differentiable, the gradient of $\boldsymbol{u}$ is the tensor field defined by

$$\{(\operatorname{grad} \boldsymbol{u})(\boldsymbol{x})\}^{\mathrm{T}}\boldsymbol{a}=\operatorname{grad}\{\boldsymbol{u}(\boldsymbol{x})\cdot\boldsymbol{a}\} \qquad \forall\, \boldsymbol{x}\in D_o \qquad \boldsymbol{a}\in\mathscr{V}. \tag{1.9.8}$$

From equation (1.9.7) it follows that,

$$\begin{aligned}\operatorname{grad}\{\boldsymbol{u}(\boldsymbol{x})\cdot\boldsymbol{a}\} &= \left(\frac{\partial \boldsymbol{u}}{\partial x^q}\cdot\boldsymbol{a}\right)\boldsymbol{e}^q \\ &= (u^p{}_{,q})(\boldsymbol{e}_p\cdot\boldsymbol{a})\boldsymbol{e}^q \\ &= (u^p{}_{,q})(\boldsymbol{e}^q\otimes\boldsymbol{e}_p)\boldsymbol{a}\end{aligned} \tag{1.9.9}$$

where use has been made of equations (1.8.5) and (1.6.47). Substituting for the right-hand side of equation (1.9.8) from equation (1.9.9) gives

$$\operatorname{grad} \boldsymbol{u} = u^p{}_{,q}\boldsymbol{e}_p \otimes \boldsymbol{e}^q \tag{1.9.10}$$

where use has been made of equation (1.6.50).

The *divergence* of a differentiable vector field $\boldsymbol{u}$, written div $\boldsymbol{u}$, is scalar-valued and is defined by

$$(\operatorname{div} \boldsymbol{u})(\boldsymbol{x})=\operatorname{tr}\{(\operatorname{grad} \boldsymbol{u})(\boldsymbol{x})\} \qquad \forall \boldsymbol{x}\in D_o. \tag{1.9.11}$$

From equation (1.9.10),

$$\operatorname{tr}\{(\operatorname{grad} \boldsymbol{u})(\boldsymbol{x})\} = u^p{}_{,q}(\boldsymbol{e}_p\boldsymbol{e}^q) = u^p{}_{,p} \tag{1.9.12}$$

where use has been made of equations (1.6.77) and (1.5.59). Substituting for the right-hand side of equation (1.9.11) from equation (1.9.12) gives

$$\operatorname{div} \boldsymbol{u} = u^p{}_{,p}. \tag{1.9.13}$$

The *curl* of a differentiable vector field $\boldsymbol{u}$, written curl $\boldsymbol{u}$, is vector-valued and is defined by

$$\{(\operatorname{curl} \boldsymbol{u})(\boldsymbol{x})\}\cdot\boldsymbol{a}=\operatorname{div}\{\boldsymbol{u}(\boldsymbol{x}) \wedge \boldsymbol{a}\} \qquad \forall\, \boldsymbol{x}\in D_o \qquad \boldsymbol{a}\in\mathscr{V}. \tag{1.9.14}$$

From equation (1.9.13)

$$\operatorname{div}\{\boldsymbol{u}(\boldsymbol{x}) \wedge \boldsymbol{a}\} = \{\pm\varepsilon^{pqr}u_r a_p\}_{,q} \tag{1.9.15}$$

from which can be obtained

$$\begin{aligned}\operatorname{div}\{\boldsymbol{u}(\boldsymbol{x}) \wedge \boldsymbol{a}\} &= \pm\varepsilon^{pqr}u_{r,q}a_p \\ &= \pm\varepsilon^{pqr}u_{r,q}\boldsymbol{e}_p\cdot\boldsymbol{a}\end{aligned} \tag{1.9.16}$$

where use has been made of equations (1.5.84) and (1.5.64). Substituting for the right-hand side of equation (1.9.14) from equation (1.9.16) gives

$$\operatorname{curl} \boldsymbol{u} = \pm\varepsilon^{pqr}u_{r,q}\boldsymbol{e}_p \tag{1.9.17}$$

where the upper or lower sign holds according to whether the basis is positive in $\mathscr{V}^+$ or in $\mathscr{V}^-$.

When a tensor field $\mathbf{A}$ is differentiable, its divergence, denoted by div $\mathbf{A}$, is the vector field defined by

$$\{(\text{div } \mathbf{A})(\boldsymbol{x})\} \cdot \boldsymbol{a} = \text{div}\{\mathbf{A}(\boldsymbol{x})\boldsymbol{a}\} \qquad \forall\, \boldsymbol{x} \in D_o \qquad \boldsymbol{a} \in \mathscr{V}. \tag{1.9.18}$$

Now, with $\mathbf{A}$ given in contravariant components

$$\begin{aligned} \mathbf{A}(\boldsymbol{x})\boldsymbol{a} &= (A^{pq}\boldsymbol{e}_p \otimes \boldsymbol{e}_q)(\boldsymbol{x})a_r\boldsymbol{e}^r \\ &= (A^{pq}a_q)\boldsymbol{e}_p \end{aligned} \tag{1.9.19}$$

where use has been made of equations (1.5.63), (1.6.47) and (1.6.60). From equations (1.9.13) and (1.9.19) it follows that

$$\begin{aligned} \text{div}(\mathbf{A}(\boldsymbol{x})\boldsymbol{a}) &= (A^{pq}a_q)_{,p} = A^{pq}{}_{,p}a_q \\ &= A^{pq}{}_{,p}(\boldsymbol{e}_q \cdot \boldsymbol{a}) = (A^{pq}{}_{,p}\boldsymbol{e}_q) \cdot \boldsymbol{a} \end{aligned} \tag{1.9.20}$$

and, substituting for div$\{\mathbf{A}(\boldsymbol{x})\boldsymbol{a}\}$ in equation (1.9.18) from equation (1.9.20) therefore gives

$$\text{div } \mathbf{A} = A^{pq}{}_{,p}\boldsymbol{e}_q. \tag{1.9.21}$$

Other component forms for div $\mathbf{A}$ follow from the use of equation (1.6.60).

1.10 INTEGRAL THEOREMS

Let R be a regular region of $\mathscr{E}$ with a boundary ∂R, and let $\boldsymbol{n}$ be the outward unit vector normal to ∂R. Take $\boldsymbol{u}$ to be a vector field which is continuous in R and continuously differentiable in the interior of R. The *divergence theorem* can be expressed in the form

$$\int_R \text{div } \boldsymbol{u} \, \mathrm{d}v = \int_{\partial R} \boldsymbol{u} \cdot \boldsymbol{n} \, \mathrm{d}a \tag{1.10.1}$$

where $\mathrm{d}v$ is an element of volume and where $\mathrm{d}a$ is an element of area. The divergence theorem can be rearranged into a form which is of particular use in continuum mechanics by replacing $\boldsymbol{u}$ by the vector field $(\boldsymbol{u} \cdot \boldsymbol{a})(\mathbf{A}\boldsymbol{b})$ where $\boldsymbol{a}$ and $\boldsymbol{b}$ are arbitrary vectors and where $\mathbf{A}$ is a tensor field which is continuous in R and continuously differentiable in the interior of R: thus

$$\int_R \text{div}\{(\boldsymbol{u} \cdot \boldsymbol{a})\mathbf{A}\boldsymbol{b}\} \, \mathrm{d}v = \int_{\partial R} \{(\boldsymbol{u} \cdot \boldsymbol{a})\mathbf{A}\boldsymbol{b}\} \cdot \boldsymbol{n} \, \mathrm{d}a. \tag{1.10.2}$$

The scalar $\{(\boldsymbol{u} \cdot \boldsymbol{a})(\mathbf{A}\boldsymbol{b})\} \cdot \boldsymbol{n}$ can be rearranged into the form

$$\boldsymbol{a} \cdot \{(\mathbf{A}\boldsymbol{b} \cdot \boldsymbol{n})\boldsymbol{u}\} = \boldsymbol{a} \cdot \{(\boldsymbol{u} \otimes \boldsymbol{n})\mathbf{A}\}\boldsymbol{b} = \boldsymbol{a} \cdot \{\boldsymbol{u} \otimes (\mathbf{A}^{\mathrm{T}}\boldsymbol{n})\}\boldsymbol{b} \tag{1.10.3}$$

where use has been made of equation (1.6.54). Also, div$\{(\boldsymbol{u} \cdot \boldsymbol{a})(\mathbf{A}\boldsymbol{b})\}$ can be rearranged: thus

$$\begin{aligned}\operatorname{div}\{(\boldsymbol{u}\cdot\boldsymbol{a})\mathbf{A}\boldsymbol{b}\} &= \mathbf{A}\boldsymbol{b}\cdot\operatorname{grad}(\boldsymbol{u}\cdot\boldsymbol{a})+(\boldsymbol{u}\cdot\boldsymbol{a})\operatorname{div}(\mathbf{A}\boldsymbol{b})\\ &= \mathbf{A}\boldsymbol{b}\cdot(\operatorname{grad}\boldsymbol{u})^{\mathrm{T}}\boldsymbol{a}+(\boldsymbol{a}\cdot\boldsymbol{u})\{(\operatorname{div}\mathbf{A})\cdot\boldsymbol{b}\}\\ &= \boldsymbol{a}\cdot\{(\operatorname{grad}\boldsymbol{u})\mathbf{A}\}\boldsymbol{b}+\boldsymbol{a}\cdot\{\boldsymbol{u}\otimes\operatorname{div}\mathbf{A}\}\boldsymbol{b}\\ &= \boldsymbol{a}\cdot[\{(\operatorname{grad}\boldsymbol{u})\mathbf{A}+\boldsymbol{u}\otimes\operatorname{div}\mathbf{A}\}\boldsymbol{b}] \qquad (1.10.4)\end{aligned}$$

where use has been made of equations (1.9.8), (1.9.18) and (1.6.42). Substituting in equation (1.10.2) from equations (1.10.3) and (1.10.4), and noting that $\boldsymbol{a}$ and $\boldsymbol{b}$ are arbitrary vectors, gives the divergence theorem in the form

$$\int_R \{\boldsymbol{u}\otimes\operatorname{div}\mathbf{A}+(\operatorname{grad}\boldsymbol{u})\mathbf{A}\}\,\mathrm{d}v=\int_{\partial R}\boldsymbol{u}\otimes(\mathbf{A}^{\mathrm{T}}\boldsymbol{n})\,\mathrm{d}a \qquad (1.10.5)$$

which finds application in the formulation of the general physical principles underlying continuum mechanics.

A special case that finds application is obtained from equation (1.10.5) by taking $\boldsymbol{u}$ to be a fixed arbitrary vector to give

$$\int_R \operatorname{div}\mathbf{A}\,\mathrm{d}v=\int_{\partial R}\mathbf{A}^{\mathrm{T}}\boldsymbol{n}\,\mathrm{d}a. \qquad (1.10.6)$$

If $\mathbf{A}=\mathbf{I}$, equation (1.10.5) reduces to

$$\int_R \operatorname{grad}\boldsymbol{u}\,\mathrm{d}v=\int_{\partial R}(\boldsymbol{u}\otimes\boldsymbol{n})\,\mathrm{d}a. \qquad (1.10.7)$$

Consider the vector identity

$$\boldsymbol{a}\wedge(\boldsymbol{b}\wedge\boldsymbol{c})=(\boldsymbol{b}\otimes\boldsymbol{c}-\boldsymbol{c}\otimes\boldsymbol{b})\boldsymbol{a}. \qquad (1.10.8)$$

Replace $\boldsymbol{b}$ by the position $\boldsymbol{x}$ of a representative point in a regular region R and set $\boldsymbol{c}=\mathbf{A}^{\mathrm{T}}\boldsymbol{n}$ which gives

$$\boldsymbol{a}\wedge(\boldsymbol{x}\wedge\mathbf{A}^{\mathrm{T}}\boldsymbol{n})=(\boldsymbol{x}\otimes\mathbf{A}^{\mathrm{T}}\boldsymbol{n}-\mathbf{A}^{\mathrm{T}}\boldsymbol{n}\otimes\boldsymbol{x})\boldsymbol{a} \qquad (1.10.9)$$

where the tensor field $\mathbf{A}$ is continuous in R and is continuously differentiable in the interior of R, it being noted that $\boldsymbol{a}$ is an arbitrary vector. Since $\boldsymbol{a}$ is an arbitrary vector, integration of both sides of equation (1.10.9) over the surface ∂R gives

$$\boldsymbol{a}\wedge\int_{\partial R}\boldsymbol{x}\wedge(\mathbf{A}^{\mathrm{T}}\boldsymbol{n})\,\mathrm{d}a=\left(\int_{\partial R}\{\boldsymbol{x}\otimes(\mathbf{A}^{\mathrm{T}}\boldsymbol{n})-(\mathbf{A}^{\mathrm{T}}\boldsymbol{n})\otimes\boldsymbol{x}\}\,\mathrm{d}a\right)\boldsymbol{a}. \qquad (1.10.10)$$

Now set $\boldsymbol{u}=\boldsymbol{x}$ in the divergence theorem in the form of equation (1.10.5) to give

$$\int_{\partial R}\boldsymbol{x}\otimes(\mathbf{A}^{\mathrm{T}}\boldsymbol{n})\,\mathrm{d}a=\int_R(\boldsymbol{x}\otimes\operatorname{div}\mathbf{A}+\mathbf{A})\,\mathrm{d}v \qquad (1.10.11)$$

it being noted that grad $\boldsymbol{x}=\mathbf{I}$ from equations (1.9.10) and (1.6.51). Hence,

$$\left(\int_{\partial R}\{\boldsymbol{x}\otimes(\mathbf{A}^{\mathrm{T}}\boldsymbol{n})-(\mathbf{A}^{\mathrm{T}}\boldsymbol{n})\otimes\boldsymbol{x}\}\,\mathrm{d}a\right)\boldsymbol{a}$$

$$=\int_R[(\boldsymbol{x}\otimes\operatorname{div}\mathbf{A}-\operatorname{div}\mathbf{A}\otimes\boldsymbol{x})\boldsymbol{a}+(\mathbf{A}-\mathbf{A}^{\mathrm{T}})\boldsymbol{a}]\,\mathrm{d}v. \quad (1.10.12)$$

Setting $\boldsymbol{b}=\boldsymbol{x}$ and $\boldsymbol{c}=\operatorname{div}\mathbf{A}$ in the identity of equation (1.10.8) gives

$$(\boldsymbol{x}\otimes\operatorname{div}\mathbf{A}-\operatorname{div}\mathbf{A}\otimes\boldsymbol{x})\boldsymbol{a}=\boldsymbol{a}\wedge(\boldsymbol{x}\wedge\operatorname{div}\mathbf{A}) \quad (1.10.13)$$

and using equation (1.6.117) in the form

$$(\mathbf{A}-\mathbf{A}^{\mathrm{T}})\boldsymbol{a}=\boldsymbol{\pi}\wedge\boldsymbol{a} \quad (1.10.14)$$

where $\boldsymbol{\pi}$ is the axial vector of $\mathbf{A}-\mathbf{A}^{\mathrm{T}}$. Substituting in equation (1.10.12) from equations (1.10.13) and (1.10.14) gives

$$\left(\int_{\partial R}\{\boldsymbol{x}\otimes(\mathbf{A}^{\mathrm{T}}\boldsymbol{n})-(\mathbf{A}^{\mathrm{T}}\boldsymbol{n})\otimes\boldsymbol{x}\}\,\mathrm{d}a\right)\boldsymbol{a}=\boldsymbol{a}\wedge\int_R(\boldsymbol{x}\wedge\operatorname{div}\mathbf{A}-\boldsymbol{\pi})\,\mathrm{d}v$$

which can, in turn, be substituted into equation (1.10.10) to give

$$\int_{\partial R}\boldsymbol{x}\wedge(\mathbf{A}^{\mathrm{T}}\boldsymbol{n})\,\mathrm{d}a=\int_R(\boldsymbol{x}\wedge\operatorname{div}\mathbf{A}-\boldsymbol{\pi})\,\mathrm{d}v \quad (1.10.15)$$

where use is being made of the fact that $\boldsymbol{a}$ is an arbitrary vector.

1.11 REPRESENTATION THEOREMS

There are physical situations in which the variables are independent of the directions considered. When a variable is independent of direction, it is described as *isotropic*. In this section consideration is given to tensor-valued variables having properties that are independent of direction, that is, isotropic tensor-valued quantities.

A tensor is defined as isotropic at a point if its components are invariant with respect to rotation of the axes of the coordinate system used to describe the tensor.

The lowest rank tensor is a tensor of rank zero, which is a scalar. A scalar by its nature must be isotropic because its value at a point cannot depend on any particular direction.

A first-order tensor is a vector, the concept of direction being basic to the definition of a vector. Using an orthonormal basis $\hat{e}=\{\hat{\boldsymbol{e}}_1,\hat{\boldsymbol{e}}_2,\hat{\boldsymbol{e}}_3\}$, a vector $\boldsymbol{a}$ is described by

$$\boldsymbol{a}=a_p\hat{\boldsymbol{e}}_p$$

and in the rotated system by,

$$\boldsymbol{a}=\bar{a}_q\bar{\boldsymbol{e}}_q$$

where

$$\bar{a}_i=a_p\bar{H}_{pi}$$

use having been made of equation $(1.5.102)_4$. For the components to be invariant, we must have $\bar{a}_i=a_i$, and hence the components a_i must satisfy the equation

$$a_i=a_p\bar{H}_{pi}$$

or

$$a_p(\delta_{pq}-\bar{H}_{pq})=0. \tag{1.11.1}$$

Equation (1.11.1) is a homogeneous set of equations for a_i, and the only solution is $a_1=a_2=a_3=0$. Thus, the only isotropic vector is the null vector $\boldsymbol{a}=\mathbf{0}$. This conclusion follows directly from §1.7.7, where it is effectively shown that the only vector whose components are unchanged by a rotation is one which is parallel to the axis of rotation. Therefore there can be no vector whose components remain unchanged by any rotation, and hence there is no isotropic vector.

An arbitrary second-order tensor **A** can be represented in the form,

$$\mathbf{A}=A_{pq}\hat{\boldsymbol{e}}_p\otimes\hat{\boldsymbol{e}}_q$$

and in the rotated system by

$$\mathbf{A}=\bar{A}_{rs}\hat{\bar{\mathbf{e}}}_r\otimes\hat{\bar{\mathbf{e}}}_s.$$

For the tensor **A** to be isotropic we must have $\bar{A}_{ij}=A_{ij}$ and thus, from equation (1.7.29)

$$A_{rs}=A_{pq}H_{pr}H_{qs} \qquad (r,s=1,2,3) \tag{1.11.2}$$

or

$$A_{pq}(\delta_{pr}\delta_{qs}-H_{pr}H_{qs})=0. \tag{1.11.3}$$

For arbitrary A_{ij} the identity of equation (1.11.3) holds only if $A_{ij}=\phi\delta_{ij}$, or in general form, if

$$\mathbf{A}=\phi\mathbf{I} \tag{1.11.4}$$

where ϕ is an arbitrary scalar. Thus the most general isotropic tensor of the second-order is the identity tensor.

A third-order, arbitrary tensor $\mathbb{P}$ can be represented in the form

$$\mathbb{P}=P_{pqr}\hat{\boldsymbol{e}}_p\otimes\hat{\boldsymbol{e}}_q\otimes\hat{\boldsymbol{e}}_r=\bar{P}_{lmn}\hat{\bar{\mathbf{e}}}_l\otimes\hat{\bar{\mathbf{e}}}_m\otimes\hat{\bar{\mathbf{e}}}_n$$

where

$$\bar{P}_{lmn} = P_{pqr} H_{pl} H_{qm} H_{rn}. \tag{1.11.5}$$

For $\mathbb{P}$ to be isotropic, $\bar{P}_{ijk} = P_{ijk}$, and hence equation (1.11.5) can be expressed in the form

$$P_{pqr} = (\delta_{pl}\delta_{qm}\delta_{rn} - H_{pl}H_{qm}H_{rn}) = 0. \tag{1.11.6}$$

Using the permutation symbol, the determinant of the matrix **A** can be expressed in the form

$$\det[A^i{}_j] = e_{pqr} A^p{}_1 A^q{}_2 A^r{}_3$$

or, equivalently

$$e_{ijk} \det \mathbf{A} = e_{pqr} A^p{}_i A^q{}_j A^r{}_k.$$

These relations, and that of equation (1.6.79) in the form $\det \mathbf{A} \det \mathbf{B} = \det(\mathbf{A}^{\mathrm{T}}\mathbf{B})$, can be used to establish the relation

$$e_{ijk}e_{lmn} = \delta_{il}(\delta_{jm}\delta_{kn} - \delta_{jn}\delta_{km}) + \delta_{im}(\delta_{jn}\delta_{kl} - \delta_{jl}\delta_{kn}) + \delta_{in}(\delta_{jl}\delta_{km} - \delta_{jm}\delta_{kl}).$$

From these relations, it can be deduced that

$$e_{pqr}e_{pqr} = 6 \qquad e_{ijk} = e_{pqr}H_{pj}H_{qi}H_{rk}$$

$$\tfrac{1}{6}e_{pqr}e_{stu}H_{sp}H_{tq}H_{ur} = 1$$

and hence it follows that the identity of equation (1.11.6) holds only if

$$P_{ijk} = \psi e_{ijk} \qquad \mathbb{P} = \psi \mathbb{E}_{(\varepsilon)} \tag{1.11.7}$$

where ψ is a scalar and where, in general, $\mathbb{E}_{(\varepsilon)}$ is the alternate or permutation tensor. Thus, the most general isotropic tensor of the third-order is the alternate or permutation tensor.

An arbitrary fourth-order tensor can be represented in the form

$$\mathsf{K} = K_{pqrs}\hat{\boldsymbol{e}}_p \otimes \hat{\boldsymbol{e}}_q \otimes \hat{\boldsymbol{e}}_r \otimes \hat{\boldsymbol{e}}_s$$

$$= \bar{K}_{klmn}\hat{\mathbf{e}}_k \otimes \hat{\mathbf{e}}_l \otimes \hat{\mathbf{e}}_m \otimes \hat{\mathbf{e}}_n$$

where

$$\bar{K}_{klmn} = K_{pqrs}H_{pk}H_{ql}H_{rm}H_{sn}. \tag{1.11.8}$$

For K to be isotropic we must have $\bar{K}_{ijkl} = K_{ijkl}$. In the summation on the right-hand side of equation (1.11.8), only three different values of p, q, r and s are possible, and hence at least two must be equal. There are three classes of solutions, depending upon which pairs of p, q, r and s are set equal.

(i) *set $r = s$ or $p = q$*: this suggests a product solution of the form $K_{pqrs} = \bar{K}_{pq}\bar{\bar{K}}_{rs}$, and equation (1.11.8) can be expressed in the form

$$\bar{K}_{kl}\bar{K}_{mn} = (\bar{K}_{pq}H_{pk}H_{ql})(\bar{\bar{K}}_{rs}H_{rm}H_{sn}). \tag{1.11.9}$$

This reduces to a product of solutions that are the same as for equation (1.11.2), and thus for this class of solutions

$$K_{ijkl}=\alpha_1\delta_{ij}\delta_{kl} \tag{1.11.10}$$

where α_1 is an arbitrary scalar multiplier.

(ii) *set* $p=r$ *or* $q=s$: for this the implied product of solutions are again the same as for equation (1.11.2) and thus for this class of solutions

$$K_{ijkl}=\alpha_2\delta_{ik}\delta_{jl} \tag{1.11.11}$$

where α_2 is an arbitrary scalar multiplier.

(iii) *set* $p=s$ *or* $q=r$: for this the implied product of solutions is also the same as for equation (1.11.2) and thus for this class of solutions

$$K_{ijkl}=\alpha_3\delta_{il}\delta_{jk} \tag{1.11.12}$$

where α_3 is an arbitrary scalar multiplier.

With this as the basis, it can be shown that the general fourth-order isotropic tensor is a linear combination of the three special solutions represented by equations (1.11.10), (1.11.11) and (1.11.12): thus

$$K_{ijkl}=\lambda_1\delta_{ij}\delta_{kl}+\lambda_2\delta_{ik}\delta_{jl}+\lambda_3\delta_{il}\delta_{jk}. \tag{1.11.13}$$

Thus, the general fourth-order isotropic tensor has been reduced to only three independent components.

A scalar-valued function of a vector may be expressed in the form

$$\phi=f(\boldsymbol{v})=f(v^i,\boldsymbol{e}_i).$$

An example of such a function is,

$$\phi=a_0+a_pv^p+A_{pq}v^pv^q+\cdots.$$

The value of ϕ remains the same when changing to a new basis, since each term is unchanged. However, with

$$\phi=a_0+\bar{a}_p\bar{v}^p+\bar{A}_{pq}\bar{v}^p\bar{v}^q+\ldots$$

where

$$\bar{a}_i=(Q^{\mathrm{T}})^p{}_ia_p$$

$$\bar{A}_{ij}=(Q^{\mathrm{T}})^q{}_i\mathrm{A}_{qr}\mathrm{Q}^r{}_j$$

the functional form of ϕ in general changes. If the coefficients a_i etc are isotropic so that $a_i=\bar{a}_i$ etc, then ϕ is called an orthogonal scalar invariant.

If $\boldsymbol{v}$ is rotated under an orthogonal transformation $\mathbf{Q}$ to a new vector $\bar{\boldsymbol{v}}$, that is $\bar{\boldsymbol{v}}=\mathbf{Q}\boldsymbol{v}$, and if

$$f(\boldsymbol{v})=f(\mathbf{Q}\boldsymbol{v}) \tag{1.11.14}$$

then $f(\boldsymbol{v})$ is called an isotropic scalar-valued vector function. A scalar-valued function of several vectors $\boldsymbol{v}_1, \boldsymbol{v}_2, \ldots, \boldsymbol{v}_n$ is isotropic if

$$f(\boldsymbol{v}_1, \boldsymbol{v}_2, \ldots, \boldsymbol{v}_n) = f(\mathbf{Q}\boldsymbol{v}_1, \mathbf{Q}\boldsymbol{v}_2, \ldots, \mathbf{Q}\boldsymbol{v}_n) \tag{1.11.15}$$

and f is called a simultaneous orthogonal invariant. A representation theorem due to Cauchy states that a scalar-valued function of vectors is a simultaneous invariant if and only if it can be expressed in the form of a function of the inner products $\boldsymbol{v}_i \cdot \boldsymbol{v}_j$ $(i, j = 1, \ldots, n)$. A scalar-valued function of vectors may also be isotropic relative to a subgroup of the full orthogonal group $\mathscr{O}$. Then equation (1.11.15) must only hold for those $\mathbf{Q}$ within the subgroup $\mathscr{G}$.

Consider an arbitrary tensor $\mathbf{A}$, and suppose

$$\boldsymbol{u} = \mathbf{A}\boldsymbol{v} \qquad \bar{\boldsymbol{u}} = \bar{\mathbf{A}}\bar{\boldsymbol{v}}.$$

If $\mathbf{Q}$ is applied to every vector so that

$$\bar{\boldsymbol{u}} = \mathbf{Q}\boldsymbol{u} \qquad \bar{\boldsymbol{v}} = \mathbf{Q}\boldsymbol{v}$$

then

$$\bar{\mathbf{A}} = \mathbf{Q}\mathbf{A}\mathbf{Q}^{\mathrm{T}} \tag{1.11.16}$$

where use has been made of equations (1.6.41) and (1.6.130).

A scalar-valued function of a tensor $\mathbf{A}$ may be expressed in the form

$$\phi = f(\mathbf{A}) = f(A^{ij}, \boldsymbol{e}_i). \tag{1.11.17}$$

An example of such a function is

$$f(\mathbf{A}) = a_0 + H_{pq}A^{pq} + K_{pqrs}A^{pq}A^{rs} + \cdots.$$

If the functional form of f is unchanged when the basis is changed then f is called an orthogonal invariant. An isotropic scalar-valued tensor function is one whose value is the same for all similar tensors, i.e. from equation (1.11.16),

$$f(\mathbf{A}) = f(\mathbf{Q}\mathbf{A}\mathbf{Q}^{\mathrm{T}}). \tag{1.11.18}$$

Applying Cauchy's theorem to the basis vectors $\boldsymbol{e}_i$ shows that an invariant ϕ must be of the form

$$\phi = f(\mathbf{A}) = f(A^{ij}, g_{ij}). \tag{1.11.19}$$

A scalar-valued tensor function of several tensors $f(\mathbf{A}_1, \mathbf{A}_2, \ldots, \mathbf{A}_n)$ is isotropic if

$$f(\mathbf{A}_1, \ldots, \mathbf{A}_n) = f(\mathbf{Q}\mathbf{A}_1\mathbf{Q}^{\mathrm{T}}, \ldots, \mathbf{Q}\mathbf{A}_n\mathbf{Q}^{\mathrm{T}}) \tag{1.11.20}$$

f being called a simultaneous invariant or simply invariant. The function f is said to be isotropic relative to $\mathscr{G}$ where $\mathscr{G}$ is a subgroup of the full orthogonal group $\mathscr{O}$ if equation (1.11.20) holds for all $\mathbf{Q}$ in $\mathscr{G}$.

A tensor-valued tensor function can be represented by

$$\mathbf{\Phi} = \mathbf{F}(\mathbf{A}) \qquad \Phi^{ij} = F^{ij}(A^{ij}, \boldsymbol{e}_i). \tag{1.11.21}$$

The function $\mathbf{F(A)}$ is said to be an isotropic tensor-valued function of a tensor if

$$\mathbf{QF(A)Q}^{\mathrm{T}}=\mathbf{F(QAQ}^{\mathrm{T}}). \tag{1.11.22}$$

For several independent tensors $\mathbf{A}_1, \mathbf{A}_2, \ldots, \mathbf{A}_n$, the function $\mathbf{F}$ is isotropic if

$$\mathbf{QF(A}_1, \ldots, \mathbf{A}_n)\mathbf{Q}^{\mathrm{T}}=\mathbf{F}(\mathbf{QA}_1\mathbf{Q}^{\mathrm{T}}, \ldots, \mathbf{QA}_n\mathbf{Q}^{\mathrm{T}}) \tag{1.11.23}$$

and $\mathbf{F}$ is isotropic relative to a subgroup $\mathscr{G}$ if equation (1.11.23) holds for every $\mathbf{Q}$ in $\mathscr{G}$.

The concepts of isotropy and isotropy relative to a subgroup $\mathscr{G}$ can be extended to functions whose values and arguments are scalars, vectors and tensors of any order. For example, a vector-valued function $\boldsymbol{v}=\boldsymbol{v}(\mathbf{A}, \boldsymbol{u})$ is isotropic if

$$\mathbf{Q}\boldsymbol{v}(\mathbf{A}, \boldsymbol{u})=\boldsymbol{v}(\mathbf{QAQ}^{\mathrm{T}}, \mathbf{Q}\boldsymbol{u}) \tag{1.11.24}$$

and isotropic relative to a subgroup $\mathscr{G}$ if equation (1.11.24) applies for all $\mathbf{Q}$ in $\mathscr{G}$.

Suppose $\mathbf{M}$ and $\mathbf{B}$ are symmetric tensors and that $\mathbf{M}$ is an isotropic tensor-valued function of $\mathbf{B}$. An example of such a function is

$$\mathbf{M(B)}=\rho_0\mathbf{I}+\rho_1\mathbf{B}+\rho_2\mathbf{B}^2 \tag{1.11.25}$$

where the ρ_k $(k=0,1,2)$ are invariants of $\mathbf{B}$ and hence can be expressed as functions

$$\rho_k=\rho_k(I_{\mathbf{B}}, II_{\mathbf{B}}, III_{\mathbf{B}}) \qquad (k=0,1,2) \tag{1.11.26}$$

of the principal invariants $I_{\mathbf{B}}$, $II_{\mathbf{B}}$ and $III_{\mathbf{B}}$ of $\mathbf{B}$. The second-order symmetric tensor $\mathbf{B}$ has the spectral representation

$$\mathbf{B}=\sum_{a=1}^{3} b_a(\boldsymbol{q}_a \otimes \boldsymbol{q}_a) \tag{1.11.27}$$

where the b_a are the proper numbers of $\mathbf{B}$ associated with the proper vectors $\boldsymbol{q}_a$. Equation (1.11.27) enables equation (1.1.25) to be expressed in the form

$$\mathbf{M}=\sum_{a=1}^{3} (\rho_0+\rho_1 b_a+\rho_2 b_a^2)(\boldsymbol{q}_a \otimes \boldsymbol{q}_a)$$

from which it is evident that every proper vector of $\mathbf{B}$ is also a proper vector of $\mathbf{M}$, and hence

$$m_a=\rho_0+\rho_1 b_a+\rho_2 b_a^2 \qquad (a=1,2,3) \tag{1.11.28}$$

where the m_a are the proper numbers of $\mathbf{M}$. For $\mathbf{M}$ to be an isotropic tensor-valued function of $\mathbf{B}$, equation (1.11.22) gives

$$\mathbf{QM(B)Q}^{\mathrm{T}}=\mathbf{M(QBQ}^{\mathrm{T}})$$

and hence, noting that $\mathbf{QB}^2\mathbf{Q}^{\mathrm{T}}=\mathbf{QBQ}^{\mathrm{T}}\mathbf{QBQ}^{\mathrm{T}}=(\mathbf{QBQ}^{\mathrm{T}})^2$, it follows, for every orthogonal $\mathbf{Q}$, that

$$\mathbf{M}(\mathbf{QBQ}^{\mathrm{T}}) = \rho_0\mathbf{I} + \rho_1(\mathbf{QBQ}^{\mathrm{T}}) + \rho_2(\mathbf{QBQ}^{\mathrm{T}})^2. \qquad (1.11.29)$$

It therefore follows that the coefficients ρ_a in equation (1.11.25) are not changed when $\mathbf{B}$ is replaced by $\mathbf{QBQ}^{\mathrm{T}}$; hence the ρ_a are invariants of $\mathbf{B}$, thus establishing the theorem, due in this form to Rivlin and Ericksen (1955) which states that, *a function* $\mathbf{M}(\mathbf{B})$ *from symmetric tensors into symmetric tensors is isotropic if and only if it has a representation of the form*

$$\mathbf{M}(\mathbf{B}) = \rho_0\mathbf{I} + \rho_1\mathbf{B} + \rho_2\mathbf{B}^2 \qquad (1.11.30)$$

where the coefficients ρ_k $(k=0,1,2)$ are of the form

$$\rho_k = \rho_k(I_{\mathbf{B}}, II_{\mathbf{B}}, III_{\mathbf{B}}). \qquad (1.11.31)$$

With regard to this representation it is to be noted that for second-order symmetric tensors, terms higher than the quadratic term do not appear because they can always be reduced using the Cayley–Hamilton theorem; thus, for example,

$$\mathbf{B}^3 = I_{\mathbf{B}}\mathbf{B}^2 - II_{\mathbf{B}}\mathbf{B} + III_{\mathbf{B}}\mathbf{I} \qquad (1.11.32)$$

$$\mathbf{B}^4 = I_{\mathbf{B}}\mathbf{B}^3 - II_{\mathbf{B}}\mathbf{B}^2 + III_{\mathbf{B}}\mathbf{B} \qquad (1.11.33)$$

with higher terms being treated in the same way. In particular, if the tensor variable $\mathbf{B}$ is restricted to be invertible, use of the Cayley–Hamilton theorem in the form

$$\mathbf{B}^2 = I_{\mathbf{B}}\mathbf{B} - II_{\mathbf{B}}\mathbf{I} + III_{\mathbf{B}}\mathbf{B}^{-1} \qquad (1.11.34)$$

gives the alternative representation theorem

$$\mathbf{M}(\mathbf{B}) = \beta_0\mathbf{I} + \beta_1\mathbf{B} + \beta_{-1}\mathbf{B}^{-1} \qquad (1.11.35)$$

where

$$\beta_0 = \rho_0 - II_{\mathbf{B}}\rho_2 \qquad \beta_1 = \rho_1 + I_{\mathbf{B}}\rho_2 \qquad \beta_{-1} = III_{\mathbf{B}}\rho_2 \qquad (1.11.36)$$

and

$$\rho_0 = \beta_0 + \frac{II_{\mathbf{B}}}{III_{\mathbf{B}}}\beta_{-1} \qquad \rho_1 = \beta_1 - \frac{I_{\mathbf{B}}}{III_{\mathbf{B}}}\beta_{-1} \qquad \rho_2 = \frac{1}{III_{\mathbf{B}}}\beta_{-1} \qquad (1.11.37)$$

it being noted that, again the β_k $(k=0, \pm 1)$ are scalar invariant functions of $I_{\mathbf{B}}$, $II_{\mathbf{B}}$, $III_{\mathbf{B}}$ of $\mathbf{B}$.

It has also been shown (Rivlin and Ericksen 1955), that an isotropic tensor function $\mathbf{K}(\mathbf{A}_1, \mathbf{A}_2)$ of two variables $\mathbf{A}_1, \mathbf{A}_2$ admits the representation,

$$\begin{aligned}\mathbf{K}(\mathbf{A}_1, \mathbf{A}_2) = {} & \alpha_0\mathbf{I} + \alpha_1\mathbf{A}_1 + \alpha_2\mathbf{A}_1^2 + \alpha_3\mathbf{A}_2 + \alpha_4\mathbf{A}_2^2 \\ & + \alpha_5(\mathbf{A}_1\mathbf{A}_2 + \mathbf{A}_2\mathbf{A}_1) + \alpha_6(\mathbf{A}_1^2\mathbf{A}_2 + \mathbf{A}_2\mathbf{A}_1^2) \\ & + \alpha_7(\mathbf{A}_1\mathbf{A}_2^2 + \mathbf{A}_2^2\mathbf{A}_1) + \alpha_8(\mathbf{A}_1^2\mathbf{A}_2^2 + \mathbf{A}_2^2\mathbf{A}_1^2) \end{aligned} \qquad (1.11.38)$$

where $\alpha_0, \alpha_1, \ldots, \alpha_8$ are functions of the joint invariants

$$\begin{array}{cccccc} \operatorname{tr} \mathbf{A}_1 & \operatorname{tr} \mathbf{A}_1^2 & \operatorname{tr} \mathbf{A}_1^3 & \operatorname{tr} \mathbf{A}_2 & \operatorname{tr} \mathbf{A}_2^2 & \operatorname{tr} \mathbf{A}_2^3 \\ \operatorname{tr} \mathbf{A}_1\mathbf{A}_2 & \operatorname{tr} \mathbf{A}_1^2\mathbf{A}_2 & \operatorname{tr} \mathbf{A}_1\mathbf{A}_2^2 & \operatorname{tr} \mathbf{A}_1^2\mathbf{A}_2^2 & & \end{array} \tag{1.11.39}$$

of $\mathbf{A}_1, \mathbf{A}_2$.

REFERENCES

Abraham R 1966 *Linear and Multilinear Algebra* (New York: Benjamin)

Bedford F W and Dwivedi T D 1970 *Vector Calculus* (New York: McGraw-Hill)

Bellman R E 1970 *Introduction to Matrix Analysis* 2nd edn (New York: McGraw-Hill)

Borisenko A I and Tarapov I E 1968 *Vector and Tensor Analysis with Applications* transl. and ed. R A Silverman (Englewood Cliffs, NJ: Prentice-Hall)

Bourne D E and Kendal P C 1977 *Vector Analysis and Cartesian Tensors* 2nd edn (New York: Academic)

Bowen R M and Wang C C 1976 *Introduction to Vectors and Tensors* (New York: Plenum)

Brillouin L 1964 *Tensors in Mechanics and Elasticity* (New York: Academic)

Buck R C 1965 *Advanced Calculus* (New York: McGraw-Hill)

Campbell H G 1977 *An Introduction to Matrices, Vectors and Linear Programming* (Englewood Cliffs, NJ: Prentice-Hall)

Chadwick P 1976 *Continuum Mechanics* (London: Allen and Unwin) pp 11–49

Chambers L G 1969 *A Course in Vector Analysis* (London: Chapman and Hall)

Chisholm J S R 1978 *Vectors in Three-Dimensional Space* (Cambridge: Cambridge University Press)

Chorlton F 1976 *Vector and Tensor Methods* (Chichester: Ellis Horwood)

Courant R 1965 *Differential and Integral Calculus* vol. II (New York: Interscience)

Eisele J A 1970 *Applied Matrix and Tensor Analysis* (New York: Wiley–Interscience)

Ericksen J L 1960 *Tensor Fields* in *Handbuch der Physik* vol. III ed. S Flügge (Berlin: Springer) pp 794–858

Graham A 1979 *Matrix Theory and Applications for Engineers and Mathematicians* (Chichester: Ellis Horwood)

Goodbody A M 1982 *Cartesian Tensors with Applications to Mechanics, Fluid Mechanics and Elasticity* (Chichester: Ellis Horwood)

Guggenheimer H W 1963 *Differential Geometry* (New York: McGraw-Hill)

Hauge B 1970 *An Introduction to Vector Analysis for Physicists and Engineers* 6th edn revised D Martin (London: Methuen)

Jeffreys H 1965 *Cartesian Tensors* (London: Cambridge University Press)

Jeffreys H and Jeffreys B S 1950 *Methods of Mathematical Physics* (London: Cambridge University Press)

Kellog O D 1929 *Foundations of Potential Theory* (Berlin: Springer)

Marsden J E and Tromba A J 1981 *Vector Calculus* 2nd edn (San Francisco: Freeman)

Martin A D and Mizel V J 1966 *Introduction to Linear Algebra* (New York: McGraw-Hill)

Morse P M and Feshbach H 1953 *Methods of Theoretical Physics* (New York: McGraw-Hill)

Pipes L A and Harvill L R 1970 *Applied Mathematics for Engineers and Physicists* (New York: McGraw-Hill)

Rivlin R S and Ericksen J L 1955 *J. Rational Mech. Anal.* vol. 4 pp 323–425

Schouten J A 1951 *Tensor Analysis for Physicists* (Oxford: Clarendon)

—— 1954 *Ricci-Calculus* (Berlin: Springer)

Thomson E M 1969 *An Introduction to Algebra of Matrices with Some Applications* (Bristol: Adam Hilger)

Tyldesley J R 1975 *An Introduction to Tensor Analysis* (London: Longman)

Wrede R C 1972 *Introduction to Vector and Tensor Analysis* (New York: Dover)

Young E C 1978 *Vector and Tensor Analysis* (New York: Marcel Dekker)

Volterra V 1930 *Theory of Functionals and Integral and Integrodifferential Equations* (Glasgow: Blackie)

2

Physical Principles

2.1 KINEMATICS

2.1.1 Bodies, Configurations and Motions

Central to any discussion of classical mechanics are the concepts of Newtonian space–time and a frame of reference. These concepts also find application in the study of continuum mechanics. In classical mechanics, physical objects are regarded as occupying *places*, which are points in a three-dimensional space, the properties of which, once defined, remain unaltered by the presence of physical objects. Any changes occurring in the physical objects are regarded as occurring at specific *instants*, which are points in a one-dimensional space, itself totally independent of the *space of places*, $\mathscr{E}$. The totality of places and instants is a topological space called the *event world*, $\mathscr{W}$. These primitive concepts find expression by way of the concept of the *Newtonian space–time*, $\mathscr{W}$, which is the union of a family of oriented three-dimensional euclidean spaces, $\{\mathscr{W}_T, T \in \mathscr{T}\}$:

$$\mathscr{W} = \bigcup_{T \in \mathscr{T}} \mathscr{W}_T \tag{2.1.1}$$

where $\mathscr{T}$, the index set of the union, is an oriented one-dimensional euclidean space. The index T is called an *instant*, and the *space of instants*, $\mathscr{T}$, is called the *Newtonian time*. The orientation in $\mathscr{T}$ assigns the set of *past instants* $\{T_0 \in \mathscr{T}, T_0 < T\}$ and the set of *future* instants $\{T_0 \in \mathscr{T}, T_0 > T\}$ relative to any *present* instant T.

Equation (2.1.1) is taken to imply that the Newtonian space–time $\mathscr{W}$ may be represented isometrically by a product space $\mathscr{E} \times \mathscr{T}$, where the space of places, $\mathscr{E}$ is an oriented three-dimensional euclidean space. The structure of the event world $\mathscr{W}$ is assumed to be such that it can be mapped homeomorphically onto the product $\mathscr{E} \times \mathscr{T}$

$$\mathscr{F}: \mathscr{W} \to \mathscr{E} \times \mathscr{T}$$

where the mapping $\mathscr{F}$ is called a *framing* and the space $\mathscr{E} \times \mathscr{T}$ is the *frame of reference*, or *observer*, onto which $\mathscr{F}$ maps $\mathscr{W}$. The orientation of the space of instants, $\mathscr{T}$ is taken to be positive in the sense of pointing from *past* to *future*. The coordinate t of an instant T is called the *time* of that instant. In practice, the unit and origin of time are regarded as being given and $\mathscr{T}$ is therefore taken to correspond to the set $\mathscr{R}$ of real numbers. Thus, the time of the instant T is the real number t and the oriented three-dimensional euclidean space $\mathscr{E}$, which represents the instantaneous spaces, is referred to as the *physical space*. The *translation space* of $\mathscr{E}$ is denoted by $\mathscr{V}$.

The physics of deformation and flow of a continuous material is here developed by way of the concept of a body $\mathscr{B}$. The properties of the body $\mathscr{B}$ will be considered by way of the concept of a *body manifold* which is an oriented three-dimensional differentiable manifold endowed with global coordinate systems. Thus a body $\mathscr{B}$ is a set whose elements, called *body points* and denoted by $X, Y, \ldots$, can be put into bijective correspondence with the points of a region B of a euclidean point space, $\mathscr{E}$. In this context, B is refered to as a *configuration* of $\mathscr{B}$. The *place* in B to which a given body-point of $\mathscr{B}$ corresponds is said to be *occupied* by that body point.

In a motion of a body, the configuration changes with the time t. If, with each value of t, there is associated a unique configuration of a body $\mathscr{B}$, the family of configurations is called a *motion* of $\mathscr{B}$. The oriented distance between instants whose times are t_1 and t_2 is $(t_2 - t_1)$. It is called the time interval, I of $\mathscr{R}$; in general, I need not be bounded. If with each value of t in I, there is associated a unique configuration B_t of a body $\mathscr{B}$, the family of configurations $\{B_t : t \in I\}$ is called a *motion* of $\mathscr{B}$. Thus, a motion is a sequence of correspondences between places and body-points.

The various physical properties characteristic of the mechanical response of a body $\mathscr{B}$, regarded as a continuous medium, are represented by scalar, vector and tensor fields defined either on a reference configuration or on the configurations constituting a motion of $\mathscr{B}$.

When defined on a reference configuration, the referential position $\boldsymbol{X}$ and the time t are the independent variables and the fields are said to be given in the referential description. In contrast, when defined on the configurations constituting a motion of $\mathscr{B}$, the independent variables are $\boldsymbol{x}$ and t, and the fields are said to be given in the spatial description. The *referential description* is obtained by adjoining the orthonormal basis $E = \{\boldsymbol{E}_1, \boldsymbol{E}_2, \boldsymbol{E}_3\}$ to the origin O and the *spatial description* by adjoining the orthonormal basis $e = \{\boldsymbol{e}_1, \boldsymbol{e}_2, \boldsymbol{e}_3\}$ to the origin o. Referential coordinates in the system (O, E) are denoted by X^α $(\alpha = 1, 2, 3)$ at a reference time $t = 0$, and spatial coordinates in the system (o, e) by x^i $(i = 1, 2, 3)$ at time t. The origin O from which X is measured need not coincide with o. Thus, the place, or *position* $\boldsymbol{x}$ is located in the particular frame of reference $\mathscr{F}$, relative to an origin o, by the relation

$$\boldsymbol{x} = \boldsymbol{o} + x^i \boldsymbol{e}_i. \tag{2.1.2}$$

Relative to the chosen coordinate system, the corresponding physical space $\mathscr{E}$ is represented isometrically by the oriented inner product space $\mathscr{R}^3$.

The term vector refers to an element

$$\boldsymbol{a} = a^k \boldsymbol{e}_k = a_k \boldsymbol{e}^k = a^\mu \boldsymbol{E}_\mu = a_\mu \boldsymbol{E}^\mu \tag{2.1.3}$$

in $\mathscr{V}$, and $|\boldsymbol{a}|$ denotes the magnitude of $\boldsymbol{a}$. The scalar, or *inner product* of two vectors $\boldsymbol{a}$ and $\boldsymbol{b}$ in $\mathscr{V}$ will be denoted by $\boldsymbol{a} \cdot \boldsymbol{b}$. Direct notation will also be used to denote linear transformations of $\mathscr{V}$ into itself, that is *tensors* over $\mathscr{V}$: thus,

$$\begin{aligned}
\mathbf{A} &= A^k{}_l \boldsymbol{e}_k \otimes \boldsymbol{e}^l = A_{kl} \boldsymbol{e}^k \otimes \boldsymbol{e}^l = A^{kl} \boldsymbol{e}_k \otimes \boldsymbol{e}_l \\
&= A^\mu{}_\nu \boldsymbol{E}_\mu \otimes \boldsymbol{E}^\nu = A^{\mu\nu} \boldsymbol{E}_\mu \otimes \boldsymbol{E}_\nu = A_{\mu\nu} \boldsymbol{E}^\mu \otimes \boldsymbol{E}^\nu \\
&= A^k{}_\mu \boldsymbol{e}_k \otimes \boldsymbol{E}^\mu = A^{k\mu} \boldsymbol{e}_k \otimes \boldsymbol{E}_\mu = A_{k\mu} \boldsymbol{e}^k \otimes \boldsymbol{E}^\mu.
\end{aligned} \tag{2.1.4}$$

The various sets of components are given by

$$a^\alpha = \boldsymbol{a} \cdot \boldsymbol{E}^\alpha \qquad a_\alpha = \boldsymbol{a} \cdot \boldsymbol{E}_\alpha \qquad a^i = \boldsymbol{a} \cdot \boldsymbol{e}^i \qquad a_i = \boldsymbol{a} \cdot \boldsymbol{e}_i \tag{2.1.5}$$

$$\begin{aligned}
&A^i{}_j = \boldsymbol{e}^i \cdot (\mathbf{A} \boldsymbol{e}_j) \qquad A^{ij} = \boldsymbol{e}^i \cdot (\mathbf{A} \boldsymbol{e}^j) \qquad A^{i\alpha} = \boldsymbol{e}^i \cdot (\mathbf{A} \boldsymbol{E}^\alpha) \\
&A^\alpha{}_\beta = \boldsymbol{E}^\alpha \cdot (\mathbf{A} \boldsymbol{E}_\beta) \qquad A^{\alpha\beta} = \boldsymbol{E}^\alpha \cdot (\mathbf{A} \boldsymbol{E}^\beta) \qquad A_{\alpha i} = \boldsymbol{E}_\alpha \cdot (\mathbf{A} \boldsymbol{e}_i).
\end{aligned} \tag{2.1.6}$$

From this it is evident that vectors may be referred to from either of the bases $\boldsymbol{e}_i$, $\boldsymbol{E}_\alpha$ and tensors to either bases or both. The expression $|\boldsymbol{x} - \boldsymbol{y}|$ will be used to denote the euclidean distance between the places $\boldsymbol{x}$ and $\boldsymbol{y}$ in $\mathscr{E}$, in accord with $(\boldsymbol{x} - \boldsymbol{y})$ being the vector in $\mathscr{V}$ that translates $\boldsymbol{y}$ into $\boldsymbol{x}$. The most general transformation $\mathbf{Q}$ of $\mathscr{V}$ that preserves the inner product itself is a linear one

$$\bar{\boldsymbol{v}} = \mathbf{Q} \boldsymbol{v} \tag{2.1.7}$$

and a linear transformation preserves the inner product if and only if the tensor $\mathbf{Q}$ is orthogonal:

$$\mathbf{Q}\mathbf{Q}^{\mathrm{T}} = \mathbf{I} \tag{2.1.8}$$

where $\mathbf{I}$ is the unit tensor.

When an origin o is selected, the points of B stand in bijective correspondence with their positions relative to o, the set $\{\boldsymbol{x}: x \in B\}$ of all such positions being denoted by B_o. In terms of the cartesian system, x^i on $\mathscr{E}$, global coordinate systems on $\mathscr{B}$, such as the mappings

$$\boldsymbol{\theta}: \mathscr{B} \to B_o \overset{(x^i)}{\longleftrightarrow} \mathscr{E}$$

and its inverse

$$\Theta: B_0 \to \mathscr{B}$$

are assumed in continuum mechanics to be orientation-preserving diffeomorphisms of $\mathscr{B}$ into $\mathscr{E}$. Such diffeomorphisms are just the configurations B of $\mathscr{B}$. The body-point X is said to *occupy* the place, i.e. the position $\boldsymbol{x}=\boldsymbol{\theta}(X)\in\mathscr{E}$ in the configuration B of $\mathscr{B}$. Thus,

$$\boldsymbol{x}=\boldsymbol{\psi}(X) \qquad X=\Psi(\boldsymbol{x}). \tag{2.1.9}$$

If with each value of t in I, there is associated a unique configuration B_t of a body $\mathscr{B}$, the family of configurations $\{B_t\colon t\in I\}$ is called a motion of $\mathscr{B}$. This definition requires the existence of functions

$$\boldsymbol{\phi}\colon \mathscr{B}\times I\rightarrow(B_t)_o \qquad \Phi\colon\{(\boldsymbol{x},t)\colon t\in I \quad \boldsymbol{x}\in(B_t)_o\}\rightarrow\mathscr{B}$$

such that

$$\boldsymbol{x}=\boldsymbol{\phi}(X,t) \qquad X=\Phi(\boldsymbol{x},t) \tag{2.1.10}$$

which identifies a motion as a sequence of correspondences between places and body-points.

In terms of the referential descriptions, a typical body-point X is identified by the position X of the place which it occupies in a particular configuration called the *reference configuration* B_r of $\mathscr{B}$. The relations of equation (2.1.9) imply the existence of functions

$$\boldsymbol{\kappa}\colon\mathscr{B}\rightarrow(B_r)_o \qquad \Pi\colon(B_r)_o\rightarrow\mathscr{B}$$

such that

$$X=\boldsymbol{\kappa}(X) \qquad X=\Pi(\boldsymbol{X}). \tag{2.1.11}$$

Equations (2.1.10) and (2.1.11) can be combined to give

$$\boldsymbol{x}=\boldsymbol{\phi}(\Pi(\boldsymbol{X}),t)=\boldsymbol{\chi}(\boldsymbol{X},t) \tag{2.1.12}$$

$$\boldsymbol{X}=\boldsymbol{\kappa}(\Phi(\boldsymbol{x},t),t)=\mathrm{X}(\boldsymbol{x},t) \tag{2.1.13}$$

which are alternative representations of a motion of $\mathscr{B}$, expressed in terms of the referential and spatial descriptions. Physically inadmissible motions are excluded by assuming that $\boldsymbol{\chi}$ and X are continuously differentiable a sufficient number of times with respect to jointly the position and time variables. This smoothness requirement implies that curves, surfaces and regions in the reference configuration B_r are carried by the motion into *material* curves, surfaces and regions in the present configuration B_t, since all such subsets of the present configuration are occupied by the same body-points at all times in I. Thus in the motion of a body $\mathscr{B}$, a representative body-point X occupies a succession of places which together form a curve called the translation path of X in euclidean point space.

The field of velocities $\boldsymbol{v}$ and the field of accelerations $\boldsymbol{a}$ are defined by taking the partial derivatives of equation (2.1.12) with respect to t: thus,

$$\boldsymbol{v}=\frac{\partial x^i}{\partial t}\,\boldsymbol{e}_i \qquad \boldsymbol{a}=\frac{\partial^2 x^i}{\partial t^2}\,\boldsymbol{e}_i. \tag{2.1.14}$$

Let φ be any general function specifying some property of a body $\mathscr{B}$. In a motion of $\mathscr{B}$ the *material derivative* of φ is the rate of change of φ with time at a fixed particle. The material derivative is denoted by $\dot{\varphi}$. Noting that the body points of $\mathscr{B}$ are identified by their positions in a reference configuration, the material derivative is calculated by differentiating φ with respect to t holding fixed the referential position $\boldsymbol{X}$. Thus, with φ given in the referential description,

$$\dot{\varphi}(\boldsymbol{X}, t) = \frac{\partial \varphi}{\partial t}(\boldsymbol{X}, t). \tag{2.1.15}$$

Or, with φ given in the spatial description

$$\dot{\varphi}(\boldsymbol{x}, t) = \frac{\partial \varphi}{\partial t}(\boldsymbol{x}, t) + [(\text{grad}\ \varphi)(\boldsymbol{x}, t)]\boldsymbol{v}(\boldsymbol{x}, t) \tag{2.1.16}$$

which is the classical relation of Euler. Identifying φ successively with the position $\boldsymbol{x}$, and with $\boldsymbol{v}$ in equation (2.1.16), gives

$$\boldsymbol{v} = \dot{\boldsymbol{x}} \qquad \boldsymbol{a} = \dot{\boldsymbol{v}} = \frac{\partial \boldsymbol{v}}{\partial t} + (\text{grad}\ \boldsymbol{v})\boldsymbol{v} = \ddot{\boldsymbol{x}} \tag{2.1.17}$$

in the spatial description.

2.1.2 Observer Transformations

Two observers are said to be equivalent if they agree about: the distance between an arbitrary pair of points in space, the orientation in space, the time elapsed between an arbitrary pair of events and the order in which two distinct events occur. An observer transformation, i.e. a change of frame, is a one-to-one mapping of space–time onto itself in which distances and time intervals are preserved. Let an event $(\boldsymbol{x}, t)$ be referred to an origin o in a frame $\mathscr{F}$ and the image event $(\boldsymbol{x}^*, t^*)$ be referred to an origin o' in a second frame $\mathscr{F}^*$. It can be shown (see for example, Noll 1964), that these two frames satisfy the above invariance requirements of equivalent observers if and only if the two events are connected by the relations

$$\boldsymbol{x}^* = \boldsymbol{c}(t) + \mathbf{Q}(t)\boldsymbol{x} \tag{2.1.18}$$

$$t^* = t - a \tag{2.1.19}$$

where $\mathbf{Q}(t)$ is an arbitrary orthogonal tensor, $\boldsymbol{c}(t)$ a point, and a is a real, arbitrary number. It is assumed that $\boldsymbol{c}(t)$ and $\mathbf{Q}(t)$ are smooth functions of t, the point $\boldsymbol{c}(t)$ depending on both the change of frame and the choice of origin, whereas the transformation $\mathbf{Q}(t)$ is uniquely determined by the change of frame. Equations (2.1.18) and (2.1.19) define a mapping of space–time onto itself which is generally referred to as the observer transform $\mathscr{F} \rightarrow \mathscr{F}^*$. It is evident from equations (2.1.18) and (2.1.19) that a change of frame, i.e. an observer transformation, can also be regarded as a time-dependent coordinate transformation which effects a change of the spatial description.

Consider two points $\boldsymbol{x}$ and $\boldsymbol{y}$ at the same time t. The change of frame denoted by equation (2.1.18) transforms these points into

$$\boldsymbol{x}^* = \boldsymbol{c} + \mathbf{Q}\boldsymbol{x} \qquad \boldsymbol{y}^* = \boldsymbol{c} + \mathbf{Q}\boldsymbol{y} \tag{2.1.20}$$

from which can be obtained the point-difference

$$\boldsymbol{x}^* - \boldsymbol{y}^* = \mathbf{Q}(\boldsymbol{x} - \boldsymbol{y}). \tag{2.1.21}$$

Identifying the point difference $\boldsymbol{x} - \boldsymbol{y}$ as the vector $\boldsymbol{u} = (\boldsymbol{x} - \boldsymbol{y})$, it follows that the vector $\boldsymbol{u}$ is transformed into

$$\boldsymbol{u}^* = \mathbf{Q}\boldsymbol{u}. \tag{2.1.22}$$

Since $\mathbf{Q}$ is orthogonal, it follows from equation (2.1.22) that a change of frame preserves not only distances but also inner products and hence angles. It follows that the unit normal vector $\boldsymbol{n}$ of a surface element must obey the transformation rule of equation (2.1.22) to give

$$\boldsymbol{n}^* = \mathbf{Q}\boldsymbol{n}. \tag{2.1.23}$$

Since it is evident that there is no one unique change of frame, $\mathscr{F}^*$ is to be regarded as an arbitrary member of the set $\mathscr{E}(\mathscr{F})$ of all frames obtainable from a given frame $\mathscr{F}$ by observer transformations. Denote by f a scalar field, defined on some arbitrary domain, referred to the frame $\mathscr{F}$ and let it be denoted by f^* when referred to the frame $\mathscr{F}^*$. Let the arbitrary vectors $\boldsymbol{a}$, $\boldsymbol{a}^*$ and arbitrary tensors $\mathbf{A}, \mathbf{A}^*$ have meanings analogous to f on the same domain. If, for all $\mathscr{F}^* \in \mathscr{E}(\mathscr{F})$,

$$f^*(\boldsymbol{x}^*, t^*) = f(\boldsymbol{x}, t) \tag{2.1.24}$$

$$\boldsymbol{a}^*(\boldsymbol{x}^*, \mathrm{t}^*) = \mathbf{Q}(t)\boldsymbol{a}(\boldsymbol{x}, t) \tag{2.1.25}$$

$$\mathbf{A}^*(\boldsymbol{x}^*, t^*) = \mathbf{Q}(t)\mathbf{A}(\boldsymbol{x}, t)\mathbf{Q}^{\mathrm{T}}(t) \tag{2.1.26}$$

the three fields are said to be objective. The transformation rules of equations (2.1.24), (2.1.25) and (2.1.26) ensure that the directions associated with the vector $\boldsymbol{a}$ and tensor $\mathbf{A}$ are unaltered by an observer transformation.

If the motion in the reference frame $\mathscr{F}$ is taken to be represented by equation (2.1.12), then it is given in $\mathscr{F}^*$ by

$$\boldsymbol{x}^* = \boldsymbol{c}(t^* + a) + \mathbf{Q}(t^* + a)\boldsymbol{\chi}(\boldsymbol{X}, t^* + a) = \boldsymbol{\chi}^*(\boldsymbol{X}, t)^* \tag{2.1.27}$$

where use has been made of equations (2.1.18) and (2.1.19). Differentiation of equation (2.1.27) with respect to t^* gives:

$$\begin{aligned} \boldsymbol{v}^*(\boldsymbol{x}^*, t^*) &= \frac{\partial \boldsymbol{\chi}^*}{\partial t^*}(\boldsymbol{X}, t^*) \\ &= \dot{\boldsymbol{c}}(t) + \dot{\mathbf{Q}}(t)\boldsymbol{x} + \mathbf{Q}(t)\boldsymbol{v}(\boldsymbol{x}, t) \\ &= \dot{\boldsymbol{c}} + \mathbf{Q}\boldsymbol{v} + \boldsymbol{\Omega}(\boldsymbol{x}^* - \boldsymbol{c}) \end{aligned} \tag{2.1.28}$$

and

$$
\begin{aligned}
\boldsymbol{a}^*(\boldsymbol{x}^*, t^*) &= \frac{\partial^2 \boldsymbol{\chi}^*}{\partial t^{*2}}(\boldsymbol{X}, t^*) \\
&= \ddot{c}(t) + \ddot{\mathbf{Q}}(t)\boldsymbol{x} + 2\dot{\mathbf{Q}}(t)\boldsymbol{v}(\boldsymbol{x}, t) + \mathbf{Q}(t)\boldsymbol{a}(\boldsymbol{x}, t) \\
&= \ddot{c} + \mathbf{Q}\boldsymbol{a} + 2\boldsymbol{\Omega}(\boldsymbol{v}^* - \dot{c}) + (\dot{\boldsymbol{\Omega}} - \boldsymbol{\Omega}^2)(\boldsymbol{x}^* - \boldsymbol{c})
\end{aligned}
\qquad (2.1.29)
$$

where the skew-symmetric tensor

$$
\boldsymbol{\Omega} = \dot{\mathbf{Q}}\mathbf{Q}^{\mathrm{T}} = -\mathbf{Q}\dot{\mathbf{Q}}^{\mathrm{T}} = -(\dot{\mathbf{Q}}\mathbf{Q}^{\mathrm{T}})^{\mathrm{T}} = -\boldsymbol{\Omega}^{\mathrm{T}} \qquad (2.1.30)
$$

represents the spin of $\mathscr{F}$ relative to $\mathscr{F}^*$. It is evident that under an observer transformation, each of the fields $\boldsymbol{v}$ and $\boldsymbol{a}$ fails to be objective.

2.1.3 The Deformation Gradient

In continuum mechanics, a body manifold may *deform* from one configuration to another one. This is in direct contrast to a configuration of a *rigid* body which is necessarily an isometry. This distinction is a significant generalisation from the theory of rigid bodies, because, in general, a diffeomorphism cannot be characterised by a finite set of parameters.

A configuration $\boldsymbol{\theta}$ of $\mathscr{B}$ is specified by identifying the coordinates, x^i of the position

$$
\boldsymbol{x} = \boldsymbol{\theta}(X) \in \mathscr{E} \qquad (2.1.31)
$$

occupied by each body-point $X \in \mathscr{B}$. Because one global coordinate system of $\mathscr{B}$ identifies uniquely the entirety of all coordinate systems of $\mathscr{B}$ by way of the family of coordinate transformations, it is sufficient to use a particular reference configuration $\boldsymbol{\kappa}$ to characterise $\mathscr{B}$; thus,

$$
\boldsymbol{\kappa}: \mathscr{B} \rightarrow B_r \qquad \boldsymbol{\kappa}(X) = X. \qquad (2.1.32)
$$

The composition of the two maps

$$
\boldsymbol{\theta}: \mathscr{B} \rightarrow B_t \overset{(x^i)}{\leftrightarrow} \mathscr{E} \qquad \boldsymbol{\kappa}: \mathscr{B} \rightarrow B_r
$$

in the form of a coordinate transformation

$$
\boldsymbol{\psi} = \boldsymbol{\theta} \circ \boldsymbol{\kappa}^{-1}: \boldsymbol{\kappa}(\mathscr{B}) \rightarrow \boldsymbol{\theta}(\mathscr{B})
$$

such that

$$
\boldsymbol{x} = \boldsymbol{\psi}(X) \qquad (2.1.33)
$$

gives rise to the important concept of a *deformation*, which in this case is the time-independent mapping $\boldsymbol{\psi}$ from configuration $\boldsymbol{\kappa}$ to the configuration $\boldsymbol{\theta}$, and hence the components of $\boldsymbol{\psi}$ are called the *deformation functions*.

Physically inadmissible deformations are excluded by assuming that equation (2.1.33) is continuously differentiable a sufficient number of times with respect to the position.

With the body manifold $\mathscr{B}$ identified with the domain $\boldsymbol{\kappa}(\mathscr{B})$ in $\mathscr{E}$, the tangent space $\mathscr{B}_X$ of $\mathscr{B}$ at any body-point $X \in \mathscr{B}$ is just the physical translation space $\mathscr{V}$. It is evident from equation (2.1.33) that for a deformation $\boldsymbol{\psi}$ from $\boldsymbol{\kappa}$ to $\boldsymbol{\theta}$ that a coordinate line of X in $\boldsymbol{\kappa}(\mathscr{B})$ is mapped into a curve in $\boldsymbol{\theta}(\mathscr{B})$, the tangent vector of this curve being

$$\frac{\partial x^i}{\partial X^\alpha}\boldsymbol{e}_i = (\text{Grad}\,\boldsymbol{\psi})(\boldsymbol{X})\boldsymbol{E}_\alpha \equiv \mathbf{F}\boldsymbol{E}_\alpha \tag{2.1.34}$$

where

$$\mathbf{F}(\boldsymbol{X}) = (\text{Grad}\,\boldsymbol{\psi})(\boldsymbol{X}) \qquad \mathbf{F} = \text{Grad}\,\boldsymbol{x}. \tag{2.1.35}$$

The field of linear maps

$$\mathbf{F}(\boldsymbol{X})\colon \mathscr{V} \to \mathscr{V} \qquad X \in \boldsymbol{\kappa}(\mathscr{B})$$

defined by equations (2.1.35) is referred to as the field of *deformation gradients*. From equations (2.1.4) and (2.1.34)

$$\mathbf{F} = F^p{}_\mu \boldsymbol{e}_p \otimes \boldsymbol{E}^\mu \qquad F^i{}_\alpha = x^i{}_{,\alpha}. \tag{2.1.36}$$

In equation (2.1.35), 'Grad' denotes the gradient operator with regard to position in the reference configuration B_r of the body $\mathscr{B}$. Since equation (2.1.33) has an inverse, the smoothness condition requiring that the mapping has continuous derivatives implies that $\mathbf{F}$ has an inverse $\mathbf{F}^{-1}$ defined by

$$\mathbf{F}^{-1}(\boldsymbol{x}) = (\text{grad}\,\boldsymbol{\psi}^{-1})(\boldsymbol{x}) \qquad \mathbf{F}^{-1} = \text{grad}\,\boldsymbol{X} \tag{2.1.37}$$

which has the component form

$$\mathbf{F}^{-1} = (F^{-1})^\mu{}_p \boldsymbol{E}_\mu \otimes \boldsymbol{e}^p \qquad (F^{-1})^\alpha{}_i = X^\alpha{}_{,i}. \tag{2.1.38}$$

In equation (2.1.37), 'grad' denotes the gradient operator with regard to position in the current configuration B_t of the body $\mathscr{B}$.

It is important to note that in observer transformations the deformation gradients transform as vectors. Taking the gradient of equation (2.1.27), with $\boldsymbol{\chi}$ replaced by $\boldsymbol{\psi}$, shows that

$$\begin{aligned}\mathbf{F}^* = (\text{Grad}\,\boldsymbol{\psi}^*)(\boldsymbol{X}, t^*) &= \mathbf{Q}(t^* + a)(\text{Grad}\,\boldsymbol{\psi})(\boldsymbol{X}, t^* + a)\\ &= \mathbf{Q}(t)\mathbf{F}(\boldsymbol{X}, t^*)\end{aligned}$$

and hence the deformation gradients in $\mathscr{F}$ and $\mathscr{F}^*$ are related by

$$\mathbf{F}^* = \mathbf{Q}\mathbf{F} \tag{2.1.39}$$

showing that the deformation gradient (which is a tensor), transforms like a vector under a change of frame at time t and thus $\mathbf{F}$ fails to be objective.

Expressing equation (2.1.35) in the equivalent form

$$\mathrm{d}\boldsymbol{x} = \mathbf{F}\,\mathrm{d}\boldsymbol{X} \tag{2.1.40}$$

the deformation gradient tensor $\mathbf{F}$ maps the neighbourhood of a *sub-body*, that is a *part* $\mathscr{P}$ of the body $\mathscr{B}$ in the reference configuration into the configuration $\boldsymbol{\psi}$, given by the jacobean

$$J = \det \mathbf{F} > 0 \tag{2.1.41}$$

which follows from regarding the deformation $\boldsymbol{\psi}$ as an orientation-preserving diffeomorphism. For mappings which have continuous derivatives, this is the necessary and sufficient condition for invertibility. Since $\mathbf{F}$ is invertible, the polar decomposition theorem (see for example Truesdell and Noll 1965), can be applied, thus giving

$$\mathbf{F} = \mathbf{R}\mathbf{U} \qquad \text{and} \qquad \mathbf{F} = \mathbf{V}\mathbf{R} \tag{2.1.42}$$

where $\mathbf{U}$ and $\mathbf{V}$ are positive definite symmetric tensors and since $J > 0$, $\mathbf{R}$ is a proper orthogonal tensor.

Consider a material curve C_t in the current configuration B_t of a body $\mathscr{B}$ with a material curve C_r in the reference configuration B_r. Let $\boldsymbol{L}$ and $\boldsymbol{l}$ be unit tangent vectors to C_r at X and to C_t at x, respectively: then

$$\mathrm{d}\boldsymbol{X} = \boldsymbol{L}\,\mathrm{d}S \qquad \mathrm{d}\boldsymbol{x} = \boldsymbol{l}\,\mathrm{d}s$$

which can be substituted into equation (2.1.40) to give, by way of the polar decomposition of equation (2.1.42)

$$\left(\frac{\mathrm{d}s}{\mathrm{d}S}\right)^2 = \boldsymbol{L}\cdot(\mathbf{C}\boldsymbol{L}) = [\boldsymbol{l}\cdot(\mathbf{B}^{-1}\boldsymbol{l})]^{-1} \tag{2.1.43}$$

where

$$\mathbf{C} \equiv \mathbf{U}^2 = \mathbf{F}^{\mathrm{T}}\mathbf{F} \tag{2.1.44}$$

$$\mathbf{B} \equiv \mathbf{V}^2 = \mathbf{F}\mathbf{F}^{\mathrm{T}} \tag{2.1.45}$$

it being noted that

$$\mathbf{V} = \mathbf{R}\mathbf{U}\mathbf{R}^{\mathrm{T}} \qquad \mathbf{B} = \mathbf{R}\mathbf{C}\mathbf{R}^{\mathrm{T}}. \tag{2.1.46}$$

The proper numbers λ_a $(a = 1, 2, 3)$ of $\mathbf{U}$ are obtained by expanding

$$\det(\mathbf{U} - \lambda_a \mathbf{I}) = 0 \tag{2.1.47}$$

to give the *characteristic equation*

$$\lambda_a^3 - \lambda_a^2 I_{\mathrm{U}} + \lambda_a II_{\mathrm{U}} - III_{\mathrm{U}} = 0 \tag{2.1.48}$$

where

$$I_{\mathrm{U}} = \operatorname{tr} \mathbf{U} = U^{\alpha}{}_{\alpha}$$

$$II_{\mathrm{U}} = \tfrac{1}{2}(\operatorname{tr} \mathbf{U})^2 - \tfrac{1}{2}\operatorname{tr} \mathbf{U}^2 = \tfrac{1}{2}(U^{\alpha}{}_{\alpha}U^{\beta}{}_{\beta} - U^{\alpha}{}_{\beta}U^{\beta}{}_{\alpha})$$

$$III_{\mathrm{U}} = \det \mathbf{U} = \varepsilon_{\alpha\beta\gamma}U^{\alpha}{}_{1}U^{\beta}{}_{2}U^{\gamma}{}_{3} \tag{2.1.49}$$

are the first, second and third invariants, respectively, of **U**. It follows directly from equation (2.1.42) that the proper vectors $\boldsymbol{q}_a$ of **V** are related to the proper vectors $\boldsymbol{p}_a$ of **U** by

$$\boldsymbol{q}_a = \mathbf{R}\boldsymbol{p}_a \tag{2.1.50}$$

and that the proper numbers of **U** and **V** are the same. In particular, it follows that

$$\frac{\mathrm{d}s}{\mathrm{d}S} = \lambda_a \tag{2.1.51}$$

from which it is concluded that the proper numbers λ_a are the extension ratios of elements in the directions represented by the proper vectors $\boldsymbol{p}_a$ of **U**. Equations (2.1.50) and (2.1.51) provide the following interpretation of equation (2.1.42). The total deformation represented by **F** can be achieved in two ways: either

(i) the initial configuration can be deformed to the final shape by imposing the stretches λ_a in the directions in B_r defined by the orthonormal vectors $\boldsymbol{p}_a$ of **U** and then rotated without change of shape to its final position by **R**, or

(ii) the initial configuration can be rotated without change of shape by **R** and then deformed to its final shape by imposing the stretches λ_a (common to **U** and **V**), in the directions in B_t defined by the orthonormal vectors $\boldsymbol{q}_a$ of **V**. With this interpretation of equation (2.1.42) as basis, **R** is called the *rotation tensor*, **U** and **V** the *right* and *left stretch tensors*, and λ_a the *principal stretches*. The principal axes associated with the proper vectors $\boldsymbol{p}_a$ and $\boldsymbol{q}_a$ are termed the *referential* and *current stretch axes*, respectively. The tensors **C** and **B** defined by equations (2.1.44) and (2.1.45) are known as the *right* and *left Cauchy–Green tensors*. It is to be noted that the referential and current stretch axes coincide with the principal axes of **C** and **B**, respectively, and the common proper numbers of **C** and **B** are the squares of the principal stretches.

From equations (2.1.39) and (2.1.45) it follows that the behaviour of the left Cauchy–Green tensor under an observer transformation is

$$\mathbf{B}^* = \mathbf{F}^*(\mathbf{F}^*)^{\mathrm{T}} = (\mathbf{QF})(\mathbf{QF})^{\mathrm{T}} = \mathbf{QFF}^{\mathrm{T}}\mathbf{Q}^{\mathrm{T}} = \mathbf{QBQ}^{\mathrm{T}}$$

from which it is evident that **B** is an objective tensor, an essential requirement for the formulation of constitutive equations.

Let λ_i $(i = 1, 2, 3)$ be the proper numbers of **U** at X, and let the principal axes of **U** at X be defined by the orthonormal basis $p = \{\boldsymbol{p}_1, \boldsymbol{p}_2, \boldsymbol{p}_3\}$, it being noted that the λ_i are the principal stretches. Since the proper numbers of **U** and **V** are the same, the λ_i are also the proper numbers of **V** at x, the principal axes of **V** at x being defined by the orthonormal basis $q = \{\boldsymbol{q}_1, \boldsymbol{q}_2, \boldsymbol{q}_3\}$. With this as basis, **U** and **V** have the spectral representations,

$$\mathbf{U} = \sum_{r=1}^{3} \lambda_r \boldsymbol{p}_r \otimes \boldsymbol{p}_r \qquad \mathbf{V} = \sum_{r=1}^{3} \lambda_r \boldsymbol{q}_r \otimes \boldsymbol{q}_r.$$

Entering the spectral representations of $\mathbf{U}$ and $\mathbf{V}$ into equations (2.1.42) respectively, gives

$$\mathbf{F}=\sum_{r=1}^{3} \lambda_r \boldsymbol{q}_r \otimes \boldsymbol{p}_r \tag{2.1.52}$$

where, in the context of equation (1.6.149), the rotation tensor is as follows $\mathbf{R}=\sum_{r=1}^{3} \boldsymbol{q}_r \otimes \boldsymbol{p}_r$.

For the more general concept of the motion of $\mathscr{B}$, the field of deformation gradients is obtained from equation (2.1.12) in the form

$$\mathbf{F}(X,t)=(\text{Grad}\,\chi)(X,t) \qquad \mathbf{F}=\text{Grad}\,\boldsymbol{x} \tag{2.1.53}$$

which is, of course, just equation (2.1.35). Thus far the discussion has been concerned with a time-independent reference frame. However, there is often a need to use a varying description as a reference frame, this invariably being identified with the present, or spatial description. In this context, the past and future are described as they appear to an observer fixed to the body point X now at the position $\boldsymbol{x}$; this description is called *relative*. Consider positions $\boldsymbol{x}$ and ξ that are values of the motion of X at the two times t and τ

$$\xi=\chi(X,\tau) \qquad \boldsymbol{x}=\chi(X,t).$$

With this as basis, it is evident that ξ is the position at time τ occupied by the body point that at the time t occupies $\boldsymbol{x}$: thus

$$\xi=\chi(\chi^{-1}(\boldsymbol{x},t),\tau)\equiv\chi_t(\boldsymbol{x},\tau) \tag{2.1.54}$$

where the function χ_t is called the *relative deformation function.* The relative deformation gradient is defined as

$$\mathbf{F}_t=\mathbf{F}_t(\boldsymbol{x},\tau)=\frac{\partial \chi_t}{\partial \boldsymbol{x}}. \tag{2.1.55}$$

If $\mathbf{F}$ is replaced by the relative deformation gradient tensor $\mathbf{F}_t$ the corresponding rotation and deformation tensors are written in the form $\mathbf{R}_t$, $\mathbf{U}_t$, $\mathbf{V}_t$, $\mathbf{C}_t$ and $\mathbf{B}_t$ and the same relations between them still apply. For example the polar decomposition of equation (2.1.42) has the same form

$$\mathbf{F}_t=\mathbf{R}_t\mathbf{U}_t \qquad \mathbf{F}_t=\mathbf{V}_t\mathbf{R}_t. \tag{2.1.56}$$

Since the value of the function χ_t is the place occupied at time τ by the body point which occupies $\boldsymbol{x}$ at time t, it follows that

$$\mathbf{F}_t(t)=\mathbf{I}. \tag{2.1.57}$$

This leads to the relations

$$\mathbf{U}_t(t)=\mathbf{V}_t(t)=\mathbf{B}_t(t)=\mathbf{C}_t(t)=\mathbf{R}_t(t)=\mathbf{I}. \tag{2.1.58}$$

2.1.4 The Velocity Gradient, Stretching and Spin

The partial derivative appearing in the last term of equation (2.1.17) is known as the velocity gradient

$$\mathbf{L} = \text{grad}\, \boldsymbol{v} = \frac{\partial v_i}{\partial x^j}\, \boldsymbol{e}_i \otimes \boldsymbol{e}^j = \text{grad}\, \dot{\boldsymbol{x}}. \tag{2.1.59}$$

The relation

$$\text{Grad}\, \dot{\boldsymbol{x}} = (\text{grad}\, \dot{\boldsymbol{x}})\mathbf{F}$$

gives, by way of equation (2.1.59)

$$\text{Grad}\, \dot{\boldsymbol{x}} = \mathbf{LF}$$

and since, in the referential description, the gradient and the material derivative commute, there follow from equation (2.1.59), with the use of equation (2.1.53), the relations

$$\mathbf{L} = \dot{\mathbf{F}}\mathbf{F}^{-1} = -\mathbf{F}(\mathbf{F}^{-1})^{\cdot} \qquad \dot{\mathbf{F}} = \mathbf{LF} \tag{2.1.60}$$

where $\dot{\mathbf{F}}$ denotes the material derivative of $\mathbf{F}$. From equations (2.1.41) and (2.1.60) there follows the relation

$$\dot{J} = \det \mathbf{F}\, \text{tr}(\dot{\mathbf{F}}\mathbf{F}^{-1}) = J\, \text{tr}\, \mathbf{L} = J\, \text{div}\, \boldsymbol{v}. \tag{2.1.61}$$

Equation (2.1.39) can be differentiated with respect to t^* to give

$$\dot{\mathbf{F}}^*(X, t^*) = \mathbf{Q}(t)\dot{\mathbf{F}}(X, t) + \dot{\mathbf{Q}}(t)\mathbf{F}(X, t) \tag{2.1.62}$$

and hence, using equation (2.1.60)

$$\mathbf{L}^* = \dot{\mathbf{F}}^*(\mathbf{F}^*)^{-1} = (\mathbf{Q}\dot{\mathbf{F}} + \dot{\mathbf{Q}}\mathbf{F})(\mathbf{QF})^{-1} = \mathbf{QLQ}^{\mathrm{T}} + \boldsymbol{\Omega} \tag{2.1.63}$$

where $\boldsymbol{\Omega}$ is defined by equation (2.1.30). It is evident from equation (2.1.63) that under an observer transformation, $\mathbf{L}$ fails to be objective.

From the polar decomposition of equation (2.1.42)

$$\dot{\mathbf{F}} = \dot{\mathbf{R}}\mathbf{U} + \mathbf{R}\dot{\mathbf{U}} = \dot{\mathbf{V}}\mathbf{R} + \mathbf{V}\dot{\mathbf{R}} \tag{2.1.64}$$

which can be substituted into equation (2.1.60) to give

$$\mathbf{L} = \mathbf{R}\dot{\mathbf{U}}\mathbf{U}^{-1}\mathbf{R}^{\mathrm{T}} + \dot{\mathbf{R}}\mathbf{R}^{\mathrm{T}} = \dot{\mathbf{V}}\mathbf{V}^{-1} + \mathbf{V}\dot{\mathbf{R}}\mathbf{R}^{\mathrm{T}}\mathbf{V}^{-1} \tag{2.1.65}$$

$$\mathbf{L}^{\mathrm{T}} = \mathbf{R}\mathbf{U}^{-1}\dot{\mathbf{U}}\mathbf{R}^{\mathrm{T}} - \dot{\mathbf{R}}\mathbf{R}^{\mathrm{T}} = \mathbf{V}^{-1}\dot{\mathbf{V}} - \mathbf{V}^{-1}\dot{\mathbf{R}}\mathbf{R}^{\mathrm{T}}\mathbf{V} \tag{2.1.66}$$

from which it follows that

$$\mathbf{L} = \mathbf{D} + \mathbf{W} \tag{2.1.67}$$

where

$$\begin{aligned}\mathbf{D} &= \tfrac{1}{2}(\mathbf{L} + \mathbf{L}^{\mathrm{T}}) \\ &= \tfrac{1}{2}\mathbf{R}(\dot{\mathbf{U}}\mathbf{U}^{-1} + \mathbf{U}^{-1}\dot{\mathbf{U}})\mathbf{R}^{\mathrm{T}} \\ &= \tfrac{1}{2}(\dot{\mathbf{V}}\mathbf{V}^{-1} + \mathbf{V}^{-1}\dot{\mathbf{V}} + \mathbf{V}\dot{\mathbf{R}}\mathbf{R}^{\mathrm{T}}\mathbf{V}^{-1} - \mathbf{V}^{-1}\dot{\mathbf{R}}\mathbf{R}^{\mathrm{T}}\mathbf{V})\end{aligned} \tag{2.1.68}$$

and

$$\begin{aligned}\mathbf{W}&=\tfrac{1}{2}(\mathbf{L}-\mathbf{L}^{\mathrm{T}})\\&=\tfrac{1}{2}\mathbf{R}(\dot{\mathbf{U}}\mathbf{U}^{-1}-\mathbf{U}^{-1}\dot{\mathbf{U}})\mathbf{R}^{\mathrm{T}}+\dot{\mathbf{R}}\mathbf{R}^{\mathrm{T}}\\&=\tfrac{1}{2}(\dot{\mathbf{V}}\mathbf{V}^{-1}-\mathbf{V}^{-1}\dot{\mathbf{V}}+\mathbf{V}\dot{\mathbf{R}}\mathbf{R}^{\mathrm{T}}\mathbf{V}^{-1}+\mathbf{V}^{-1}\dot{\mathbf{R}}\mathbf{R}^{\mathrm{T}}\mathbf{V}).\end{aligned}\tag{2.1.69}$$

When the reference configuration implicit in the definitions of **F**, **U**, **V** and **R** is chosen to be the current configuration B_t of $\mathscr{B}$, then equations (2.1.57), (2.1.58) and equations (2.1.65) to (2.1.69) give

$$\mathbf{D}=\dot{\mathbf{U}}_t(t)=\dot{\mathbf{V}}_t(t) \qquad \mathbf{W}=\dot{\mathbf{R}}_t(t).\tag{2.1.70}$$

Equations (2.1.67), (2.1.68) and (2.1.69) show that **D** and **W** are the symmetric and skew parts, respectively, of the velocity gradient **L**. The fundamental equation (2.1.67) is known as the *Cauchy–Stokes decomposition* of the velocity gradient. Because the symmetric part of the velocity gradient is the rate of change of the stretch as the body passes through its current configuration, **D** is called the *stretching* or *rate of deformation tensor*. The skew-symmetric part of **L** is the rate of change of the rotation as a body passes through its current configuration and hence **W** is called the *spin tensor*.

The components of the stretching tensor **D** are given by

$$D^i{}_j=\boldsymbol{e}_i\cdot(\mathbf{D}\boldsymbol{e}^j).$$

Consider a material curve C_t in the current configuration B_t of a body $\mathscr{B}$ with a material curve C_r in the reference configuration B_r. Let $\boldsymbol{L}$ and $\boldsymbol{l}$ be unit tangent vectors to C_r at X and to C_t at x, respectively. From equations (2.1.40) and (2.1.67), together with $\mathrm{d}\boldsymbol{X}=\boldsymbol{L}\,\mathrm{d}s$ and $\mathrm{d}\boldsymbol{x}=\boldsymbol{l}\,\mathrm{d}s$

$$\begin{aligned}(d\boldsymbol{x})^{\boldsymbol{\cdot}}&=\dot{\mathbf{F}}\,\mathrm{d}\boldsymbol{X}=\mathbf{L}\mathbf{F}\,\mathrm{d}\boldsymbol{X}=\mathbf{L}\,\mathrm{d}\boldsymbol{x}=\mathbf{L}\boldsymbol{l}\,\mathrm{d}s\\&=\boldsymbol{l}(\mathrm{d}s)^{\boldsymbol{\cdot}}+\dot{\boldsymbol{l}}\,\mathrm{d}s\end{aligned}$$

and hence

$$\begin{aligned}\frac{(\mathrm{d}s)^{\boldsymbol{\cdot}}}{\mathrm{d}s}&=\boldsymbol{l}\cdot(\mathbf{D}\boldsymbol{l})+\boldsymbol{l}\cdot(\mathbf{W}\boldsymbol{l})=\boldsymbol{l}\cdot(\mathbf{D}\boldsymbol{l})\\ \dot{\boldsymbol{l}}&=\mathbf{L}\boldsymbol{l}-[\boldsymbol{l}\cdot(\mathbf{L}\boldsymbol{l})]\boldsymbol{l}\end{aligned}\tag{2.1.71}$$

from which it follows that the component $\boldsymbol{l}\cdot(\mathbf{D}\boldsymbol{l})$ is the rate of extension per unit length of a material line element which, in the current configuration B_t, is situated at x and momentarily aligned with the direction defined by the base vector $\boldsymbol{e}_i$. Since **D** is symmetric, it has the spectral representation

$$\mathbf{D}=\sum_{a=1}^{3}v_a(\boldsymbol{d}_a\otimes\boldsymbol{d}_a)$$

and hence

$$\frac{(\mathrm{d}s_a)^{\cdot}}{\mathrm{d}s_a} = v_a$$

where the proper numbers v_a $(a = 1, 2, 3)$ are called the *principal stretchings* and the associated set of orthonormal proper vectors $\boldsymbol{d}_a$ define the *principal axes of stretching*.

Consider the intersection of a material curve D_t in the current configuration B_t of a body $\mathscr{B}$, with the material curve C_t. Let $\boldsymbol{m}$ be the unit tangent vector on D_t at x, and let θ be the angle of intersection, so that $\cos\theta = \boldsymbol{l}\cdot\boldsymbol{m}$ from which can be obtained

$$-\sin\theta\dot{\theta} = \dot{\boldsymbol{l}}\cdot\boldsymbol{m} + \boldsymbol{l}\cdot\dot{\boldsymbol{m}}.$$

From equation (2.1.71)

$$\dot{\boldsymbol{l}} = \mathbf{L}\boldsymbol{l} - [\boldsymbol{l}\cdot(\mathbf{L}\boldsymbol{l})]\boldsymbol{l}$$

and hence, similarly

$$\dot{\boldsymbol{m}} = \mathbf{L}\boldsymbol{m} - [\boldsymbol{m}\cdot(\mathbf{L}\boldsymbol{m})]\boldsymbol{m}.$$

Thus one can obtain, for the condition that $\boldsymbol{l}$ and $\boldsymbol{m}$ are orthogonal unit vectors, the condition

$$\boldsymbol{l}\cdot(\mathbf{D}\boldsymbol{m}) = -\tfrac{1}{2}\dot{\theta}$$

from which it follows that the component $\boldsymbol{l}\cdot(\mathbf{D}\boldsymbol{m})(i \neq j)$ is half the rate of decrease of the angle between a pair of material line elements which intersect at x and are instantaneously in the directions represented by $\boldsymbol{e}_i$ and $\boldsymbol{e}^j$.

From equation (2.1.71) it follows that, since $\boldsymbol{l}\cdot(\mathbf{W}\boldsymbol{l}) = 0$, applying $\mathbf{W}$ alone will produce no change in the distance between material points and must therefore give rise to a rigid body rotation.

These considerations lead to the following general statement: relative motion in the locality of the point x consists of a tri-axial stretching, represented by $\mathbf{D}$, superimposed upon a rigid rotation specified by $\mathbf{W}$.

Use of equations (2.1.63) and (2.1.68) gives

$$\begin{aligned}\mathbf{D}^* &= \tfrac{1}{2}I(\mathbf{L}^* + \mathbf{L}^{*\mathrm{T}})\\ &= \tfrac{1}{2}(\mathbf{QLQ}^{\mathrm{T}} + \mathbf{QL}^{\mathrm{T}}\mathbf{Q}^{\mathrm{T}} + \boldsymbol{\Omega} + \boldsymbol{\Omega}^{\mathrm{T}})\end{aligned}$$

where $\boldsymbol{\Omega}$ is defined by equation (2.1.30), and hence

$$\mathbf{D}^* = \mathbf{QDQ}^{\mathrm{T}}$$

from which it is evident that under an observer transformation, $\mathbf{D}$ is an objective tensor, an essential requirement for the formulation of constitutive equations.

2.1.5 The Corotational and Convected Time Derivatives

Following from equation (2.1.26)

$$\dot{\mathbf{A}}^* = \mathbf{Q}\dot{\mathbf{A}}\mathbf{Q}^{\mathrm{T}} + \dot{\mathbf{Q}}\mathbf{A}\mathbf{Q}^{\mathrm{T}} + \mathbf{Q}\mathbf{A}\dot{\mathbf{Q}}^{\mathrm{T}} \tag{2.1.72}$$

it is evident that under an observer transformation the material derivative fails to be objective due to the presence of the last two terms on the right-hand side. The spin tensor can be expressed in the form

$$\mathbf{W}^* = \mathbf{Q}\mathbf{W}\mathbf{Q}^{\mathrm{T}} + \boldsymbol{\Omega} \tag{2.1.73}$$

where $\boldsymbol{\Omega}$ is defined by equation (2.1.30). From equation (2.1.73)

$$\dot{\mathbf{Q}} = \mathbf{W}^*\mathbf{Q} - \mathbf{Q}\mathbf{W}. \tag{2.1.74}$$

Substituting for $\dot{\mathbf{Q}}$ in equation (2.1.72) from equation (2.1.74) and rearranging gives

$$\dot{\mathbf{A}}^* - \mathbf{W}^*\mathbf{A}^* + \mathbf{A}^*\mathbf{W}^* = \mathbf{Q}(\dot{\mathbf{A}} - \mathbf{W}\mathbf{A} + \mathbf{A}\mathbf{W})\mathbf{Q}^{\mathrm{T}} \tag{2.1.75}$$

which can be expressed in the form

$$(\mathbf{A}^*)^{\mathrm{r}} = \mathbf{Q}\mathbf{A}^{\mathrm{r}}\mathbf{Q}^{\mathrm{T}} \tag{2.1.76}$$

where

$$\mathbf{A}^{\mathrm{r}} = \dot{\mathbf{A}} - \mathbf{W}\mathbf{A} + \mathbf{A}\mathbf{W} \tag{2.1.77}$$

is referred to as the *corotational time derivative*. The corotational derivative is seen to be equal to the material derivative of $\mathbf{A}$ as it would appear to an observer in a frame of reference fixed within the body and rotating with it at an angular velocity equal to the instantaneous value of the angular velocity $\mathbf{W}$ of the body.

The addition of the objective quantity $\mathbf{DA} + \mathbf{AD}$ to $\mathbf{A}^{\mathrm{r}}$ gives the *convected time derivative*

$$\mathbf{A}^{\mathrm{c}} = \mathbf{A}^{\mathrm{r}} + \mathbf{D}\mathbf{A} + \mathbf{A}\mathbf{D} \tag{2.1.78}$$

or, alternatively, using equation (2.1.67)

$$\mathbf{A}^{\mathrm{c}} = \dot{\mathbf{A}} + \mathbf{L}^{\mathrm{T}}\mathbf{A} + \mathbf{A}\mathbf{L}. \tag{2.1.79}$$

The uses and limitations of the convected derivative have been extensively considered in relation to the formulation of constitutive equations by Oldroyd (1950, 1958).

2.1.6 Isochoric Motions and Incompressibility

Let an element of volume $\mathrm{d}V = (\mathrm{d}X \wedge \delta X)\cdot\Delta X$ at X be carried into the element of volume

$$\mathrm{d}v = (\mathrm{d}\boldsymbol{x} \wedge \delta\boldsymbol{x})\cdot\Delta\boldsymbol{x} = [(\mathbf{F}\,\mathrm{d}\boldsymbol{X}) \wedge (\mathbf{F}\,\delta\boldsymbol{X})]\cdot(\mathbf{F}\,\Delta\boldsymbol{X})$$

at x, where x is an interior point of a material region R_t in the current configuration B_t of a body $\mathscr{B}$, the particle currently situated at x occupying the body-point X in the reference configuration B_r. Using the general relation (see §1.6.4),

$$\mathbf{A}^+(\boldsymbol{a} \wedge \boldsymbol{b}) = (\mathbf{A}\boldsymbol{a}) \wedge (\mathbf{A}\boldsymbol{b}) \tag{2.1.80}$$

$$\mathbf{A}^{-1} = (\det \mathbf{A}^{\mathrm{T}})^{-1}\mathbf{A}^{\mathrm{T}+} \tag{2.1.81}$$

where $\mathbf{A}^+$ is the adjugate of the tensor $\mathbf{A}$ and $\boldsymbol{a}, \boldsymbol{b}$ are arbitrary vectors, the element of volume $\mathrm{d}v$ can be rearranged into the form

$$\begin{aligned} \mathrm{d}v &= \mathbf{F}^+(\mathrm{d}X \wedge \delta X)\cdot(\mathbf{F}\,\Delta X) \\ &= J(\mathbf{F}^{-1})^{\mathrm{T}}(\mathrm{d}X \wedge \delta X)\cdot(\mathbf{F}\,\Delta X) \\ &= J\,\mathrm{d}V \end{aligned} \tag{2.1.82}$$

from which it follows that $J > 0$ in accord with equation (2.1.41).

A motion of a body $\mathscr{B}$ is called *isochoric* if the volume of the shape of each part of $\mathscr{B}$ remains constant in time. In this case the volume elements of equation (2.1.82) have equal content so that $\mathrm{d}V = \mathrm{d}v$ and hence equation (2.1.82) reduces to

$$J = 1 \tag{2.1.83}$$

on each configuration in $\{B_t\}$. Since the volume of the shape of each part of $\mathscr{B}$ remains constant in time, it follows from equations (2.1.61) and (2.1.68) that

$$\operatorname{tr}\mathbf{L} = \operatorname{tr}\mathbf{D} = \operatorname{div}\boldsymbol{v} = 0. \tag{2.1.84}$$

Either one of the conditions of equations (2.1.83) and (2.1.84) is necessary and sufficient for isochoric motion.

A material body is said to be *incompressible* if it can undergo only isochoric motions. Thus, for an incompressible material, it follows from equations (2.1.41) and (2.1.83) that

$$\det \mathbf{F} = 1 \qquad III_{\mathbf{B}} = III_{\mathbf{C}} = 1 \tag{2.1.85}$$

where $III_{\mathbf{B}}$, $III_{\mathbf{C}}$ are the third invariants of the Cauchy–Green deformation tensors defined by equations (2.1.44) and (2.1.45).

In a motion of a body $\mathscr{B}$ the material derivative of the left Cauchy–Green tensor $\mathbf{B}$ can be expressed, using equation (2.1.45), in the form

$$\dot{\mathbf{B}} = \mathbf{L}\mathbf{B} + \mathbf{B}\mathbf{L}^{\mathrm{T}} \tag{2.1.86}$$

where use has been made of equations (2.1.60). From equations (2.1.77) and (2.1.79),

$$2\mathbf{B}^{\mathrm{r}} - \mathbf{B}^{\mathrm{c}} = \dot{\mathbf{B}} - (\mathbf{L}^{\mathrm{T}} + 2\mathbf{W})\mathbf{B} + \mathbf{B}(2\mathbf{W} - \mathbf{L}). \tag{2.1.87}$$

Substituting for $\mathbf{W}$ and for $\dot{\mathbf{B}}$ in equation (2.1.87) from equations (2.1.67) and

(2.1.86), respectively, gives

$$2\mathbf{B}^{\mathrm{r}} - \mathbf{B}^{\mathrm{c}} = \mathbf{O}. \tag{2.1.88}$$

Similarly, there can be obtained

$$(\mathbf{B}^{-1})^{\cdot} = -\mathbf{L}^{\mathrm{T}}\mathbf{B}^{-1} - \mathbf{B}^{-1}\mathbf{L} \tag{2.1.89}$$

which, used with equation (2.1.79) gives

$$(\mathbf{B}^{-1})^{\mathrm{c}} = (\mathbf{B}^{-1})^{\cdot} + \mathbf{L}^{\mathrm{T}}\mathbf{B}^{-1} + \mathbf{B}^{-1}\mathbf{L} = \mathbf{O}. \tag{2.1.90}$$

The material derivatives of $I_{\mathbf{B}}$ and $I_{\mathbf{B}^{-1}}$ follow from equations (2.1.86) and (2.1.89) in the form

$$\dot{I}_{\mathbf{B}} = 2\,\mathrm{tr}(\mathbf{BD}) \qquad \dot{I}_{\mathbf{B}^{-1}} = -2\,\mathrm{tr}(\mathbf{B}^{-1}\mathbf{D}) \tag{2.1.91}$$

where $I_{\mathbf{B}}$ and $I_{\mathbf{B}^{-1}}$ are the first invariants of $\mathbf{B}$ and $\mathbf{B}^{-1}$ respectively. Substituting for $\mathbf{B}$ in equation (2.1.88) from the relation

$$\mathbf{B}' = \mathbf{B} - \tfrac{1}{3}(\mathrm{tr}\,\mathbf{B})\mathbf{I} \tag{2.1.92}$$

gives

$$2(\mathbf{B}')^{\mathrm{r}} - (\mathbf{B}')^{\mathrm{c}} = \tfrac{2}{3}I_{\mathbf{B}}\mathbf{D}' - \tfrac{2}{3}\mathbf{I}\,\mathrm{tr}(\mathbf{BD}') \tag{2.1.93}$$

where $\mathbf{B}'$ and $\mathbf{D}'$ are the deviators of $\mathbf{B}$ and $\mathbf{D}$ respectively. Similarly, substituting for $\mathbf{B}^{-1}$ in equation (2.1.90) from the relation

$$(\mathbf{B}^{-1})' = \mathbf{B}^{-1} - \tfrac{1}{3}(\mathrm{tr}\,\mathbf{B}^{-1})\mathbf{I} \tag{2.1.94}$$

gives

$$[(\mathbf{B}^{-1})']^{\mathrm{c}} = -\tfrac{2}{3}I_{\mathbf{B}^{-1}}\mathbf{D}' + \tfrac{2}{3}\mathbf{I}\,\mathrm{tr}(\mathbf{B}^{-1}\mathbf{D}') \tag{2.1.95}$$

where $(\mathbf{B}^{-1})'$ is the deviator of $\mathbf{B}^{-1}$. Equations (2.1.93) and (2.1.95) can be rearranged and combined to give

$$\mathbf{D}' = 2\alpha_1(\mathbf{B}')^{\mathrm{r}} + \alpha_{-1}[(\mathbf{B}^{-1})']^{\mathrm{c}} - \alpha_1(\mathbf{B}')^{\mathrm{c}} \tag{2.1.96}$$

where

$$\alpha_1 = (3\dot{I}_{\mathbf{B}^{-1}} + 2I_{\mathbf{D}}I_{\mathbf{B}^{-1}})/2(\dot{I}_{\mathbf{B}}I_{\mathbf{B}^{-1}} + I_{\mathbf{B}}\dot{I}_{\mathbf{B}^{-1}}) \tag{2.1.97}$$

$$\alpha_{-1} = -(3\dot{I}_{\mathbf{B}} - 2I_{\mathbf{D}}I_{\mathbf{B}})/2(\dot{I}_{\mathbf{B}}I_{\mathbf{B}^{-1}} + I_{\mathbf{B}}\dot{I}_{\mathbf{B}^{-1}}). \tag{2.1.98}$$

In a motion of a body $\mathscr{B}$ the material derivative of the right Cauchy–Green tensor $\mathbf{C}$ can be expressed in the form

$$\dot{\mathbf{C}} = (\mathbf{F}^{\mathrm{T}})^{\cdot}\mathbf{F} + \mathbf{F}^{\mathrm{T}}\dot{\mathbf{F}} \tag{2.1.99}$$

where use has been made of equation (2.1.44). Equation (2.1.99) can be rearranged using equation (2.1.60) to give

$$\mathbf{D} = \tfrac{1}{2}(\mathbf{F}^{\mathrm{T}})^{-1}\dot{\mathbf{C}}\mathbf{F}^{-1}. \tag{2.1.100}$$

2.2 PHYSICAL PRINCIPLES OF CONTINUUM MECHANICS

2.2.1 Balance Principles: I, Mass and Momentum

In classical mechanics the resistance of a material body to change of movement is characterised by a quantity which is called its mass. The mass of a body is a measure of the amount of material contained in an arbitrary sub-body, that is part $\mathscr{P}$ of a body $\mathscr{B}$. Since a body $\mathscr{B}$ is the sum of its parts, the mass of $\mathscr{B}$ is the sum of the masses of its parts. This concept of mass implies that the mass of a body $\mathscr{B}$ is a non-negative, time-independent scalar quantity which is also generally independent of the size of the configuration occupied by the arbitrary part $\mathscr{P}$ of the body. Let the body manifold $\mathscr{B}$ be represented by a reference configuration B_r, and let the mass associated with $\mathscr{P}$ be denoted by the non-negative number M. The mass $\mathbf{M}(\mathscr{P})$ possesses the following properties

(i) $M(\mathscr{P}_1 \cup \mathscr{P}_2) = M(\mathscr{P}_1) + M(\mathscr{P}_2)$ for all pairs $\mathscr{P}_1, \mathscr{P}_2$ of disjoint subsets of $\mathscr{B}$
(ii) $M(\mathscr{P}) \to 0$ as the volume $V \to 0$.

Property (ii) excludes concentrated masses in accord with the essential assumption of continuum physics that mass be absolutely continuous with respect to the euclidean volume measure on B_r. Hence the properties specified by (i) and (ii) imply the existence of a scalar field ρ_r, defined on B_r. Thus, for any sub-body $\mathscr{P}$ of $\mathscr{B}$,

$$M(\mathscr{P}) = \int_{\mathscr{P}} \rho_r \, \mathrm{d}V \tag{2.2.1}$$

where ρ_r, the mass per unit volume, is the density in the reference configuration $B_r(\mathscr{P})$ of the material of which the sub-body, $\mathscr{P}$ of $\mathscr{B}$, is composed.

With regard to the motion of a material body, the mass of a body $\mathscr{B}$ possesses one other property:

(iii) The mass $M(\mathscr{R})$ of the material $\mathscr{R}$ occupying an arbitrary material region R_t at time t in the current configuration B_t is independent of the motion $\{B_t: t \in I\}$ of the body $\mathscr{B}$.

The time independence of the mass $M(\mathscr{R})$ gives, by way of equation (2.2.1), the *equation of mass balance*:

$$\frac{\mathrm{d}}{\mathrm{d}t} \int_{R_t(\mathscr{P})} \rho \, \mathrm{d}v = 0 \qquad (\mathscr{P} \subset \mathscr{B}) \tag{2.2.2}$$

where ρ is now the density in the current configuration $B_t(\mathscr{P})$ of $\mathscr{P}$. It follows from equations (2.2.1) and (2.2.2), with the use of equation (2.1.82), that the conservation of mass can be expressed in the form

$$\rho_r = J\rho \tag{2.2.3}$$

where ρ_r and ρ are the mass densities in the reference and current configurations, respectively. A motion in which $\rho = \rho_r$ is called an isochoric, or volume-preserving motion (see equation (2.1.83)), while if the density is uniform in both space and time it is called a *homochoric motion.*

The linear momentum $\boldsymbol{p}(R_t)$ of the material occupying R_t is defined by the relation

$$\boldsymbol{p}(R_t) = \int_{R_t(\mathscr{P})} \rho \boldsymbol{v} \, \mathrm{d}v \qquad (\mathscr{P} \subset \mathscr{B}) \tag{2.2.4}$$

where $\boldsymbol{v}$ is the velocity of the material particle at the position $\boldsymbol{x}$. The angular momentum $\boldsymbol{H}(R_t)$ of the same material is defined by the relation

$$\boldsymbol{H}(R_t) = \int_{R_t(\mathscr{P})} (\boldsymbol{x} \wedge \boldsymbol{v}) \rho \, \mathrm{d}v \qquad (\mathscr{P} \subset \mathscr{B}) \tag{2.2.5}$$

where $\boldsymbol{x}$ is the position of a representative point of R_t relative to an origin o, $\boldsymbol{H}(R_t)$ being the angular momentum with respect to o.

Euler's equations of motion can now be stated in the form of two principles:

(i) The rate of change of linear momentum $\boldsymbol{p}$ is equal to the total applied force $\boldsymbol{F}$ acting on the body; i.e.

$$\dot{\boldsymbol{p}} = \boldsymbol{F} \tag{2.2.6}$$

(ii) The rate of change of angular momentum $\boldsymbol{H}$ is equal to the total applied torque $\boldsymbol{\Gamma}$; i.e.

$$\dot{\boldsymbol{H}} = \boldsymbol{\Gamma}. \tag{2.2.7}$$

Two types of external force are assumed to act on a body $\mathscr{B}$

(i) body forces, such as gravity, which act on every material element throughout the body and can be described by a vector field $\boldsymbol{b}$ which is referred to as the body force per unit mass and which is defined on the configurations of $\mathscr{B}$;

(ii) surface forces, such as friction, which act on the surface elements of area and can be described by a vector $\boldsymbol{t}$ which is referred to as the surface traction per unit area and which is defined on the surface S_t of the material occupying that part of B_t which is acted upon by the surface traction.

The total force $\boldsymbol{F}$ is defined as

$$\boldsymbol{F}(\mathscr{P}) = \int_{R_t(\mathscr{P})} \rho \boldsymbol{b} \, \mathrm{d}v + \int_{\partial R_t(\mathscr{P})} \boldsymbol{t} \, \mathrm{d}a \tag{2.2.8}$$

where the second integral on the right-hand side of equation (2.2.8) represents the contribution to $\boldsymbol{F}$ from the contact force acting on the boundary ∂R_t of the arbitrary material region R_t. Equation (2.2.8) can be expressed by way of equations (2.2.4) and (2.2.6) in the form

$$\frac{\mathrm{d}}{\mathrm{d}t}\int_{R_t(\mathscr{P})} \rho \boldsymbol{v}\, \mathrm{d}v = \int_{R_t(\mathscr{P})} \rho \boldsymbol{b}\, \mathrm{d}v + \int_{\partial R_t(\mathscr{P})} \boldsymbol{t}\, \mathrm{d}a \tag{2.2.9}$$

which is just the integral form of the equation of linear momentum balance. For this case it is evident that the body force $\boldsymbol{b}$ acts as an internal source of linear momentum while the surface traction promotes a flux of momentum across the surface on which it acts.

The total torque about the origin o is defined as

$$\boldsymbol{\Gamma}(\mathscr{P}) = \int_{R_t(\mathscr{P})} \rho(\boldsymbol{x} \wedge \boldsymbol{b})\, \mathrm{d}v + \int_{\partial R_t(\mathscr{P})} (\boldsymbol{x} \wedge \boldsymbol{t})\, \mathrm{d}a \tag{2.2.10}$$

where the second integral on the right-hand side of equation (2.2.10) represents the contribution to $\boldsymbol{\Gamma}$ of the contact forces acting on the boundary ∂R_t of the arbitrary material region R_t. If the origin o is moved a distance $\boldsymbol{d}$ then the torque $\boldsymbol{\Gamma}^*$ about the new origin o^* can be expressed, by way of equations (2.2.8) and (2.2.10), in the form

$$\begin{aligned}\boldsymbol{\Gamma}^* &= \int_{R_t(\mathscr{P})} (\boldsymbol{x}-\boldsymbol{d}) \wedge \boldsymbol{b}\rho\, \mathrm{d}v + \int_{\partial R_t(\mathscr{P})} (\boldsymbol{x}-\boldsymbol{d}) \wedge \boldsymbol{t}\, \mathrm{d}a \\ &= \Gamma - \boldsymbol{d} \wedge \boldsymbol{F}.\end{aligned} \tag{2.2.11}$$

The angular momentum will also change in such a change of origin to $\boldsymbol{H}^*$ where

$$\boldsymbol{H}^* = \int_{R_t(\mathscr{P})} (\boldsymbol{x}-\boldsymbol{d}) \wedge \boldsymbol{v}\rho\, \mathrm{d}v = \boldsymbol{H} - \boldsymbol{d} \wedge \boldsymbol{p}. \tag{2.2.12}$$

Using equation (2.2.7) allows equations (2.2.11) and (2.2.12) to be rearranged to give, by way of equation (2.2.6), the relation

$$\boldsymbol{\Gamma}^* = \dot{\boldsymbol{H}}^* - \boldsymbol{d} \wedge (\boldsymbol{F} - \dot{\boldsymbol{p}}) = \dot{\boldsymbol{H}}^* \tag{2.2.13}$$

from which it is evident that if the principles (i) and (ii) hold for one choice of origin they also hold for any other choice of origin, provided it is at rest relative to the first origin. Using equations (2.2.5) and (2.2.7), equation (2.2.10) can be expressed in the form

$$\frac{\mathrm{d}}{\mathrm{d}t}\int_{R_t(\mathscr{P})} \rho(\boldsymbol{x} \wedge \boldsymbol{v})\, \mathrm{d}v = \int_{R_t(\mathscr{P})} \rho(\boldsymbol{x} \wedge \boldsymbol{b})\, \mathrm{d}v + \int_{\partial R_t(\mathscr{P})} (\boldsymbol{x} \wedge \boldsymbol{t})\, \mathrm{d}a \tag{2.2.14}$$

which is just the integral form of the equation of angular momentum balance. For this case it is evident that $(\boldsymbol{x} \wedge \boldsymbol{b})$ is the supply of angular momentum per unit mass and $(\boldsymbol{x} \wedge \boldsymbol{t})$ is the influx of angular momentum per unit area in B_t.

If, following Dahler and Scriven (1963), allowance is made for a body couple density $\boldsymbol{G}$ and a surface couple density $\boldsymbol{c}(\boldsymbol{n})$ to act, then the rate of change of total angular momentum density $\boldsymbol{h}$ is obtained by extending Euler's laws to give

$$\frac{\mathrm{d}}{\mathrm{d}t}\int_{R_t(\mathscr{P})} \rho \boldsymbol{h}\,\mathrm{d}v = \int_{R_t(\mathscr{P})} (\boldsymbol{x} \wedge \boldsymbol{b} + \boldsymbol{G})\rho\,\mathrm{d}v + \int_{\partial R_t(\mathscr{P})} [(\boldsymbol{x} \wedge \boldsymbol{t}(\boldsymbol{n})) + \boldsymbol{c}(\boldsymbol{n})]\,\mathrm{d}a \tag{2.2.15}$$

which is the integral form of the equation of total angular momentum balance ($\boldsymbol{h}$ is considered further in §2.2.5).

It is to be noted that all the equations relating to the discussion of the balance of mass and momentum have been stated in terms of an arbitrary sub-body $\mathscr{P}$ of the body $\mathscr{B}$. Since $\mathscr{B}$ is a sub-body of itself, all the equations remain valid when $\mathscr{P}$ is replaced by $\mathscr{B}$. The discussion has been concerned with an arbitrary sub-body $\mathscr{P} \subset \mathscr{B}$ because in the case of a deformable body manifold $\mathscr{B}$, the different sub-bodies of $\mathscr{B}$ will, in general, deform independently relative to one another.

2.2.2 Transport Formulae

Consider a material surface S_t in the current configuration B_t of a body $\mathscr{B}$ with a material surface S_r in the reference configuration B_r. Let the element of area $\mathrm{d}A$ at X, defined by the vectors $\mathrm{d}\boldsymbol{X}, \delta\boldsymbol{X}$, be carried into the element of surface $\mathrm{d}a$ at $x \in S_t$ defined by the two vectors $\mathrm{d}\boldsymbol{x}$ and $\delta\boldsymbol{x}$. Thus,

$$\boldsymbol{N}\,\mathrm{d}A = \mathrm{d}\boldsymbol{X} \wedge \delta\boldsymbol{X} \qquad \boldsymbol{n}\,\mathrm{d}a = \mathrm{d}\boldsymbol{x} \wedge \delta\boldsymbol{x}$$

where $\boldsymbol{N}$ and $\boldsymbol{n}$ are unit vectors normal to S_r and S_t at X and x, respectively. Using the general relations of equations (2.1.80) and (2.1.81), there follows the relation

$$\boldsymbol{n}\,\mathrm{d}a = (\mathbf{F}\,\mathrm{d}\boldsymbol{X}) \wedge (\mathbf{F}\,\delta\boldsymbol{X}) = \mathbf{F}^{+}(\mathrm{d}\boldsymbol{X} \wedge \delta\boldsymbol{X})$$
$$= J(\mathbf{F}^{-1})^{\mathrm{T}}\boldsymbol{N}\,\mathrm{d}A$$

and hence

$$(\mathrm{d}a/\mathrm{d}A)^2 = J^2\boldsymbol{N}\cdot(\mathbf{C}^{-1}\boldsymbol{N}) = J^2[\boldsymbol{n}\cdot(\mathbf{B}\boldsymbol{n})]^{-1} \tag{2.2.16}$$

where $\mathbf{C}$ and $\mathbf{B}$ are defined by equations (2.1.44) and (2.1.45).

Using the notation $\mathrm{d}\boldsymbol{s}$ for a directed element of arc length, the required set of transport formulae can be expressed as follows.

Let ψ be a continuously differentiable scalar field representing some property of a body $\mathscr{B}$. The transport formulae can be expressed, by way of equations (2.1.40), (2.1.60), (2.1.61), (2.1.82) and (2.2.16) in the form:

$$\frac{\mathrm{d}}{\mathrm{d}t}\int_{C_t} \psi\,\mathrm{d}\boldsymbol{s} = \int_{C_t} (\dot{\psi}\,\mathrm{d}\boldsymbol{s} + \psi\mathbf{L}\,\mathrm{d}\boldsymbol{s}) \tag{2.2.17}$$

$$\frac{\mathrm{d}}{\mathrm{d}t}\int_{S_t} \psi\boldsymbol{n}\,\mathrm{d}a = \int_{S_t} [(\dot{\psi} + \psi\,\mathrm{tr}\,\mathbf{L})\boldsymbol{n} - \psi\mathbf{L}^{\mathrm{T}}\boldsymbol{n}]\,\mathrm{d}a \tag{2.2.18}$$

$$\frac{\mathrm{d}}{\mathrm{d}t}\int_{R_t}\psi\,\mathrm{d}v=\int_{R_t}(\dot{\psi}+\psi\,\mathrm{tr}\,\mathbf{L})\,\mathrm{d}v \tag{2.2.19}$$

where C_t, S_t and R_t denote a material curve, a material surface, and a material region respectively in the current configuration B_t of a body $\mathscr{B}$. Similarly, for a continuously differentiable vector field $\boldsymbol{c}$,

$$\frac{\mathrm{d}}{\mathrm{d}t}\int_{C_t}\boldsymbol{c}\cdot\mathrm{d}\boldsymbol{s}=\int_{C_t}(\dot{\boldsymbol{c}}+\mathbf{L}^{\mathrm{T}}\boldsymbol{c})\cdot\mathrm{d}\boldsymbol{s} \tag{2.2.20}$$

$$\frac{\mathrm{d}}{\mathrm{d}t}\int_{S_t}\boldsymbol{c}\cdot\boldsymbol{n}\,\mathrm{d}a=\int_{S_t}(\dot{\boldsymbol{c}}+\boldsymbol{c}\,\mathrm{tr}\,\mathbf{L}-\mathbf{L}\boldsymbol{c})\cdot\boldsymbol{n}\,\mathrm{d}a \tag{2.2.21}$$

$$\frac{\mathrm{d}}{\mathrm{d}t}\int_{R_t}\boldsymbol{c}\,\mathrm{d}v=\int_{R_t}(\dot{\boldsymbol{c}}+\boldsymbol{c}\,\mathrm{tr}\,\mathbf{L})\,\mathrm{d}v. \tag{2.2.22}$$

2.2.3 Cauchy's Postulate and the Stress Principle

It is the interaction of the atoms and molecules of which the material is composed at the sub-continuum level that determines the cohesion of a body and the transmission throughout the body of the effects of forces acting on its surface. Because the strength of the interatomic forces usually decays rapidly with increasing atomic separation, an atom only interacts strongly with its nearest neighbours. Thus any small volume element of a material is influenced by the rest of the material primarily through forces acting on its surface. This aspect of continuum mechanics finds expression by way of Cauchy's postulate that the action of the rest of the material upon any volume element of it is of the same form as distributed surface forces. The concept of contact forces finds mathematical expression by way of the *stress principle* of Cauchy which can be stated as follows:

The action of the material occupying the part of B_t exterior to a closed surface S on the material occupying the interior part is equivalent to a distribution of force vectors $\boldsymbol{t}$ per unit area, defined on S.

The force vector per unit area $\boldsymbol{t}$ is called the surface traction or stress vector and is associated with an element of area $\mathrm{d}\boldsymbol{a}$ having a unit normal $\boldsymbol{n}$. Thus, at each point in space there is a doubly infinite set of stress vectors corresponding to the different possible directions of the surface element $\mathrm{d}\boldsymbol{a}$. When the dependence of the stress vector on direction is to be emphasised it will be written in the form $\boldsymbol{t}(\boldsymbol{n})$. Since Cauchy's stress principle admits only contact forces, it is said to define *non-polar materials*, thus recognising that contact torques are excluded.

Setting $\psi=\rho$ in equation (2.2.19) and using equations (2.1.61) and (2.2.2) gives

$$\int_{R_t} (\dot{\rho} + \rho \operatorname{div} \boldsymbol{v}) \, dv = 0. \tag{2.2.23}$$

Setting $\boldsymbol{c} = \rho \boldsymbol{v}$ in equation (2.2.22) and using equation (2.2.23) gives

$$\frac{d}{dt} \int_{R_t} \rho \boldsymbol{v} \, dv = \int_{R_t} \rho \boldsymbol{a} \, dv. \tag{2.2.24}$$

Substituting the left-hand side of equation (2.2.24) for the left-hand side of equation (2.2.9) and rearranging gives

$$\int_{R_t} \rho(\boldsymbol{a} - \boldsymbol{b}) \, dv = \int_{\partial R_t} \boldsymbol{t} \, da \tag{2.2.25}$$

which can be applied to a sub-body $\mathscr{P}$ of $\mathscr{B}$ such that $B_t(\mathscr{P})$ is a tetrahedron with vertices at (x^1, x^2, x^3), $(x^1 + \varepsilon/n^1, x^2, x^3)$, $(x^1, x^2 + \varepsilon/n^2, x^3)$, $(x^1, x^2, x^3 + \varepsilon/n^3)$, where (n^1, n^2, n^3) are the components of a fixed unit vector $\boldsymbol{n}$, and where ε is a small parameter. In this way by use of the mean value theorem of integral calculus, the following can be obtained

$$\begin{aligned}\frac{\varepsilon^3}{6n^1n^2n^3}[\rho(\boldsymbol{a} - \boldsymbol{b})](\boldsymbol{x}, t) = {}& \frac{\varepsilon^2}{2n^2n^3} \boldsymbol{t}(\boldsymbol{x}_1, t, -\boldsymbol{e}_1) + \frac{\varepsilon^2}{2n^1n^3} \boldsymbol{t}(\boldsymbol{x}_2, t, -\boldsymbol{e}_2) \\ & + \frac{\varepsilon^2}{2n^1n^2} \boldsymbol{t}(\boldsymbol{x}_3, t, -\boldsymbol{e}_3) + \frac{\varepsilon^2}{2n^1n^2n^3} \boldsymbol{t}(\hat{\boldsymbol{x}}, t, \boldsymbol{n})\end{aligned} \tag{2.2.26}$$

where $\boldsymbol{x}$ is a particular point in the tetrahedron, and where $\boldsymbol{x}_1, \boldsymbol{x}_2, \boldsymbol{x}_3$ and $\hat{\boldsymbol{x}}$ are particular points on the appropriate boundary triangles of the tetrahedron. Dividing equation (2.2.26) by ε^2, and then letting $\varepsilon \to 0$, there follows the relation

$$\boldsymbol{t}(\boldsymbol{x}, t, \boldsymbol{n}) = -\boldsymbol{t}(\boldsymbol{x}, t, -\boldsymbol{e}_i) n^i \tag{2.2.27}$$

where use has been made of the continuity condition on $\boldsymbol{t}$ and the fact that $\boldsymbol{x}_1, \boldsymbol{x}_2, \boldsymbol{x}_3$ and $\hat{\boldsymbol{x}}$ all approach $\boldsymbol{x}$ as $\varepsilon \to 0$. Equation (2.2.27) is taken to imply that $\boldsymbol{t}(\boldsymbol{x}, t, \cdot)$ may be extended into a linear function on $\mathscr{V}$, since the coefficients of n_i on the right-hand side of equation (2.2.27) depend only on $(\boldsymbol{x}, t)$. Thus, there is a tensor field

$$\mathbf{T} = \mathbf{T}(\boldsymbol{x}, t): \mathscr{V} \to \mathscr{V} \qquad \boldsymbol{x} \in B_t(\mathscr{P}) \qquad t \in \mathbf{I} \tag{2.2.28}$$

such that

$$\boldsymbol{t} = \mathbf{T}^{\mathrm{T}} \boldsymbol{n} \tag{2.2.29}$$

for all unit vectors $\boldsymbol{n}$ at each $(\boldsymbol{x}, t)$. The tensor $\mathbf{T}$, called the *Cauchy stress tensor*, is defined on the configurations of the material body under consideration and does not depend upon $\boldsymbol{n}$.

Setting $\boldsymbol{n}=\boldsymbol{e}_j$ in equation (2.2.27) gives

$$\boldsymbol{t}(\boldsymbol{n})=-\boldsymbol{t}(-\boldsymbol{n}) \tag{2.2.30}$$

and hence equation (2.2.27) can be expressed in the form

$$\boldsymbol{t}(\boldsymbol{x},t,\boldsymbol{n})=\boldsymbol{t}(\boldsymbol{x},t,\boldsymbol{e}^i)n_i=T_j{}^i(\boldsymbol{x},t)n_i\boldsymbol{e}^j \tag{2.2.31}$$

where

$$\boldsymbol{t}(\boldsymbol{x},t,\boldsymbol{e}^i)=T_j{}^i(\boldsymbol{x},t)\boldsymbol{e}^j \qquad (i=1,2,3). \tag{2.2.32}$$

The component form given by equation (2.2.31) for $\boldsymbol{t}(\boldsymbol{x},t,\boldsymbol{n})$ implies

$$T_i{}^j=\boldsymbol{e}_i\cdot(\mathbf{T}\boldsymbol{e}^j)=\boldsymbol{e}_j\cdot(\mathbf{T}^{\mathrm{T}}\boldsymbol{e}^i). \tag{2.2.33}$$

At a given point x, there exists in general three mutually perpendicular directions $\boldsymbol{q}_a$ $(a=1,2,3)$ such that

$$(\mathbf{T}-\sigma\mathbf{I})\boldsymbol{q}=0 \tag{2.2.34}$$

where σ_a $(a=1,2,3)$, the proper numbers of $\mathbf{T}$, are called the *principal stresses.* The stress tensor and the stress vector at x now have the representations

$$\mathbf{T}=\sum_{a=1}^{3}\sigma_a\boldsymbol{q}_a\otimes\boldsymbol{q}_a \qquad \boldsymbol{t}=\sum_{a=1}^{3}\sigma_a n_a\boldsymbol{q}_a \tag{2.2.35}$$

where $\boldsymbol{n}$ is the unit normal vector of equation (2.2.29). Thus the principal stresses act along the normal to the surface elements. The principal stresses σ_a $(a=1,2,3)$ of $\mathbf{T}$ are obtained by expanding

$$\det(\mathbf{T}-\sigma_a\mathbf{I})=0$$

to give the characteristic equation

$$\sigma_a{}^3-\sigma_a{}^2I_{\mathrm{T}}+\sigma_a II_{\mathrm{T}}-III_{\mathrm{T}}=0 \tag{2.2.36}$$

where

$$I_{\mathrm{T}}=\operatorname{tr}\mathbf{T} \tag{2.2.37}$$

$$II_{\mathrm{T}}=\tfrac{1}{2}(\operatorname{tr}\mathbf{T})^2-\tfrac{1}{2}\operatorname{tr}\mathbf{T}^2 \tag{2.2.38}$$

$$III_{\mathrm{T}}=\det\mathbf{T}=\tfrac{1}{6}(\operatorname{tr}\mathbf{T})^3-\tfrac{1}{2}\operatorname{tr}\mathbf{T}\operatorname{tr}\mathbf{T}^2+\tfrac{1}{3}\operatorname{tr}\mathbf{T}^3 \tag{2.2.39}$$

are the first, second and third invariants, respectively of $\mathbf{T}$. With the coordinate axes taken as the principal directions

$$[T_i{}^j]=\begin{bmatrix}\sigma_1 & 0 & 0\\ 0 & \sigma_2 & 0\\ 0 & 0 & \sigma_3\end{bmatrix} \tag{2.2.40}$$

from which it follows that

$$I_T = \sigma_1 + \sigma_2 + \sigma_3 \tag{2.2.41}$$

$$II_T = \sigma_1\sigma_2 + \sigma_2\sigma_3 + \sigma_3\sigma_1 \tag{2.2.42}$$

$$III_T = \sigma_1\sigma_2\sigma_3. \tag{2.2.43}$$

The deviator $\mathbf{T}'$ of the stress tensor $\mathbf{T}$ is defined as

$$\mathbf{T}' = \mathbf{T} - \tfrac{1}{3}(\operatorname{tr}\mathbf{T})\mathbf{I}. \tag{2.2.44}$$

The expression $T = \frac{1}{3}\operatorname{tr}\mathbf{T}$ is called the mean normal stress.

The stress vector can be resolved into a component $\boldsymbol{t}_{(n)}$ normal to the surface and a component $\boldsymbol{t}_{(s)}$ directed tangentially to S.

The component $\boldsymbol{t}_{(n)}$ is called the *normal stress* and is said to be *tensile* when positive and *compressive* when negative. The magnitude $\sigma_{(n)}$ of $\boldsymbol{t}_{(n)}$ is given by

$$\sigma_{(n)} = \boldsymbol{t}\cdot\boldsymbol{n} = \boldsymbol{n}\cdot(\mathbf{T}\boldsymbol{n}). \tag{2.2.45}$$

If the principal directions are taken as the coordinate axes it follows from equations (2.2.40) and (2.2.45) that

$$\sigma_{(n)} = \sigma_1 n_1^2 + \sigma_2 n_2^2 + \sigma_3 n_3^2. \tag{2.2.46}$$

The component $\boldsymbol{t}_{(s)}$ is called the *shearing stress*. By its definition the shearing stress $\boldsymbol{t}_{(s)}$ can be written

$$\boldsymbol{t}_{(s)} = \boldsymbol{t} - \boldsymbol{t}_{(n)}$$

and its magnitude $\tau_{(s)}$ is given by

$$\tau_{(s)} = |\boldsymbol{t} - (\boldsymbol{t}\cdot\boldsymbol{n})\boldsymbol{n}| = (|\boldsymbol{t}|^2 - \sigma_{(n)}^2)^{1/2} \tag{2.2.47}$$

it being noted that $\boldsymbol{t}_{(s)}$ and $\boldsymbol{t}_{(n)}$ are perpendicular. It follows that $\tau_{(s)}$ can be expressed in the form

$$\tau_{(s)}^2 = \sigma_1^2 n_1^2 + \sigma_2^2 n_2^2 + \sigma_3^2 n_3^2 - (\sigma_1 n_1^2 + \sigma_2 n_2^2 + \sigma_3 n_3^2)^2. \tag{2.2.48}$$

Noting that $n_1^2 + n_2^2 + n_3^2 = 1$, the stationary values of $\tau_{(s)}$ can be obtained by differentiating equation (2.2.48), it being found that $\tau_{(s)}$ possesses minimum values along the principal directions and maximum values on lines in the plane of two principal directions but bisecting the angle between them. For example, with

$$n_1 = 0 \qquad n_2^2 = n_3^2 = \tfrac{1}{2}$$

substitution in equation (2.2.48) gives

$$\tau_{(s)} = \tfrac{1}{2}|\sigma_2 - \sigma_3|. \tag{2.2.49}$$

The values of $\tau_{(s)}$ satisfying such relations are called the *principal shearing stresses*. The maximum shearing stress is thus equal to half the difference between the maximum and minimum principal stresses and acts on a plane which makes an angle of $\pi/8$ with the directions of these principle stresses.

Cauchy's stress tensor refers the stress to the instantaneous area of the surface element on which the stress vector acts.

A measure of stress can be defined by referring the force to unit initial area to give the *nominal stress*, $\mathbf{T}^0$, which finds application in the form of the *engineering stress* in tensile tests. Let the force on an element of area $\mathrm{d}a$ in the present configuration B_t of a body $\mathscr{B}$ be $\mathrm{d}\boldsymbol{F}$. This is related to the Cauchy stress tensor by

$$\mathrm{d}\boldsymbol{F}=\mathbf{T}^{\mathrm{T}}\boldsymbol{n}\,\mathrm{d}a=\mathbf{T}^{\mathrm{T}}\,\mathrm{d}\boldsymbol{a}. \tag{2.2.50}$$

By analogy with equation (2.2.50) a stress tensor $\mathbf{T}^0$ can be constructed which refers to the initial area $\mathrm{d}A$ in the reference configuration B_r: thus,

$$\mathrm{d}\boldsymbol{F}=(\mathbf{T}^0)^{\mathrm{T}}\mathrm{d}\boldsymbol{A}=(\mathbf{T}^0)^{\mathrm{T}}\boldsymbol{N}\,\mathrm{d}A. \tag{2.2.51}$$

Equations (2.2.50) and (2.2.51) can be combined to give the ratio $\mathrm{d}a/\mathrm{d}A$ which can be substituted for from equation (2.2.16) to give

$$\mathbf{T}^0=J\mathbf{F}^{-1}\mathbf{T} \tag{2.2.52}$$

where $\mathbf{T}^0$, the nominal stress tensor, is not, in general, a symmetric tensor, and it is therefore difficult to incorporate into constitutive relations between stress and strain.

The first Piola–Kirchhoff stress tensor is stated as

$$\mathbf{T}_{\mathrm{R}}=(\mathbf{T}^0)^{\mathrm{T}} \tag{2.2.53}$$

and the second Piola–Kirchhoff stress tensor as

$$\begin{aligned}\tilde{\mathbf{T}}&=J\mathbf{F}^{-1}\mathbf{T}(\mathbf{F}^{-1})^{\mathrm{T}}\\&=\mathbf{F}^{-1}\mathbf{T}_{\mathrm{R}}=\mathbf{T}^0(\mathbf{F}^{-1})^{\mathrm{T}}.\end{aligned} \tag{2.2.54}$$

It is evident from equation (2.2.54) that $\tilde{\mathbf{T}}=\tilde{\mathbf{T}}^{\mathrm{T}}$, i.e. $\tilde{\mathbf{T}}$ is symmetric if $\mathbf{T}$ is symmetric, and hence $\tilde{\mathbf{T}}$ is suitable for the formulation of constitutive equations.

2.2.4 Balance Principles: II, Energy and Entropy

Substituting for $\boldsymbol{t}$ in equation (2.2.25) from equation (2.2.29) and using the divergence theorem gives

$$\int_{R_t(\mathscr{P})}[\rho(\boldsymbol{a}-\boldsymbol{b})-\operatorname{div}\mathbf{T}]\,\mathrm{d}v=0. \tag{2.2.55}$$

The condition that equation (2.2.55) holds for all arbitrary material regions R_t, however small, leads to the relation

$$\rho\dot{\boldsymbol{v}}=\operatorname{div}\mathbf{T}+\rho\boldsymbol{b} \tag{2.2.56}$$

Multiplying equation (2.2.56) by $\boldsymbol{v}$ and noting that

$$\operatorname{div}(\mathbf{T}\boldsymbol{v}) = \boldsymbol{v} \cdot \operatorname{div} \mathbf{T} + \operatorname{tr}(\mathbf{TL})$$

gives

$$\tfrac{1}{2}\rho(\boldsymbol{v} \cdot \boldsymbol{v})^{\cdot} = \operatorname{div}(\mathbf{T}\boldsymbol{v}) - \operatorname{tr}(\mathbf{TL}) + \rho\boldsymbol{v} \cdot \boldsymbol{b}. \tag{2.2.57}$$

Integrating equation (2.2.57) throughout an arbitrary material region R_t and noting that $(\rho\, \mathrm{d}v)^{\cdot}$ is zero gives

$$\frac{\mathrm{d}}{\mathrm{d}t} K(R_t) = P(R_t) - \int_{R_t(\mathscr{P})} \operatorname{tr}(\mathbf{TL})\, \mathrm{d}v \tag{2.2.58}$$

where

$$K(R_t) = \int_{R_t(\mathscr{P})} \tfrac{1}{2}\rho\boldsymbol{v} \cdot \boldsymbol{v}\, \mathrm{d}v \tag{2.2.59}$$

defines the kinetic energy of the material occupying R_t and

$$P(R_t) = \int_{R_t(\mathscr{P})} \rho\boldsymbol{v} \cdot \boldsymbol{b}\, \mathrm{d}v + \int_{\partial R_t(\mathscr{P})} \boldsymbol{v} \cdot \boldsymbol{t}\, \mathrm{d}a \tag{2.2.60}$$

is the mechanical power, i.e. the rate that work is being done on the element. If $\mathbf{T}$ is symmetric it follows from equation (2.1.67) that

$$\operatorname{tr}(\mathbf{TL}) = \operatorname{tr}(\mathbf{TD}) + \operatorname{tr}(\mathbf{TW}) = \operatorname{tr}(\mathbf{TD}) \tag{2.2.61}$$

and this term is called the stress power per unit volume in B_t. Substituting for $\operatorname{tr}(\mathbf{TL})$ in equation (2.2.58) from equation (2.2.61) gives

$$\frac{\mathrm{d}}{\mathrm{d}t} K(R_t) = P(R_t) - M(R_t) \tag{2.2.62}$$

where

$$\mathrm{M}(R_t) = \int_{R_t(\mathscr{P})} \operatorname{tr}(\mathbf{TD})\, \mathrm{d}v. \tag{2.2.63}$$

Equation (2.2.62) is called the mechanical energy balance equation.

The fundamental postulate that all forms of energy are convertible from one form into another finds expression in the *first law of thermodynamics*,

$$\frac{\mathrm{d}}{\mathrm{d}t}[K(R_t) + U(R_t)] = H(R_t) + P(R_t) \tag{2.2.64}$$

where $U(R_t)$ is the internal energy and $H(R_t)$ is the rate at which thermal energy is being fed into the element. The internal energy $U(R_t)$ is an additive set function like mass and therefore has the properties (i) and (ii) given in §2.2.1. These conditions ensure the existence of a scalar field ε; hence $U(R_t)$ can be expressed in the form

$$U(R_t)=\int_{R_t(\mathscr{P})} \rho\varepsilon \,\mathrm{d}v \tag{2.2.65}$$

where ε is now seen to be the internal energy per unit mass. It is assumed that

$$H(R_t)=\int_{R_t(\mathscr{P})} \rho r \,\mathrm{d}v+\int_{\partial R_t(\mathscr{P})} h \,\mathrm{d}a \tag{2.2.66}$$

where h is the rate of flow of thermal energy per unit area into the material occupying the region R_t across the boundary ∂R_t, and r is a heat supply or source per unit mass and time. Substituting in equation (2.2.64) for $K(R_t)$, $P(R_t)$, $U(R_t)$ and $H(R_t)$ from equations (2.2.59), (2.2.60), (2.2.65) and (2.2.66), respectively gives, for the equation of total energy balance in integral form, the relation

$$\frac{\mathrm{d}}{\mathrm{d}t}\int_{R_t(\mathscr{P})} \rho(\varepsilon+\tfrac{1}{2}\boldsymbol{v}\cdot\boldsymbol{v}) \,\mathrm{d}v=\int_{R_t(\mathscr{P})} \rho(\boldsymbol{v}\cdot\boldsymbol{b}+r) \,\mathrm{d}v+\int_{\partial R_t(\mathscr{P})} (\boldsymbol{v}\cdot\boldsymbol{t}+h) \,\mathrm{d}a. \tag{2.2.67}$$

The difference between the mechanical energy balance equation (2.2.62) and the total energy balance equation (2.2.67), i.e.

$$\frac{\mathrm{d}}{\mathrm{d}t} U(R_t)=H(R_t)+M(R_t) \tag{2.2.68}$$

is referred to as the equation of thermal energy balance. Application of the thermal energy balance (equation (2.2.68)) to a tetrahedral subregion of B_t in a way similar to that described in §2.2.3, establishes the existence of a heat flux vector $\boldsymbol{q}$, which relates to the unit normal $\boldsymbol{n}$ according to the relation

$$h=\boldsymbol{q}\cdot\boldsymbol{n}. \tag{2.2.69}$$

The heat vector $\boldsymbol{q}$ is directed opposite to the flux of heat. Equation (2.2.69) can be regarded as the formulation of a concept which is as basic to heat flow as the concept of the stress vector is to the theory of stress. Substituting in equation (2.2.68) for $M(R_t)$, $U(R_t)$ and $H(R_t)$ from equations (2.2.63), (2.2.65) and (2.2.66), h being replaced from equation (2.2.69), gives

$$\frac{\mathrm{d}}{\mathrm{d}t}\int_{R_t(\mathscr{P})} \rho\varepsilon \,\mathrm{d}v=\int_{R_t(\mathscr{P})} [\rho r+\mathrm{tr}(\mathbf{TD})] \,\mathrm{d}v+\int_{\partial R_t(\mathscr{P})} \boldsymbol{q}\cdot\boldsymbol{n} \,\mathrm{d}a \tag{2.2.70}$$

which is the integral form of the equation of thermal energy balance.

When all the information necessary to characterise a material element is available it is said that the state of the system is known and the set of characterising variables are called state variables. If the state variables of a homogeneous system are independent of time the system is said to be in thermodynamic equilibrium.

In concluding this section, reference must be made to the two fundamental, but undefined concepts of absolute temperature θ and entropy S which together characterise the thermal state of a system in thermodynamic equilibrium.

Entropy has the two properties:

(i) The entropy of a system is the sum of the entropies of its parts (i.e. entropy, like mass, is an additive set function, the set being the material points).
(ii) The entropy of a system can change either by interacting with its surroundings or by changes within the system.

Thus,

$$\frac{\mathrm{d}S}{\mathrm{d}t}=\frac{dS_{\mathrm{I}}}{\mathrm{d}t}+\frac{\mathrm{d}S_{\mathrm{E}}}{\mathrm{d}t} \tag{2.2.71}$$

where the term on the left-hand side is the rate of increase of the total entropy, the first term on the right-hand side is the rate of entropy increase due to internal causes, and the second term is the rate of entropy increase due to interaction with the surroundings. The rate of entropy increase due to interaction with the surroundings is taken to be

$$\frac{\mathrm{d}S_{\mathrm{E}}}{\mathrm{d}t}=\int_{R_t(\mathscr{P})}\rho\frac{r}{\theta}\,\mathrm{d}v+\int_{\partial R_t(\mathscr{P})}\frac{\boldsymbol{q}\cdot\boldsymbol{n}}{\theta}\,\mathrm{d}a. \tag{2.2.72}$$

If η is the total entropy per unit mass and γ the entropy production per unit mass, then

$$S(\mathscr{P})=\int_{R_t(\mathscr{P})}\rho\eta\,\mathrm{d}v \qquad \frac{\mathrm{d}S_{\mathrm{I}}}{\mathrm{d}t}=\int_{R_t(\mathscr{P})}\rho\gamma\,\mathrm{d}v. \tag{2.2.73}$$

Substituting in equation (2.2.71) from equations (2.2.72) and (2.2.73) gives

$$\frac{\mathrm{d}}{\mathrm{d}t}\int_{R_t(\mathscr{P})}\rho\eta\,\mathrm{d}v=\int_{R_t(\mathscr{P})}\rho\gamma\,\mathrm{d}v+\int_{R_t(\mathscr{P})}\rho\frac{r}{\theta}\,\mathrm{d}v+\int_{\partial R_t(\mathscr{P})}\frac{\boldsymbol{q}\cdot\boldsymbol{n}}{\theta}\,\mathrm{d}a. \tag{2.2.74}$$

The *second law of thermodynamics* states that

$$\frac{\mathrm{d}S_{\mathrm{I}}}{\mathrm{d}t}\geqslant 0 \qquad \text{and hence } \gamma\geqslant 0. \tag{2.2.75}$$

If $\mathrm{d}S_{\mathrm{I}}/\mathrm{d}t$ or γ are zero the flow process is said to be reversible and if they are positive the process is said to be irreversible.

2.2.5 Field Equations

In §§ 2.2.1 and 2.2.4 six general balance principles which govern the motions of all body manifolds have been formulated. These principles are stated in integral forms by equations (2.2.2), (2.2.9), (2.2.14), (2.2.15), (2.2.62), (2.2.67) and (2.2.70). The differential forms of the balance principles will be derived in this section. These forms are called the *field equations.*

Since equation (2.2.23) holds for all arbitrary material regions R_t, however small, it follows that

$$\dot{\rho}+\rho \operatorname{div} \boldsymbol{v}=0 \tag{2.2.76}$$

which is the field equation form of the conservation of mass. From equation (2.1.16)

$$\dot{\rho}=\frac{\partial \rho}{\partial t}+(\operatorname{grad} \rho)\cdot \boldsymbol{v} \tag{2.2.77}$$

which can be used to rearrange equation (2.2.76) into the form

$$\frac{\partial \rho}{\partial t}+\operatorname{div}(\rho \boldsymbol{v})=0. \tag{2.2.78}$$

Equations (2.2.76) and (2.2.78) are generally referred to as the *equations of continuity*. The balance law of equation (2.2.2) and the field equation (2.2.76) are equivalent expressions of the *principle of mass conservation.* It is evident from equation (2.1.82) that the conservation of mass can be expressed in the form

$$\rho_{\mathrm{r}}=J\rho \tag{2.2.79}$$

where ρ_{r} and ρ are the mass densities in the reference and current configuration, respectively.

Substituting for $\boldsymbol{t}$ in equation (2.2.25) from equation (2.2.29) and using the divergence theorem gives

$$\int_{R_t(\mathscr{P})} [\rho(\dot{\boldsymbol{v}}-\boldsymbol{b})-\operatorname{div}\mathbf{T}]\,\mathrm{d}v=0. \tag{2.2.80}$$

The condition that equation (2.2.80) holds for all arbitrary material regions R_t, however small, leads to the field equation form of the balance of linear momentum:

$$\rho\dot{\boldsymbol{v}}=\operatorname{div}\mathbf{T}+\rho\boldsymbol{b}=\rho\boldsymbol{a} \tag{2.2.81}$$

which is Cauchy's equation of motion. Equation (2.2.76) can be used to give the alternative form

$$(\rho\boldsymbol{v})^{\boldsymbol{\cdot}}+\rho\boldsymbol{v}\operatorname{div}\boldsymbol{v}=\operatorname{div}\mathbf{T}+\rho\boldsymbol{b}. \tag{2.2.82}$$

Use of equation (2.2.79) allows equation (2.2.82) to be expressed in terms of the referential description to give

$$\rho_{\mathrm{r}}\dot{\boldsymbol{v}}=J\operatorname{div}\mathbf{T}+\rho_{\mathrm{r}}\boldsymbol{b} \tag{2.2.83}$$

where ρ_{r} is the mass density in the reference configuration.

Substituting from equation (2.2.29) for $\boldsymbol{t}$ in equation (2.2.14) and rearranging by way of equations (2.2.22) and (2.2.76) gives

$$\int_{R_t(\mathscr{P})} \rho \boldsymbol{x} \wedge (\dot{\boldsymbol{v}} - \boldsymbol{b})\, \mathrm{d}v = \int_{\partial R_t(\mathscr{P})} \boldsymbol{x} \wedge (\mathbf{T}^{\mathrm{T}}\boldsymbol{n})\, \mathrm{d}a. \tag{2.2.84}$$

From equation (2.2.84) the following relation can be obtained

$$\int_{R_t(\mathscr{P})} \rho[\boldsymbol{x}\otimes(\dot{\boldsymbol{v}} - \boldsymbol{b}) - (\dot{\boldsymbol{v}} - \boldsymbol{b})\otimes\boldsymbol{x}]\, \mathrm{d}v = \int_{\partial R_t(\mathscr{P})} [\boldsymbol{x} \otimes (\mathbf{T}^{\mathrm{T}}\boldsymbol{n}) - (\mathbf{T}^{\mathrm{T}}\boldsymbol{n}) \otimes \boldsymbol{x}]\, \mathrm{d}a \tag{2.2.85}$$

where use has been made of the vector identity $\boldsymbol{a} \wedge (\boldsymbol{b} \wedge \boldsymbol{c}) = (\boldsymbol{b} \otimes \boldsymbol{c} - \boldsymbol{c} \otimes \boldsymbol{b})\boldsymbol{a}$. Using the divergence theorem in the form

$$\int_{\partial R_t(\mathscr{P})} \boldsymbol{u} \otimes (\mathbf{T}^{\mathrm{T}}\boldsymbol{n})\, \mathrm{d}a = \int_{R_t(\mathscr{P})} [\boldsymbol{u} \otimes \operatorname{div} \mathbf{T} + (\operatorname{grad} \boldsymbol{u})\mathbf{T}]\, \mathrm{d}v$$

$$\int_{\partial R_t(\mathscr{P})} (\mathbf{T}^{\mathrm{T}}\boldsymbol{n})\, \mathrm{d}a = \int_{R_t(\mathscr{P})} \operatorname{div} \mathbf{T}\, \mathrm{d}v$$

where $\boldsymbol{u}$ is a fixed arbitrary vector, equation (2.2.85) can be rearranged into the form

$$\int_{R_t(\mathscr{P})} \{\boldsymbol{x} \otimes [\rho(\dot{\boldsymbol{v}} - \boldsymbol{b}) - \operatorname{div} \mathbf{T}] - [\rho(\dot{\boldsymbol{v}} - \boldsymbol{b}) - \operatorname{div} \mathbf{T}] \otimes \boldsymbol{x} + \mathbf{T} - \mathbf{T}^{\mathrm{T}}\}\, \mathrm{d}v = \mathbf{O}. \tag{2.2.86}$$

The condition that equation (2.2.86) holds for all arbitrary material regions R_t, however small, gives by way of equation (2.2.81), the condition

$$\mathbf{T} = \mathbf{T}^{\mathrm{T}} \tag{2.2.87}$$

$$T^{ij} = T^{ji} \tag{2.2.88}$$

thus establishing the symmetry of the stress matrix in general without recourse to any assumption of equilibrium or of uniformity of the stress distribution.

In equation (2.2.15), the couple stress vector $\boldsymbol{c}$ at point $\boldsymbol{x}$ is a linear transformation of the unit normal $\boldsymbol{n}$; thus

$$\boldsymbol{c} = \mathbf{M}^{\mathrm{T}}\boldsymbol{n} \tag{2.2.89}$$

where $\mathbf{M}$ is referred to as the *couple stress tensor*. The concepts underlying the formulation of equation (2.2.89) paralleling those underlying those of the stress vector. Substituting in equation (2.2.15) from equations (2.2.29) and (2.2.89) for $\boldsymbol{t}$ and $\boldsymbol{c}$, respectively and setting

$$\boldsymbol{h} = \boldsymbol{m} + \boldsymbol{x} \wedge \boldsymbol{v} \tag{2.2.90}$$

allows equation (2.2.15) to be rearranged, by way of equations (2.2.22) and

(2.2.76), to give

$$\int_{R_t(\mathscr{P})} \rho \boldsymbol{x} \wedge (\dot{\boldsymbol{v}} - \boldsymbol{b})\, \mathrm{d}v + \int_{R_t(\mathscr{P})} \rho(\dot{\boldsymbol{m}} - \boldsymbol{G})\, \mathrm{d}v = \int_{\partial R_t(\mathscr{P})} \boldsymbol{x} \wedge (\mathbf{T}^{\mathrm{T}}\boldsymbol{n})\, \mathrm{d}a + \int_{\partial R_t(\mathscr{P})} \mathbf{M}^{\mathrm{T}}\boldsymbol{n}\, \mathrm{d}a \tag{2.2.91}$$

which is to be compared with equation (2.2.84). Equation (2.2.91) can be rearranged in the same way as equation (2.2.84), followed by use of the divergence theorem, to give

$$\rho\dot{\boldsymbol{m}} = \operatorname{div} \mathbf{M} + \rho \boldsymbol{G} - 2\boldsymbol{\pi} \tag{2.2.92}$$

where $\boldsymbol{\pi}$ is the axial vector of $\mathbf{T} - \mathbf{T}^{\mathrm{T}}$, and

$$\boldsymbol{m} = \boldsymbol{h} - \boldsymbol{x} \wedge \boldsymbol{v} \tag{2.2.93}$$

is termed the internal angular momentum. The terms $\mathbf{M}$ and $\boldsymbol{G}$ are usually of electromagnetic origin and in most engineering applications are very much smaller than the stress $\mathbf{T}$. Thus, provided the coupling between the moment of linear momentum and the internal angular momentum is weak, the stress tensor will usually be very nearly symmetric.

Equation (2.2.70) can be rearranged by way of equations (2.2.19), (2.2.76) and the divergence theorem to give

$$\int_{R_t(\mathscr{P})} [\rho\dot{\varepsilon} - \rho r - \operatorname{tr}(\mathbf{TD}) - \operatorname{div} \boldsymbol{q}]\, \mathrm{d}v = 0$$

which, since it must hold for all arbitrary material regions R_t, however small, leads to the relation

$$\rho\dot{\varepsilon} = \operatorname{tr}(\mathbf{TD}) + \operatorname{div} \boldsymbol{q} + \rho r \tag{2.2.94}$$

which is the field equation form of the balance of thermal energy.

Equation (2.2.74) can be rearranged by way of equations (2.2.19) and (2.2.76) to give

$$\int_{R_t(\mathscr{P})} \left(\rho\dot{\eta} - \rho\frac{r}{\theta} - \rho\gamma\right) \mathrm{d}v = \int_{\partial R_t(\mathscr{P})} \frac{\boldsymbol{q} \cdot \boldsymbol{n}}{\theta}\, \mathrm{d}a \tag{2.2.95}$$

which can be further rearranged by way of the divergence theorem to give

$$\int_{R_t(\mathscr{P})} \left(\rho\dot{\eta} - \rho\frac{r}{\theta} - \rho\gamma - \operatorname{div}\left(\frac{\boldsymbol{q}}{\theta}\right)\right) \mathrm{d}v = 0. \tag{2.2.96}$$

As the integral must be zero for any arbitrary material region R_t, the integrand must be zero, giving

$$\rho\gamma = \rho\dot{\eta} - \rho\frac{r}{\theta} - \operatorname{div}\left(\frac{\boldsymbol{q}}{\theta}\right) \tag{2.2.97}$$

which is the field equation form of equation (2.2.74). But by the second law of thermodynamics $\gamma \geqslant 0$, and thus

$$\rho\dot{\eta} - \rho\frac{r}{\theta} - \operatorname{div}\left(\frac{\boldsymbol{q}}{\theta}\right) \geqslant 0. \tag{2.2.98}$$

This relation is called the Clausius–Duhem inequality after Clausius (1854) and Duhem (1901). It is to be noted, however, that reservations regarding the generality of the Clausius–Duhem inequality have been expressed by Green and Naghdi (1977). Eliminating ρr between equations (2.2.94) and (2.2.97) gives

$$\rho\gamma\theta \equiv \rho\theta\dot{\eta} - \rho\dot{\varepsilon} + \operatorname{tr}(\mathbf{TD}) + \frac{1}{\theta}\boldsymbol{q}\cdot \operatorname{grad}\theta \geqslant 0. \tag{2.2.99}$$

2.2.6 The General Balance Equation

It is evident from the preceeding sections that the physical principles of continuum physics considered in these sections all take the form of integral equations. From these balance principles there follow, by way of suitable conditions of continuity, the corresponding field equations, and in this respect the integral equation form of the physical balance principles is to be regarded as the fundamental form.

From equations (2.2.2), (2.2.9), (2.2.14), (2.2.15), (2.2.62), (2.2.67) and (2.2.70), it is evident that all the physical balance principles can be accommodated within the general balance equation

$$\frac{\mathrm{d}}{\mathrm{d}t}\int_{R_t(\mathscr{P})} \rho\psi\,\mathrm{d}v = \int_{R_t(\mathscr{P})} \rho s\,\mathrm{d}v - \int_{\partial R_t(\mathscr{P})} \boldsymbol{j}\cdot\boldsymbol{n}\,\mathrm{d}a \tag{2.2.100}$$

where ψ is some physical quantity per unit mass, which may be a scalar, vector or tensor over the current configuration B_t of a body $\mathscr{B}$, s the supply of ψ per unit mass, and $-\boldsymbol{j}\cdot\boldsymbol{n}$ the influx of ψ per unit area in B_t. The various balance equations can be obtained using the forms of $-\boldsymbol{j}\cdot\boldsymbol{n}$ and s given in table 2.1.

Taking the material region R_t which corresponds at time t to the configuration of the material region R_r and using Green's theorem gives equation (2.2.100) in the form

$$\int_{R_t(\mathscr{P})} (\rho\psi\,\mathrm{d}v)^{\boldsymbol{\cdot}} = \int_{R_t(\mathscr{P})} (\rho s - \operatorname{div}\boldsymbol{j})\,\mathrm{d}v \tag{2.2.101}$$

and noting that the conservation of mass requires $(\rho\,\mathrm{d}v)^{\boldsymbol{\cdot}} = 0$, together with the condition that equation (2.2.101) holds for all material regions R_r, however small, leads to the field equation form of the general balance equation

$$\rho\dot{\psi} = \rho s - \operatorname{div}\boldsymbol{j}. \tag{2.2.102}$$

Table 2.1. The flux and source terms appearing in the general balance equation.

Physical quantity	Physical quantity per unit mass ψ	Supply of ψ per unit mass s	Influx of ψ per unit area in B_t $-\boldsymbol{j}\cdot\boldsymbol{n}$
Mass	1	0	0
Linear momentum	$\boldsymbol{v}$	$\boldsymbol{b}$	$\boldsymbol{t}$
Angular momentum	$\boldsymbol{x}\wedge\boldsymbol{v}$	$\boldsymbol{x}\wedge\boldsymbol{b}$	$\boldsymbol{x}\wedge\boldsymbol{t}$
Total angular momentum	$\boldsymbol{x}\wedge\boldsymbol{v}+\boldsymbol{m}$	$\boldsymbol{x}\wedge\boldsymbol{b}+\boldsymbol{G}$	$\boldsymbol{x}\wedge\boldsymbol{t}(\boldsymbol{n})+\boldsymbol{c}(\boldsymbol{n})$
Energy	$\varepsilon+\frac{1}{2}\mathrm{v}\cdot\boldsymbol{v}$	$\boldsymbol{v}\cdot\boldsymbol{b}+r$	$\boldsymbol{v}\cdot\boldsymbol{t}+h$
Entropy	η	$\gamma+r/\theta$	$(\boldsymbol{q}\cdot\boldsymbol{n})/\theta$

2.3 PRINCIPLES OF CONSTITUTIVE THEORY

2.3.1 General Constitutive Equation

For a given body there must exist some relation between the dynamical state of the body at some instant, the kinematical state of the body at the same instant, and possibly the kinematical state of the body at all instants of the body's past history. The equations which express the relation between the kinematical and dynamical variables are called constitutive equations. In formulating constitutive equations certain general principles must be satisfied. The most important of them are the principles of determinism, local action and objectivity.

Regarding the body force $\boldsymbol{b}$ and heat supply r as known, the field equations (2.2.76), (2.2.81) and (2.2.94), together with equation (2.2.87), provide eight scalar relations between the seventeen scalar fields consisting of ρ, ε and the components of $\boldsymbol{v}$, $\boldsymbol{q}$ and $\mathbf{T}$. Hence, in general, the field equations do not by themselves determine the behaviour of a material body.

Consider a motion of a material body $\mathscr{B}$ to be described by equation (2.1.12). Let the density in some reference configuration be prescribed and consider only a symmetric state of stress $\mathbf{T}$. In this way the spatial continuity equation (2.2.76) and the symmetry condition of equation (2.2.87) are satisfied. By appropriate choice of the body force $\boldsymbol{b}$ it can be ensured that the equation (2.2.81) of motion also holds. There remains only the energy equation (2.2.94) to be satisfied. This implies that the way to obtain the further relations which must be adjoined to the field equations to give a complete system is to specify $\mathbf{T}$, ε and $\boldsymbol{q}$. However, if the prescription of these fields depends only on a knowledge of the motion of $\mathscr{B}$ then the field equations would eventually furnish five scalar relations for only four scalar fields identified as the density and the components of $\boldsymbol{v}$. Hence, it is evident that $\mathbf{T}$, ε and $\boldsymbol{q}$ are not, in general, fully determined by the motion of $\mathscr{B}$ but depend also upon a scalar field which is not present as a dependent variable in the field equations. Having regard to the physical significance of ε and $\boldsymbol{q}$, the additional scalar field is expected to measure the heat content and is identified as the temperature θ defined on the configurations of $\mathscr{B}$. Thus, for a simple thermomechanical process, experience has been used to identify the motion of the material points of a body and their temperatures, that is

$$\boldsymbol{x}=\boldsymbol{\chi}(\boldsymbol{X},t) \qquad \theta=\theta(\boldsymbol{X},t) \tag{2.3.1}$$

as the independent constitutive variables, and the energy equation (2.2.99) identifies $\mathbf{T}$, $\boldsymbol{q}$, ε and η as the dependent constitutive variables.

These considerations give rise to the following basic constitutive statement:

For a simple thermomechanical process, the *principle of determinism* asserts that at the point x occupied at time t by an arbitrary material point of a body $\mathscr{B}$, the thermomechanical constitutive functions $\mathbf{T}$, $\boldsymbol{q}$, ε and η are uniquely determined by the history of the motion and temperature of the body $\mathscr{B}$.

The principle of determinism applied to a simple thermomechanical process gives the four constitutive equations:

$$\mathbf{T}(\boldsymbol{X},t)=\mathbf{G}[\boldsymbol{\chi}^t(s),\theta^t(s),\boldsymbol{X},t] \tag{2.3.2}$$

$$\boldsymbol{q}(\boldsymbol{X},t)=\boldsymbol{L}[\boldsymbol{\chi}^t(s),\theta^t(s),\boldsymbol{X},t] \tag{2.3.3}$$

$$\varepsilon(\boldsymbol{X},t)=\mathscr{E}[\boldsymbol{\chi}^t(s),\theta^t(s),\boldsymbol{X},t] \tag{2.3.4}$$

$$\eta(\boldsymbol{X},t)=\mathscr{H}[\boldsymbol{\chi}^t(s),\theta^t(s),\boldsymbol{X},t] \tag{2.3.5}$$

where $\mathbf{G}$ and $\boldsymbol{L}$ are tensor-valued and vector-valued *functionals*, respectively, and $\mathscr{E}$ and $\mathscr{H}$ are scalar-valued functionals defined over the fields of the real functions $\boldsymbol{\chi}^t(s)$ and $\theta^t(s)$. For the present time t, the function

$$\boldsymbol{\chi}^t(s)=\boldsymbol{\chi}(t-s) \tag{2.3.6}$$

describes the fixed history of the entire motion of the body and

$$\theta^t(s)=\theta(t-s)$$

is the function describing the fixed history of the entire temperature field of the body. The principle of determinism excludes the dependence of the material behaviour at $\boldsymbol{X}$ on any point outside the body and any future events. That is, equations (2.3.2) to (2.3.5) are only applicable to all $\boldsymbol{X}\in\mathscr{B}$ and only valid for all $\tau\leqslant t$. The domain $(0\leqslant s<\infty)$ of the elapsed time s can, if required, be indicated by using the more explicit notation

$$\mathbf{T}(\boldsymbol{X},t)=\underset{s=0}{\overset{\infty}{\mathbf{G}}}[\boldsymbol{\chi}^t(s),\theta^t(s),\boldsymbol{X},t] \tag{2.3.7}$$

due to Volterra (1930).

Continuum mechanics is essentially concerned with contact forces which are determined by the conditions in the immediate neighbourhood of the point of application, and therefore to exclude possible actions at a distance from constitutive equations it is proposed to consider the response at a body point X. This procedure gives rise to the *principle of local action*, according to which the response at a body point X is determined if the conditions are known in an arbitrarily small neighbourhood $\mathcal{N}(X)$ of X, the motion outside being disregarded. Consider a small neighbourhood $\mathcal{N}(X)$ of X and let Z denote a typical body point in $\mathcal{N}(X)$. The principal of local action requires equation (2.3.2) to be expressed in the form

$$\mathbf{T}(t)=\mathbf{G}[\boldsymbol{\chi}(Z,\tau),\theta(Z,\tau),\boldsymbol{X},t] \tag{2.3.8}$$

where $\tau=t-s$.

Of particular interest are the *simple materials* considered by Noll (1958) which are characterised by the absence of a microstructure. Noll develops the general theory of simple materials by way of two basic assertions of invariance. These are the *principle of objectivity* and various conditions relating to *material symmetry*.

2.3.2 Principle of Objectivity

Measurements of distances and time intervals can only be made relative to a frame of reference which is given for all time. Material properties are intrinsic properties of a body and therefore effects produced in a material by a given set of causes must be independent of the observer, i.e. the frame of reference to which the effects are referred. These physical considerations give rise to the principal of material objectivity according to which material properties are invariant to observer transformations.

The principle of objectivity requires that the form of the functional $\mathbf{G}$ should be the same as any two objectively equivalent motions. Thus, for a process $(\boldsymbol{\chi}^*,\mathbf{T}^*)$ equivalent to $(\boldsymbol{\chi},\mathbf{T})$

$$\mathbf{T}(t)=\mathbf{G}[\boldsymbol{\chi}(Z,\tau)\,\boldsymbol{X},t] \tag{2.3.9}$$

and

$$\mathbf{T}^*(t^*)=\mathbf{G}[\boldsymbol{\chi}^*(Z,\tau^*),\boldsymbol{X},t^*] \tag{2.3.10}$$

for all $\mathbf{Q}(t)$, $\boldsymbol{c}(t)$ and a such that

$$\boldsymbol{x}^*(Z,\tau^*)=\boldsymbol{c}(t)+\mathbf{Q}(\tau)\boldsymbol{x}(Z,\tau) \tag{2.3.11}$$

$$\tau^*=\tau-a \tag{2.3.12}$$

where τ is now being used as the time variable and all variable body points Z lie within the body $\mathcal{B}$.

The restrictions on the form of the functional $\mathbf{G}$ imposed by the principle of objectivity can be examined by way of three special transformations of the frame $\boldsymbol{x}$. These special transformations enable the position, the time and the orientation of a moving observer to be transformed away.

Rigid translation of the spatial frame. Consider a small neighbourhood $\mathcal{N}(X)$ of X and let Z denote a typical body point in $\mathcal{N}(X)$. Let $\boldsymbol{\kappa}$ be any configuration of $\mathcal{N}(X)$. The vector-valued function $\boldsymbol{\kappa}_X$ defined by the body point difference

$$\boldsymbol{Z}=\boldsymbol{\kappa}_X(Z)=\boldsymbol{\kappa}(Z)-\boldsymbol{\kappa}(X) \tag{2.3.13}$$

is called the *localisation* in $\mathcal{N}(X)$ of the configuration $\boldsymbol{\kappa}$, it being noted that $\boldsymbol{Z}$ is the position vector of the body point Z when the position of X is taken as origin. Replacing $\boldsymbol{\kappa}$ by the configuration at time τ in a motion $\boldsymbol{\chi}$ gives rise to the concept of localised motion $\boldsymbol{\chi}_X$ which describes the motion of a neighbourhood of X as it appears to an observer moving with the body point X, $\boldsymbol{\chi}_X$ being defined by the relation

$$\boldsymbol{\chi}_X(Z,\tau)=\boldsymbol{\chi}(Z,\tau)-\boldsymbol{\chi}(X,\tau). \tag{2.3.14}$$

The particular conditions for which $\mathbf{Q}(\tau)=\mathbf{I}$, $a=0$, $\boldsymbol{c}(\tau)=-\boldsymbol{\chi}(X,t)$ determine a rigid translation of the spatial frame of reference in which the fixed body point X at time t remains at the origin. Inserting these conditions into equations (2.3.11) and (2.3.12) gives

$$\begin{aligned}\boldsymbol{\chi}^*(Z,\tau)&=\boldsymbol{\chi}(Z,\tau)-\boldsymbol{\chi}(X,\tau)\\ \tau^*=\tau \qquad & \mathbf{T}^*(t)=\mathbf{T}(t)\end{aligned} \tag{2.3.15}$$

from which it follows that $\boldsymbol{\chi}^*$ is just the localisation $\boldsymbol{\chi}_X$ at X of the motion $\boldsymbol{\chi}$ as defined by equation (2.3.14). Insertion of these conditions into equation (2.3.10) gives the constitutive equation (2.3.9) in the reduced form

$$\begin{aligned}\mathbf{T}(t)&=\mathbf{G}[\boldsymbol{\chi}(Z,\tau)-\boldsymbol{\chi}(X,\tau),\boldsymbol{X},t]\\ &=\mathbf{G}[\boldsymbol{\chi}_X(Z,\tau),\boldsymbol{X},t].\end{aligned} \tag{2.3.16}$$

Shift of time. Select $\mathbf{Q}(\tau)=\mathbf{I}$, $\boldsymbol{c}(\tau)=\boldsymbol{o}$ and $a=t$. This particular shift of timescale makes the present time t the reference time after the change of frame; i.e. setting $t=a$ in equations (2.1.19) and (2.3.12) gives

$$\begin{aligned}\tau^*=\tau-t=-s \qquad t^*=t-t=0 \qquad \mathbf{T}^*(0)=\mathbf{T}(t)\\ \boldsymbol{\chi}^*(Z,\tau^*)=\boldsymbol{\chi}(Z,t+\tau^*).\end{aligned} \tag{2.3.17}$$

Insertion of these conditions into equation (2.3.10) gives

$$\mathbf{T}(t)=T^*(0)=\mathbf{G}[\boldsymbol{\chi}(Z,t+\tau^*),\boldsymbol{X},0] \tag{2.3.18}$$

from which it is evident that $\mathbf{G}$ is independent of t. Combining this result with that of equation (2.3.16) gives

$$\mathbf{T}(t)=\mathbf{G}[\boldsymbol{\chi}(Z,t+\tau^*)-\boldsymbol{\chi}(X,t+\tau^*);\boldsymbol{X}] \tag{2.3.19}$$

where it is to be noted that, for all $s \geqslant 0$,

$$\chi_X^*(Z, \tau^*) = \chi_X(Z, \tau) = \chi_X^t(Z, s). \tag{2.3.20}$$

From equation (2.3.19) it can be concluded that **G** is a functional in the history of the relative motions from the present.

Having regard to the principle of local action, for the two equivalent motions $\boldsymbol{\chi}$ and $\boldsymbol{\chi}^*$ to coincide in some neighbourhood $\mathcal{N}(X)$ for all times $\tau \leqslant t$, it is evident from equation (2.3.19) that **G** is a functional in the history of the localisation $\boldsymbol{\chi}_X^t$ in $\mathcal{N}(X)$ of the motion $\boldsymbol{\chi}$ for $s \geqslant 0$. Hence, from equation (2.3.19), there can be obtained the approximation

$$\boldsymbol{\chi}_X^t(\boldsymbol{Z}, s) \approx \mathbf{F}^t(s)\boldsymbol{Z} = \mathbf{F}^t(s)\,\mathrm{d}\boldsymbol{X} \tag{2.3.21}$$

where the approximation can be made as precise as is required by confining the position vector $\boldsymbol{Z}$, defined by equation (2.3.13), to a small enough neighbourhood of the origin. Thus, application of the principle of objectivity reduces the constitutive equation (2.3.19) for a simple material to the form

$$\mathbf{T} = \mathbf{G}[\mathbf{F}^t(s), \boldsymbol{X}, \mathrm{d}\boldsymbol{X}] \tag{2.3.22}$$

from which it is evident that dependence on $\boldsymbol{\chi}^t$ has been eliminated. The fact that **G** may depend on the past values of **F** is taken to imply that in general a simple material may have *memory effects.*

Rigid rotations of the spatial frame. Let $\boldsymbol{c}(\tau) = \boldsymbol{o}$, $a = 0$, and take $\mathbf{Q}(t)$ to be arbitrary. This represents an arbitrary time-dependent rotation of the spatial frame of reference. In this rotation, it follows from equation (2.1.26) that the stress transforms as

$$\mathbf{T}^*(\boldsymbol{X}, t) = \mathbf{Q}(t)\mathbf{T}(\boldsymbol{X}, t)\mathbf{Q}(t)^{\mathrm{T}} \tag{2.3.23}$$

which when applied to equation (2.3.19) gives the identity

$$\mathbf{Q}(0)\mathbf{G}[\boldsymbol{\chi}_X^t(Z, s), \boldsymbol{X}]\mathbf{Q}(0)^{\mathrm{T}} = \mathbf{G}[\mathbf{Q}(s)\boldsymbol{\chi}_X^t(Z, s), \boldsymbol{X}] \tag{2.3.24}$$

for any history of rotations

$$\mathbf{Q}(s) \in \mathscr{E}\mathscr{O}(\mathscr{V}) \qquad s \in [0, \infty].$$

The condition represented by equation (2.3.24) is the restriction that must be imposed on the functional **G** in order that it is objective.

Thus, the restrictions on the form of the functionals imposed by the principal of objectivity gives equation (2.3.22), subject to the condition of equation (2.3.24), as the most general constitutive equation for the stress in purely mechanical theories of non-polar continuous materials.

2.3.3 Material Symmetry

All directional vectors $\mathrm{d}\boldsymbol{X}$ originating from $\boldsymbol{X}$ in three dimensions can be decomposed along three vectors $\boldsymbol{K}_\alpha$ ($\alpha = 1, 2, 3$) where $\boldsymbol{K}_\alpha$ is a material frame characterising the local reference configuration B_r. Hence the constitutive equation (2.3.22) can be expressed in the general form

$$\mathbf{T}(t) = \mathbf{G}[\mathbf{F}^t(s), X, \boldsymbol{K}_\alpha]. \tag{2.3.25}$$

Consider a body $\mathscr{B}$ which is subjected to a given deformation in relation to a given reference configuration. Now apply the same deformation in relation to an alternative choice of reference configuration and it follows from equation (2.3.25) that the stress produced at a representative body point X will generally be different, the appearance of $\boldsymbol{K}_\alpha$ in the constitutive equation serving to stress this possible directional dependence of the material properties at a material point X, that is the anisotropic properties of the material.

If the intended reference configuration is clearly identified, then the explicit dependence of $\mathbf{G}$ upon $\boldsymbol{K}_\alpha$ can be replaced by an implicit dependence, the functional being replaced, for example, by $\mathbf{G}_{\boldsymbol{\kappa}}$, such that

$$\mathbf{T}(t) = \mathbf{G}_{\boldsymbol{\kappa}}(\mathbf{F}^t(s)) \tag{2.3.26}$$

where the subscript $\boldsymbol{\kappa}$ denotes the reference configuration B_r spanned by $\boldsymbol{K}_\alpha$.

Structural ordering of the material points in a body gives rise to certain crystallographic symmetries in the material properties. These crystallographic symmetries in the material properties imply the existence of particular changes of reference configuration which leave the stress at X arising from an arbitrary deformation invariant. The collection of such transformations is a measure of the degree of symmetry possessed by the material. Constitutive equations must be form invariant with respect to material symmetry transformations.

As a basis for developing the concepts underlying the symmetry properties of a material, Noll (1958) considered the effect of a change of reference configuration B_r on the deformation gradient $\mathbf{F}$ and hence upon the general constitutive equation. With this as basis, it follows from equation (2.3.26) that for a simple material in the absence of thermal effects,

$$\mathbf{T}(t) = \mathbf{G}_{\boldsymbol{\kappa}}(\mathbf{F}^t(s)) \qquad \mathbf{T}(t) = \mathbf{G}_{\hat{\boldsymbol{\kappa}}}(\hat{\mathbf{F}}^t(s)) \tag{2.3.27}$$

where $\boldsymbol{\kappa}$ and $\hat{\boldsymbol{\kappa}}$ imply an arbitrary pair of reference configurations B_r and $\hat{B}_r$ of a body $\mathscr{B}$ and the deformation gradients $\mathbf{F}$ and $\hat{\mathbf{F}}$ relate the current configuration B_t of $\mathscr{B}$ to B_r and $\hat{B}_r$. Equation (2.1.36) gives

$$\mathbf{F} = F^i{}_\alpha \boldsymbol{e}_i \otimes \boldsymbol{E}^\alpha \qquad \hat{\mathbf{F}} = \hat{F}^i{}_\alpha \boldsymbol{e}_i \otimes \hat{\boldsymbol{E}}^\alpha \tag{2.3.28}$$

and hence

$$\begin{aligned}\mathbf{F} &= \hat{F}^i{}_\alpha \frac{\partial \hat{X}^\beta}{\partial X^\alpha} \boldsymbol{e}_i \otimes \boldsymbol{E}^\alpha \\ &= (\hat{F}^i{}_\beta \boldsymbol{e}_i \otimes \hat{\boldsymbol{E}}^\beta)\left(\frac{\partial \hat{X}_\gamma}{\partial X_\alpha} \hat{\boldsymbol{E}}_\gamma \otimes \boldsymbol{E}^\alpha\right) = \hat{\mathbf{F}}\mathbf{P}\end{aligned} \tag{2.3.29}$$

where

$$\mathbf{P}=\frac{\partial\hat{X}}{\partial X}=\frac{\partial\lambda(X)}{\partial X} \qquad \hat{X}=\lambda(X) \tag{2.3.30}$$

and $\mathbf{P}$ is the deformation gradient associated with the mapping $B_r\rightarrow\hat{B}_r$. It follows from equation (2.3.29) that the constitutive equation (2.3.27) can be expressed in the form

$$\mathbf{T}=\mathbf{G}_{\kappa}(\mathbf{F}^t(s))=\mathbf{G}_{\hat{\kappa}}(\mathbf{F}^t(s),\mathbf{P}^{-1}). \tag{2.3.31}$$

Equation (2.3.31) is the relation which must hold for all $\mathbf{F}$ between functionals for different reference configurations.

If for any two neighbourhoods $\mathcal{N}(X)$ and $\mathcal{N}(\hat{X})$ there exist reference configurations B_r and $\hat{B}_r$, respectively, in which the properties of the neighbourhoods are identical, i.e.

$$\rho_r=\hat{\rho}_r \qquad \mathbf{G}_{\kappa}=\mathbf{G}_{\hat{\kappa}} \tag{2.3.32}$$

then the two body points X and $\hat{X}$ are said to be materially isomorphic.

For the body as a whole to be homogeneous there must exist a compatible reference configuration B_r for the whole body such that the response functional $\mathbf{G}_{\kappa}$ is the same at each point.

Inhomogeneity is a property of the body as a whole, relating specifically to the way in which the material neighbourhoods are pieced together in contrast to ρ and $\mathbf{G}$ which characterise properties of the material.

The requirement that the response functional $\mathbf{G}$ with respect to B_r shall coincide with the response functional with respect to $\hat{B}_r$ implies that $\mathbf{G}_{\hat{\kappa}}$ is the same functional of the argument $\hat{\mathbf{F}}^t(s)=\mathbf{F}^t(s)\mathbf{P}^{-1}$ as $\mathbf{G}_{\kappa}$ is of $\mathbf{F}^t(s)$; i.e.

$$\mathbf{G}_{\hat{\kappa}}[\mathbf{F}^t(s)\mathbf{P}^{-1}]=\mathbf{G}_{\kappa}[\mathbf{F}^t(s)]. \tag{2.3.33}$$

It is evident from equations (2.3.31) and (2.3.33) that the determination of the stress at the representative body point X is independent of which of the response functionals $\mathbf{G}_{\kappa}$ and $\hat{\mathbf{G}}_{\kappa}$ is used and is also independent of which of the deformation gradients $\mathbf{F}$ and $\hat{\mathbf{F}}$ is used.

In equation (2.3.33), $\mathbf{P}$ is a unimodular tensor (that is, det $\mathbf{P}=\pm 1$), since B_r and $\hat{B}_r$ have the same density. For convenience, set $\mathbf{P}^{-1}=\mathbf{H}$, giving

$$\mathbf{G}(\mathbf{F}^t(s))=\mathbf{G}(\mathbf{F}^t(s)\mathbf{H}) \tag{2.3.34}$$

where $\mathbf{H}$ represents a unimodular mapping that leaves the response functional $\mathbf{G}$ invariant.

The property represented by equation (2.3.34) does not hold for all materials for all unimodular $\mathbf{H}$. For a given class of materials the set of all unimodular tensors $\mathbf{H}$ for which equation (2.3.34) does hold forms a group called the isotropy group or material symmetry group $\mathcal{G}_{\kappa}$ of the material at the body point whose position in the reference configuration B_r is X. It is a subgroup of the group $\mathcal{U}$ of all unimodular transformations.

Since $\mathbf{H}$ belongs to $\mathscr{G}_\kappa$, $\mathbf{F}^t(s)$ in equation (2.3.34) can be replaced by $\mathbf{F}^t(s)\mathbf{P}$ to give

$$\mathbf{G}_{\hat{\kappa}}(\mathbf{F}^t(s)\mathbf{P}) = \mathbf{G}_{\hat{\kappa}}(\mathbf{F}^t(s)\mathbf{PH}). \tag{2.3.35}$$

Use of equation (2.3.31) now gives

$$\mathbf{G}_{\hat{\kappa}}(\mathbf{F}^t(s)\mathbf{PHP}^{-1}) = \mathbf{G}_{\hat{\kappa}}(\mathbf{F}^t(s)) \tag{2.3.36}$$

and hence, since $\mathbf{H} \in \mathscr{G}_\kappa$, $\mathbf{PHP}^{-1}$ is an element of the symmetry group $\mathscr{G}_{\hat{\kappa}}$ of the material relative to $\hat{B}_r$ from which it follows that

$$\mathscr{G}_{\hat{\kappa}} = \mathbf{P}\mathscr{G}_\kappa\mathbf{P}^{-1} \tag{2.3.37}$$

as stated by Noll (1958).

An orthogonal transformation $\mathbf{Q}$ belongs to the isotropy group $\mathscr{G}_\kappa$ if and only if

$$\mathbf{QG}(\mathbf{F}^t(s))\mathbf{Q}^{\mathrm{T}} = \mathbf{G}(\mathbf{QF}^t(s)\mathbf{Q}^{\mathrm{T}}). \tag{2.3.38}$$

If $\mathbf{Q}$ belongs to $\mathscr{G}_\kappa$. then so does its inverse $\mathbf{Q}^{\mathrm{T}} = \mathbf{Q}^{-1}$. Equation (2.3.38) can be established by replacing $\mathbf{F}$ by $\mathbf{QF}$ and $\mathbf{H}$ by $\mathbf{Q}^{\mathrm{T}}$ in equation (2.3.34) to give

$$\mathbf{G}(\mathbf{QF}^t(s)\mathbf{Q}^{\mathrm{T}}) = \mathbf{G}(\mathbf{QF}^t(s)) \tag{2.3.39}$$

where $\mathbf{Q}$ is time independent. The restriction imposed by the principle of objectivity on the functional $\mathbf{G}$ can be expressed by way of equation (2.3.24) in the form

$$\mathbf{G}(\mathbf{Q}^t(s)\mathbf{F}^t(s)) = \mathbf{Q}(t)\mathbf{G}(\mathbf{F}^t(s))\mathbf{Q}(t)^{\mathrm{T}} \tag{2.3.40}$$

for all orthogonal tensor histories $\mathbf{Q}^t(s)$. Setting $\mathbf{Q}^t(s) \equiv \mathbf{Q}$ independent of s, gives

$$\mathbf{G}(\mathbf{QF}^t(s)) = \mathbf{QG}(\mathbf{F}^t(s))\mathbf{Q}^{\mathrm{T}} \tag{2.3.41}$$

which combined with equation (2.3.39) gives equation (2.3.38). Thus equation (2.3.38) is the functional equation to be solved to determine the orthogonal members $\mathbf{Q}$ of $\mathscr{G}_\kappa$.

It is evident from equation (2.3.38) that since $\mathscr{G}_\kappa$ is a group it contains both the identity transformation $\mathbf{I}$ and the inverse transformation $-\mathbf{I}$. Hence, by the definition of a group, if $\mathscr{G}_\kappa$ contains a certain proper orthogonal transformation $\mathbf{Q}$ it also contains the improper transformation $-\mathbf{Q}$.

For the condition that the material at a body point X is indistinguishable in its response both before and after it has been deformed into a new configuration, the group of all the possible static density-preserving deformations of the reference configuration B_r constitute the isotropy group $\mathscr{G}_\kappa$. If $\mathbf{H} = \mathrm{Grad}\ \boldsymbol{\lambda}$, where $\mathbf{H}$ is a member of the isotropy group $\mathscr{G}_\kappa$, then $\boldsymbol{\lambda}$ is the motion which carries one configuration into another in which the material is indistinguishable, and $\mathbf{H}$ which is unidmoular may, or may not, be orthogonal. A material which possesses a local reference configuration B_r such that the isotropy group contains the full orthogonal group $\mathcal{O}$ is said to be isotropic, and B_r is called an undistorted state of the material. For an isotropic material the response in any test to a deformation from an undistorted state is the same irrespective of whether the material is given an arbitrary rotation before the test. This says no more than that there are no preferred orientations of the material relative to the directions associated with the test.

A material is said to be anisotropic (or aelotropic), if its isotropy group relative to an undistorted state is a proper subgroup of the full orthogonal group $\mathcal{O}$.

2.3.4 Internal Constraints

Materials may be subjected to geometrical and internal constraints arising from symmetries other than material ones. These internal constraints must be regarded as part of the constitutive equations.

The particular internal constraints of present interest can be represented by equations of the form

$$\tilde{\eta}(\mathbf{F})=0 \tag{2.3.42}$$

where $\tilde{\eta}$ is a scalar-valued function. The principle of objectivity in the form of equation (2.1.39) can be applied to equation (2.3.42) for the special choice $\mathbf{Q}=\mathbf{R}^{\mathrm{T}}$ to give by way of equation (2.1.42)

$$\tilde{\eta}(\mathbf{R}^{\mathrm{T}}\mathbf{R}\mathbf{U})=\tilde{\eta}(\mathbf{F})=0 \tag{2.3.43}$$

from which it is evident that $\tilde{\eta}$ depends upon the deformation gradient through $\mathbf{U}$ and hence through the right Cauchy–Green deformation tensor $\mathbf{C}$ defined by equation (2.1.44). Hence, application of the principle of objectivity shows that the internal constraints can be represented by the condition

$$\lambda(\mathbf{C})=0 \tag{2.3.44}$$

where

$$\lambda(\mathbf{C})=\lambda(\mathbf{U}^2)\equiv\tilde{\eta}(\mathbf{U}).$$

A material body $\mathscr{B}$ must be able to accommodate its motion to an internal constraint by way of suitably introduced contact forces. Hence the general constitutive equation governing the stress response of the body $\mathscr{B}$ must be modified to allow these forces to act. This is achieved by amplifying the statement of the principle of determinism given in §2.3.1. The required modification centres on the assumption that the part of the stress arising from the effect of the constraint does no work and is not determined by the motion.

For simple materials subject to internal constraints, the required amplification of the basic constitutive statement can be expressed as follows: the stress at time t is uniquely determined by the history of the deformation gradient only to within a stress $\mathbf{N}$ that does no work in any motion compatible with the constraints.

From equation (2.2.94) it follows that the condition that equation (2.3.44) shall be a workless constraint can be expressed in the form

$$\operatorname{tr}(\mathbf{ND})=0 \tag{2.3.45}$$

for all allowable motions.

Differentiation of equation (2.3.44) gives

$$\dot{\lambda}=\frac{\partial\lambda}{\partial C^{\alpha\beta}}\dot{C}^{\alpha\beta}=\operatorname{tr}(\lambda_{\mathrm{C}}\dot{\mathbf{C}})=0 \tag{2.3.46}$$

where

$$\lambda_{\mathrm{C}}=\frac{\partial\lambda}{\partial C^{\alpha\beta}}\boldsymbol{E}^{\alpha}\otimes\boldsymbol{E}^{\beta} \tag{2.3.47}$$

Substituting for $\dot{\mathbf{C}}$ in equation (2.3.46) from equation (2.1.100) gives

$$\operatorname{tr}(\lambda_{\mathrm{C}}\mathbf{F}^{\mathrm{T}}\mathbf{DF})=\operatorname{tr}(\mathbf{F}\lambda_{\mathrm{C}}\mathbf{F}^{\mathrm{T}}\mathbf{D})=0 \tag{2.3.48}$$

and hence, by comparison with equation (2.3.45)

$$\mathbf{N}=q\mathbf{F}\lambda_{\mathrm{C}}\mathbf{F}^{\mathrm{T}} \tag{2.3.49}$$

for all stretching tensors $\mathbf{D}$ satisfying equation (2.3.45), q being a scalar multiplier. Thus, $\mathbf{N}$ is determined, to within a scalar multiplier, by the expression of equation (2.3.44) for the constraint.

If there are several constraints represented by the equation

$$\lambda_m(\mathbf{C})=0 \qquad \text{for } m=1,2,\ldots,p$$

then

$$\mathbf{N}=\sum_{m=1}^{p} q_m\mathbf{F}(\lambda_m)_{\mathrm{C}}\mathbf{F}^{\mathrm{T}}. \tag{2.3.50}$$

The tensor $\mathbf{T}+\mathbf{N}$ is called the extra stress $\mathbf{T}_{\mathrm{E}}$; i.e.

$$\mathbf{T}_{\mathrm{E}}=\mathbf{T}+\sum_{m=1}^{p} q_m\mathbf{F}(\lambda_m)_{\mathrm{C}}\mathbf{F}^{\mathrm{T}}. \tag{2.3.51}$$

For an incompressible material, equations (2.1.41) and (2.1.83) give $J(\tau)=|\det\mathbf{F}(\tau)|=1$ and hence for isochoric motions $\det\mathbf{C}(\tau)=1$, from which it is evident that an appropriate form of equation (2.3.44) is

$$\lambda(\mathbf{C})=\det\mathbf{C}-1=0. \tag{2.3.52}$$

Differentiation of equation (2.3.52) gives

$$\dot{\lambda}=\det\mathbf{C}\operatorname{tr}(\dot{\mathbf{C}}\mathbf{C}^{-1})=\operatorname{tr}(\mathbf{C}^{-1}\dot{\mathbf{C}}) \tag{2.3.53}$$

which compared with equation (2.3.46) gives

$$\lambda_{\mathrm{C}}=\mathbf{C}^{-1}=\mathbf{F}^{-1}(\mathbf{F}^{-1})^{\mathrm{T}} \tag{2.3.54}$$

from which it follows, by way of equation (2.3.49), that the constraint stress

$$\mathbf{N}=P\mathbf{I} \tag{2.3.55}$$

and hence the extra stress

$$\mathbf{T}_{\mathrm{E}}=\mathbf{T}+P\mathbf{I} \tag{2.3.56}$$

the disposable scalar $P(\equiv q)$ here being identified as pressure.

2.3.5 Reduced Constitutive Equations

From the definition of a simple material given in §2.3.1 and from equation (2.3.22) it is evident that, in the absence of thermal effects, the functional definition of a simple material takes the form

$$\mathbf{T}(X,t)=\mathbf{G}(\mathbf{F}^t(s),X,t). \tag{2.3.57}$$

Under arbitrary rotations, the principle of objectivity imposes the restriction that $\mathbf{G}$ must satisfy the relation

$$\mathbf{Q}(0)\mathbf{G}(\mathbf{F}^t(s),X,t)\mathbf{Q}(0)^{\mathrm{T}}=\mathbf{G}(\mathbf{Q}(s)\mathbf{F}^t(s),X,t) \tag{2.3.58}$$

where use has been made of equation (2.1.39). Replacing $\mathbf{F}$ by its right polar decomposition $\mathbf{RU}$ and setting $\mathbf{Q}=\mathbf{R}^{\mathrm{T}}$, equation (2.3.58) becomes

$$\begin{aligned}\mathbf{G}(\mathbf{R}(t)^{\mathrm{T}}\mathbf{F}^t(s),X,t)&=\mathbf{R}(t)^{T}\mathbf{G}(\mathbf{F}^t(s),X,t)\mathbf{R}(t)\\&=\mathbf{G}(\mathbf{U}^t(s),X,t)\end{aligned}$$

which can be transformed to

$$\mathbf{G}(\mathbf{F}^t(s),X,t)=\mathbf{R}(t)\mathbf{G}(\mathbf{U}^t(s),X,t)\mathbf{R}(t)^{\mathrm{T}}. \tag{2.3.59}$$

Hence the functional equation (2.3.57) for a simple material admits the reduced representation

$$\mathbf{T}(X,t)=\mathbf{R}(t)\mathbf{G}(\mathbf{U}^t(s),X,t)\mathbf{R}(t)^{\mathrm{T}} \tag{2.3.60}$$

it being noted that in the transition from equation (2.3.57) to equation (2.3.60) the effects of stretch and rotation on the stress at a typical body point have been separated out. This important result shows that the stress at time t is affected only by the present value of the rotation $\mathbf{R}(t)$, and not by the past history of the rotation $\mathbf{R}^t(s)$ for $s>0$. Thus, the stress at a body point X at time t is the stress at time t that would result from a pure stretch history $\mathbf{U}^t(s)$ relative to a reference configuration B_{r}, followed by a rotation $\mathbf{R}(t)$ relative to B_{r} at time t.

Although $\mathbf{R}$ and $\mathbf{U}$ have direct physical significance, neither can be simply calculated from the motion $\boldsymbol{\chi}(X,t)$. However, $\mathbf{F}$ and therefore the deformation tensors $\mathbf{C}$ and $\mathbf{B}$, defined by equations (2.1.44) and (2.1.45), respectively, can be readily calculated. Hence, defining another response function $\mathbf{p}$ by way of the relation

$$\mathbf{p}(\mathbf{U}^2)=\mathbf{G}(\mathbf{U})=\mathbf{p}(\mathbf{C}) \tag{2.3.61}$$

it follows that equation (2.3.60) can be expressed in the alternative form

$$\mathbf{T}(X,t)=\mathbf{R}(t)\mathbf{p}(\mathbf{C}^t(s),X,t)\mathbf{R}(t)^{\mathrm{T}}. \tag{2.3.62}$$

This result shows that the stress, while initially assumed to depend upon the history $\mathbf{F}^t(s)$ of the deformation gradient, can only depend on the history of one of the strain measures, such as the right stretch $\mathbf{U}$, the deformation tensor $\mathbf{C}$, or the strain $\mathbf{E}=\frac{1}{2}(\mathbf{C}-\mathbf{I})$. The strain $\mathbf{E}$ is the classical Green–St Venant measure of finite strain.

REFERENCES

Dahler J S and Scriven L E 1963 *Proc. R. Soc. Lond.* **A275** 504–27
Duhem P 1901 *C.R. Acad. Sci. (Paris)* **132** 117–20
Green A E and Naghdi P M 1977 *Proc. R. Soc. Lond.* **A357** 253–70
Noll W 1958 *Arch. Rational Mech. Anal.* **2** 197–226
—— 1964 *Am. Math. Monthly* **71** 129–44
Oldroyd J G 1950 *Proc. R. Soc. Lond.* **A200** 523–41
—— 1958 *Proc. R. Soc. Lond.* **A245** 278–97

GENERAL REFERENCES

Billington E W and Tate A 1981 *The Physics of Deformation and Flow* (New York: McGraw-Hill)
Chadwick P 1976 *Continuum Mechanics* (London: Allen and Unwin)
Fung Y C 1965 *Foundations of Solid Mechanics* (Englewood Cliffs, NJ: Prentice-Hall)
Green A E and Adkins J E 1960 *Large Elastic Deformation and Non-Linear Continuum Mechanics* (Oxford: Clarendon)
Gurtin M E 1981 *Continuum Mechanics* (New York: Academic)
Hunter S C 1983 *Mechanics of Continuous Media* 2nd edn (Chichester: Ellis Horwood)
Jaunzemis W 1967 *Continuum Mechanics* (New York: Macmillan)
Malvern L E 1969 *Introduction to the Mechanics of a Continuous Medium* (Englewood Cliffs, NJ: Prentice-Hall)
Prager W 1961 *Introduction to Mechanics of Continua* (Boston: Ginn)
Spencer A J M 1980 *Continuum Mechanics* (London: Longman)
Truesdell C and Noll W 1965 *The Non-linear Field Theories of Mechanics* in *Handbuch der Physik* vol. III/3 ed. S. Flügge (Berlin: Springer)
Truesdell C and Toupin R A 1960 *The Classical Field Theories* in *Handbuch der Physik* vol. III/1 ed. S Flügge (Berlin: Springer) pp 226–858
Volterra V 1930 *Theory of Functionals and Integral and Integrodifferential Equations* (Glasgow: Blackie)
Wang C-C and Truesdell C 1973 *Introduction to Rational Elasticity* (Leyden: Noordhoff)

3

Initial Yield

3.1 GENERALISED ISOTROPIC YIELD CRITERION

3.1.1 Basic Assumptions

Deformation of solid bodies beyond the elastic range is characterised either by permanent dimensional changes, resulting from the mechanisms of slip, or from dislocations at the atomic level, the latter irreversible deformation being known as plastic deformation. Plastic deformations occur only at stress intensities above a certain threshold value is known as the elastic limit. Or, in terms of an idealised theory of plasticity, this threshold is known as the initial yield stress, denoted as Y for observations made in a state of uniaxial stress and as k for observations made in pure shear.

Almost all theories of continuum plasticity assume that in general the criterion for yield can be interpreted geometrically in terms of a surface

$$f(\mathbf{T})=0 \tag{3.1.1}$$

where f is the yield function and $\mathbf{T}$ is the Cauchy stress tensor.

Any yield criterion for an isotropic material can be expressed in the form

$$f(I_{\mathrm{T}}, II_{\mathrm{T}}, III_{\mathrm{T}})=0 \tag{3.1.2}$$

where the three invariants I_{T}, II_{T}, III_{T} of the stress tensor $\mathbf{T}$ are expressed in terms of the principal stresses, σ_a $(a=1,2,3)$. The condition that an isotropic material depends only on the magnitude of the three principal applied stresses, and not on their direction, is satisfied by specifying that, for an isotropic material, f is to be a symmetric function of the principal stresses σ_a $(a=1,2,3)$.

It is further assumed in the classical theory of plasticity that yield is unaffected by hydrostatic stress. This assumption is based on the studies of Bridgman (1923, 1952) who has shown, to a good degree of approximation, that hydrostatic pressure does not affect either the initial yield or the post-yield deformation behaviour of a wide range of metals. The observations of

Crossland (1954) are in accord with these conclusions. With this as a basis it is possible to formulate the yield criterion in terms of the stress deviator $\mathbf{T}'$, where the prime denotes a deviator. For an isotropic, homogeneous, incompressible material, the assumption of isotropy implies that the plastic yielding can depend only on the magnitude of the deviators of the three principal stresses, and not on their directions and hence the yield criterion reduces to the form

$$f(J'_2, J'_3) = 0 \tag{3.1.3}$$

where the invariants

$$J'_2 = \tfrac{1}{2}\operatorname{tr} T'^2 \qquad J'_3 = \det \mathbf{T}' \tag{3.1.4}$$

of the stress deviator $\mathbf{T}'$ are defined by the coefficient of the characteristic equation

$$J'_3 + \sigma'_a J'_2 - \sigma'^3_a = 0 \tag{3.1.5}$$

and where the proper numbers σ'_a $(a = 1, 2, 3)$ of $\mathbf{T}'$ are the deviators of the principal stresses, σ_a $(a = 1, 2, 3)$. The condition that $I_{\mathbf{T}'} = J'_1 = 0$ is in accord with the assumption that yield is unaffected by hydrostatic stress.

To ensure that the yield stress is the same in tension and compression, it is necessary to impose the further restriction that

$$f(-\mathbf{T}) = f(\mathbf{T}). \tag{3.1.6}$$

In applying the equation (3.1.6) condition and taking equation (3.1.3) into account, it is noted from equation (3.1.4) that $J'_2(-\mathbf{T}) = J'_2(\mathbf{T})$ whereas $J'_3(-\mathbf{T}) = -J'_3(\mathbf{T})$.

The additional assumptions formulated by way of equations (3.1.2), (3.1.3) and (3.1.6), although limiting the generality of the yield function of equation (3.1.1), are all included in those theories of plasticity most widely used to describe the mechanical response of metals.

In general, the yield function may contain several parameters characterising the state of the material. These material parameters will be expected to vary with temperature or other thermodynamic variables. As formulated, the general yield condition relates only to the state of stress at a point in space and, irrespective of whether the mechanical response is elastic or plastic, does not depend on the stress at neighbouring points. For this reason the yield condition contains no dependence on the stress gradient.

3.1.2 A Geometrical Representation of Stress

The yield criterion can be interpreted for an isotropic material in terms of a geometrical representation of the stress state obtained by taking the principal stresses as coordinates, as shown in figure 3.1 (see, for example, Haigh 1920,

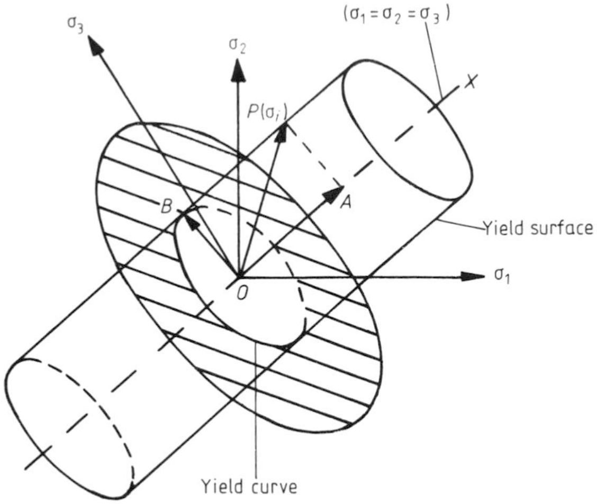

Figure 3.1 Yield surface in the principal stress space showing the yield curve in the π-plane.

Westergaard 1920, 1952). Every point in this principal stress space corresponds to a state of stress, and the position vector ***OP*** of any such point P $(\sigma_1, \sigma_2, \sigma_3)$ may be resolved into a component ***OA*** along the line ***OX***, which makes equal angles with the coordinate axes, together with a component ***OB*** in a plane known as the π-plane which is perpendicular to ***OX*** and passes through the origin. In stress space the yield criterion defines a surface which is generally referred to as the yield surface. The yield surfaces are parallel to the hydrostatic line ***OA*** characterised by the condition

$$\sigma_1 = \sigma_2 = \sigma_3 \quad \text{(hydrostatic line } OA\text{)}. \tag{3.1.7}$$

This follows from the assumption that yield is independent of the spherical part of the stress which requires that if $(\sigma_1, \sigma_2, \sigma_3)$ is on the yield surface, then so is $(\sigma_1 + OA, \sigma_2 + OA, \sigma_3 + OA)$ for any value of OA. This in turn implies that the yield surfaces are in general to be regarded as prisms with elements perpendicular to the deviatoric plane characterised by the condition

$$\sigma_1 + \sigma_2 + \sigma_3 = 0 \quad \text{(deviatoric plane)}. \tag{3.1.8}$$

Stress points that lie inside the yield surface represent elastic stress states, while those which lie on the yield surface represent incipient plastic stress states.

The intersection of the yield surface with the deviatoric, that is the π-plane, is generally referred to as the yield curve. Viewed toward the origin O and along OX, the principal stress axes are seen from figure 3.2 to be symmetrically placed $2\pi/3$ apart in the deviatoric plane. The assumption of isotropy requires that f be a symmetric function of $\sigma_1, \sigma_2, \sigma_3$ such that interchanging any two of the three arguments leaves f unchanged, from which it follows that the locus in

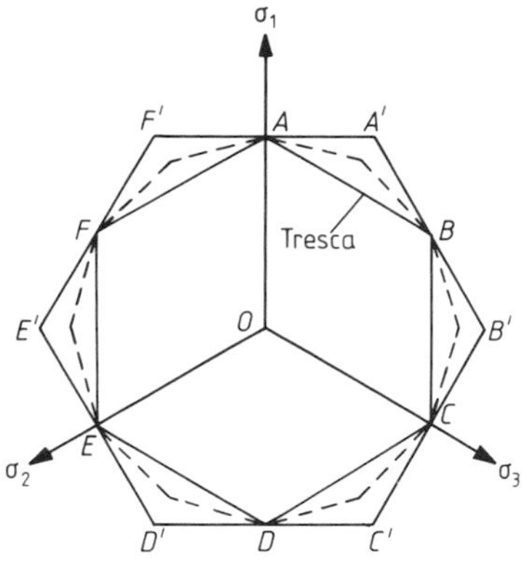

Figure 3.2 Yield surface in the deviatoric plane for three piecewise continuous yield functions.

the deviatoric plane must be symmetric with respect to each of the three projected axes. The location in the π-plane of the arbitrary point P can be determined by noting that each of the stress space axes make an angle $\cos^{-1}\sqrt{\frac{2}{3}}$ with the π-plane, and hence the projected deviatoric components are $(\sqrt{\frac{2}{3}}\sigma_1, \sqrt{\frac{2}{3}}\sigma_2, \sqrt{\frac{2}{3}}\sigma_3)$.

3.1.3 Formulation of the Yield Criterion

Consider the yield function

$$f = J_2' H(J_2', J_3') - k^2 = 0 \tag{3.1.9}$$

where k is the maximum shear stress at yielding in a state of pure shear.

Following Prager (1945), the stress dependence of H will be taken to be of the form

$$H = H(J_2', J_3') = H(\omega) \tag{3.1.10}$$

to give equation (3.1.9) in the form

$$f = J_2' H(\omega) - k^2 = 0 \qquad H(0) = 1 \tag{3.1.11}$$

where

$$\omega = \frac{27}{4}\frac{J_3'^2}{J_2'^3} = \mu^2 \frac{(\mu^2-9)^2}{(3+\mu^2)^3} \qquad (0 \leqslant \omega \leqslant 1) \tag{3.1.12}$$

was used by Fromm (1933) in the form of a parameter $(\omega/54)^{1/2}$ and where the stress parameter

$$\mu = \frac{3\sigma_1'}{\sigma_3' - \sigma_2'} \tag{3.1.13}$$

was introduced by Lode (1926). The condition of equation (3.1.10) reduces the general form of the yield function defined by equation (3.1.9) to the restricted form of equation (3.1.11) which satisfies the condition of equation (3.1.6).

Further consideration of the formulation of the yield criterion is restricted to those yield criteria for which the yield surface is convex. Convexity of the yield surface imposes a restriction on the possible forms of the yield function. In stress space the condition of convexity of the yield surface is taken to imply that any straight line in the deviatoric plane may pierce the yield curve in at most two points. With this as basis, the criterion for convexity of the yield surface can be formulated in terms of the yield curve in the deviatoric plane by way of the parameter

$$\bar{\mu} = -\sqrt{3}\,\mathrm{d}v/\mathrm{d}u \tag{3.1.14}$$

where

$$u = \sqrt{\tfrac{3}{2}}\,\sigma_1' \qquad v = (\sigma_3' - \sigma_2')/\sqrt{2}. \tag{3.1.15}$$

For the yield function defined by equation (3.1.11),

$$\bar{\mu} = \mu\left[1 - \frac{(3+\mu^2)}{\mu^2}\left(1 - \frac{H}{\Lambda}\right)\right] \tag{3.1.16}$$

where

$$\begin{aligned}\Lambda &= H\left[1 - \tfrac{1}{6}(3+\mu^2)\frac{\mu}{H}\frac{\mathrm{d}H}{\mathrm{d}\mu}\right] = H\frac{(3+\mu^2)}{(3+\mu\bar{\mu})}\\ &= H\left[1 - 3(-1)^n \tan\left(\frac{(-1)^n}{6}\cos^{-1}(1-2\omega) + \frac{n(3-n)}{6}\pi\right)\right.\\ &\qquad\left.\frac{[\omega(1-\omega)]^{1/2}}{H}\frac{\mathrm{d}H}{\mathrm{d}\omega}\right]\end{aligned} \tag{3.1.17}$$

and where (having solved equation (3.1.12) for μ^2 in terms of ω)

$$\mu = \pm\sqrt{3}\tan\left[\frac{(-1)^n}{6}\cos^{-1}(1-2\omega) + \frac{n(3-n)}{6}\pi\right] \qquad (n=0,1,2)$$

it being noted that

$$n = \begin{cases} 0 \\ 1 \\ 2 \end{cases} \quad \text{for} \quad \begin{cases} (0 \leqslant \mu^2 \leqslant 1) & (0 \leqslant \omega \leqslant 1) \\ (1 \leqslant \mu^2 \leqslant 9) & (1 \geqslant \omega \geqslant 0). \\ (9 \leqslant \mu^2 \leqslant \infty) & (0 \leqslant \omega \leqslant 1) \end{cases} \tag{3.1.18}$$

The parameter Λ is important in relation to the formulation of constitutive equations. Convexity of the yield surface requires that

$$\mathrm{d}\bar{\mu}/\mathrm{d}\mu \geqslant 0.$$

Equation (3.1.16) can be rearranged using equation (3.1.17) and integrated to give

$$H = \exp\left[-2\int_0^\theta \tan(c-\phi)\,\mathrm{d}c\right] \tag{3.1.19}$$

where

$$\mu = \pm\sqrt{3}\tan\theta \qquad \bar{\mu} = \pm\sqrt{3}\tan\phi \tag{3.1.20}$$

and where, from equation (3.1.12)

$$\theta = [(-1)^n \cos^{-1}(1-2\omega) + n(3-n)\pi]/6 \tag{3.1.21}$$

it being noted that n takes the values given in equation (3.1.18). Provided a chosen form for $\bar{\mu}$ satisfies the conditions implied by convexity of the yield surface, then the resulting yield function, obtained by way of equations (3.1.11) and (3.1.19), must also satisfy the condition of convexity for the yield surface.

It is evident from equation (3.1.18) that, in terms of the parameter μ, there are twleve $\pi/6$ segments in the deviatoric, or π-plane. It is further concluded that for an isotropic condition of incompressible yielding, the shape of the yield locus in each of the twelve $\pi/6$ segments is the same, apart from reflections. Thus, it is only necessary to consider stress states whose vectors lie in just one of the twelve $\pi/6$ segments. This implies that for the purpose of observing how a material satisfies the conditions at initial yield, it is sufficient to apply only those stress systems whose stress vectors lie in a selected segment. In this context, selection of an appropriate $\pi/6$ segment can be inferred from Lode's (1926) stress parameter μ defined by equation (3.1.13). That this is so can be appreciated by considering two limiting values of μ.

The second and third invariants of the stress deviator $\mathbf{T}'$, as defined in equation (3.1.4), can be expressed in the form,

$$J_2' = \tfrac{1}{2}(\sigma_1'^2 + \sigma_2'^2 + \sigma_3'^2) \qquad J_3' = \sigma_1'\sigma_2'\sigma_3' \tag{3.1.22}$$

and then rearranged to give

$$J_2' = \frac{1}{12}(3+\mu^2)(\sigma_3'-\sigma_2')^2 \qquad J_3' = -\frac{1}{108}\mu(9-\mu^2)(\sigma_3'-\sigma_2')^3 \tag{3.1.23}$$

or, alternatively

$$J_2' = \frac{3}{4}\frac{(3+\mu^2)}{\mu^2}\sigma_1'^2 \qquad J_3' = -\frac{1}{4}\frac{(9-\mu^2)}{\mu^2}\sigma_1'^3 \tag{3.1.24}$$

where μ is defined by equation (3.1.13).

Substituting for J_2' in equation (3.1.11) from equation (3.1.23) and rearranging gives

$$\tfrac{1}{2}(\sigma_3 - \sigma_2) = \frac{\sqrt{3}}{(3+\mu^2)^{1/2}}\frac{1}{H^{1/2}}k. \tag{3.1.25}$$

For $\mu=0$, equation (3.1.12) gives $\omega=0$, and equation (3.1.20) gives $\theta=0$. Setting $\theta=0$

$$H=1 \qquad \text{for } \omega=0,\ \mu=0. \tag{3.1.26}$$

Thus, when $\mu=0$, the limiting conditions of equation (3.1.26) reduce equation (3.1.25) to

$$k=\tfrac{1}{2}(\sigma_3-\sigma_2) \tag{3.1.27}$$

and from the discussion given in §2.2.3, in particular equation (2.2.49), it is evident that the limiting condition $\mu=0$ identifies k as the maximum shear stress at yielding in a state of pure shear.

Enter the value of J_2' given by equation (3.1.24) into equation (3.1.11) and rearrange to give

$$2\sigma_1-(\sigma_2+\sigma_3)=2\sqrt{3}\,\frac{\mu}{(3+\mu^2)^{1/2}}\,\frac{1}{H^{1/2}}\,k. \tag{3.1.28}$$

Setting $\mu=-1$ in equation (3.1.13) gives $\sigma_1=\sigma_2$. Thus, the limiting value $\mu=-1$ reduces equation (3.1.28) to the form

$$\sigma_2-\sigma_3=-\sqrt{3}\,k/[H^{1/2}]_{\mu=-1} \tag{3.1.29}$$

where $[H^{1/2}]_{\mu=-1}$ is the value of $H^{1/2}$ corresponding to $\mu=-1$. For the simple uniaxial extension of a solid rod, the only non-vanishing component of stress at yield is the initial, tensile yield stress Y. Hence for simple extension, with regard to the conditions relating to $\mu=-1$, set $\sigma_1=\sigma_2=0$ and identify Y with σ_3. These limiting conditions reduce equation (3.1.29) to

$$Y=\sqrt{3}\,k/[H^{1/2}]_{\mu=-1} \tag{3.1.30}$$

where Y is the initial yield stress in simple uniaxial tension.

From equation (3.1.21) it is evident that the values $\mu=0$, $\mu=-1$ are the limits of one of the $\pi/6$ segments. The limiting condition $\mu=-1$ can be obtained in practice from measurements of the initial yield stress of a solid rod deformed in simple uniaxial extension, the other limiting condition for which $\mu=0$ having been obtained from measurements of the shear stress at initial yield using a thin-walled tube twisted in pure torsion. The whole of the intermediate range $(0<\mu^2<1)$ can be obtained from the combined stress measurements of a thin-walled tube deformed in combined tension and torsion. An example of this approach to determining the conditions at initial yield are the classical combined stress measurements of Taylor and Quinney (1931).

Another method of covering the range is to stress a thin-walled cylindrical tube under combined axial tension and internal pressure. An example of this approach to determining the conditions at initial yield is the classical combined stress measurements of Lode (1926).

Entering the values for the difference $(\sigma_3-\sigma_2)$ given by equation (3.1.25) into equation (3.1.15) gives

$$u=\frac{1}{\sqrt{3}}\mu v \qquad v=\frac{\sqrt{3}}{(3+\mu^2)^{1/2}H^{1/2}}(\sqrt{2}\,k) \tag{3.1.31}$$

where use has been made of equation (3.1.13).

3.2 PIECEWISE CONTINUOUS YIELD CRITERIA

3.2.1 Tresca Yield Criterion

The studies of Tresca (1868) would appear to be the first investigation of the conditions at initial yield. The formulation of the Tresca yield criterion is based on the assumption that yielding occurs when the maximum shear stress reaches the prescribed value k. Thus the Tresca yield criterion can be expressed by way of equation (2.2.49) in the form

$$f=(\sigma_1-\sigma_2)-2k=0. \tag{3.2.1}$$

However, since the maximum shear stress yield equality is satisfied by any one of the six linear equations

$$\tfrac{1}{2}(\sigma_1-\sigma_2)=\pm k \qquad \tfrac{1}{2}(\sigma_2-\sigma_3)=\pm k \qquad \tfrac{1}{2}(\sigma_3-\sigma_1)=\pm k \tag{3.2.2}$$

it is evident that the Tresca yield criterion consists of six linear sections. For isotropic incompressible yielding, symmetry requires that these six linear sections form a regular hexagonal prism whose six plane faces are defined by equation (3.2.2).

It is evident from equation (3.2.2) that the Tresca yield criterion can be expressed in the more general form,

$$f=[(\sigma_1-\sigma_2)^2-4k^2][(\sigma_2-\sigma_3)^2-4k^2][(\sigma_3-\sigma_1)^2-4k^2]=0 \tag{3.2.3}$$

which is now symmetric with respect to the principal stresses. For uniaxial tension, where the only non-zero component of stress at initial yield is taken to be Y $(\equiv\sigma_3)$, so that $\sigma_1=\sigma_2=0$, equation (3.2.3) gives

$$Y=2k. \tag{3.2.4}$$

The Tresca yield criterion can also be expressed in the form (Reuss 1930)

$$f=J_2'(1-\omega)^{1/3}-\left[4-3\left(\frac{J_2'}{k^2}\right)\right]^{2/3}k^2=0 \tag{3.2.5}$$

which can be rearranged into the form of equation (3.1.11) with

$$H=\cos^2[\tfrac{1}{6}\cos^{-1}(1-2\omega)] \tag{3.2.6}$$

where ω is defined by equation (3.1.12).

Entering the form for H given by equation (3.2.6) into equation (3.1.31) gives

$$\frac{u}{(\sqrt{2}\,k)}=\frac{1}{\sqrt{3}}\mu\frac{v}{(\sqrt{2}\,k)} \qquad \frac{v}{(\sqrt{2}\,k)}=\frac{\sec[\frac{1}{6}\cos^{-1}(1-2\omega)]}{(1+\frac{1}{3}\mu^2)^{1/2}}. \tag{3.2.7}$$

Using equations (3.2.7) and (3.1.12) gives the yield locus in the deviatoric, i.e. the π-plane, which is seen from figure 3.2 to be a regular hexagon.

Equation (3.2.7) gives $v/(\sqrt{2}\,k)=1$ for all μ in the range $(0\leqslant\mu\leqslant 1)$, that is $n=0$, and hence from equation (3.1.14), $\bar{\mu}=0$ for $n=0$. Thus, a regular hexagon must have

$$\left.\begin{aligned}\bar{\mu}&=0\\ \bar{\mu}&=\pm 3\end{aligned}\right\} \quad \text{for} \quad \begin{cases}(0\leqslant\mu^2\leqslant 1) & (n=0)\\ (1\leqslant\mu^2\leqslant\infty) & (n=1,2)\end{cases}. \tag{3.2.8}$$

For $\mu=-1$, equation (3.1.12) gives $\omega=1$, and hence equations (3.2.6), (3.2.8) and (3.1.17) give

$$[H]_{\mu=-1}=\tfrac{3}{4} \qquad [\Lambda]_{\mu=-1}=\begin{cases}1\\ \frac{1}{2}\end{cases} \quad \text{for} \quad \bar{\mu}=\begin{cases}0\\ -3\end{cases}. \tag{3.2.9}$$

Substituting for H in equation (3.1.30) from equation (3.2.9) gives equation (3.2.4).

Setting $n=0$ in equation (3.1.21) and substituting the resulting form for θ into equation (3.1.20) gives

$$\mu=\pm\sqrt{3}\tan[\tfrac{1}{6}\cos^{-1}(1-2\omega)] \qquad (n=0). \tag{3.2.10}$$

Hence from equation (3.2.6)

$$H=3/(3+\mu^2) \qquad (n=0) \tag{3.2.11}$$

for all μ^2 in the range $(0\leqslant\mu^2\leqslant 1)$. Equation $(3.1.20)_1$ and (3.1.21) give

$$|\mu|=\sqrt{3}\tan\left(\frac{\pi}{3}-\frac{1}{6}\cos^{-1}(1-2\omega)\right) \qquad (n=1)$$

$$|\mu|=\sqrt{3}\tan\left(\frac{\pi}{3}+\frac{1}{6}\cos^{-1}(1-2\omega)\right) \qquad (n=2)$$

and hence use of equation (3.2.6) gives

$$H=\frac{3}{4}\frac{(1+|\mu|)^2}{(3+\mu^2)} \qquad (n=1,2). \tag{3.2.12}$$

From equation (3.2.6)

$$3\frac{\sqrt{\omega(1-\omega)}}{H}\frac{\mathrm{d}H}{\mathrm{d}\omega}=-\tan\alpha \tag{3.2.13}$$

which can be entered into equation (3.1.17) to give

$$\Lambda=\cos^2\alpha\left[1+(-1)^n\tan\alpha\tan\left((-1)^n\alpha+\frac{n(3-n)}{6}\pi\right)\right] \quad (3.2.14)$$

where

$$\alpha=\tfrac{1}{6}\cos^{-1}(1-2\omega) \qquad (0\leqslant\omega\leqslant 1). \quad (3.2.15)$$

Equation (3.2.14) can be rearranged into the form

$$\Lambda=\{1-(-1)^n\tan\alpha\tan[n(3-n)\pi/6]\}^{-1} \qquad (n=0,1,2) \quad (3.2.16)$$

which for $n=0$ gives $\Lambda=1$ for all μ in the range $(0\leqslant\mu^2\leqslant 1)$.

Equations (3.2.6) and (3.2.16) can be used, to give a relation for the ratio H/Λ which can be entered into equation (3.1.16) to give the conditions of equation (3.2.8), use being made of equations (3.2.11) and (3.2.12).

3.2.2 Rotated Tresca Yield Criterion

Consider the yield criterion,

$$f=J_2'\omega^{1/3}-3[\tfrac{1}{2}-\tfrac{1}{2}(J_2'/k^2)]^{2/3}k^2=0 \quad (3.2.17)$$

which can be rearranged into the form of equation (3.1.11) with

$$H=\tfrac{4}{3}\cos^2[\tfrac{1}{6}\cos^{-1}(2\omega-1)] \quad (3.2.18)$$

where ω is defined by equation (3.1.12).

Entering the form for H given by equation (3.2.18) into equation (3.1.31) gives

$$\frac{u}{(\sqrt{2}\,k)}=\frac{1}{\sqrt{3}}\mu\frac{v}{(\sqrt{2}\,k)} \qquad \frac{v}{(\sqrt{2}\,k)}=\frac{\sqrt{3}}{2}\frac{\sec[\tfrac{1}{6}\cos^{-1}(2\omega-1)]}{(1+\tfrac{1}{3}\mu^2)^{1/2}}. \quad (3.2.19)$$

Using equations (3.2.19) and (3.1.12) gives the yield locus in the deviatoric, that is the π-plane, and is shown in figure 3.2 as the regular hexagon $A'B'C'D'E'F'A'$.

Equation (3.2.19) gives $u/(\sqrt{2}\,k)=\sqrt{3}/2$ for all μ in the range $(9\leqslant\mu^2\leqslant\infty)$, that is $n=3$, and hence from equation(3.1.14), $\bar{\mu}=\pm\infty$ for $n=3$. Thus, for the regular hexagon $A'B'C'D'E'F'A'$ shown in figure 3.2 must have

$$\bar{\mu}=\begin{cases}\pm 1\\ \pm\infty\end{cases} \quad \text{for} \begin{cases}(0\leqslant\mu^2\leqslant 9)\\ (9\leqslant\mu^2\leqslant\infty)\end{cases}. \quad (3.2.20)$$

For $\mu=-1$, equation (3.1.12) gives $\omega=1$, and hence equations (3.2.18), (3.2.20) and (3.1.17) give

$$[H]_{\mu=-1}=\tfrac{4}{3} \qquad [\Lambda]_{\mu=-1}=\tfrac{4}{3}. \quad (3.2.21)$$

Substituting for H in equation (3.1.30) from equation (3.2.21) gives

$$Y=\tfrac{3}{2}k \quad (3.2.22)$$

which is to be compared with equation (3.2.4).

Equations (3.1.20) and (3.1.21) give

$$|\mu| = \sqrt{3}\tan[\tfrac{1}{6}\cos^{-1}(1-2\omega)] \qquad (n=0)$$
$$|\mu| = \sqrt{3}\tan[(\pi/3) - \tfrac{1}{6}\cos^{-1}(1-2\omega)] \qquad (n=1) \qquad (3.2.23)$$
$$|\mu| = \sqrt{3}\tan[(\pi/3) + \tfrac{1}{6}\cos^{-1}(1-2\omega)] \qquad (n=2)$$

which can be rearranged and combined with equation (3.2.18) to give

$$H = \frac{(3+|\mu|)^2}{3(3+\mu^2)} \qquad (n=0,1)$$
$$H = \frac{4\quad \mu^2}{3(3+\mu^2)} \qquad (n=2). \qquad (3.2.24)$$

From equation (3.2.18)

$$3\frac{\sqrt{\omega(1-\omega)}}{H}\frac{\mathrm{d}H}{\mathrm{d}\omega} = \tan\gamma \qquad (3.2.25)$$

which can be entered into equation (3.1.17) to give,

$$\Delta = \tfrac{4}{3}\cos^2\gamma\left[1-(-1)^n\tan\gamma\tan\left((-1)^n\alpha + \frac{n(3-n)}{6}\pi\right)\right] \qquad (3.2.26)$$

where

$$\gamma = \tfrac{1}{6}\cos^{-1}(2\omega-1) = (\pi/6) - \alpha \qquad (3.2.27)$$

it being noted that α is defined by equation (3.2.15). For $n=0,1,2$ equation (3.2.26) reduces to

$$\Lambda = 1 + (\tan\alpha/\sqrt{3}) = 1 + \tfrac{1}{3}|\mu| \qquad (n=0)$$
$$\Lambda = \frac{2\quad(\sqrt{3}+\tan\alpha)}{\sqrt{3}(1+\sqrt{3}\tan\alpha)} = 1 + \tfrac{1}{3}|\mu| \qquad (n=1) \qquad (3.2.28)$$
$$\Lambda = 0 \qquad (n=2).$$

Entering the forms for H and Λ given by equations (3.2.24) and (3.2.28), respectively, into equation $(3.1.17)_2$ and rearranging gives the values of $\bar{\mu}$ given by equation (3.2.20).

Equations (3.1.14), (3.1.16) and (3.2.20) imply that the yield criterion defined by equation (3.1.11) with H given by equation (3.2.18) consists of six linear sections. For isotropic incompressible yielding, symmetry requires that these six linear sections form a regular hexagonal prism. Thus the yield criterion defined by equation (3.1.11) with H given by equation (3.2.18) is also characterised by a hexagonal yield locus in the π-plane, but which differs by $\pi/6$ in orientation from the hexagonal yield locus characteristic of the Tresca yield criterion. Hence the term rotated Tresca yield criterion. In figure 3.2, the regular hexagonal surface $ABCDEFA$ corresponds to the Tresca yield criterion and the regular hexagonal surface $A'B'C'D'E'F'A'$ corresponds to the rotated Tresca yield criterion.

Symmetry considerations imply that no piecewise smooth curve passing through A, may lie outside the straight line $F'AA'$, and still form part of a convex yield surface. Consequently, with reference to figure 3.2 all possible yield surfaces satisfying isotropy, convexity, independence from the effect of hydrostatic stress, and for which the yield stress in tension and compression is the same, must lie between the regular hexagonal surfaces $ABCDEFA$ and $A'B'C'D'E'F'A'$. However, it is to be emphasised that not all surfaces lying between these two hexagonal surfaces are admissible. The convexity condition of the yield surface determines whether they are admissible.

3.2.3 Twelve Linear Sections

In the context of equation (3.1.18) it has been concluded that, in terms of μ, there are twelve $\pi/6$ segments in the π-plane. This observation is taken to imply the possible existence of yield criteria characterised by a yield locus in the π-plane consisting of twelve linear sections.

It is evident from equation (3.1.20) that the condition $\bar{\mu}=\bar{\mu}_n=$constant for all values of μ gives $\phi=$constant for all θ, where θ is defined by equation (3.1.21). Substituting $\phi=$constant in equation (3.1.19) gives

$$H=\frac{H(\sqrt{3}\cos\theta+\bar{\mu}_n\sin\theta)^2}{[(\sqrt{3}\cos\theta+\bar{\mu}_n\sin\theta)^2]_{\omega=0}} \qquad (n=0,1,2) \qquad (3.2.29)$$

where θ is defined by equation (3.1.21) and where the significance of the subscript n can be identified from equation (3.1.18). The values of $\bar{\mu}_n$ can be obtained from the relation

$$|\bar{\mu}_n|=|\bar{\mu}_0|+\frac{(3-2|\bar{\mu}_0|-\bar{\mu}_0^2)}{(1+|\bar{\mu}_0|)}n(2-n)+\frac{(3+\bar{\mu}_0^2)}{(1-|\bar{\mu}_0|)}\frac{n(n-1)}{2} \qquad (3.2.30)$$

where, for $n=0$

$$\bar{\mu}_0=-\lim_{\sigma_1'\to 0}\frac{\mathrm{d}}{\mathrm{d}\sigma_1'}(\sigma_3'-\sigma_2') \qquad (3.2.31)$$

it being noted that at

$$\omega=0 \qquad n=\begin{cases}1,2\\0\end{cases} \qquad \mu^2=\begin{cases}9\\0\end{cases}. \qquad (3.2.32)$$

Equation (3.1.11) with H given by equation (3.2.29) defines a yield condition which is characterised by twelve linear sections in the π-plane. A typical example is shown as the broken line polygon in figure 3.2 which corresponds to $\bar{\mu}_0=-0.5$.

Equation (3.1.16) can be rearranged to give

$$\Lambda = \frac{(1+\frac{1}{3}\mu\bar{\mu}_n)}{[(\cos\theta + \bar{\mu}_n \sin\theta/\sqrt{3})^2]_{\omega=0}}. \tag{3.2.33}$$

For the Tresca yield criterion, and the range of μ^2 for which $n=0$, $\bar{\mu}_0=0$ and equation (3.2.33) gives $\Lambda=1$ for all μ in the range $(0 \leqslant \mu^2 \leqslant 1)$, a result which has also been shown to follow from equation (3.2.16).

3.2.4 Yield Criterion in Terms of Lode's Stress Parameter

Thus far the yield criteria have been considered in terms of the parameter ω defined by equation (3.1.12). It is of interest to express H in terms of μ.

Equation (3.2.29) can be expressed in the form (Billington 1981)

$$H = \frac{(3+\bar{\mu}_n\mu)^2}{(3+\mu^2)}\left[\frac{(3+\mu^2)}{(3+\bar{\mu}_n\mu)^2}\right]_{\omega=0} \qquad (n=0,1,2) \tag{3.2.34}$$

where $\bar{\mu}_n$ is to be obtained from equation (3.2.30) and where use has been made of equation (3.1.20).

Entering the form for H given by equation (3.2.34) into equation (3.1.17) gives

$$\Lambda = (3+\bar{\mu}_n\mu)\left[\frac{(3+\mu^2)}{(3+\bar{\mu}_n\mu)^2}\right]_{\omega=0} \qquad (n=0,1,2). \tag{3.2.35}$$

The condition of convexity of the yield surface imposes upper and lower limits on $\bar{\mu}_0$.

At the lower limit, $\bar{\mu}_0=0$ and equation (3.2.34) gives

$$H = \frac{3}{(3+\mu^2)} \qquad (n=0) \tag{3.2.36}$$

$$H = \frac{3}{4}\frac{(1+|\mu|)^2}{(3+\mu^2)} \qquad (n=1,2) \tag{3.2.37}$$

which when substituted into equation (3.1.11) gives the classical Tresca yield criterion. Equations (3.1.23) and (3.1.24) can be used together with equation (3.1.11) to give the parameters u and v of equation (3.1.15) in the form

$$u = \frac{\mu}{(3+\mu^2)^{1/2}H^{1/2}}\sqrt{2}\,k \qquad v = \frac{\sqrt{3}}{(3+\mu^2)^{1/2}H^{1/2}}\sqrt{2}\,k. \tag{3.2.38}$$

For $\bar{\mu}_0=0$, $v=$constant for all μ in the range $(0 \leqslant \mu^2 \leqslant 1)$, and hence equation (3.2.38) reduces to equation (3.2.36), it being noted that for $\mu=0$, $\omega=0$ and $H=1$.

For the upper limit, $|\bar{\mu}_0|=1$ and equation (3.2.34) gives

$$H=\frac{(3+|\mu|)^2}{3(3+\mu^2)} \qquad (n=0,1) \tag{3.2.39}$$

$$H=\frac{4}{3}\frac{\mu^2}{(3+\mu^2)} \qquad (n=2) \tag{3.2.40}$$

which when substituted into equation (3.1.11) gives the rotated Tresca yield criterion. The condition $|\bar{\mu}_2|=\infty$, gives u=constant for all μ in the range $(9\leqslant\mu^2\leqslant\infty)$ and hence equation (3.2.38) reduces to equation (3.2.40), use being made of the condition that for $\mu^2=9$, $\omega=0$ and $H=1$.

3.2.5 Limitation of a Piecewise Continuous Yield Condition

The characteristic feature of a piecewise continuous yield condition is the existence of corners (singular regimes), on the yield surface. A characteristic feature of a vertex on the yield surface is the existence of a fan of normals to the contributing surfaces, which in turn means that the normal vectors at these corners are not unique with regard to their directions.

In the case of the Tresca yield criterion, the existence of a vertex on the yield surface at $\mu=-1$ renders this criterion unsuitable for use in studies of the uniaxial tensile deformation behaviour of a material.

Similarly, in the case of a rotated Tresca yield criterion, the existence of a vertex on the yield surface at $\mu=0$, renders this criterion unsuitable for use in studies of a material deformed in simple shear.

Koiter (1953) has generalised the concept of a singular yield surface by regarding the yield surface to be made up of several smooth surfaces, defined by several functions, $f_1(\mathbf{T}), f_2(\mathbf{T}), \ldots, f_n(\mathbf{T})$. For elastic states all the functions are negative, while at yield at least one must equal zero. If more than one yield function is zero, the stress point is at a vertex on the yield surface.

From the above discussion, this particular limitation has physical significance in relation to the characterisation of a materials mechanical properties by way of stress–strain measurements, the significant quantity in this latter context being the parameter Λ defined by equation (3.1.17).

3.3 CONTINUOUSLY DIFFERENTIABLE YIELD CRITERIA

3.3.1 Yield Criteria of the von Mises Type

A particularly simple yield function is obtained by setting $\theta=\phi$ in equations (3.1.19), (3.1.20) and (3.1.17) to give

$$H=1 \qquad \Lambda=1 \qquad \mu=\bar{\mu} \tag{3.3.1}$$

for all μ. The condition $H=1$ reduces equation (3.1.11) to

$$f=J_2'-k^2=0 \tag{3.3.2}$$

which is the classical yield criterion of von Mises (1913). Substituting $H=1$ in equation (3.1.30) gives

$$Y=\sqrt{3}\,k \tag{3.3.3}$$

for the ratio of the initial yield stress in simple uniaxial extension to the initial yield stress in pure torsion. For the von Mises yield criterion, the yield surface is the right circular cylinder defined by equation (3.3.2), such that the locus in the π-plane is symmetric with respect to each of the three projected axes. That this is so can be seen as follows. For $H=1$, equations (3.2.38) can be combined to give

$$(u^2+v^2)^{1/2}=\sqrt{2}\,k=\sqrt{\tfrac{2}{3}}\,Y \tag{3.3.4}$$

where use has been made of equation (3.3.3). From equation (3.3.4), it follows that the locus in the π-plane is the von Mises circle of radius $\sqrt{\frac{2}{3}}\,Y$, which is shown in figure 3.3, where it is compared with the Tresca yield criterion which is shown as a broken line hexagon. The von Mises yield criterion is the simplest continuously differentiable isotropic condition of incompressible yielding.

Consider the particular case for which

$$\bar{\mu}=\mu\left(1-\frac{9c(3+\mu^2)(1-\mu^2)(9-\mu^2)}{(3+\mu^2)^3-8c\mu^4(9-\mu^2)}\right) \qquad (c>0) \tag{3.3.5}$$

where c is a constant characteristic of material properties. Substituting for ϕ in equation (3.1.19) using equations (3.1.20), (3.1.21) and (3.3.5) gives on integration

$$H=1-c\omega \tag{3.3.6}$$

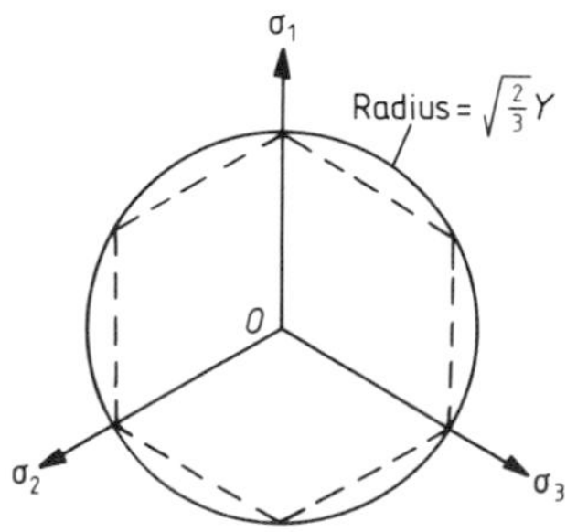

Figure 3.3 Yield curves for the von Mises circle and Tresca hexagon as seen in the deviatoric plane, and viewed towards O, along OX of figure 3.1.

which can be entered into equation (3.1.11) to give the yield function

$$f = J_2'(1 - c\omega) - k^2 = 0. \tag{3.3.7}$$

Setting $c = 0$ reduces equation (3.3.7) to the von Mises yield condition. A limiting value of c can be found from the convexity condition of the yield surface which, in the context of equation (3.1.14), requires that the slope of the $(\bar{\mu}, \mu)$ relation must be positive or zero, but cannot be negative. Differentiation of equation (3.3.5) with respect to μ produces, at $\mu = 0$, the condition $c = \frac{1}{9}$ for $\mathrm{d}\bar{\mu}/\mathrm{d}\mu = 0$, and hence $c = \frac{1}{9}$ is the limiting value. Thus the condition of convexity of the yield surface restricts c to the range $(0 < c \leqslant \frac{1}{9})$. Use of equations (3.2.38), (3.1.30) and (3.3.6) gives

$$(u^2 + v^2)^{1/2} = \sqrt{2}\, k/(1 - c\omega)^{1/2} \qquad (0 < c \leqslant 1/9) \tag{3.3.8}$$

it being noted that for $\mu = -1$, equation (3.1.12) gives $\omega = 1$. From a comparison of equations (3.3.4) and (3.3.8), it is evident that for the permitted range of c, the deviation (as seen in the π-plane) of the yield function defined by equation (3.3.7) from the von Mises yield criterion is $\sim c\omega/2$ $(c\omega \ll 1)$. The variation of $\bar{\mu}$ with μ for $c = 0.06$ and $c = \frac{1}{9}$ is shown in figure 3.4, where they are compared with the von Mises yield criterion in the form of the broken straight line, $\mu = \bar{\mu}$. Entering the form for H given by equation (3.3.6) into equation (3.1.17) gives

$$\Lambda = 1 - [8\mu^2/(9 - \mu^2)]c\omega. \tag{3.3.9}$$

The yield function of equation (3.3.7) is also continuously differentiable and is generally referred to as a modified von Mises yield criterion (Drucker 1949, Freudenthal and Gou 1969, Prager 1945).

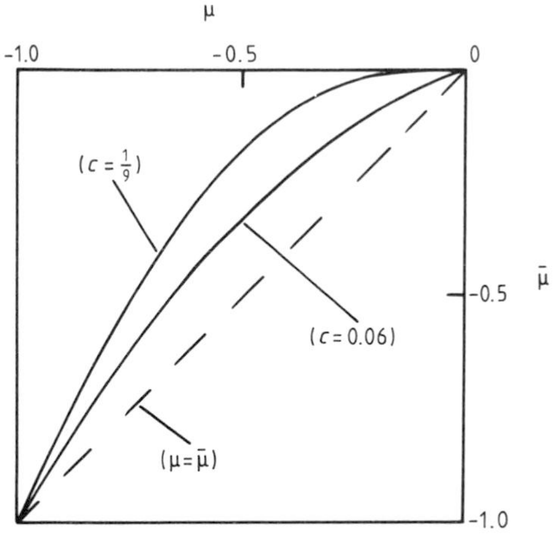

Figure 3.4 The $(\bar{\mu}, \mu)$ relation for a continuously differentiable yield function of the modified von Mises type.

3.3.2 Modified Tresca Yield Criterion

One way in which the Tresca yield criterion can be made continuously differentiable is to replace ω in equation (3.2.6) by $a\omega$ to give

$$H=\cos^2[\tfrac{1}{6}\cos^{-1}(1-2a\omega)] \qquad (0\leqslant a\leqslant 1) \qquad (3.3.10)$$

where a is a constant. The yield function obtained using equation (3.1.11) with H given by equation (3.3.10) is continuously differentiable for all a in the range $(0\leqslant a\leqslant 1)$. It is of interest to note that the limiting value $a=0$ gives $H=1$ for all ω which is the condition of equation (3.3.1), and hence this limiting value of a reduces the modified Tresca yield criterion to the von Mises yield criterion. Thus, the yield function obtained using equations (3.1.11) and (3.3.10) includes as its limiting forms the Tresca yield criterion for which $a=1$, and the von Mises yield criterion for which $a=0$.

The parameter Λ can be obtained from equations (3.1.17) and (3.3.10) in the form

$$\Lambda=\cos^2\hat{\alpha}\left[1+(-1)^n\left(\frac{a(1-\omega)}{(1-a\omega)}\right)^{1/2}\tan\hat{\alpha}\tan\left((-1)^n\alpha+\frac{n(3-n)}{6}\pi\right)\right] \qquad (3.3.11)$$

where

$$\alpha=\tfrac{1}{6}\cos^{-1}(1-2\omega) \qquad (0\leqslant\omega\leqslant 1) \qquad (3.3.12)$$

$$\hat{\alpha}=\tfrac{1}{6}\cos^{-1}(1-2a\omega) \qquad (0\leqslant a\leqslant 1). \qquad (3.3.13)$$

Equations (3.3.10) and (3.3.11) can be used to give a relation for the ratio H/Λ which can be entered into equation (3.1.16) to give values of $\bar{\mu}$. It is evident from equations (3.3.10) to (3.3.13), together with equation (3.1.16), that

$$\left.\begin{aligned}[\Lambda]_{\omega=1}&=[H]_{\omega=1}\\ [\Lambda]_{\omega=0}&=[H]_{\omega=0}\end{aligned}\right\} \quad \text{and} \quad \bar{\mu}=\mu \qquad (3.3.14)$$

and hence the modified Tresca yield criterion is continuously differentiable for all μ.

The variation of $\bar{\mu}$ with μ is shown in figure 3.5 for several values of a and for μ in the range $(-1\leqslant\mu\leqslant 0)$.

By letting a approximate closely to unity the modified yield function, defined by equations (3.1.11) and (3.3.10), can be made to approximate the Tresca yield criterion to any required degree of accuracy, whilst retaining the condition of being continuously differentiable through the corners.

3.3.3 Modified Form of a Yield Criterion having Twelve Linear Sections

The twelve-sided, linear, piecewise continuous yield condition defined by equations (3.1.11) and (3.2.29) can be made continuously differentiable

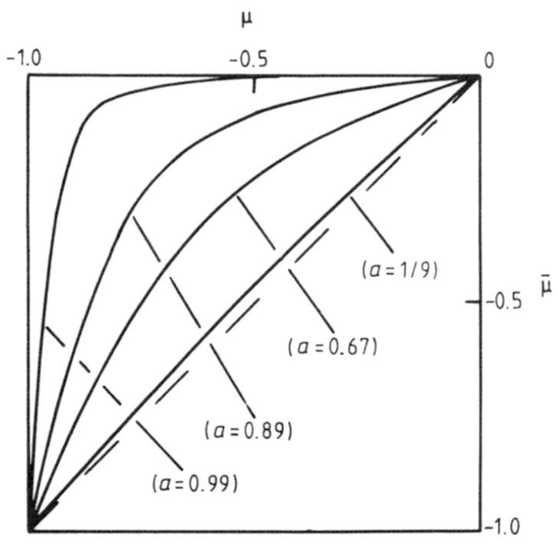

Figure 3.5. The $(\bar{\mu}, \mu)$ relation for a modified Tresca yield criterion.

through the corners at $(\mu^2=1, \omega=1)$, and $(\mu^2=9, \omega=0)$ by taking,

$$|\bar{\mu}|=|\bar{\mu}_0|+\frac{(3-2|\bar{\mu}_0|-\bar{\mu}_0^2)}{1+|\bar{\mu}_0|+2\exp[(1-|\mu|)/\varepsilon]}$$

$$+\frac{8|\bar{\mu}_0|}{(1+|\bar{\mu}_0|)\{1-|\bar{\mu}_0|+(1+|\bar{\mu}_0|)\exp[(3-|\mu|)/\varepsilon]\}} \tag{3.3.15}$$

where ε is an adjustable constant, and where $\bar{\mu}_0$ is defined by equation (3.2.31). By adjusting ε, the modified yield function approximates the piecewise yield condition to any required degree of accuracy, whilst retaining the condition of being continuously differentiable through the corners.

3.4. COMPOSITE YIELD FUNCTION

A composite yield function can be obtained by taking H to be of the form

$$H=\phi H_{(1)}+(H_{(2)}-\phi H_{(1)})U \tag{3.4.1}$$

where U is some appropriate form of step function, $H_{(1)}$ is restricted to be of the form of equation (3.2.29) with $\bar{\mu}$ given by equation (3.3.15) and where $H_{(2)}$ can be any one of the forms given by equations (3.3.1), (3.3.6) and (3.3.10). In the context of equation (3.1.18), for each value of $n=0,1,2$, there is a particular value of $\mu=\mu^*$ at which the yield function changes from that obtained using equation (3.1.11) with $H=H_{(2)}$, for which $U=1$, to that obtained using equation (3.1.11) with $H=\phi H_{(1)}$, for which $H=0$. Thus, at the discontinuous change from $U=0$ to $U=1$,

$$[H]_{\mu=\mu^*}=[H_{(2)}]_{\mu=\mu^*}=[\phi H_{(1)}]_{\mu=\mu^*} \qquad (3.4.2)$$

it being noted that ϕ is solely a function of μ^*.

With $H_{(2)}=1$ and $U=1$, the yield function defined by equation (3.1.11) is just that of von Mises.

For $n=0$, μ^2 lies in the range $(0\leqslant\mu^2\leqslant 1)$ and hence

$$U=\begin{cases}1 & (0\leqslant\mu^2\leqslant\mu^{*2})\\ 0 & (\mu^{*2}\leqslant\mu^2\leqslant 1)\end{cases} \qquad (\mu^*\equiv\bar{\mu}_0) \qquad (3.4.3)$$

and

$$\phi=3/(3+\mu^{*2}) \qquad (\mu^*\equiv\bar{\mu}_0) \qquad (3.4.4)$$

it being noted that for $n=0$, $\mu^*\equiv\bar{\mu}_0$.

The yield function defined by equation (3.1.11), with H given by equation (3.4.1) is continuously differentiable for all values of μ.

3.5 CONCLUDING REMARKS

Some materials, initially loaded in tension, are observed to yield at a much reduced stress when reloaded in compression. This is known as the Bauschinger effect. Materials which exhibit the Bauschinger effect do not satisfy the restriction formulated in equation (3.1.6).

Because the process of slip on a crystal plane is directional, plastic deformation on a microscopic scale must be regarded as being physically anisotropic. For this reason continuum plastic deformation will often, though by no means always, be accompanied by the progressive development of anisotropy. Thus, for some materials it is to be expected that isotropy which is initially present will usually be destroyed by continuous plastic deformation, thus invalidating equation (3.1.2). Hill (1948) has proposed a generalisation of the von Mises quadratic yield condition for states of anisotropy possessing three mutually orthogonal planes of symmetry, when the reference Cartesian coordinate planes are chosen parallel to the three symmetry planes. This approach has been applied for a particular type of anisotropy associated with the behaviour of sheet metal, where orthotropic symmetry is to be expected with symmetry axes perpendicular to the sheet, and parallel and perpendicular to the rolling direction used in manufacture of the sheet (see also Hill 1950).

Throughout the present chapter, it has been assumed that yield is unaffected by hydrostatic stress, a situation generally in accord with experiment for a wide variety of materials when tested at moderate stress levels. However, there is an increasing body of experimental evidence showing that yield and flow stress of certain materials are sensitive to hydrostatic pressure, but that work hardening and the magnitude of the strength differential (SD) effect between compression and tension are relatively insensitive (see, for example, Drucker 1973, Spitzig *et al* 1975). Measurements of the volume changes occurring during plastic deformation at atmospheric pressure for these materials show that the plastic volume expansion was larger than that measured in materials which show a smaller SD effect. The SD effect can, in principle, be accounted for by way of a yield function which involves odd powers of J'_3 (Freudenthal and Gou 1969; see also Ainbinder *et al* 1965, Pae and Bhateja 1975).

Although the above remarks indicate that many materials of technical use do not satisfy the basic assumptions set out in §3.1, the remaining discussion of this book will be restricted to materials which do satisfy these assumptions. This approach recognises the fact that the complex behaviour of materials of technical interest are likely to be resolved only when a physically acceptable theory has been developed for, and shown to predict correctly, the observed behaviour of materials which do satisfy these basic assumptions. Thus, the remaining discussion is restricted in a number of important ways. It is concerned solely with isotropic, homogeneous, incompressible materials characterised by the existence of a universal stress–strain relation. This latter restriction requires the Bauschinger effect to be neglected thus ensuring that the yield stress is the same in tension and compression. No account is taken of thermodynamic restrictions, the proposed constitutive equations being purely mechanical. These restrictions, though limiting the generality of the approach, are all included in those theories of plasticity most widely used to describe the mechanical response of many metals.

REFERENCES

Ainbinder S B, Laka M G and Maiors I Y 1965 *Polymer Mechanics* **1** 50–5
Billington E W 1981 *J. Phys D: Appl. Phys.* **14** 1071–85
Bridgman P W 1923 *Proc. Am. Acad. Art. Sci.* **58** 163–242
Crossland B 1954 *Inst. Mech. Eng. Proc.* **168** 935–46
Drucker D C 1949 *J. Appl. Mech.* **16** 349–57
—— 1973 *Met. Trans.* **4** 667–73
Freudenthal A M and Gou P F 1969 *Acta Mechan.* **8** 34–52
Fromm H 1933 *Ing-Arch.* **4** 432–66
Haigh B P 1920 *Engineering* **109** 158–60
Hill R 1948 *Proc. R. Soc.* A **193** 281–97
Koiter W T 1953 *Quart. Appl. Math.* **11** 350–4
Lode W 1926 *Z. Phys.* **36** 913–39
Pae K D and Bhateja S K 1975 *J. Macromol. Sci.—Revs. Macromol. Chem.* **C13** 1–75
Prager W 1945 *J. Appl. Phys.* **16** 837–40
Reuss E 1930 *Z. Angew. (Math.) Mech.* **10** 266–74
Spitzig W A, Sober R J and Richmond O 1975 *Acta Metall.* **23** 885–93

Taylor G I and Quinney H 1931 *Phil. Trans. R. Soc.* A **230** 323–62
Tresca H 1868 *Comptes Rendu Acad. Sci. Paris* **59** 754–8
von Mises R 1913 *Gottinger Nachrichten, Math.–Phys. Klasse* 582–92
Westergaard H M 1920 *J. Franklin Inst.* **189** 627–40

GENERAL REFERENCES

Bridgman P W 1952 *Studies in Large Plastic Flow and Fracture* (New York: McGraw-Hill)
Drucker D C 1967 *Introduction to the Mechanics of Deformable Solids* (New York: McGraw-Hill)
Freudenthal A M and Geiringer H 1958 *The Inelastic Continuum* in *Handbuch der Physik* vol. 4, ed. S Flügge (Berlin: Springer) pp 229–433
Fung Y C 1965 *Foundations of Solid Mechanics* (Englewood Cliffs, NJ: Prentice-Hall)
Hill R 1950 *Plasticity* (Oxford: Clarendon)
Lubahn J D and Felgar R P 1961 *Plasticity and Creep of Metals* (New York: Wiley)
Nadai A 1950 *Theory of Flow and Fracture of Solids* (New York: McGraw-Hill)
Prager W 1959 *An Introduction to Plasticity* (Reading, MA: Addison Wesley)
—— 1961 *Introduction to Mechanics of Continua* (Boston: Ginn)
Westergaard H M 1952 *Theory of Elasticity and Plasticity* (Cambridge, MA: Harvard University Press)

4

Purely Elastic Materials

4.1 SIMPLE ELASTIC MATERIALS

4.1.1 Constitutive Equation

A *simple solid* is a simple material for which at X there is a reference configuration B_r such that the associated isotropy group $\mathscr{G}$ is a subgroup of the orthogonal group $\mathscr{O}$; i.e.

$$\mathscr{G} \subset \mathscr{O} \qquad \text{(simple solid)}.$$

The preferred reference configurations satisfying this condition are identified as the undistorted states of the solid.

It has already been noted in §2.3.3 that a simple material is called *isotropic* at a particle X if it possesses at least one local configuration such that for the response functional from that configuration the isotropy group $\mathscr{G}$ contains the full orthogonal group $\mathscr{O}$; i.e.

$$\mathscr{O} \subset \mathscr{G} \qquad \text{(isotropic material)}.$$

The preferred reference configurations satisfying this condition are identified as the undistorted states of the isotropic material.

Thus, the definitions of a simple solid and an isotropic material may be combined to give a definition of an *undistorted state* of a simple material:

an undistorted state of a simple material is a reference configuration for which the isotropy group contains the full orthogonal group or is a subgroup of it.

A theorem of group theory states that there is no group $\mathscr{G}$ which is a proper subgroup of the full unimodular group $\mathscr{U}$ and which in turn contains the full orthogonal group $\mathscr{O}$ as a proper subgroup (Brauer 1965, Noll 1965). Thus, if $\mathscr{O} \subset \mathscr{G} \subset \mathscr{U}$, then either $\mathscr{O} = \mathscr{G}$ or else $\mathscr{G} = \mathscr{U}$; i.e. the orthogonal group is maximal in the unimodular group. It can therefore be concluded that isotropic simple materials must be either simple fluids or isotropic simple solids. Thus, for any undistorted state, if $\mathscr{G}=\mathscr{U}$ then the simple material is said to be an *isotropic simple solid*.

In equation (2.3.22) the response functional **G** is a material property that determines the stress for any choice of histories. Thus, for an isotropic elastic material, in the absence of thermal effects, the functional of the general constitutive equation (2.3.22) of a simple material reduces to a function. For these conditions the constitutive equation which characterises a *simple elastic material* can be obtained from equation (2.3.22) in the form

$$\mathbf{T}=\mathbf{g}(\mathbf{F}) \tag{4.1.1}$$

where **g** is a symmetric tensor-valued function. The appearance of the deformation gradient **F** in equation (4.1.1) implies a dependence of the stress on the choice of a given reference configuration. Since the choice of reference configuration is arbitrary, the constitutive equation (4.1.1) implies that changes of stress at a representative point of a simple elastic material arise in response to configuration changes regardless of the intermediate states occupied by the material in passing from one configuration to another.

From equations (2.1.39) and (2.3.24) it follows that the constitutive equation (4.1.1) is compatible with the principle of objectivity if and only if

$$\mathbf{g}(\mathbf{QF})=\mathbf{Qg}(\mathbf{F})\mathbf{Q}^{\mathrm{T}} \tag{4.1.2}$$

for all orthogonal tensors **Q**. Replacing **F** by its right polar decomposition in the form of the left-hand relation of equation (2.1.42), gives equation (4.1.2) in the form

$$\mathbf{g}(\mathbf{QRU})=\mathbf{Qg}(\mathbf{F})\mathbf{Q}^{\mathrm{T}} \tag{4.1.3}$$

where **R** is the orthogonal rotation tensor and where **U** is the right stretch tensor. By making the special choice $\mathbf{Q}=\mathbf{R}^{\mathrm{T}}$, equation (4.1.3) can be transformed to (cf equation (2.3.59))

$$\mathbf{g}(\mathbf{F})=\mathbf{Rg}(\mathbf{U})\mathbf{R}^{\mathrm{T}}. \tag{4.1.4}$$

Hence the elastic stress relation of equation (4.1.1) admits the reduced representation (cf equation (2.3.60)),

$$\mathbf{T}=\mathbf{Rg}(\mathbf{U})\mathbf{R}^{\mathrm{T}}. \tag{4.1.5}$$

From equation (4.1.5) it follows that the response function **g** can be determined from observations involving pure deformations alone relative to a reference configuration B_{r} since the effect of rotation is merely to rotate the stress by means of the rotation **R** relative to B_{r}.

An elastic material is isotropic if and only if its response function **g(F)**, taken relative to an undistorted state as reference configuration B_{r}, satisfies the relation (Noll 1958)

$$\mathbf{g}(\mathbf{FQ})=\mathbf{g}(\mathbf{F}) \tag{4.1.6}$$

for all orthogonal $\mathbf{Q}$. Replacing $\mathbf{F}$ by its left polar decomposition in the form of the right-hand relation of equation (2.1.42) and setting $\mathbf{Q}=\mathbf{R}^{\mathrm{T}}$ on the left-hand side of equation (4.1.6) gives $\mathbf{g}(\mathbf{F})=\mathbf{g}(\mathbf{V})$, where $\mathbf{V}$ is the left stretch tensor. Hence, using equation (2.1.46), the constitutive equation (4.1.1) can be expressed in the form

$$\mathbf{T}=\mathbf{g}(\mathbf{V}) \tag{4.1.7}$$

or, alternatively, since equation (2.1.45) gives $\mathbf{B}=\mathbf{V}^2$, it follows that for an isotropic simple elastic material, the constitutive equation (4.1.1) can be expressed in the form

$$\mathbf{T}=\mathbf{M}(\mathbf{B}) \tag{4.1.8}$$

where $\mathbf{M}$ is a symmetric tensor-valued function. Thus, it follows that the stress in an isotropic simple elastic material is wholly determined by the stretch measured from an undistorted state.

The restriction placed on the response function $\mathbf{M}$ by the principle of objectivity is

$$\mathbf{M}(\mathbf{Q}\mathbf{B}\mathbf{Q}^{\mathrm{T}})=\mathbf{Q}\mathbf{M}(\mathbf{B})\mathbf{Q}^{\mathrm{T}} \tag{4.1.9}$$

where use has been made of equation (4.1.2) and the identity $\mathbf{g}(\mathbf{F})=\mathbf{M}(\mathbf{B})$. Equation (4.1.9) is just a form of equation (1.11.22), from which it follows that the response function $\mathbf{M}$ is isotropic, and hence the representation theorem in the form of equation (1.11.30) can be applied to give

$$\mathbf{T}=\mathbf{M}(\mathbf{B})=a_0\mathbf{I}+a_1\mathbf{B}+a_2\mathbf{B}^2 \tag{4.1.10}$$

where the response coefficients

$$a_k=a_k(I_{\mathbf{B}}, II_{\mathbf{B}}, III_{\mathbf{B}}) \qquad (k=0,1,2) \tag{4.1.11}$$

are scalar functions of the principal invariants $I_{\mathbf{B}}$, $II_{\mathbf{B}}$ and $III_{\mathbf{B}}$ of $\mathbf{B}$. It follows from equation (4.1.10) that the stress tensor $\mathbf{T}$ and the deformation tensor $\mathbf{B}$ are coaxial in every configuration of an isotropic simple elastic material and that the stress in an undistorted state of such a material is spherical.

For the case that the argument tensor has an inverse, it follows from the Cayley–Hamilton theorem in the form of equation (1.11.34) that an equivalent form of equation (4.1.10) is the constitutive equation

$$\mathbf{T}=\beta_0\mathbf{I}+\beta_1\mathbf{B}+\beta_{-1}\mathbf{B}^{-1} \tag{4.1.12}$$

where again the β_a $(a=0, \pm 1)$ are scalar invariant functions of $I_{\mathbf{B}}$, $II_{\mathbf{B}}$ and $III_{\mathbf{B}}$ of $\mathbf{B}$.

The a_k $(k=0,1,2)$ and the β_a $(a=0, \pm 1)$ are in general connected by the relations

$$a_0=\beta_0+\frac{II_{\mathbf{B}}}{III_{\mathbf{B}}}\beta_{-1} \qquad a_1=\beta_1-\frac{I_{\mathbf{B}}}{III_{\mathbf{B}}}\beta_{-1} \qquad a_2=\frac{1}{III_{\mathbf{B}}}\beta_{-1} \tag{4.1.13}$$

$$\beta_0 = a_0 - II_{\mathbf{B}} a_2 \qquad \beta_1 = a_1 + I_{\mathbf{B}} a_2 \qquad \beta_{-1} = III_{\mathbf{B}} a_2. \tag{4.1.14}$$

For incompressible materials, the stress $\mathbf{T}$ in equation (4.1.1) must be replaced by the extra stress $\mathbf{T}_E = \mathbf{T} + P\mathbf{I}$ of equation (2.3.56), where P is an indeterminate pressure. Hence, for an incompressible isotropic, simple elastic material the constitutive equations (4.1.10) and (4.1.12) are replaced by

$$\mathbf{T} = -P_a\mathbf{I} + a_1\mathbf{B} + a_2\mathbf{B}^2 \tag{4.1.15}$$

$$\mathbf{T} = -P_\beta\mathbf{I} + \beta_1\mathbf{B} + \beta_{-1}\mathbf{B}^{-1} \tag{4.1.16}$$

where now the response coefficients

$$a_k = a_k(I_{\mathbf{B}}, II_{\mathbf{B}}) \qquad (k = 1, 2) \tag{4.1.17}$$

$$\beta_a = \beta_a(I_{\mathbf{B}}, II_{\mathbf{B}}) \qquad (a = \pm 1) \tag{4.1.18}$$

are functions only of $I_{\mathbf{B}}$ and $II_{\mathbf{B}}$, since for an incompressible material only isochoric deformations are possible, and therefore the conditions of equation (2.1.85) apply. In equations (4.1.15) and (4.1.16), the spherical terms $a_0\mathbf{I}$ and $\beta_0\mathbf{I}$ have been absorbed into the constraint stress. The condition $III_{\mathbf{B}} = 1$ reduces equations (4.1.13) and (4.1.14) to the relations

$$a_1 = \beta_1 - I_{\mathbf{B}}\beta_{-1} \qquad a_2 = \beta_{-1} \qquad \beta_1 = a_1 + I_{\mathbf{B}} a_2. \tag{4.1.19}$$

4.1.2 Response Function

Let b_i be the proper numbers of $\mathbf{B}$, the principal axes being defined by the orthonormal basis $\{\boldsymbol{q}_1, \boldsymbol{q}_2, \boldsymbol{q}_3\}$. Then $\mathbf{B}$ has the spectral representation

$$\mathbf{B} = \sum_{r=1}^{3} b_r \boldsymbol{q}_r \otimes \boldsymbol{q}_r. \tag{4.1.20}$$

Entering this form for $\mathbf{B}$ into the constitutive equation (4.1.12) gives

$$\mathbf{T} = \sum_{r=1}^{3} \sigma_r \boldsymbol{q}_r \otimes \boldsymbol{q}_r \tag{4.1.21}$$

where

$$\sigma_i = \beta_0 + \beta_1 b_i + \beta_{-1} b_i^{-1} \qquad (i = 1, 2, 3). \tag{4.1.22}$$

The constitutive equations (4.1.10), (4.1.12) and (4.1.16) can be expressed in the form

$$\begin{aligned}\mathbf{T}' &= \rho_0\mathbf{I} + \rho_1\mathbf{B}' + \rho_2\mathbf{B}'^2 \\ &= \beta_1\mathbf{B}' + \beta_{-1}(\mathbf{B}^{-1})' = \mathbf{M}'(\mathbf{B}')\end{aligned} \tag{4.1.23}$$

where a prime denotes a deviator, and where

$$\rho_0 = -\tfrac{1}{3}\rho_2 \operatorname{tr}\mathbf{B}'^2 \qquad \rho_1 = a_1 + \tfrac{2}{3} I_{\mathbf{B}} a_2 \qquad \rho_2 = a_2 \tag{4.1.24}$$

$$\rho_1 = \beta_1 - \tfrac{1}{3}\frac{I_{\mathbf{B}}}{III_{\mathbf{B}}}\beta_{-1} \qquad \rho_2 = \frac{1}{III_{\mathbf{B}}}\beta_{-1}$$

Equations (4.1.20) and (4.1.21) can be used to express equation (4.1.23) in the form

$$\begin{aligned}\sigma_i' &= \rho_0 + \rho_1 b_i' + \rho_2 b_i'^2 \\ &= \beta_1 b_i' + \beta_{-1}(b^{-1})_i' = m_i' \qquad (i=1,2,3)\end{aligned} \tag{4.1.25}$$

where the proper numbers σ_i' $(i=1,2,3)$ of $\mathbf{T}'$ are the deviators of the principal stresses σ_i $(i=1,2,3)$ and similarly the proper numbers b_i', $(b^{-1})_i'$ $(i=1,2,3)$ of $\mathbf{B}'$ and $(\mathbf{B}^{-1})'$ respectively, are the deviators of the proper numbers b_i and b_i^{-1} of $\mathbf{B}$ and $\mathbf{B}^{-1}$ respectively. The m_i' $(i=1,2,3)$ are the proper numbers of the response function $\mathbf{M}'(\mathbf{B}')$.

Lode's (1926) stress parameter,

$$\mu = 3\sigma_1'/(\sigma_3' - \sigma_2') \tag{4.1.26}$$

has been introduced in §3.1.3. It is important to note that the order in which the indices $(i=1,2,3)$ of the σ_i' appear in the definition of μ are determined by the inequalities existing between the σ_i $(i=1,2,3)$. Thus, the definition of Lode's stress parameter given by equation (4.1.26) relates to the inequalities $(\sigma_3 > \sigma_1 > \sigma_2$.

In a similar way, Lode (1926) introduced a strain parameter ν, defined, however, in the context of the incremental theory of plasticity. For present purposes, a deformation parameter κ based on Lode's (1926) strain parameter, will be defined by the relation

$$\kappa = \frac{3b_1'}{b_3' - b_2'} \tag{4.1.27}$$

where the b_i' $(i=1,2,3)$ are the proper numbers of $\mathbf{B}'$.

Entering the values of σ_i' given by equation (4.1.25) into equation (4.1.26) gives the identity

$$\mu = \frac{3\sigma_1'}{\sigma_3' - \sigma_2'} = \frac{3m_1'}{m_3' - m_2'} = \bar{\nu} \tag{4.1.28}$$

where

$$\bar{\nu} = \kappa\left(1 + \frac{3}{2}\frac{(1-\kappa^2)}{\kappa^2}\frac{1}{1-(\rho_1/\rho_2 b_1')}\right) \tag{4.1.29}$$

and where κ is defined by equation (4.1.27).

Equation (4.1.29) can be rearranged into the form

$$\mu = \kappa\left(1 + \frac{(1-\kappa^2)}{2\kappa^2}\frac{(3b_1 - I_{\mathbf{B}})}{(b_1 + III_{\mathbf{B}}q)}\right) \tag{4.1.30}$$

where

$$q = -\beta_1/\beta_{-1}. \tag{4.1.31}$$

The relation of equation (4.1.30) provides a possible way of determining whether a particular material satisfies the constitutive equation of a simple elastic material in the form of equation (4.1.23) using only information obtained from pure stretches.

4.1.3 Initial Yield

Equation (4.1.23) can be expressed in the form

$$J_2' = \tfrac{1}{2}\operatorname{tr}\mathbf{T}'^2 = \tfrac{1}{2}\operatorname{tr}\mathbf{M}'^2 \qquad (4.1.32)$$

or, alternatively,

$$J_2' - \varkappa = 0 \qquad (4.1.33)$$

where

$$\varkappa = \tfrac{1}{2}\operatorname{tr}\mathbf{M}'^2. \qquad (4.1.34)$$

Since $\varkappa$ is a function only of the deformation tensor $\mathbf{B}$, it is evident that equation (4.1.33) is of the form of equation (3.3.2). It is thus concluded that the constitutive equations (4.1.12) and (4.1.16) of a simple elastic material apply only to those materials which satisfy von Mises yield criterion.

Associated with the theories of crystal dislocation mechanics are various possible ways in which plastic deformation can occur at the crystal dislocation level. These various sub-continuum mechanisms of deformation are taken to imply the possible existence of materials which satisfy yield conditions other than the von Mises yield criterion. The existence of such materials requires the formulation of a more general form of constitutive equation.

4.2 NON-SIMPLE ELASTIC MATERIALS

4.2.1 Loading Function

The basic requirement is that the non-simple elastic material will, at initial yield, satisfy the generalised isotropic yield criterion discussed in Chapter 3. A simple way of satisfying this requirement is by way of two assumptions.

For initial and post yield deformation, both the elastic part of an element of the material and the associated plastic part are subject to identically the same stress system as that acting on the element as a whole. With this as a basis, it is assumed that there exists a stress intensity function

$$h = J_2' H(\omega) - a = 0 \qquad H(0) = 1 \qquad (4.2.1)$$

which defines the stress systems that determine a given mode of deformation over the entire range of deformation. Equation (4.2.1) is just equation (3.1.11) with k^2 replaced by a, it being noted that ω is defined by equation (3.1.12).

Thus, for initial and post yield conditions of deformation, h is identical to the yield function, the term *stress intensity function* being introduced so as to emphasise that it is now assumed to apply also to the purely elastic range of deformation which precedes initial yield.

The stress intensity function h is assumed to determine a loading function $\mathbf{G}'$ by way of the relation

$$\mathbf{G}' = \frac{\partial h}{\partial \mathbf{T}}. \tag{4.2.2}$$

Substituting for h in equation (4.2.2) from equation (4.2.1) gives

$$\mathbf{G}' = \delta_0 \mathbf{I} + \delta_1 \mathbf{T}' + \delta_2 \mathbf{T}'^2 \tag{4.2.3}$$

where the loading coefficients δ_a $(a = 0, 1, 2)$ are given by

$$\begin{aligned} \delta_0 &= -\tfrac{2}{3} J_2' \, \partial h / \partial J_3' = -\tfrac{2}{3} J_2' \delta_2 \\ \delta_1 &= \frac{\partial h}{\partial J_2'} = H\left(1 - 3 \frac{\omega}{H} \frac{\mathrm{d}H}{\mathrm{d}\omega}\right) \\ \delta_2 &= \frac{\partial h}{\partial J_3'} = 2H\left(\frac{J_2'}{J_3'}\right) \frac{\omega}{H} \frac{\mathrm{d}H}{\mathrm{d}\omega} \end{aligned} \tag{4.2.4}$$

having noted that the invariants J_2' and J_3' are defined by equation (3.1.4), and that H can take any of the forms discussed in Chapter 3.

Interpreting $\mathbf{G}'$ as a deviator implies the existence of a function

$$\mathbf{G} = c_0 \mathbf{I} + c_1 \mathbf{T} + c_2 \mathbf{T}^2 \tag{4.2.5}$$

where

$$\delta_0 = -\tfrac{2}{3} J_2' c_2 \qquad \delta_1 = c_1 + \tfrac{2}{3} I_{\mathrm{T}} c_2 \qquad \delta_2 = c_2 \tag{4.2.6}$$

having noted that I_{T} is the first invariant of the Cauchy stress tensor $\mathbf{T}$.

4.2.2 The Constitutive Equation

With the above assumptions as the basis, the constitutive equation for a class of general elastic materials will be taken to be

$$\mathbf{G}(\mathbf{T}) = \mathbf{g}(\mathbf{F}) \tag{4.2.7}$$

where, in the immediate context of equation (4.2.7), $\mathbf{G}$ is a tensor-valued function of $\mathbf{T}$, and where $\mathbf{g}$ is a tensor-valued function of the deformation gradient tensor $\mathbf{F}$. The constitutive equation (4.2.7) satisfies the principle of objectivity if and only if

$$\mathbf{G}(\mathbf{Q}\mathbf{T}\mathbf{Q}^{\mathrm{T}}) = \mathbf{Q}\mathbf{G}(\mathbf{T})\mathbf{Q}^{\mathrm{T}} \tag{4.2.8}$$

$$\mathbf{g}(\mathbf{Q}\mathbf{F}) = \mathbf{Q}\mathbf{g}(\mathbf{F})\mathbf{Q}^{\mathrm{T}} \tag{4.2.9}$$

for all orthogonal tensors $\mathbf{Q}$.

Equation (4.2.7) can be expressed in several different forms, reduced so as to satisfy the principle of objectivity. Thus, replacing $\mathbf{F}$ by its right polar decomposition $\mathbf{RU}$ on the left-hand side of equation (4.2.9) and setting $\mathbf{Q}=\mathbf{R}^{\mathrm{T}}$ gives equation (4.2.7) in the form

$$\mathbf{G}(\mathbf{T})=\mathbf{R}\mathbf{g}(\mathbf{U})\mathbf{R}^{\mathrm{T}} \tag{4.2.10}$$

where $\mathbf{U}$ is the right stretch tensor and where $\mathbf{R}$ is the rotation tensor.

An elastic material is isotropic if and only if its response function $\mathbf{g}(\mathbf{F})$, taken relative to an undistorted state as reference configuration B_{r}, satisfies the relation (Noll 1958)

$$\mathbf{g}(\mathbf{FQ})=\mathbf{g}(\mathbf{F}) \tag{4.2.11}$$

for all orthogonal $\mathbf{Q}$. Replacing $\mathbf{F}$ by its left polar decomposition $\mathbf{VR}$ and setting $\mathbf{Q}=\mathbf{R}^{\mathrm{T}}$ on the left-hand side of equation (4.2.11) gives $\mathbf{g}(\mathbf{F})=\mathbf{g}(\mathbf{V})$, where $\mathbf{V}$ is the left stretch tensor. From equation (2.1.45), $\mathbf{B}=\mathbf{V}^2$, and it therefore follows that for an isotropic elastic material, the constitutive equation (4.2.7) can be expressed in the form

$$\mathbf{G}(\mathbf{T})=\mathbf{M}(\mathbf{B}) \tag{4.2.12}$$

where $\mathbf{M}$ is a symmetric tensor-valued function, it being noted in relation to equation (4.2.10) that $\mathbf{V}=\mathbf{RUR}^{\mathrm{T}}$. The restriction placed on the response function $\mathbf{M}$ by the principle of objectivity is

$$\mathbf{M}(\mathbf{QBQ}^{\mathrm{T}})=\mathbf{QM}(\mathbf{B})\mathbf{Q}^{\mathrm{T}} \tag{4.2.13}$$

where use has been made of equation (4.2.9) and the identity $\mathbf{g}(\mathbf{F})=\mathbf{M}(\mathbf{B})$. Equation (4.2.13) is just a form of equation (1.11.22), from which it follows that the response function $\mathbf{M}$ is isotropic, and hence the representation theorem in the form of equation (1.11.30) can be applied to give

$$\mathbf{G}(\mathbf{T})=\phi_0\mathbf{I}+\phi_1\mathbf{B}+\phi_2\mathbf{B}^2 \tag{4.2.14}$$

where the response coefficients,

$$\phi_a=\phi_a(I_{\mathbf{B}}, II_{\mathbf{B}}, III_{\mathbf{B}}) \qquad (a=0, 1, 2) \tag{4.2.15}$$

are scalar functions of the principal invariants $I_{\mathbf{B}}$, $II_{\mathbf{B}}$ and $III_{\mathbf{B}}$ of $\mathbf{B}$. It follows from equation (4.2.14) that the loading function $\mathbf{G}$, and the deformation tensor $\mathbf{B}$ are coaxial in every configuration of an isotropic, general elastic material and that the loading function in an undistorted state of such a material is spherical.

Since equation (4.2.8) is just a form of equation (1.11.22), it follows that the loading function $\mathbf{G}$ is isotropic and hence the representation theorem in the form of equation (1.11.30) can be applied to give

$$\mathbf{G}(\mathbf{T})=c_0\mathbf{I}+c_1\mathbf{T}+c_2\mathbf{T}^2 \tag{4.2.16}$$

which is just equation (4.2.5), having noted that the coefficients

$$c_a = c_a(I_{\mathrm{T}}, II_{\mathrm{T}}, III_{\mathrm{T}}) \qquad (a=0,1,2) \tag{4.2.17}$$

are scalar functions of the principal invariants I_{T}, II_{T} and III_{T} of $\mathbf{T}$.

Substituting for $\mathbf{G}$ in equation (4.2.14) from equation (4.2.16) gives

$$\begin{aligned}\mathbf{G}(\mathbf{T}) = c_0\mathbf{I} + c_1\mathbf{T} + c_2\mathbf{T}^2 = \mathbf{M}(\mathbf{B}) &= \phi_0\mathbf{I} + \phi_1\mathbf{B} + \phi_2\mathbf{B}^2 \\ &= \beta_0\mathbf{I} + \beta_1\mathbf{B} + \beta_{-1}\mathbf{B}^{-1}\end{aligned} \tag{4.2.18}$$

which is the required form of the constitutive equation. Using the Cayley–Hamilton theorem, the constitutive equation (4.2.18) can be rearranged into the alternative form

$$\mathbf{G}(\mathbf{T}) = c_0\mathbf{I} + c_1\mathbf{T} + c_2\mathbf{T}^2 = \mathbf{M}(\bar{\mathbf{E}}) = \tilde{\alpha}_0\mathbf{I} + \tilde{\alpha}_1\bar{\mathbf{E}} + \tilde{\alpha}_2\bar{\mathbf{E}}^2 \tag{4.2.19}$$

where the strain tensor

$$\bar{\mathbf{E}} = \tfrac{1}{4}(\mathbf{B} - \mathbf{B}^{-1}) \tag{4.2.20}$$

and where

$$\tilde{\alpha}_0 = \phi_0 + \frac{1}{4}\frac{II_{\mathrm{B}}}{III_{\mathrm{B}}}\tilde{\alpha}_1 + \frac{1}{16}\left(2 + \frac{I_{\mathrm{B}}}{III_{\mathrm{B}}} - \frac{II_{\mathrm{B}}^2}{III_{\mathrm{B}}^2}\right)\tilde{\alpha}_2 \tag{4.2.21}$$

$$\tilde{\alpha}_1 = 4\frac{(II_{\mathrm{B}} + III_{\mathrm{B}}^2)[\phi_1 + \phi_2(I_{\mathrm{B}} + III_{\mathrm{B}})]}{[1 + II_{\mathrm{B}} + (I_{\mathrm{B}} + III_{\mathrm{B}})III_{\mathrm{B}}]} - 4III_{\mathrm{B}}\phi_2 \tag{4.2.22}$$

$$\tilde{\alpha}_2 = 16\frac{III_{\mathrm{B}}[\phi_1 + \phi_2(I_{\mathrm{B}} + III_{\mathrm{B}})]}{[1 + II_{\mathrm{B}} + (I_{\mathrm{B}} + III_{\mathrm{B}})III_{\mathrm{B}}]}. \tag{4.2.23}$$

With regard to equations (4.2.18) and (4.2.19), it is important to note that the response function $\mathbf{M}$ of equation (4.2.19) is identically equivalent to the response function $\mathbf{M}$ of equation (4.2.18). This follows from realising that the representation in equation (4.2.19) is no more than a rearrangement of the representation in equation (4.2.18) which has been obtained solely through the Cayley–Hamilton theorem.

The constitutive equation (4.2.18) differs from the constitutive equation (4.1.10) by an additional term which is quadratic in the stress. Although the constitutive equation now contains a term which is quadratic in stress it is important to note that the assumptions formulated by way of equations (4.2.1) and (4.2.2) ensure that the dimensions of the loading function remain those of stress. This follows from noting that both ω and H are dimensionless and hence the loading coefficient δ_1 defined by equation (4.2.4) is dimensionless in contrast with δ_2 whose dimensions are seen from equation (4.2.4) to be those of the ratio J'_2/J'_3. From equation (3.1.24)

$$\frac{J'_2}{J'_3} = -3\frac{(3+\mu^2)}{(9-\mu^2)}\frac{1}{\sigma'_1} \tag{4.2.24}$$

and since μ is dimensionless, it follows that the ratio J_2'/J_3' has the dimensions of reciprocal stress and hence δ_2 also has the dimensions of reciprocal stress. Thus, using equation (4.2.6), these conditions on δ_1 and δ_2 make

$$c_1 = \delta_1 - \tfrac{2}{3} I_{\mathrm{T}} \delta_2$$

dimensionless, and the product $c_2 \mathbf{T}^2$ must have the dimensions of stress, from which it follows that the loading function $\mathbf{G}$ defined by equation (4.2.5) must have the dimensions of stress.

For incompressible materials, the loading function $\mathbf{G}$ in equations (4.2.18) and (4.2.19) must be replaced by the extra stress $\mathbf{T}_{\mathrm{E}} = \mathbf{G} + P\mathbf{I}$, where P is an indeterminate pressure. Hence, for an incompressible isotropic non-simple elastic material the constitutive equations (4.2.18) and (4.2.19) are replaced by

$$\begin{aligned} \mathbf{G}(\mathbf{T}) &= -P_\phi \mathbf{I} + \phi_1 \mathbf{B} + \phi_2 \mathbf{B}^2 \\ &= -P_\alpha \mathbf{I} + \tilde{\alpha}_1 \bar{\mathbf{E}} + \tilde{\alpha}_2 \bar{\mathbf{E}}^2 \end{aligned} \tag{4.2.25}$$

where now the response coefficients,

$$\phi_k = \phi_k(I_{\mathbf{B}}, II_{\mathbf{B}}) \qquad \tilde{\alpha}_k = \tilde{\alpha}_k(I_{\mathbf{B}}, II_{\mathbf{B}}) \qquad (k = 1, 2) \tag{4.2.26}$$

are functions only of $I_{\mathbf{B}}$ and $II_{\mathbf{B}}$, since for an incompressible material, only isochoric deformations are possible, and therefore the conditions of equation (2.1.85) apply. In equation (4.2.25), the spherical terms $\phi_0 \mathbf{I}$ and $\tilde{\alpha}_0 \mathbf{I}$ have been absorbed into the corresponding constraint stress.

Equations (4.2.18), (4.2.19) and (4.2.25) can be expressed in the form

$$\mathbf{G}'(\mathbf{T}') = \delta_0 \mathbf{I} + \delta_1 \mathbf{T}' + \delta_2 \mathbf{T}'^2 = \mathbf{M}'(\mathbf{B}') = 2\bar{G}\bar{\mathbf{M}}'(\bar{\mathbf{E}}') \tag{4.2.27}$$

where

$$\bar{G} = \bar{G}(K_2', K_3')$$

is a non-negative scalar factor of proportionality, and where

$$K_2' = \tfrac{1}{2} \operatorname{tr} \bar{\mathbf{E}}'^2 \qquad K_3' = \det \bar{\mathbf{E}}'. \tag{4.2.28}$$

4.2.3 Response Function

Entering the form for $\mathbf{B}$ given by equation (4.1.20) into the constitutive equation (4.2.18) gives

$$\mathbf{G} = \sum_{r=1}^{3} g_r \boldsymbol{q}_r \otimes \boldsymbol{q}_r = \sum_{r=1}^{3} (c_0 + c_1 \sigma_r + c_2 \sigma_r^2) \boldsymbol{q}_r \otimes \boldsymbol{q}_r \tag{4.2.29}$$

where

$$\begin{aligned} g_i &= \phi_0 + \phi_1 b_i + \phi_2 b_i^2 \\ &= c_0 + c_1 \sigma_i + c_2 \sigma_i^2 \qquad (i = 1, 2, 3) \end{aligned} \tag{4.2.30}$$

and where the g_i are the proper numbers of the symmetric tensor $\mathbf{G}$. In the same way there can be obtained from equation (4.2.19) the relation

$$g_i = c_0 + c_1\sigma_i + c_2\sigma_i^2 = \tilde{\alpha}_0 + \tilde{\alpha}_1\bar{\varepsilon}_i + \tilde{\alpha}_2\bar{\varepsilon}_i^2 \qquad (i=1,2,3) \qquad (4.2.31)$$

where the principal strains

$$\bar{\varepsilon}_i = [b_i - (b^{-1})_i]/4 \qquad (i=1,2,3)$$

are the proper numbers of $\bar{\mathbf{E}}$.

With the definition of Lode's (1926) stress parameter defined by equation (4.1.26) as the basis, the following useful parameter can be defined

$$\bar{\mu} = 3g_1'/(g_3' - g_2') \qquad (4.2.32)$$

where the g_i' $(i=1,2,3)$ are the proper numbers of $\mathbf{G}'$. It follows from equation (4.2.27) that $\mathbf{G}'$ and $\mathbf{T}'$ are coaxial in every loading state of an isotropic material. Hence, entering the spectral representations of the symmetric tensors $\mathbf{G}'$ and $\mathbf{T}'$ into equation (4.2.27) gives

$$g_i' = \delta_0 + \delta_1\sigma_i' + \delta_2\sigma_i'^2 \qquad (i=1,2,3). \qquad (4.2.33)$$

Substituting for the g_i' in equation (4.2.32) from equation (4.2.33) and rearranging gives

$$\bar{\mu} = \mu\left(1 + \frac{(1-\mu^2)}{2\mu^2}\,\frac{3}{1-(\delta_1/\delta_2\sigma_1')}\right) \qquad (4.2.34)$$

where use has been made of equation (4.1.26), and where, for the stress intensity function h of equation (4.2.1), the δ_1 and δ_2 are defined by equations (4.2.4). It is of particular interest to note that the assumptions formulated by way of equations (4.2.1) and (4.2.2) identify the $\bar{\mu}$ of equation (4.2.32) with the $\bar{\mu}$ of equation (3.1.16) it being further noted that these same assumptions relate the Λ of equation (3.1.17) to δ_1 and δ_2 by way of the relations

$$\Lambda = \delta_1 - \delta_2\sigma_1' = H\left[1 - \tfrac{1}{6}(3+\mu^2)\,\frac{\mu}{H}\,\frac{\mathrm{d}H}{\mathrm{d}\mu}\right] \qquad (4.2.35)$$

where use has been made of equations (4.2.4) and (4.2.33).

Using equation (4.2.29), the spectral representation of $\mathbf{G}'$ can be entered into equation (4.2.27) to give

$$g_i' = m_i' = 2\bar{G}\bar{m}_i' \qquad (i=1,2,3) \qquad (4.2.36)$$

where the g_i' are given by equation (4.2.33), and the $m_i', \bar{m}_i'$ are the proper numbers of the symmetric tensors

$$\mathbf{M}' = \sum_{r=1}^{3} m_r'\boldsymbol{q}_r \otimes \boldsymbol{q}_r \qquad \bar{\mathbf{M}}' = \sum_{r=1}^{3} \bar{m}_r'\boldsymbol{q}_r \otimes \boldsymbol{q}_r. \qquad (4.2.37)$$

Substituting for the g_i' in equation (4.2.32) from equation (4.2.36) gives the *generalised Lode relation,*

$$\bar{\mu}=\frac{3g_1'}{g_3'-g_2'}\equiv\frac{3m_1'}{m_3'-m_2'}\equiv\frac{3\bar{m}_1'}{\bar{m}_3'-\bar{m}_2'}=\bar{v} \tag{4.2.38}$$

where, in the context of the deformation behaviour of the material, the parameter

$$\bar{v}=\frac{3m_1'}{m_3'-m_2'}=\frac{3\bar{m}_1'}{\bar{m}_3'-\bar{m}_2'} \tag{4.2.39}$$

is the counterpart to $\bar{\mu}$. There are two cases of interest, depending upon which representation of the response function $\mathbf{M}'$ is used.

(i) $\mathbf{M}'=\mathbf{M}'(\mathbf{B}')$: for this representation, equation (4.2.27) can be expressed in the form

$$\begin{aligned}\mathbf{G}'(\mathbf{T}')=\mathbf{M}'(\mathbf{B}')&=\rho_0\mathbf{I}+\rho_1\mathbf{B}'+\rho_2\mathbf{B}'^2\\&=\beta_1\mathbf{B}'+\beta_{-1}(\mathbf{B}^{-1})'\end{aligned} \tag{4.2.40}$$

where

$$\begin{aligned}\rho_0=-\tfrac{1}{3}\rho_2\,\mathrm{tr}\,\mathbf{B}'^2 \qquad \rho_1=\phi_1+\tfrac{2}{3}I_{\mathbf{B}}\phi_2 \qquad \rho_2=\phi_2\\ \beta_1=\rho_1+\tfrac{1}{3}I_{\mathbf{B}}\rho_2 \qquad \beta_{-1}=III_{\mathbf{B}}\rho_2\end{aligned} \tag{4.2.41}$$

Equations (4.1.20) and (4.2.29) can be used to express equation (4.2.40) in the form

$$g_i'=m_i'=\rho_0+\rho_1 b_i'+\rho_2 b_i'^2=\beta_1 b_i'+\beta_{-1}(b^{-1})_i' \qquad (i=1,2,3) \tag{4.2.42}$$

which is to be compared with equation (4.1.25). Substituting in equation (4.2.39) for the m_i' from equation (4.2.42) gives,

$$\bar{v}=\kappa\left(1+\frac{(1-\kappa^2)}{2\kappa^2}\,\frac{3}{1-(\rho_1/\rho_2 b_1')}\right) \tag{4.2.43}$$

which is just equation (4.1.29), κ being defined by equation (4.1.27).

(ii) $\mathbf{M}'=2\bar{G}\bar{\mathbf{M}}'(\bar{\mathbf{E}}')$: for this representation, equation (4.2.27) can be expressed in the form

$$\mathbf{G}'(\mathbf{T}')=\mathbf{M}'(\bar{\mathbf{E}}')=2\bar{G}\bar{\mathbf{M}}'(\bar{\mathbf{E}}')=2\bar{G}(\varphi_0\mathbf{I}+\varphi_1\bar{\mathbf{E}}'+\varphi_2\bar{\mathbf{E}}'^2) \tag{4.2.44}$$

where

$$\varphi_0=-\tfrac{2}{3}K_2'\varphi_2 \qquad \varphi_1=(\tilde{\alpha}_1+\tfrac{2}{3}I_{\bar{\mathbf{E}}}\tilde{\alpha}_2)/(2\bar{G}) \qquad \varphi_2=\tilde{\alpha}_2/(2\bar{G}) \tag{4.2.45}$$

use having been made of the Cayley–Hamilton theorem. Entering the spectral representations of the symmetric tensors $\bar{\mathbf{M}}'$ and

$$\bar{\mathbf{E}}'=\sum_{r=1}^{3}\bar{\varepsilon}_r'\boldsymbol{q}_r\otimes\boldsymbol{q}_r \qquad \bar{\varepsilon}_i'=[b_i'-(b^{-1})_i']/4 \qquad (i=1,2,3)$$

into equation (4.2.44) gives

$$g_i' = \delta_0 + \delta_1 \sigma_i' + \delta_2 \sigma_i'^2 = m_i' = 2\bar{G}(\varphi_0 + \varphi_1 \bar{\varepsilon}_i' + \varphi_2 \bar{\varepsilon}_i'^2) = 2\bar{G}\bar{m}_i'. \quad (4.2.46)$$

Substituting for the $\bar{m}_i'$ in equation (4.2.39) from equation (4.2.46) and rearranging gives

$$\bar{v} = v\left(1 + \frac{(1-v^2)}{2v^2} \frac{3}{1-(\varphi_1/\varphi_2 \bar{\varepsilon}_1')}\right) \quad (4.2.47)$$

where

$$v = 3\bar{\varepsilon}_1'/(\bar{\varepsilon}_3' - \bar{\varepsilon}_2') \quad (4.2.48)$$

can be regarded as a modified form of Lode's (1926) strain parameter.

It is evident from equation (4.2.38) that both $\bar{v}$ and $\bar{\mu}$ are independent of which form of representation is used for $\mathbf{M}'$. Thus, the only parameters which distinguish the two representations for $\mathbf{M}'$ are the κ and v defined by equations (4.1.27) and (4.2.48). The two representations for $\mathbf{M}'$ are related by the response coefficients of equation (4.2.45), where use of equation (4.2.41) gives the $\tilde{\alpha}_i$ $(i=0, 1, 2)$ of equations (4.2.21), (4.2.22) and (4.2.23) in the form

$$\tilde{\alpha}_0 = \beta_0 + (\beta_1 + \beta_{-1}) \frac{(I_{\mathbf{B}} + II_{\mathbf{B}} III_{\mathbf{B}} + 2III_{\mathbf{B}})}{(1 + I_{\mathbf{B}} III_{\mathbf{B}} + II_{\mathbf{B}} + III_{\mathbf{B}}^2)} \quad (4.2.49)$$

$$\tilde{\alpha}_1 = 4 \frac{(\beta_1 + \beta_{-1})(II_{\mathbf{B}} + III_{\mathbf{B}}^2)}{(1 + I_{\mathbf{B}} III_{\mathbf{B}} + II_{\mathbf{B}} + III_{\mathbf{B}}^2)} - 4\beta_{-1} \quad (4.2.50)$$

$$\tilde{\alpha}_2 = \frac{16(\beta_1 + \beta_{-1}) III_{\mathbf{B}}}{(1 + I_{\mathbf{B}} III_{\mathbf{B}} + II_{\mathbf{B}} + III_{\mathbf{B}}^2)} \quad (4.2.51)$$

and hence the strain response coefficients,

$$\varphi_0 = -\tfrac{2}{3} K_2' \varphi_2 \qquad \varphi_2 = \tilde{\alpha}_2/(2\bar{G})$$

$$\varphi_1 = \tfrac{2}{3}(\beta_1 + \beta_{-1}) \frac{[3(II_{\mathbf{B}} + III_{\mathbf{B}}^2) + 2(I_{\mathbf{B}} - II_{\mathbf{B}}) III_{\mathbf{B}}]}{(1 + I_{\mathbf{B}} III_{\mathbf{B}} + II_{\mathbf{B}} + III_{\mathbf{B}}^2)\bar{G}} - 2\frac{\beta_{-1}}{\bar{G}} \quad (4.2.52)$$

where use has been made of equation (4.2.45). The deformation response coefficients β_i $(i = \pm 1)$ can be obtained from equation (4.2.52) in the form,

$$\beta_1 = \tfrac{1}{2}\bar{G}[\varphi_1 + \tfrac{1}{12}(3 + I_{\mathbf{B}} III_{\mathbf{B}} + 2II_{\mathbf{B}} III_{\mathbf{B}})\varphi_2] \quad (4.2.53)$$

$$\beta_{-1} = -\tfrac{1}{2}\bar{G}[\varphi_1 - \tfrac{1}{12}(3III_{\mathbf{B}}^2 + 2I_{\mathbf{B}} III_{\mathbf{B}} + 3II_{\mathbf{B}} - 2II_{\mathbf{B}} III_{\mathbf{B}})\varphi_2]. \quad (4.2.54)$$

For the purposes of reference, case (i) will be referred to as a deformation type response function or simply as the *deformation response function* and case (ii) will be referred to as a strain type of response function or simply as the *strain response function.*

4.3 THE RESPONSE COEFFICIENTS

4.3.1 Deformation Response Function

By direct analogy with the formulation of the concept of a stress intensity function, it is assumed that there exists a *deformation intensity function*

$$m = L_2' K(\Gamma) - c = 0 \qquad K(0) = 1 \tag{4.3.1}$$

where

$$\Gamma = \frac{27}{4}\frac{L_3'^2}{L_2'^3} = \kappa^2 \frac{(\kappa^2 - 9)^2}{(3+\kappa^2)^3} \qquad (0 \leqslant \Gamma \leqslant 1) \tag{4.3.2}$$

and where

$$L_2' = \tfrac{1}{2}\operatorname{tr}\mathbf{B}'^2 \qquad L_3' = \det \mathbf{B}' \tag{4.3.3}$$

it being noted that c is to be regarded as a constant for a given state of loading. The further assumption is made that

$$\mathbf{M}'(\mathbf{B}') = 2\tilde{G}\frac{\partial m}{\partial \mathbf{B}} = 2\tilde{G}(\hat{\rho}_0 \mathbf{I} + \hat{\rho}_1 \mathbf{B}' + \hat{\rho}_2 \mathbf{B}'^2) \tag{4.3.4}$$

which is to be compared with equation (4.2.2). Substituting for m in equation (4.3.4) from equation (4.3.1) and comparing with equation (4.2.40) gives the deformation response coefficients,

$$\rho_0 = 2\tilde{G}\hat{\rho}_0 = -\tfrac{2}{3}L_2'\rho_2 \qquad \rho_1 = 2\tilde{G}\hat{\rho}_1 = 2\tilde{G}K\left(1 - 3\frac{\Gamma\,\mathrm{d}K}{K\,\mathrm{d}\Gamma}\right)$$
$$\rho_2 = 2\tilde{G}\hat{\rho}_2 = -12\tilde{G}\frac{K}{b_1'}\frac{(3+\kappa^2)}{(9-\kappa^2)}\frac{\Gamma}{K}\frac{\mathrm{d}K}{\mathrm{d}\Gamma} \tag{4.3.5}$$

which is to be compared with equation (4.2.4). Thus, given a form for K, the deformation response coefficients ρ_a $(a = 0, 1, 2)$ can be evaluated from equation (4.3.5). Entering the form for the ρ_p $(p = 1, 2)$ given by equation (4.3.5) into equation (4.2.43) and rearranging gives

$$\bar{\nu} = \kappa\left[1 - \frac{(3+\kappa^2)}{\kappa^2}\left(1 - \frac{K}{\psi}\right)\right] \tag{4.3.6}$$

where

$$\psi = \hat{\rho}_1 - \hat{\rho}_2 b_1' = K\left[1 - \frac{(3+\kappa^2)}{6}\frac{\kappa}{K}\frac{\mathrm{d}K}{\mathrm{d}\kappa}\right] \tag{4.3.7}$$

use having been made of equation (4.3.5).

4.3.2 Strain Response Function

By direct analogy with the formulation of the concept of a stress intensity function, it is assumed that there exists a *strain intensity function*

$$g = K_2' N(\Omega) - b = 0 \qquad N(0) = 1 \tag{4.3.8}$$

where

$$\Omega = \frac{27}{4} \frac{K_3'^2}{K_2'^3} = v^2 \frac{(v^2-9)^2}{(3+v^2)^3} \qquad (0 \leqslant \Omega \leqslant 1) \tag{4.3.9}$$

and where b has to be regarded as a constant for a given state of loading. The further assumption is made that,

$$\bar{\mathbf{M}}' = \partial g / \partial \bar{\mathbf{E}} \tag{4.3.10}$$

which is to be compared with equation (4.2.2). Substituting for g in equation (4.3.10) from equation (4.3.8) and comparing with equation (4.2.44) gives the strain response coefficients

$$\begin{aligned} \varphi_0 &= -\tfrac{2}{3} K_2' \frac{\partial g}{\partial K_3'} = -\tfrac{2}{3} K_2' \varphi_2 \\ \varphi_1 &= \frac{\partial g}{\partial K_2'} = N\left(1 - 3 \frac{\Omega}{N} \frac{\mathrm{d}N}{\mathrm{d}\Omega}\right) \\ \varphi_2 &= \frac{\partial g}{\partial K_3'} = 2N\left(\frac{K_2'}{K_3'}\right) \frac{\Omega}{N} \frac{\mathrm{d}N}{\mathrm{d}\Omega} \end{aligned} \tag{4.3.11}$$

which is to be compared with equation (4.2.4). Thus, given a form for N, the strain response coefficients φ_a $(a = 0, 1, 2)$ can be evaluated from equation (4.3.11). Entering the form for the φ_p $(p = 1, 2)$ given by equation (4.3.11) into equation (4.2.47) and rearranging gives

$$\bar{v} = v\left[1 - \frac{(3+v^2)}{v^2}\left(1 - \frac{N}{\Psi}\right)\right] \tag{4.3.12}$$

where

$$\Psi = \varphi_1 - \varphi_2 \bar{\varepsilon}_1' = N\left(1 - \frac{(3+v^2)}{6} \frac{v}{N} \frac{\mathrm{d}N}{\mathrm{d}v}\right). \tag{4.3.13}$$

There are three forms for N of present interest.

(i) ***Continuously differentiable.*** Taking

$$N = 1 - \bar{c}\Omega \qquad (\bar{c} = \text{constant}) \tag{4.3.14}$$

gives

$$\varphi_0 = -\tfrac{2}{3} K_2' \varphi_2 \qquad \varphi_1 = 1 + 2\bar{c}\Omega \qquad \varphi_2 = 6 \frac{(3+v^2)}{(9-v^2)} \frac{\bar{c}\Omega}{\bar{\varepsilon}_1'} \tag{4.3.15}$$

which when entered into equation (4.3.12) gives

$$\bar{\nu}=\nu\left(1-\frac{9\bar{c}(3+\nu^2)(1-\nu^2)(9-\nu^2)}{(3+\nu^2)^3-8\bar{c}\nu^4(9-\nu^2)}\right)\equiv\bar{\mu} \tag{4.3.16}$$

where use has been made of equation (4.3.13).

(ii) ***Piecewise linear.*** For a given state of loading, the following relation can be obtained from equation (4.3.8)

$$b=\tfrac{1}{12}(3+\nu^2)(\bar{\varepsilon}_3'-\bar{\varepsilon}_2')^2N(\Omega)$$

which can be differentiated with respect to $\bar{\varepsilon}_1'$ and then rearranged using equation (4.2.47) and (4.2.48) to give the condition,

$$\bar{\nu}=-\frac{\mathrm{d}}{\mathrm{d}\bar{\varepsilon}_1'}(\bar{\varepsilon}_3'-\bar{\varepsilon}_2'). \tag{4.3.17}$$

An interesting form for the strain intensity function results from taking $\bar{\nu}=$ constant in equation (4.3.17), integrating and rearranging to give

$$\tfrac{4}{3}K_2'\frac{(3+\bar{\nu}\nu)^2}{(3+\nu^2)}-A_{(2)}^2=0 \qquad (A_{(2)}=\text{constant}) \tag{4.3.18}$$

which when compared with equation (4.3.8) is taken to imply

$$N=\frac{(3+\bar{\nu}_n\nu)^2}{(3+\nu^2)}\left(\frac{(3+\nu^2)}{(3+\bar{\nu}_n\nu)^2}\right)_{\Omega=0} \qquad (n=0,1,2) \tag{4.3.19}$$

where,

$$|\bar{\nu}_n|=|\bar{\nu}_0|+\frac{(3-2|\bar{\nu}_0|-\bar{\nu}_0^2)}{(1+|\bar{\nu}_0|)}n(2-n)+\frac{(3+\bar{\nu}_0^2)\,n(n-1)}{(1-|\bar{\nu}_0|\quad 2} \qquad (n=0,1,2) \tag{4.3.20}$$

and where, for $n=0$, equation (4.3.17) gives

$$\bar{\nu}_0=-\lim_{\bar{\varepsilon}_1'\to 0}\frac{\mathrm{d}}{\mathrm{d}\bar{\varepsilon}_1'}(\bar{\varepsilon}_3'-\bar{\varepsilon}_2') \tag{4.3.21}$$

it being noted that,

$$n=\begin{cases}0\\1\\2\end{cases}\quad\text{for}\quad\begin{cases}(0\leqslant\nu^2\leqslant 1) & (0\leqslant\Omega\leqslant 1)\\(1\leqslant\nu^2\leqslant 9) & (1\geqslant\Omega\geqslant 0).\\(9\leqslant\nu^2\leqslant\infty) & (0\leqslant\Omega\leqslant 1)\end{cases} \tag{4.3.22}$$

(iii) ***Composite strain intensity function.*** A composite strain intensity function can be obtained by taking N to be of the form

$$N=\hat{\phi}\hat{N}_{(1)}+(\hat{N}_{(2)}-\hat{\phi}\hat{N}_{(1)})U \tag{4.3.23}$$

where U is some appropriate form of step function. $\hat{N}_{(1)}$ is restricted to be of the form of equation (4.3.19) and $\hat{N}_{(2)}$ is restricted to be of the form of equation (4.3.14). In the context of equation (4.3.22), for each value of $n=0, 1, 2$, there is a particular value of $v=v^*$ at which the strain intensity function changes from that obtained using equation (4.3.8) with $N=\hat{N}_{(2)}$, for which $U=1$, to that obtained using equation (4.3.8) with $N=\hat{\phi}\hat{N}_{(1)}$, for which $U=0$. Thus, at the discontinuous change from $U=0$ to $U=1$

$$[N]_{v=v^*}=[\hat{N}_{(2)}]_{v=v^*}=[\hat{\phi}\hat{N}_{(1)}]_{v=v^*} \tag{4.3.24}$$

it being noted that $\hat{\phi}$ is solely a function of v^*.

Since there is a direct analogy between the formulation of the strain intensity function g and the stress intensity function h, which itself depends upon the formulation of the initial yield function, further details regarding g can be obtained by way of Chapter 3.

4.4 RESTRICTIONS ON THE FORM OF THE STRESS AND STRAIN INTENSITY FUNCTIONS

In the context of the spectral representation,

$$\bar{\mathbf{E}}'=\sum_{r=1}^{3} \bar{\varepsilon}_r' \boldsymbol{q}_r \otimes \boldsymbol{q}_r \tag{4.4.1}$$

let the principal axes of $\bar{\mathbf{E}}'$ be oriented so that they are equally inclined to a plane Π, through the origin o of the system (o, q). Each of the deviatoric strain space axes makes an angle $\cos^{-1}\sqrt{\frac{2}{3}}$ with the Π-plane, and hence the projected deviatoric components are $\sqrt{\frac{2}{3}}\,\bar{\varepsilon}_i'$ $(i=1,2,3)$. The $(\tilde{u}, \tilde{v})$ curve, where

$$\tilde{u}=\sqrt{\tfrac{3}{2}}\,\bar{\varepsilon}_1'=v\tilde{v}/\sqrt{3} \qquad \tilde{v}=\frac{(\bar{\varepsilon}_3'-\bar{\varepsilon}_2')}{\sqrt{2}}=\frac{\sqrt{3}}{(3+v^2)^{1/2}N^{1/2}}\sqrt{2b} \tag{4.4.2}$$

constitutes a geometrical representation in the Π-plane of any strain system defined by the strain intensity function g. It is of particular interest to note that

$$\bar{v}=-\sqrt{3}\,\frac{d\tilde{v}}{d\tilde{u}}=-\frac{d}{d\bar{\varepsilon}_1'}(\bar{\varepsilon}_3'-\bar{\varepsilon}_2')\equiv-\frac{d}{d\sigma_1'}(\sigma_3'-\sigma_2')=-\sqrt{3}\,\frac{dv}{du}=\bar{\mu} \tag{4.4.3}$$

where use has been made of equations (3.1.14), (3.1.15) and (4.2.38). Thus, for $v=\bar{v}$ for all v, the $(\tilde{u}, \tilde{v})$ curve in the Π-plane is a circle. Deviations from a circle in the Π-plane are obtained by taking N to be given by equation (4.3.14). For a piecewise linear, continuous, strain intensity function, the $(\tilde{u}, \tilde{v})$ curve in the Π-plane consists of twelve linear sections, it being noted that $\bar{v}\equiv v_n=$ constant for all v in each of the three ranges of v^2 identified by the three values of n in equation (4.3.22).

Thus far the stress and strain parameters, μ and ν, have been considered separately. However, in terms of the constitutive equation (4.2.27), the principal axes of $\bar{\mathbf{M}}'$ and $\bar{\mathbf{E}}'$ are also those of $\mathbf{G}'$ and $\mathbf{T}'$, and hence the π-plane of §3.1.2 must be indistinguishable from the Π-plane, this common plane being the deviatoric plane of §3.1.2. This conclusion leads to the interesting result that, in principle, the strain systems defined by the strain intensity function g, can have geometrical representations in the deviatoric plane which differ from that for the corresponding stress system. Compatibility of the two different geometrical representations is ensured, at least in principle, by the identity of equation (4.2.38), it being noted that

$$\mu^2 = \bar{\mu}^2 \equiv \bar{\nu}^2 = \nu^2 = \begin{cases} 0, 9 \\ 1, \infty \end{cases} \qquad \text{for} \quad \omega = \Omega = \begin{cases} 0 \\ 1 \end{cases} \tag{4.4.4}$$

where ω is defined by equation (3.1.12).

The identity of equation (4.2.38) does, however, exclude the possible use of a continuously differentiable stress intensity function with a piecewise *linear* continuous, strain intensity function. This is because in the case of a continuously differentiable stress intensity function, $\bar{\mu}$ varies continuously in each of the three ranges of $n = 0, 1, 2$, whereas $\bar{\nu} \equiv \bar{\nu}_n =$ constant for each value of n. Similarly, the identity of equation (4.2.38) excludes the use of a continuously differentiable strain intensity function, with a piecewise linear, continuous, stress intensity function.

The strain intensity function g is subject to a further restriction.

From equation (4.4.2),

$$\tfrac{1}{2}(\bar{\varepsilon}_3' - \bar{\varepsilon}_2') = \frac{\sqrt{3}}{[N(3+\nu^2)]^{1/2}} \sqrt{b}. \tag{4.4.5}$$

For $\nu = 0$, equation (4.3.9) gives $\Omega = 0$, and hence equation (4.3.8) gives

$$[K_2']_{\nu=0} = b \qquad (N(0) = 1). \tag{4.4.6}$$

These conditions, when entered into equation (4.4.5) give

$$\sqrt{b} = \tfrac{1}{2}(\bar{\varepsilon}_3 - \bar{\varepsilon}_2)(\equiv \gamma_1). \tag{4.4.7}$$

From the constitutive equation (4.2.46),

$$\sigma_3 - \sigma_2 = 2\left[\bar{G}\frac{\Psi}{\Lambda}\right]_{\nu=0} (\bar{\varepsilon}_3 - \bar{\varepsilon}_2) \tag{4.4.8}$$

and hence at initial yield

$$k = 2\left[\bar{G}\frac{\Psi}{\Lambda}\right]_{\nu=0} [\sqrt{b}]_k = 2\left[\bar{G}\frac{\Psi}{\Lambda}\right]_{\nu=0} [\gamma_1]_k \tag{4.4.9}$$

where use has been made of equation (3.1.27), and where $[\gamma_1]_k$ is the value of $\gamma_1(\equiv\sqrt{b})$ corresponding to the maximum shear stress k at initial yield in a state of pure shear. From equation (4.4.2)

$$\bar{\varepsilon}_1' = \frac{2v}{\sqrt{3}\,[N(3+v^2)]^{1/2}} \sqrt{b}. \tag{4.4.10}$$

For the particular case $v = -1$, equation (4.2.48) gives

$$\bar{\varepsilon}_1' = \bar{\varepsilon}_2' = -\tfrac{1}{2}\bar{\varepsilon}_3' \qquad (v = -1). \tag{4.4.11}$$

Entering these conditions into equation (4.4.10) gives,

$$\bar{\varepsilon}_3' = \frac{2}{\sqrt{3}\,[N^{1/2}]_{v=-1}} \sqrt{b}. \tag{4.4.12}$$

For the conditions $\mu = -1, \sigma_1 = \sigma_2 = 0$ the constitutive equation (4.2.46) gives

$$\sigma_3 = 3\left[\frac{\bar{G}\Psi}{\Lambda}\right]_{\mu=v=-1} \bar{\varepsilon}_3' \tag{4.4.13}$$

and hence at initial yield,

$$Y = 3\left[\frac{\bar{G}\Psi}{\Lambda}\right]_{\mu=v=-1} [\bar{\varepsilon}_3']_Y \tag{4.4.14}$$

where $[\bar{\varepsilon}_3']_Y$ is the value of $\bar{\varepsilon}_3'$ corresponding to the initial yield stress $\sigma_3 (\equiv Y)$ in a state of simple uniaxial tension. Thus, at initial yield, equations (4.4.7), (4.4.9), (4.4.12) and (4.4.14) give,

$$[\bar{\varepsilon}_3']_Y = \frac{1}{\sqrt{3}\,[N^{1/2}]_{v=-1}} [2\gamma_1]_k \tag{4.4.15}$$

which is the counterpart, in terms of the kinematics of the system, to equation (3.1.30). Consider for example

(i) ***Modified Tresca form for*** g.

$$N = \cos^2[\tfrac{1}{6}\cos^{-1}(1 - 2a\Omega)] \qquad (0 \leqslant a \leqslant 1). \tag{4.4.16}$$

For $v = -1, \Omega = 1$, equations (4.4.15) and (4.4.16) therefore give

$$[\bar{\varepsilon}_3']_Y = \frac{1}{\cos[\frac{1}{6}\cos^{-1}(1-2a)]} \left[\frac{2}{\sqrt{3}}\gamma_1\right]_k. \tag{4.4.17}$$

Hence for the Tresca form for g, for which $a = 1$

$$[\bar{\varepsilon}_3']_Y = \frac{2}{\sqrt{3}} \left[\frac{2}{\sqrt{3}}\gamma_1\right]_k \qquad (a = 1). \tag{4.4.18}$$

(ii) ***Modified rotated Tresca form for*** g.

$$N = \tfrac{4}{3}\cos^2[\tfrac{1}{6}\cos^{-1}(2a\Omega - 1)] \qquad (0 \leqslant a \leqslant 1). \tag{4.4.19}$$

For $v=-1$, $\Omega=1$, equations (4.4.15) and (4.4.19) therefore give

$$[\bar{\varepsilon}_3']_Y=\frac{1}{\cos[\frac{1}{6}\cos^{-1}(2a-1)]}\left[\frac{2}{\sqrt{3}}\gamma_1\right]_k. \tag{4.4.20}$$

Thus for the rotated Tresca form for g, for which $a=1$

$$[\bar{\varepsilon}_3']_Y=\frac{\sqrt{3}}{2}\left[\frac{2}{\sqrt{3}}\gamma_1\right]_k. \tag{4.4.21}$$

Equations (4.4.18) and (4.4.21) give,

$$[\bar{\varepsilon}_3']_Y>\left[\frac{2}{\sqrt{3}}\gamma_1\right]_k \qquad \text{(Tresca)} \tag{4.4.22}$$

$$[\bar{\varepsilon}_3']_Y<\left[\frac{2}{\sqrt{3}}\gamma_1\right]_k \qquad \text{(rotated Tresca).} \tag{4.4.23}$$

It is evident from equations (4.4.22) and (4.4.23) that in the context of the stress–strain relations of equations (4.48) and (4.4.13), the conditions at initial yield impose a restriction on the form for N, that is on the form for the strain intensity function g.

It is of interest to note that for the modified Tresca form for g, with v in the range $(-1\leqslant v\leqslant 0)$

$$v\geqslant\bar{v} \qquad N=\cos^2[\tfrac{1}{6}\cos^{-1}(1-2a\Omega)]$$

whereas for the modified, rotated Tresca form of g, with v in the range $(-1\leqslant v\leqslant 0)$

$$v\leqslant\bar{v} \qquad N=\tfrac{4}{3}\cos^2[\tfrac{1}{6}\cos^{-1}(2a\Omega-1)].$$

Hence, if a modified Tresca form for the stress intensity function h is used in conjunction with a modified Tresca form for g, then it is possible, at least in principle, to satisfy both the generalised Lode relation $\bar{\mu}\equiv\bar{v}$ and the classical Lode relation $\mu=v$. However, use of a modified Tresca form of the stress intensity function h with a modified rotated Tresca form for g excludes all possibility of satisfying the classical Lode relation $\mu=v$ (except at $\mu^2(=v^2)=0$, 1, 9, ∞), it is possible however, to satisfy the fundamental identity $\bar{\mu}=\bar{v}$, that is, the generalised Lode relation.

4.5 INFINITESIMAL STRAIN

The Almansi–Hamel strain is defined by (Almansi 1911, Hamel 1912),

$$\mathbf{e}=\tfrac{1}{2}(\mathbf{I}-\mathbf{B}^{-1}). \tag{4.5.1}$$

For classical infinitesimal linearised elasticity, the assumption that the displacements and displacement gradient components are small compared to unity implies the use of the same reference axes for both $\boldsymbol{x}$ and $\boldsymbol{X}$, such that the displacement vector

$$\boldsymbol{u}=\boldsymbol{x}-\boldsymbol{X}=\boldsymbol{u}(\boldsymbol{X},t). \tag{4.5.2}$$

This approximation allows the introduction of the displacement gradient

$$\tilde{\mathbf{S}}=\mathbf{F}-\mathbf{I}. \tag{4.5.3}$$

Substituting for $\mathbf{F}$ in equation (2.1.45) from equation (4.5.3) gives

$$\mathbf{B}^{-1}=(\mathbf{I}+\tilde{\mathbf{S}}^{\mathrm{T}})^{-1}(\mathbf{I}+\tilde{\mathbf{S}})^{-1}. \tag{4.5.4}$$

To a first order in $\tilde{\mathbf{S}}$, equation (4.5.4) gives

$$\mathbf{B}^{-1}=\mathbf{I}-(\tilde{\mathbf{S}}+\tilde{\mathbf{S}}^{\mathrm{T}}). \tag{4.5.5}$$

Substituting for $\mathbf{B}^{-1}$ in equation (4.5.1) from equation (4.5.5) reduces $\mathbf{e}$ to the small strain tensor,

$$\tilde{\mathbf{e}}=\tfrac{1}{2}(\tilde{\mathbf{S}}+\tilde{\mathbf{S}}^{\mathrm{T}}). \tag{4.5.6}$$

Thus, for sufficiently small strains

$$2\tilde{\mathbf{e}}=\mathbf{I}-\mathbf{B}^{-1}\equiv\mathbf{I}-\tilde{\mathbf{B}}^{-1}. \tag{4.5.7}$$

From the definition of $\mathbf{B}$, it follows that

$$\tilde{\mathbf{B}}=\mathbf{I}+(\tilde{\mathbf{S}}+\tilde{\mathbf{S}}^{\mathrm{T}})=\mathbf{I}+2\tilde{\mathbf{e}}. \tag{4.5.8}$$

Substituting for $\mathbf{B}$ and $\mathbf{B}^{-1}$ in the constitutive equation (4.1.12) from equations (4.5.7) and (4.5.8) gives

$$\mathbf{T}=(\beta_0+\beta_1+\beta_{-1})\mathbf{I}+2(\beta_1-\beta_{-1})\tilde{\mathbf{e}}. \tag{4.5.9}$$

The substitutions

$$\left.\begin{aligned}\beta_0=-G_0I_{\tilde{e}}\qquad & \beta_1=\tfrac{1}{2}G_0+\tfrac{1}{2}(\lambda+G_0)I_{\tilde{e}}\\ \beta_{-1}=-\tfrac{1}{2}G_0 & +\tfrac{1}{2}(\lambda+G_0)I_{\tilde{e}}\end{aligned}\right\} \tag{4.5.10}$$

reduce equation (4.5.9) to

$$\mathbf{T}=\lambda I_{\tilde{e}}\mathbf{I}+2G_0\tilde{\mathbf{e}} \tag{4.5.11}$$

which is the stress–strain relationship of the classical infinitesimal theory of elasticity, it being noted that $I_{\tilde{e}}$ is the first invariant of $\tilde{\mathbf{e}}$ and that λ and G_0 are the Lamé elastic constants.

Substituting for $\mathbf{B}$ and $\mathbf{B}^{-1}$ in equation (4.2.20) from equations (4.5.7) and (4.5.8) gives, for sufficiently small strains,

$$\bar{\mathbf{E}}=\tfrac{1}{2}(\tilde{\mathbf{S}}+\tilde{\mathbf{S}}^{\mathrm{T}})=\tilde{\mathbf{e}}. \tag{4.5.12}$$

Entering this form for $\bar{\mathbf{E}}$ in equation (4.2.19) and neglecting the quadratic term in $\tilde{\mathbf{e}}$ gives

$$\mathbf{G}(\mathbf{T}) = \tilde{\alpha}_0\mathbf{I} + \tilde{\alpha}_1\tilde{\mathbf{e}}. \tag{4.5.13}$$

Entering the form for the β_a ($a = 0, \pm 1$) given by equation (4.5.10) into equations (4.2.21), (4.2.49) and (4.2.50) reduces the $\tilde{\alpha}_0$ and $\tilde{\alpha}_1$ for sufficiently small strains to

$$\tilde{\alpha}_0 = \lambda I_{\tilde{\mathbf{e}}} \qquad \tilde{\alpha}_1 = 2G_0 \tag{4.5.14}$$

use being made of the relations

$$I_{\mathbf{B}} \approx 3 + 2I_{\tilde{\mathbf{e}}} \qquad II_{\mathbf{B}} \approx 3 + 4I_{\tilde{\mathbf{e}}} \qquad III_{\mathbf{B}} \approx 1 + 2I_{\tilde{\mathbf{e}}}. \tag{4.5.15}$$

Entering the limiting form for the $\tilde{\alpha}_0$ and $\tilde{\alpha}_1$ given by equation (4.5.14) into equation (4.5.13) gives

$$\mathbf{G}(\mathbf{T}) = \lambda I_{\tilde{\mathbf{e}}}\mathbf{I} + 2G_0\tilde{\mathbf{e}}. \tag{4.5.16}$$

For the von Mises type stress intensity function, $\mathbf{G} = \mathbf{T}$ and equation (4.5.16) reduce to equation (4.5.11), from which it can be concluded that the constitutive equation of a simple elastic material and the constitutive equation of the classical theory of infinitesimal elasticity both apply only to materials which satisfy the von Mises type of stress intensity function.

In the context of equation (4.2.27), the limiting condition of equation (4.5.14) requires that,

$$\lim_{\bar{\mathbf{E}}' \to 0} \bar{G} = G_0 \tag{4.5.17}$$

where G_0 is the classical, or ground state shear modulus.

4.6 THE GROUND STATE

The limiting conditions

$$\begin{aligned} &\delta_0 = 0, \quad \delta_1 = 1, \quad \delta_2 = 0, \quad \mathbf{G}' = \mathbf{T}', \quad H = 1, \quad \Lambda = 1 \\ &\varphi_0 = 0, \quad \varphi_1 = 1, \quad \varphi_2 = 0, \quad \bar{\mathbf{M}}' = \bar{\mathbf{E}}', \quad N = 1, \quad \Psi = 1 \end{aligned} \tag{4.6.1}$$

reduce the constitutive equation (4.2.44) to

$$\mathbf{T}' = 2\bar{G}\bar{\mathbf{E}}' \tag{4.6.2}$$

and equation (4.2.46) to

$$\sigma_i' = 2\bar{G}\bar{\varepsilon}_i' \qquad (i = 1, 2, 3). \tag{4.6.3}$$

These in turn reduce equation (4.2.38) to

$$\mu = \bar{\mu} \equiv \bar{\nu} = \nu. \tag{4.6.4}$$

The identity of equation (4.6.4), first stated by Lode (1926) in the context of classical plasticity theory in the form $\mu = \nu$, will be referred to as the Lode relation.

Setting, $\bar{G} = G_0$ in equation (4.6.2) gives,

$$\mathbf{T}' = 2G_0\bar{\mathbf{E}}' \tag{4.6.5}$$

which will be referred to as the ground state form of the constitutive equation for an isotropic material, the limiting conditions of equation (4.6.1) together with the limiting condition $\bar{G} = G_0$ being referred to as the ground state conditions.

It should be noted that materials whose mechanical response are characterised by the ground state conditions, satisfy only stress and strain intensity functions of the von Mises type.

REFERENCES

Almansi E 1911 *Rend. Lincei.* (5A) **20** 705–14

Brauer R 1965 *Arch. Ration. Mech. Anal.* **18** 97–9

Hamel G 1912 *Elementare Mechanik* (Leipzig: Teubner) (Engl. Transl. 1949 New York: Springer)

Lode W 1926 *Z. Phys.* **36** 913–39

Noll W 1958 *Arch. Ration. Mech. Anal.* **2** 197–226

—— 1965 *Arch. Ration. Mech. Anal.* **18** 100–2

GENERAL REFERENCES

Chadwick P 1976 *Continuum Mechanics* (London: Allen and Unwin)

Gurtin M E 1981 *Continuum Mechanics* (New York: Academic)

Jaunzemis W 1967 *Continuum Mechanics* (New York: Macmillan)

Truesdell C and Noll W 1965 *The Non-linear Field Theories of Mechanics* in *Handbuch der Physik* vol. III/3 ed. S Flügge (Berlin: Springer)

Truesdell C and Toupin R A 1960 *The Classical Field Theories* in *Handbuch der Physik* vol. III/1 ed. S Flügge (Berlin: Springer) pp 226–858

Wang C-C and Truesdell C 1973 *Introduction to Rational Elasticity* (Leyden: Noordhoff)

5

Elastic–Plastic Materials

5.1 TOTAL ELASTIC–PLASTIC DEFORMATION

5.1.1 Constitutive Equation for Loading

Bridgman (1923, 1952) has shown to a good degree of approximation that hydrostatic pressure does not affect either the initial yield or the post-yield deformation behaviour of a wide range of metals and the observations of Crossland (1954) are in accord with these conclusions. With this as a basis, the present discussion is restricted to incompressible materials.

A typical one-dimensional (i.e. true tensile stress–engineering strain) curve (σ, e), is shown schematically in figure 5.1. The yield stress Y, i.e. at the point P, is seen to separate the characteristic stress–strain curve into an elastic range and a plastic range. The range of response from O to P is entirely elastic. In this initial elastic range OP, both the loading and the unloading paths coincide. Having unloaded from the point P, further deformation on reloading requires an increase in stress, a condition referred to as *work-hardening* or *strain-hardening*. A specimen initially loaded in tension is observed to yield at a much reduced stress when reloaded in compression. This is known as the *Bauschinger* effect. For the purposes of developing an idealised theory of plasticity it will, however, be assumed that the ideal plastic body does not exhibit a Bauschinger effect. This restriction permits the yield stress to be taken to be the same in tension and compression. The classical three-dimensional theory of plastic behaviour can be regarded as a generalisation of the three particular idealisations of the one-dimensional stress–strain curve, illustrated schematically in figure 5.2.

Work-hardening materials are characterised by a yield surface which changes with continued plastic deformation, that is continued loading beyond initial yield. Such changes in the yield surface arising from work-hardening are accounted for by generalising the initial yield function, f, defined by equation (3.1.11). A basic requirement is that the constitutive equation (4.2.27) for a non-

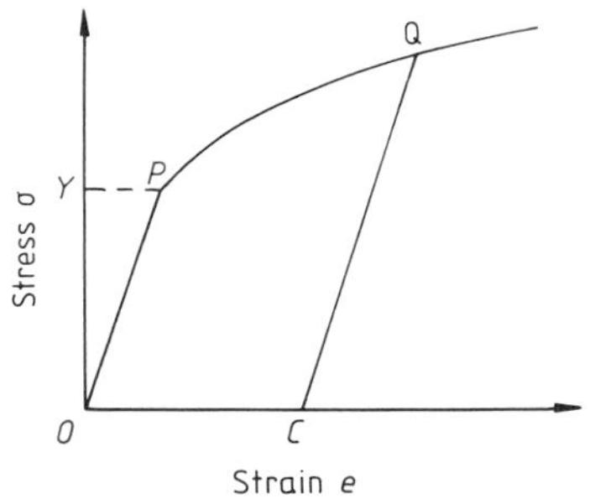

Figure 5.1 Schematic representation of a uniaxial stress–strain curve for a work-hardening material.

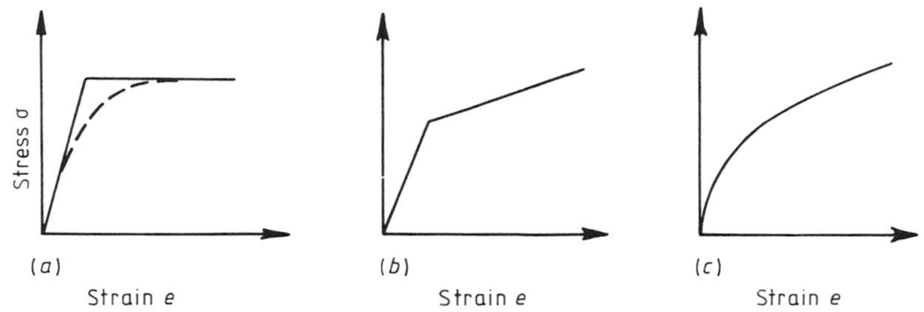

Figure 5.2 (*a*) Idealised, elastic–perfectly plastic stress–strain curve; (*b*) idealised, elastic–linear work-hardening stress–strain curve; (*c*) idealised, annealed, elastic–plastic stress–strain curve.

simple elastic material will correctly predict the mechanical response of a non-simple elastic–plastic material over the initial, purely elastic, range of deformation. A way of satisfying this requirement has been discussed in §4.2.1 in the context of two basic assumptions.

For initial and post-yield deformation, both the elastic part of an element of the material and the associated plastic part are subject to identically the same stress system as that acting on the element as a whole. With this as the basis it is assumed that there exists a stress intensity function

$$h = J_2' H(\omega) - a = 0 \qquad H(0) = 1 \tag{5.1.1}$$

which defines the stress systems that determine a given mode of deformation over the entire range of deformation. Equation (5.1.1) is just equation (3.1.11) with k^2 replaced by a, it being noted that ω is defined by equation (3.1.12). Thus, for initial and post-yield conditions of deformation, h is identical to the yield function, the term *stress intensity function* being introduced to emphasise that it is now assumed to apply also to the purely elastic range of deformation which precedes initial yield.

The stress intensity function h is assumed to determine a loading function $\mathbf{G}'$ by way of the relation,

$$\mathbf{G}' = \partial h / \partial \mathbf{T}.$$

The mechanical response of some fully annealed metals approximates closely to the idealised stress–strain behaviour shown schematically in figure 5.2(c). These metals are characterised by a smoothly continuous transition from the initial purely elastic range of deformation, to the total elastic–plastic, work-hardening range of deformation. For this type of elastic–plastic material, the (total) mechanical response can be described by the constitutive equation (4.2.27) expressed in the form,

$$\mathbf{G}'(\mathbf{T}') = 2\bar{G}\bar{\mathbf{M}}'(\bar{\mathbf{E}}') = 2\bar{G}(\varphi_0 \mathbf{I} + \varphi_1 \bar{\mathbf{E}}' + \varphi_2 \bar{\mathbf{E}}'^2) \tag{5.1.2}$$

where the apparent strain

$$\bar{\mathbf{E}} = \tfrac{1}{4}(\mathbf{B} - \mathbf{B}^{-1}) \qquad (\mathbf{B} = \mathbf{F}\mathbf{F}^{\mathrm{T}}). \tag{5.1.3}$$

Here

$$\bar{G} = \bar{G}(K_2', K_3') \tag{5.1.4}$$

is a non-negative scalar factor of proportionality, having noted that

$$K_2' = \tfrac{1}{2}\mathrm{tr}\, \bar{\mathbf{E}}'^2 \qquad K_3' = \det \bar{\mathbf{E}}' \tag{5.1.5}$$

and that

$$\varphi_a = \varphi_a(K_2', K_3') \qquad (a = 0, 1, 2) \tag{5.1.6}$$

are the response coefficients. The way in which the response coefficients are evaluated by way of the concept of a strain intensity function has been discussed in §4.3.2.

The constitutive equation (5.1.2) can be rearranged into the form

$$\sigma'' = 3\frac{[N^{1/2}]_{\nu=-1}}{[H^{1/2}]_{\mu=-1}} G\varepsilon'' \qquad G = \bar{G}\left[\frac{\Psi(3+\mu\bar{\mu})}{\Lambda(3+\nu\bar{\nu})}\right]^{1/2} \tag{5.1.7}$$

where use has been made of the generalised Lode relation of equation (4.2.38) to eliminate the terms involving $\bar{\mu}$ and $\bar{\nu}$. Here the *effective stress* is

$$\sigma'' = \frac{(3J_2'H)^{1/2}}{[H^{1/2}]_{\mu=-1}} = \frac{[H(3+\mu^2)]^{1/2}}{2[H^{1/2}]_{\mu=-1}}(\sigma_3 - \sigma_2) \tag{5.1.8}$$

and the *effective strain* is

$$\varepsilon'' = \frac{(\tfrac{4}{3}K_2'N)^{1/2}}{[N^{1/2}]_{\nu=-1}} = \frac{[N(3+\nu^2)]^{1/2}}{3[N^{1/2}]_{\nu=-1}}(\varepsilon_3 - \varepsilon_2) \tag{5.1.9}$$

use having been made of the relations,

$$\Lambda = H\frac{(3+\mu^2)}{(3+\mu\bar{\mu})} \qquad \Psi = N\frac{(3+v^2)}{(3+v\bar{v})} \tag{5.1.10}$$

which follows from equations (3.1.17) and (4.3.12).

For many materials which exhibit the behaviour of elastic–perfectly plastic materials, the mechanical response approximates closely to the continuous transition identified in figure 5.2(*a*) by the broken curve; see for example the stress–strain curves shown in figure 4 of Lianis and Ford (1957). This behaviour can be described by taking the generalised shear modulus, G, to be of the form,

$$G = G_0 \tanh(\phi\varepsilon'')/\phi\varepsilon'' \tag{5.1.11}$$

where ϕ is a constant characteristic of material properties, and where ε'' is defined by equation (5.1.9), having noted that G_0 is the ground state, i.e. the classical, shear modulus. Entering the form of $\bar{G}$, given by equation (5.1.11), into equation (5.1.7) gives,

$$\sigma'' = \sigma_0 \tanh(\phi\varepsilon'') \tag{5.1.12}$$

where,

$$\sigma_0 = 3\frac{[N^{1/2}]_{v=-1}}{[H^{1/2}]_{\mu=-1}}\frac{G_0}{\phi}\left(\frac{\Psi(3+\mu\bar{\mu})}{\Lambda(3+v\bar{v})}\right)^{1/2}. \tag{5.1.13}$$

A particular form of equation (5.1.12) was first proposed by Prager (1938, 1942). For a more general form of equation (5.1.11), see §7.1.4.

The behaviour shown schematically in figure 5.2(*b*) can be described by the constitutive equation (5.1.2) by formulating G in such a way that it changes discontinuously at some particular value of the total strain.

5.1.2 Constitutive Equation for Unloading

Referring to figure 5.1, unloading from the point Q in the plastic range of strain is seen to result in the stress–strain point (σ, e), i.e. the state point, following the path QC which is substantially parallel to the initial elastic portion of the curve. At C, for which $\sigma = 0$, there remains a permanent plastic strain. Thus the total strain resulting from loading to some arbitrary point in the elastic–plastic range of strain consists of two parts, one an elastic component and the other a plastic component. The question thus arises of how to identify the elastic and plastic behaviour of the material separately.

The elastic and plastic components of any quantity α will be denoted by α^{e} and α^{p} respectively.

Following Lee (1969), who assumed that the total deformation gradient tensor $\mathbf{F}$ can be decomposed into a matrix product,

$$\mathbf{F} = \mathbf{F}^{\mathrm{e}}\mathbf{F}^{\mathrm{p}} \tag{5.1.14}$$

of the local elastic $\mathbf{F}^{e}$, and plastic $\mathbf{F}^{p}$ deformation gradient tensors, various attempts have been made to develop theories for elastic–plastic solids with large strains (see, for example, Freund 1970, Hahn 1974, Hahn and Jaunzemis 1971, Lee and Germain 1974). Implicit in the formulation of equation (5.1.14) is the concept of an intermediate reference state. Thus, at any instant in time t, the motion of the body $\mathscr{B}$ is regarded, as a rigid–plastic deformation of the initial reference configuration B_r into an intermediate configuration B^{p} superimposed on a recoverable deformation of the intermediate configuration into the configuration B_t. The intermediate description is obtained by adjoining to an origin $\bar{o}$ the orthonormal basis $\bar{e}=\{\bar{e}_1, \bar{e}_2, \bar{e}_3\}$. Intermediate description coordinates in the system $(\bar{o}, \bar{e})$ are denoted by $\bar{x}^{\bar{r}}$ $(\bar{r}=1, 2, 3)$. Thus, from equation (2.1.36)

$$\mathbf{F}^{p}=(F^{p})^{\bar{r}}{}_{\mu}\bar{\boldsymbol{e}}_{\bar{r}}\otimes \boldsymbol{E}^{\mu} \qquad (F^{p})^{\bar{m}}{}_{\alpha}=\bar{x}^{\bar{m}}{}_{,\alpha}. \tag{5.1.15}$$

Entering the compatibility condition of equation (5.1.14) into equation (5.1.3) gives,

$$\bar{\mathbf{E}}=\tfrac{1}{4}\{\mathbf{F}^{e}\mathbf{B}^{p}(\mathbf{F}^{e})^{T}-[\mathbf{F}^{e}\mathbf{B}^{p}(\mathbf{F}^{e})^{T}]^{-1}\}. \tag{5.1.16}$$

For the initial, purely elastic range of deformation, the plastic strain is zero, and hence $\mathbf{B}^{p}=\mathbf{I}$, which condition, when entered into equation (5.1.16), gives the apparent elastic strain

$$\bar{\mathbf{E}}^{e}=\tfrac{1}{4}[\mathbf{B}^{e}-(\mathbf{B}^{e})^{-1}] \tag{5.1.17}$$

which can be entered into equation (5.1.2) to give the constitutive equation for loading and *unloading* in the purely elastic range of strain.

As already noted, unloading from the arbitrary point Q in the total elastic–plastic range of strain is seen from figure 5.1 to result in the stress–strain point (σ, e), following the path QC which is substantially parallel to the initial elastic portion of the curve. Thus, it is concluded that unloading from the arbitrary point Q is to be described by the same constitutive equation as that used to describe the initial, purely elastic range of strain. That is unloading is to be described by equation (5.1.2) with $\bar{\mathbf{E}}$ replaced by the $\bar{\mathbf{E}}^{e}$ of equation (5.1.17), or equivalently by setting $\mathbf{B}^{p}=\mathbf{I}$ in equation (5.1.16).

With regard to the arbitrary point Q, it is important to note that for neutral, that is sustained loading at Q, there is no way of determining whether the point Q was attained by loading irreversibly along the path OPQ or whether it was attained by recoverable loading along the path CQ. This says no more than that the material has *no memory* of the plastic states irreversibly traversed during loading along the path OPQ, whereas (ideally), it has a *perfect memory* of all the reversible elastic states it traverses, irrespective of how the material attains the arbitrary point Q.

5.2 PURELY VISCOUS SOLIDS

5.2.1 Constitutive Equation for a Purely Viscous Solid

The actual changes in material properties associated with the phenomenon of work-hardening—that is the strain-hardening arising from the cold-working involved in plastic deformation—imply that the stress is a history-dependent function of the strain. This is because after each unloading it is a different material which is responding to reloading. History dependence is a general feature characteristic of plastic deformation and is the principal physical difference between elastic and plastic deformation.

For plastic deformation, restricted descriptions can be obtained by introducing irreversible parameters in the form of time-like parameters which are specified to keep on increasing, thus recognising that the plastic response of a material is an irreversible process. Assuming an isothermal deformation process, the yield criterion in the form of equation (3.1.1) can be generalised to include a dependence upon work hardening by introducing the stress intensity function

$$f(\mathbf{T}, \boldsymbol{\varepsilon}^{\mathrm{p}}, \bar{a}) = 0 \tag{5.2.1}$$

which in addition to the Cauchy stress $\mathbf{T}$, also depends upon the plastic component $\boldsymbol{\varepsilon}^{\mathrm{p}}$ of some, as yet undefined, measure of total finite strain $\boldsymbol{\varepsilon}$, and the work-hardening characteristics represented by the parameter $\bar{a}$ which is dependent upon the history of the deformation. Set

$$\boldsymbol{\varepsilon}^{\mathrm{p}} = \int \mathbf{D}^{\mathrm{p}}\,\mathrm{d}t \tag{5.2.2}$$

where $\mathbf{D}^{\mathrm{p}}$ is the plastic component of the total stretching tensor $\mathbf{D}$ defined by equation (2.1.68). Equation (5.2.2) implies decomposition of the total stretching tensor into elastic and plastic parts $\mathbf{D}^{\mathrm{e}}$ and $\mathbf{D}^{\mathrm{p}}$, respectively, it being noted, however, that in general the $\mathbf{D}^{\mathrm{e}}$ and $\mathbf{D}^{\mathrm{p}}$ need not satisfy rate of deformation compatibility conditions. The following relation can be obtained from equations (5.2.1) and (5.2.2)

$$\dot{f} = \mathrm{tr}(\mathbf{G}'\dot{\mathbf{T}}) + \mathrm{tr}(\boldsymbol{\Sigma}^{\mathrm{T}}\mathbf{D}^{\mathrm{p}}) + \frac{\partial f}{\partial \bar{a}}\mathrm{tr}(\boldsymbol{\Pi}^{\mathrm{T}}\mathbf{D}^{\mathrm{p}}) = 0 \tag{5.2.3}$$

where

$$\mathbf{G}' = \partial \mathbf{f}/\partial \mathbf{T} \tag{5.2.4}$$

and where

$$\boldsymbol{\Sigma} = \partial f/\partial \boldsymbol{\varepsilon}^{\mathrm{p}} \qquad \boldsymbol{\Pi} = \partial \bar{a}/\partial \boldsymbol{\varepsilon}^{\mathrm{p}}. \tag{5.2.5}$$

Equation (5.2.3) is satisfied by the constitutive equation

$$\mathbf{D}^{\mathrm{p}} = \mathbf{G}'\dot{m} \tag{5.2.6}$$

where

$$\dot{m}=-\operatorname{tr}(\mathbf{G}'\dot{\mathbf{T}})/\operatorname{tr}[(\boldsymbol{\Sigma}^{\mathrm{T}}+\boldsymbol{\Pi}^{\mathrm{T}}\,\partial f/\partial\bar{a})\mathbf{G}']. \tag{5.2.7}$$

This constitutive assumption states that the rate of straining is determined by the present value of $\mathbf{G}'$. The quantity $\mathbf{G}'$ is, effectively, just the loading function discussed in §4.2.1. In the context of these equations, f is known as the plastic potential (see, for example, Hill 1950).

Substituting for f in equation (5.2.4) from equation (3.1.11) gives

$$\mathbf{G}'=\delta_0\mathbf{I}+\delta_1\mathbf{T}'+\delta_2\mathbf{T}'^2=\bar{\eta}\mathbf{D}^{\mathrm{p}} \qquad (\bar{\eta}=1/\dot{m}) \tag{5.2.8}$$

where the loading coefficients δ_a $(a=0,1,2)$ are given by

$$\begin{gathered}\delta_0=-\tfrac{2}{3}J_2'\delta_2\\ \delta_1=H\left(1-3\frac{\omega}{H}\frac{\mathrm{d}H}{\mathrm{d}\omega}\right) \qquad \delta_2=2H\left(\frac{J_2'}{J_3'}\right)\frac{\omega}{H}\frac{\mathrm{d}H}{\mathrm{d}\omega}\end{gathered} \tag{5.2.9}$$

having noted that the invariants J_2' and J_3' are defined by equation (3.1.4), and that H can take any one of the forms discussed in Chapter 3.

The conditions for the simplest continuously differentiable isotropic condition of incompressible yielding, that is the von Mises yield criterion, are given in equation (3.3.1), having noted that $H=1$. Entering the condition $H=1$ into equation (5.2.9) gives,

$$\delta_0=0, \qquad \delta_1=1, \qquad \delta_2=0 \qquad \Lambda=1, \qquad \mathbf{G}'=\mathbf{T}' \qquad (H=1) \tag{5.2.10}$$

which reduce equation (5.2.8) to

$$\mathbf{T}'=2\eta\mathbf{D}^{\mathrm{p}} \tag{5.2.11}$$

which is just the constitutive equation of a purely viscous (non-Newtonian), fluid (see for example, Billington and Tate 1981).

It has been shown (Reiner 1945, Rivlin 1948, 1949), that the constitutive equation of a viscous fluid can be represented in the more general form

$$\mathbf{T}'=\eta_0\mathbf{I}+\eta_1\mathbf{D}'+\eta_2\mathbf{D}'^2 \tag{5.2.12}$$

where the coefficients

$$\eta_a=\eta_a(II_{\mathrm{D}}, III_{\mathrm{D}}) \qquad (a=0,1,2) \tag{5.2.13}$$

are scalar functions of the principal invariants II_{D} and III_{D} of $\mathbf{D}$ only, since for an incompressible fluid $I_{\mathrm{D}}=0$.

In the context of equations (5.2.11) and (5.2.12), it is assumed that the constitutive equation of a purely viscous solid can be represented in the more general form

$$\mathbf{G}'(\mathbf{T}')=\eta_0\mathbf{I}+\eta_1\mathbf{D}^{\mathrm{p}}+\eta_2(\mathbf{D}^{\mathrm{p}})^2 \tag{5.2.14}$$

where

$$\eta_a = \eta_a(II_{\mathbf{D}^p}, III_{\mathbf{D}^p}) \qquad (a=0,1,2) \tag{5.2.15}$$

having noted that $I_{\mathbf{D}}=0$ for an incompressible, viscous solid.

5.2.2 Stable Plastic Material

The loading surface is the yield surface for $f=0$, whereas $f<0$ is a surface in the elastic region inside the yield surface for which $d\varepsilon^p=0$. Since $\bar{a}$ is associated with the plastic strains, the work-hardening increment $d\bar{a}$ is also zero. Since $f>0$ is outside the yield surface, this particular condition has no physical significance. These considerations give rise to the three loading conditions:

$$\begin{aligned} \text{Unloading: } f&=0 \qquad \operatorname{tr}(\mathbf{G}'\dot{\mathbf{T}})<0 \\ \text{Neutral loading: } f&=0 \qquad \operatorname{tr}(\mathbf{G}'\dot{\mathbf{T}})=0. \\ \text{Loading: } f&=0 \qquad \operatorname{tr}(\mathbf{G}'\dot{\mathbf{T}})>0 \end{aligned} \tag{5.2.16}$$

It is evident that loading can occur only if the material work-hardens, for which condition the yield surface simply moves outward.

Drucker (1951) has proposed a generalisation of the work-hardening concept (see also Drucker 1959, 1960, 1967, Hill 1950, 1958, 1968). To formulate this generalisation. Drucker considers the work expended by an external agency which slowly applies a set of self-equilibrating forces and then slowly removes it from a body which is initially in a state of equilibrium. The external agency is regarded as being quite distinct from the agency causing the state of stress which existed initially. Upon completion of the cycle, the stress is returned to its initial equilibrium value, whereas the body may or may not return to its original configuration. Drucker proceeds to define a stable work-hardening material as one which satisfies both of the following conditions:

(i) for all such added sets of stresses, the plastic work done by the external agency during the application of the additional stresses is positive;
(ii) the net total work performed by the external agency during the cycle of adding or removing stresses is either zero or positive.

The physical significance of this important concept can be examined by way of the following loading cycle. Loading is from an existing state of stress $\mathbf{T}^*$ on or inside a loading surface in nine-dimensional stress space, to $\mathbf{T}$ where $\mathbf{T}^*$ is less than the yield stress. This loading is followed by incremental loading from $\mathbf{T}$ to $\mathbf{T}+d\mathbf{T}$ where the positive increment of stress $d\mathbf{T}$ produces an increment of plastic strain followed in turn by unloading to $\mathbf{T}^*$. Drucker's postulate can now be expressed in the form

$$\operatorname{tr}[(\mathbf{T}-\mathbf{T}^*)^{\mathrm{T}}\mathbf{D}^p)] \geqslant 0 \qquad \operatorname{tr}(\dot{\mathbf{T}}^{\mathrm{T}}\mathbf{D}^p) \geqslant 0 \tag{5.2.17}$$

where $\mathbf{D}^p$ is the plastic component of the stretching tensor, it being noted that the equality sign holds only during neutral loading (see equation (5.2.16)). The inequality on the right-hand side of equation (5.2.17) is often referred to as the condition of uniqueness. The fact that Drucker's definition of work-hardening implies the inequalities of equation (5.2.17) can be shown from the following considerations of the net work done by the external agency during the loading cycle of present interest. At time $t=0$, let the existing state of stress be $\mathbf{T}^*$. The external agency changes the stress to $\mathbf{T}^{(1)}$ on the yield surface and then to a neighbouring point $(\mathbf{T}^{(1)}+\delta\mathbf{T})$ outside or on the initial yield surface at time $(t_1+\delta t)$ followed by unloading back to $\mathbf{T}^*$ at time t^*. The net work done by the external agency during the cycle is $\Delta W_{\text{ext}}=\delta W_{\text{T}}-\delta W_0$ where δW_{T} is the total work done during the cycle and δW_0 is the work that would be done during the same deformation by the initial stress $\mathbf{T}^*$ if it were held constant. Noting that the net elastic work during the cycle is zero,it follows that

$$\partial W_{\text{T}}=\int_0^{t_1}\text{tr}(\mathbf{TD}^{e})\,dt+\int_{t_1}^{t_1+\delta t}[\text{tr}(\mathbf{TD}^{e})+\text{tr}(\mathbf{TD}^{p})]\,dt+\int_{t_1+\delta t}^{t^*}\text{tr}(\mathbf{TD}^{e})\,dt$$

$$=\int_{t_1}^{t_1+\delta t}\text{tr}(\mathbf{TD}^{p})\,dt \tag{5.2.18}$$

$$\delta W_0=\int_{t_1}^{t_1+\delta t}\text{tr}(\mathbf{T}^*\mathbf{D}^{p})\,dt \tag{5.2.19}$$

and therefore

$$\Delta W_{\text{ext}}=\int_{t_1}^{t_1+\delta t}\text{tr}[(\mathbf{T}-\mathbf{T}^*)\mathbf{D}^{p}]\,dt. \tag{5.2.20}$$

$\text{tr}(\mathbf{TD})$ is the stress power per unit volume in the current configuration B_t of the body $\mathscr{B}$ (see equation (2.2.61)) and $\mathbf{D}^e$ is the elastic component of the stretching tensor in equation (5.2.18).

Applying the second part of Drucker's definition, which requires ΔW_{ext} to be either zero or positive for any plastic deformation during an arbitrarily short δt following a loading to an arbitrary point $\mathbf{T}$ on the yield surface, it is evident that the integrand in equation (5.2.20) must either be zero or positive. This in turn establishes the inequality on the left-hand side of equation (5.2.17).

The first part of Drucker's definition can now be applied. For an initial

$$\mathbf{T}^*\equiv\mathbf{T}^{(1)} \qquad d\mathbf{T}=\mathbf{T}-\mathbf{T}^*$$

and the requirement that

$$\text{tr}[(\mathbf{T}-\mathbf{T}^*)\mathbf{D}^{p}]>0$$

then implies that

$$\text{tr}(\dot{\mathbf{T}}\mathbf{D}^{p})>0$$

which is simply the inequality on the right-hand side of equation (5.2.17).

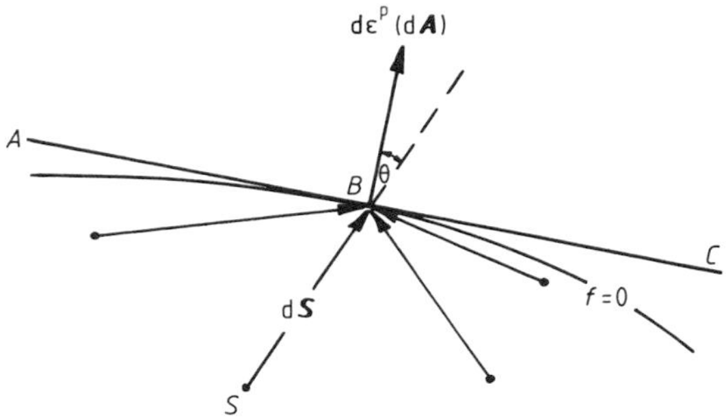

Figure 5.3 Schematic representation of the convex yield surface in stress space.

Following Naghdi (1960), a useful concept is one which considers an increment of stress $d\mathbf{T}$ as a vector $d\boldsymbol{S}$ in the nine-dimensional stress space, and the corresponding increment of plastic strain $d\boldsymbol{\varepsilon}^{p}$ as the components of a vector $d\boldsymbol{A}$ in the same stress space. From equation (5.2.17) it then follows that

$$d\boldsymbol{S}\cdot d\boldsymbol{A}=|d\boldsymbol{S}||d\boldsymbol{A}|\cos\theta\geqslant 0 \tag{5.2.21}$$

which implies that

$$-\pi/2\leqslant\theta\leqslant\pi/2 \tag{5.2.22}$$

from which it is concluded that the angle between $d\boldsymbol{S}$ and $d\boldsymbol{A}$ is acute. Figure 5.3 shows a schematic representation of the plastic strain vector $d\boldsymbol{A}$ at a regular point B on the yield surface in stress space. The line ABC in figure 5.3 represents a hyperplane normal to $d\boldsymbol{A}$. From equations (5.2.21) and (5.2.22) it is evident that all interior points, i.e. the vectors $d\boldsymbol{S}$, must lie to one side of the hyperplane ABC and opposite to the side from which $d\boldsymbol{A}$ points away. The loading vectors $d\boldsymbol{S}$ are outward vectors whose directions are bounded by the plane of the yield surface, and therefore the hyperplane ABC must be tangent to the loading surface at the point B. Hence, since $d\boldsymbol{A}$ is normal to the hyperplane it must also be normal to the yield surface at the point B. Because the tangent plane at a regular point on a surface is unique, it follows that the hyperplane ABC must be unique at a regular point of the yield surface. Hence, the direction of $d\boldsymbol{A}$ normal to the hyperplane must also be unique. From this follows the conclusion that at a smooth point of the yield surface, the direction of $d\boldsymbol{\varepsilon}^{p}$ is independent of the direction of $d\mathbf{T}$. These considerations imply that through every such point B on the yield surface there is a plane such that all points interior to the surface lie on the same side of the plane; i.e. the *yield surface is convex.*

5.3 SIMPLE ELASTIC–PLASTIC MATERIALS

Simple elastic–plastic materials are simple materials subject to the following two constitutive assumptions:

(i) the defining response functional of equation (2.3.7), expressed in terms of the notation of equation (2.3.21), satisfies the identity

$$\mathbf{T}(t) = \underset{s=0}{\overset{\infty}{\mathbf{G}}} (\mathbf{F}^t(s)) = \underset{s=0}{\overset{\infty}{\mathbf{G}}} (\mathbf{F}^t[\sigma(s)]) \tag{5.3.1}$$

for every $\mathbf{F}^t(s)$ in the domain of $\mathbf{G}$ and every increasing function $\sigma(s)$ subject to the limiting conditions

$$\sigma(0) = 0 \qquad \lim_{s\to\infty} \sigma(s) = \infty \tag{5.3.2}$$

(ii) there is a tensor function $\mathbf{g}$ such that

$$\dot{\mathbf{T}} = \mathbf{g}(\mathbf{T}, \mathbf{L}) \tag{5.3.3}$$

for every process possible in the material. The function $\mathbf{g}$ is assumed to be continuously differentiable near $\mathbf{L} = \mathbf{O}$.

The physical significance of the constitutive assumption (i) is that the stress $\mathbf{T}(t)$ at time t depends on the order in which the body $\mathscr{B}$ has occupied its past configurations only and not on the time-rate at which these past configurations were traversed. This restriction is of particular relevance to work-hardening materials in so far as the time can be regarded as a convenient parameter for determining the sequence of events associated with the irreversible behaviour of plastic deformation, thus emphasising the irreversible nature of the underlying mechanisms of deformation. The function $\sigma(s)$ represents a change of time scale, and equation (5.3.1) asserts the invariance of the present stress under such a change.

The constitutive assumption (ii) states that the instantaneous rate of stress, $\dot{\mathbf{T}}$ is determined by the present velocity gradient $\mathbf{L}$ and the present stress $\mathbf{T}$. Thus, the stressing of a simple elastic–plastic material subject to a given stress is determined uniquely by the stretching and spin to which it is instantaneously subjected. Let the deformation gradient $\mathbf{F}(\tau)$ be known for all times τ between some initial time 0 and the present time t. In the event that the initial stress $\mathbf{T}(0)$ is also known, then the differential equation (5.3.3) uniquely determines the present stress $\mathbf{T}(t)$. Hence, to determine $\mathbf{T}(t)$, all that needs to be known about the kinematical history previous to the initial instant 0 is the effect of this previous history on the initial stress.

The variables entering the equations of simple elastic–plastic materials are all quantities defined in the present configuration B_t of the body $\mathscr{B}$.

Having regard to equations (2.1.59), (2.1.60), (2.1.67) and (2.1.68) it is evident that equation (5.3.3) is a constitutive equation of the rate-type. In order to reduce the constitutive equation (5.3.3) so as to satisfy the principle of objectivity, it is convenient to express the constitutive equation (2.3.60) in the more general form

$$\mathbf{T}(t)=\mathbf{R}(t)\,\underset{s=0}{\overset{\infty}{\mathbf{G}}}\,(\mathbf{U}(t-s))\mathbf{R}(t)^{\mathrm{T}} \tag{5.3.4}$$

which can in turn be expressed in the form (Noll 1958)

$$\mathbf{T}(t)=\mathbf{R}_{t'}(t)\mathbf{R}(t')\,\underset{s=0}{\overset{\infty}{\mathbf{H}}}\,(\mathbf{U}^*_{t'}(t-s);\,\mathbf{U}(t'))\mathbf{R}(t')^{\mathrm{T}}\mathbf{R}_{t'}(t)^{\mathrm{T}} \tag{5.3.5}$$

where $\mathbf{R}_t$ is the relative rotation tensor

$$\mathbf{U}^*_{t'}(t)=\mathbf{R}(t)^{\mathrm{T}}\mathbf{U}_t(\tau)\mathbf{R}(t) \tag{5.3.6}$$

and $\mathbf{H}$ is a functional of the function

$$\mathbf{U}^*_{t'}(t-s) \qquad s\geqslant 0$$

and a function of the tensor parameter $\mathbf{U}(t')$. Setting

$$\mathbf{T}^*_{t'}(t)=\mathbf{R}(t')^{\mathrm{T}}\mathbf{R}_{t'}(t)^{\mathrm{T}}\mathbf{T}(t)\mathbf{R}_{t'}(t)\mathbf{R}(t') \tag{5.3.7}$$

allows equation (5.3.5) to be expressed in the form

$$\mathbf{T}^*_{t'}(t)=\underset{s=0}{\overset{\infty}{\mathbf{H}}}\,[\mathbf{U}^*_{t'}(t-s);\,\mathbf{U}(t')] \tag{5.3.8}$$

from which it is evident that the function $\mathbf{T}^*_{t'}(t)$ is completely determined for all times t if the function

$$\mathbf{U}^*_{t'}(\tau)=\mathbf{R}(t')^{\mathrm{T}}\mathbf{U}_{t'}(\tau)\mathbf{R}(t') \tag{5.3.9}$$

is prescribed for all τ. For the material to be of the rate-type, it follows that the functions $\mathbf{T}^*_{t'}(t)$ and $\mathbf{U}^*_{t'}(t)$ must be related by a differential equation of the form

$$\dot{\mathbf{T}}^*_{t'}(t)=\mathbf{f}[\mathbf{T}^*_{t'}(t),\,\dot{\mathbf{T}}^*_{t'}(t);\,\mathbf{U}^*_{t'}(t),\,\dot{\mathbf{U}}^*_{t'}(t)]. \tag{5.3.10}$$

Since t' is arbitrary, one can set $t'=t$ without loss of generality. Use of equation (5.3.9) together with the relation

$$\mathbf{D}(t)=\frac{\mathrm{D}}{\mathrm{D}t}\,\mathbf{U}_t(\tau)\bigg|_{\tau=t} \tag{5.3.11}$$

gives

$$\dot{\mathbf{U}}^*_t(t)=\mathbf{R}(t)^{\mathrm{T}}\mathbf{D}(t)\mathbf{R}(t) \tag{5.3.12}$$

where $\mathbf{D}$ is the stretching tensor. The tensor

$$\mathbf{T}^{\mathrm{r}}=\frac{\mathrm{D}}{\mathrm{D}t}\,[\mathbf{R}_{t'}(t)^{\mathrm{T}}\mathbf{T}(t)\mathbf{R}_{t'}(t)]_{t'=t} \tag{5.3.13}$$

is called the corotational stress rate (see equation (2.1.77)). It follows from equations (5.3.7) and (5.3.13) that equation (5.3.10), for $t'=t$, takes the form (Noll 1958)

$$(\mathbf{T}^*)^r=\mathbf{f}(\mathbf{T}^*,(\mathbf{T}^*)^r;\mathbf{D}^*,\mathbf{U}) \tag{5.3.14}$$

where

$$(\mathbf{T}^*)^r=\mathbf{R}^T\mathbf{T}^r\mathbf{R}. \tag{5.3.15}$$

Equation (5.3.14) is a form of the general constitutive equation of the rate-type.

With this as the basis, it is evident that equation (5.3.3) is a constitutive equation of the rate-type, the reduced equation being of the form

$$\mathbf{R}^T\mathbf{T}^r\mathbf{R}=\mathbf{g}(\mathbf{R}^T\mathbf{T}\mathbf{R},\mathbf{R}^T\mathbf{D}\mathbf{R};\mathbf{U}) \tag{5.3.16}$$

all quantities being taken with respect to the fixed reference configuration B_r. Equation (5.3.3) shows that $\dot{\mathbf{T}}$, and hence $\mathbf{T}^r$, must be determined independently of the choice of B_r. Having regard to the polar decomposition of equation (2.1.42), given any orthogonal $\mathbf{R}$ and any symmetric $\mathbf{U}$, it is always possible to find a reference configuration such that the present deformation gradient is given by $\mathbf{F}(t)=\mathbf{R}\mathbf{U}$. Therefore equation (5.3.16) must be an identity in $\mathbf{R}$ and $\mathbf{U}$. Taking $\mathbf{U}=\mathbf{I}$ and $\mathbf{R}=\mathbf{I}$, equation (5.3.16) becomes

$$\mathbf{T}^r=\mathbf{g}(\mathbf{T},\mathbf{D}). \tag{5.3.17}$$

Taking $\mathbf{U}=\mathbf{I}$ and $\mathbf{R}$ an arbitrary orthogonal tensor, equations (5.3.16) and (5.3.17) are taken to imply that $\mathbf{g}$ obeys the identity

$$\mathbf{R}^T\mathbf{g}(\mathbf{T},\mathbf{D})\mathbf{R}=\mathbf{g}(\mathbf{R}^T\mathbf{T}\mathbf{R},\mathbf{R}^T\mathbf{D}\mathbf{R}) \tag{5.3.18}$$

which implies that $\mathbf{g}$ is an isotropic tensor function of its two symmetric tensor arguments.

Consider two kinematical histories defined by $\mathbf{F}^t(s)=\mathbf{F}(t-s)$ and $\tilde{\mathbf{F}}^t(s)=\tilde{\mathbf{F}}(t-s)=\mathbf{F}(t-\sigma(s))$. The constitutive restriction formulated by equation (5.3.1) states that these two histories should determine the same present stress $\mathbf{T}(t)$. The deformation gradient $\tilde{\mathbf{F}}$ corresponds to

$$\tilde{\mathbf{D}}(t)=\dot{\sigma}(0)\mathbf{D}(t) \qquad \tilde{\mathbf{W}}(t)=\dot{\sigma}(0)\mathbf{W} \qquad \dot{\tilde{\mathbf{T}}}(t)=\dot{\sigma}(0)\dot{\mathbf{T}}(t). \tag{5.3.19}$$

Use of equation (5.3.17) to describe the process corresponding to $\tilde{\mathbf{F}}$ gives

$$\tilde{\mathbf{T}}^r=\alpha\mathbf{T}^r=\mathbf{h}(\mathbf{T},\tilde{\mathbf{D}})=\mathbf{h}(\mathbf{T},\alpha\mathbf{D}) \tag{5.3.20}$$

where $\alpha=\dot{\sigma}(0)$. Combination of equations (5.3.17) and (5.3.20) shows that

$$\mathbf{h}(\mathbf{T},\alpha\mathbf{D})=\alpha\mathbf{h}(\mathbf{T},\mathbf{D}). \tag{5.3.21}$$

Given any positive constant α, it is possible to choose $\sigma(s)=\alpha(s)$, $\dot{\sigma}(0)=\alpha$. Therefore, equation (5.3.21) holds for every positive α, which means that $\mathbf{h}(\mathbf{T},\mathbf{D})$ is positively homogeneous in $\mathbf{D}$.

These considerations enable the constitutive equation of a simple elastic–plastic material to be expressed in the form

$$\mathbf{T}^r=\mathsf{H}(\mathbf{T})[\mathbf{D}] \tag{5.3.22}$$

where the tensor function $\mathbf{H}(\mathbf{T})[\mathbf{D}]$ is isotropic in $\mathbf{T}$ and $\mathbf{D}$.

Equation (5.3.22) can be expressed in terms of the convected stress rate $\mathbf{T}^c$; thus

$$\mathbf{T}^c = \bar{\mathbf{H}}(\mathbf{T})[\mathbf{D}] \tag{5.3.23}$$

where $\bar{\mathbf{H}}$ differs from $\mathbf{H}$ by the term $(\mathbf{DT}+\mathbf{TD})$ (see equation (2.1.78)).

In accord with §2.3.4, in particular equation (2.3.56), incompressible elastic–plastic materials are defined by constitutive equations of the form of equations (5.3.22) and (5.3.23), except that $\mathbf{T}$ is replaced by $(\mathbf{T}+P\mathbf{I})$ to give, for example

$$\mathbf{T}^r = -\dot{P}\mathbf{I} + \mathbf{H}(\mathbf{T}+P\mathbf{I})[\mathbf{D}] \tag{5.3.24}$$

where only isochoric motions characterised by $\operatorname{tr}\mathbf{D}=0$ are to be considered. Equation (5.3.24) is the constitutive equation of an incompressible, elastic–plastic material as formulated by Truesdell (1955a,b) (see also Green 1956, Thomas 1955, Truesdell and Noll 1965).

The response function $\mathbf{H}$ can be represented in the form of equation (1.11.38). The resulting constitutive equation of the rate-type for an incompressible material can then be rearranged into the form

$$\tilde{\varphi}_0\mathbf{I} + \tilde{\varphi}_1\mathbf{D}' + \tilde{\varphi}_2\mathbf{D}'^2 = \frac{\mathscr{D}}{\mathscr{D}t}\mathbf{T}' + \tilde{\alpha}_1\mathbf{T}' \tag{5.3.25}$$

where a prime denotes a deviator, and where

$$\frac{\mathscr{D}}{\mathscr{D}t}\mathbf{T}' = (\mathbf{T}')^r - [(\mathbf{T}'\tilde{\boldsymbol{\Psi}} + \tilde{\boldsymbol{\Psi}}\mathbf{T}') - \tfrac{2}{3}\mathbf{I}\operatorname{tr}(\mathbf{T}'\tilde{\boldsymbol{\Psi}})]$$
$$- [(\mathbf{T}'^2\tilde{\boldsymbol{\Phi}} + \tilde{\boldsymbol{\Phi}}\mathbf{T}'^2) - \tfrac{2}{3}\mathbf{I}\operatorname{tr}(\tilde{\boldsymbol{\Phi}}\mathbf{T}'^2)] \tag{5.3.26}$$

and

$$\tilde{\boldsymbol{\Psi}} = \tilde{\psi}_5\mathbf{D}' + \tilde{\psi}_7\mathbf{D}'^2 + \tfrac{2}{3}(\operatorname{tr}\mathbf{T})\tilde{\boldsymbol{\Phi}} \qquad \tilde{\boldsymbol{\Phi}} = \tilde{\psi}_6\mathbf{D}' + \tilde{\psi}_8\mathbf{D}'^2 \tag{5.3.27}$$

$$\tilde{\varphi}_0 = -\tfrac{2}{3}\tilde{K}'_2\tilde{\varphi}_2 \qquad \tilde{\varphi}_1 = \tilde{\psi}_3 + \tfrac{2}{3}\tilde{\psi}_5(\operatorname{tr}\mathbf{T}) + \tfrac{2}{9}\tilde{\psi}_6(\operatorname{tr}\mathbf{T})^2 \tag{5.3.28}$$

$$\tilde{\varphi}_2 = \tilde{\psi}_4 + \tfrac{2}{3}\tilde{\psi}_7(\operatorname{tr}\mathbf{T}) + \tfrac{2}{9}\tilde{\psi}_8(\operatorname{tr}\mathbf{T})^2 \tag{5.3.29}$$

$$\tilde{a}_1 = -\tilde{\psi}_1 - \tfrac{2}{3}\tilde{\psi}_2(\operatorname{tr}\mathbf{T}) \tag{5.3.30}$$

$$\tilde{K}'_2 = \tfrac{1}{2}\operatorname{tr}\mathbf{D}'^2. \tag{5.3.31}$$

It must be noted that the $\tilde{\psi}_a$ $(a=1,\ldots,8)$ are functions of the basic invariants given by equation (1.11.39), subject to the condition $\operatorname{tr}\mathbf{D}=0$, since only isochoric motions are to be considered.

The particular conditions

$$\begin{gathered}\tilde{\varphi}_0 = 0 \qquad \tilde{\varphi}_2 = 0 \qquad \tilde{\psi}_4 = 0 \qquad \tilde{\psi}_6 = 0 \\ \tilde{\psi}_7 = 0 \qquad \tilde{\psi}_8 = 0 \qquad \tilde{\boldsymbol{\Phi}} = \mathbf{O}\end{gathered} \tag{5.3.32}$$

reduce equation (5.3.25) to the form

$$\mathbf{D}' = \frac{1}{\tilde{\varphi}_1} \frac{\mathscr{D}}{\mathscr{D}t} \mathbf{T}' + \dot{m}\mathbf{T}' \tag{5.3.33}$$

where

$$\tilde{\varphi}_1 = \tilde{\psi}_3 + \tfrac{2}{3}\tilde{\psi}_5(\operatorname{tr}\mathbf{T}) \qquad \dot{m} = \tilde{a}_1/\tilde{\varphi}_1. \tag{5.3.34}$$

Also, equation (5.3.26) now has the form

$$\frac{\mathscr{D}}{\mathscr{D}t}\mathbf{T}' = (\mathbf{T}')^{\mathrm{r}} - \tilde{\psi}_5\{[(\mathbf{T}')^{\mathrm{c}} - (\mathbf{T}')^{\mathrm{r}}] - \tfrac{1}{3}\mathbf{I}\operatorname{tr}[(\mathbf{T}')^{\mathrm{c}} - (\mathbf{T}')^{\mathrm{r}}]\} \tag{5.3.35}$$

which is the form of the loading rate proposed by Spencer and Ferrier (1973).

5.4 NON-SIMPLE ELASTIC–PLASTIC MATERIALS

5.4.1 Constitutive Equation of the Rate-Type

Consider the constitutive equation

$$\mathbf{G}^{\mathrm{r}} = -\dot{P}\mathbf{I} + \mathsf{H}(\mathbf{G} + P\mathbf{I})[\mathbf{D}] \tag{5.4.1}$$

where P is the pressure, $\mathbf{D}$ the stretching tensor and where $\mathbf{G}^{\mathrm{r}}$ is the corotational derivative of the loading function $\mathbf{G}$ defined by equation (4.2.5). Equation (5.4.1) is just equation (5.3.24) with $\mathbf{T}$ replaced by $\mathbf{G}$.

The response function H can be represented in the form of equation (1.11.38), the resulting constitutive equation of the rate-type for an incompressible non-simple elastic–plastic material then being rearranged into the form

$$\varphi_0\mathbf{I} + \varphi_1\mathbf{D}' + \varphi_2\mathbf{D}'^2 = \frac{\mathscr{D}}{\mathscr{D}t}\mathbf{G}' + a_0\mathbf{I} + a_1\mathbf{G}' + a_2\mathbf{G}'^2 \tag{5.4.2}$$

where

$$\frac{\mathscr{D}}{\mathscr{D}t}\mathbf{G}' = (\mathbf{G}')^{\mathrm{r}} - [(\mathbf{G}'\mathbf{\Psi} + \mathbf{\Psi}\mathbf{G}') - \tfrac{2}{3}\mathbf{I}\operatorname{tr}(\mathbf{G}'\mathbf{\Psi})]$$
$$- [(\mathbf{G}'^2\mathbf{\Phi} + \mathbf{\Phi}\mathbf{G}'^2) - \tfrac{2}{3}\mathbf{I}\operatorname{tr}(\mathbf{\Phi}\mathbf{G}'^2)] \tag{5.4.3}$$

and

$$\mathbf{\Psi} = \psi_5\mathbf{D}' + \psi_7\mathbf{D}'^2 + \tfrac{2}{3}(\operatorname{tr}\mathbf{G})\mathbf{\Phi} \qquad \mathbf{\Phi} = \psi_6\mathbf{D}' + \psi_8\mathbf{D}'^2 \tag{5.4.4}$$

$$\varphi_0 = -\tfrac{2}{3}\tilde{K}'_2\varphi_2 \qquad \varphi_1 = \psi_3 + \tfrac{2}{3}\psi_5(\operatorname{tr}\mathbf{G}) + \tfrac{2}{9}\psi_6(\operatorname{tr}\mathbf{G})^2$$
$$\varphi_2 = \psi_4 + \tfrac{2}{3}\psi_7(\operatorname{tr}\mathbf{G}) + \tfrac{2}{9}\psi_8(\operatorname{tr}\mathbf{G})^2 \tag{5.4.5}$$

$$a_0 = \tfrac{2}{3}L'_2\psi_2 \qquad a_1 = -\psi_1 - \tfrac{2}{3}\psi_2(\operatorname{tr}\mathbf{G}) \qquad a_2 = -\psi_2 \tag{5.4.6}$$

$$\tilde{K}'_2 = \tfrac{1}{2}\,\mathrm{tr}\,\mathbf{D}'^2 \qquad L'_2 = \tfrac{1}{2}\,\mathrm{tr}\,\mathbf{G}'^2. \tag{5.4.7}$$

It must be noted that the ψ_a $(a = 1, \ldots, 8)$ are functions of the basic invariants given by equation (1.11.39), subject to the condition, $\mathrm{tr}\,\mathbf{D} = 0$, since only isochoric motions are to be considered.

For a material which satisfies von Mises yield criterion, for which $\mathbf{G}' = \mathbf{T}'$ (see equation (5.2.10)), the assumption is made that equation (5.4.2) must reduce to equation (5.3.33). The particular class of materials characterised by the conditions

$$\begin{array}{llll} a_0 = 0 & a_2 = -\psi_2 = 0 & a_1 = -\psi_1 & \\ \varphi_0 = 0 & \varphi_2 = 0 & \psi_4 = 0 & \psi_6 = 0 \\ \boldsymbol{\Phi} = \mathbf{0} & \psi_7 = 0 & \psi_8 = 0 & \end{array} \tag{5.4.8}$$

satisfy the required assumption.

The conditions of equation (5.4.8) reduce equation (5.4.2) to the constitutive equation

$$\mathbf{D}' = \frac{1}{\varphi_1}\frac{\mathscr{D}}{\mathscr{D}t}\mathbf{G}' + \dot{m}\mathbf{G}' \tag{5.4.9}$$

where

$$\varphi_1 = \psi_3 + \tfrac{2}{3}\psi_5(\mathrm{tr}\,\mathbf{G}) \qquad \dot{m} = -\psi_1/\varphi_1 \tag{5.4.10}$$

and where equation (5.4.3) now has the form

$$\frac{\mathscr{D}}{\mathscr{D}t}\mathbf{G}' = (\mathbf{G}')^{\mathrm{r}} - \psi_5\{[(\mathbf{G}')^{\mathrm{c}} - (\mathbf{G}')^{\mathrm{r}}] - \tfrac{1}{3}\mathbf{I}\,\mathrm{tr}[(\mathbf{G}')^{\mathrm{c}} - (\mathbf{G}')^{\mathrm{r}}]\} \tag{5.4.11}$$

it being noted that $(\mathbf{G}')^{\mathrm{c}}$ is the convected derivative of $\mathbf{G}'$. The quantity $\mathscr{D}\mathbf{G}'/\mathscr{D}t$, defined by equation (5.4.11), will be referred to as a *general loading rate*. In this connection, although it is widely recognised that there exist very many forms of loading rate (see for example Cotter and Rivlin 1955, Green 1956, Havner 1982, Hill 1968, Jaumann 1911, Spencer and Ferrier 1973, Truesdell 1955a,b), it should be noted that terms which represent the difference between such possible loading rates are objective quantities and may therefore appear in constitutive equations, as for example the term $(\mathbf{DA} + \mathbf{AD})$ of equation (2.1.78).

5.4.2 Incremental-Type Constitutive Equation

Denoting by $\boldsymbol{\psi}$ the deformation of a body $\mathscr{B}$, that is the time-dependent mapping $\boldsymbol{\psi}$ from reference configuration $\boldsymbol{\kappa}$ to the current configuration $\boldsymbol{\theta}$, then from §2.1.3

$$\boldsymbol{x} = \boldsymbol{\psi}(X)$$

which is just equation (2.1.33), $\boldsymbol{x}$ being the current position occupied by each body-point $X \in \mathscr{B}$. Let the deformation be changed to $\boldsymbol{\psi}^*$ where now

$$\boldsymbol{x}^* = \boldsymbol{\psi}^*(X)$$

is the new current position of the body-point $X \in \mathscr{B}$. The displacement of a body-point due to this change is,

$$\boldsymbol{x}^* - \boldsymbol{x} = \boldsymbol{\psi}^*(X) - \boldsymbol{\psi}(X)$$

which can be written in the form,

$$\delta\boldsymbol{x} = \delta\boldsymbol{\psi}(X)$$

where the operator δ is defined by, $\delta\boldsymbol{\psi} = \boldsymbol{\psi}^* - \boldsymbol{\psi}$.

The present discussion is concerned only with those displacements $\delta\boldsymbol{x}$ which are sufficiently small for each $X \in \mathscr{B}$ that terms of order $|\delta\boldsymbol{x}|^2$ can be neglected in comparison with those of order $|\delta\boldsymbol{x}|$. In the context of this definition of smallness, $\delta\boldsymbol{\psi}$ is referred to as an *incremental deformation* from the configuration described by $\boldsymbol{\psi}$.

In terms of the concept of an incremental deformation the deformation gradient tensor $\mathbf{F}$, defined by equation (2.1.35), changes as follows:

$$\delta\mathbf{F}(\boldsymbol{X}) = \delta(\mathrm{Grad}\ \boldsymbol{\psi})(\boldsymbol{X}) = \mathrm{Grad}\ \delta\boldsymbol{\psi}(\boldsymbol{X})$$

from which it is evident that the operators δ and Grad commute. It must be noted that this relation is exact, and in particular is valid even if $\delta\boldsymbol{\psi}(\boldsymbol{X})$ is not incremental in the sense of the definition of an incremental deformation given above.

In order to obtain incremental stress–strain relations, the rates of stress and rates of strain occurring in rate equations may be replaced by the corresponding increments, having noted that the resulting incremental equations are approximations.

Substituting for $\mathbf{D}'$ in equation (5.4.9) from equation (2.1.96) gives

$$\alpha_1[2(\mathbf{B}')^{\mathrm{r}} - (\mathbf{B}')^{\mathrm{c}}] + \alpha_{-1}[(\mathbf{B}^{-1})']^{\mathrm{c}} = \frac{1}{\varphi_1}\frac{\mathscr{D}}{\mathscr{D}t}\mathbf{G}' + \dot{m}\mathbf{G}' \tag{5.4.12}$$

where the α_i $(i = \pm 1)$ are given by equations (2.1.97) and (2.1.98).

For infinitesimal deformations from an arbitrary configuration at time t, to one at $t + \varepsilon$, the constitutive equation (5.4.12) reduces to the form

$$\alpha_1\ \delta\mathbf{B}' + \alpha_{-1}\ \delta(\mathbf{B}^{-1})' = (1/\varphi_1)\ \delta\mathbf{G}' + \mathbf{G}'\ \delta m + \mathrm{O}(\varepsilon). \tag{5.4.13}$$

Thus, when the error term, $\mathrm{O}(\varepsilon)$ is omitted, the constitutive equation (5.4.9) reduces to the incremental-type constitutive equation

$$\alpha_1\ \delta\mathbf{B}' + \alpha_{-1}\ \delta(\mathbf{B}^{-1})' = (1/\varphi_1)\ \delta\mathbf{G}' + \mathbf{G}'\ \delta m \tag{5.4.14}$$

where the α_i, defined by equations (2.1.97) and (2.1.98) now have the form

$$\alpha_1 = \frac{3q}{2(qI_{\mathbf{B}} + II_{\mathbf{B}})} = -q\alpha_{-1} \qquad (I_{\mathbf{B}^{-1}} = II_{\mathbf{B}}) \tag{5.4.15}$$

$$q = \lim_{\varepsilon \to 0} (\dot{II}_{\mathbf{B}}/\dot{I}_{\mathbf{B}}). \tag{5.4.16}$$

From the constitutive equation (4.2.21) one can obtain the incremental-type constitutive equation,

$$\delta\mathbf{G}' = \beta_1\, \delta\mathbf{B}' + \beta_{-1}\, \delta(\mathbf{B}^{-1})' + \mathbf{B}'\, \delta\beta_1 + (\mathbf{B}^{-1})'\, \delta\beta_{-1}. \tag{5.4.17}$$

Setting

$$\beta_i = \alpha_i/m = 2\hat{A}\alpha_i \qquad (i = \pm 1) \tag{5.4.18}$$

where

$$\hat{A} = \hat{A}(I_{\mathbf{B}}, II_{\mathbf{B}}) \tag{5.4.19}$$

allows equation (5.4.14) to be rearranged into the form

$$\begin{aligned}\delta\mathbf{G}' = {} & \beta_1\, \delta\mathbf{B}' + \beta_{-1}\, \delta(\mathbf{B}^{-1})' + \mathbf{B}'\, \delta\beta_1 + (\mathbf{B}^{-1})'\, \delta\beta_{-1} \\ & + (1 - 2\hat{A}/\varphi_1)\, \delta\mathbf{G}' - 2\hat{A}[\mathbf{B}'\, \delta\alpha_1 + (\mathbf{B}^{-1})'\, \delta\alpha_{-1}]\end{aligned} \tag{5.4.20}$$

from which it is evident, by comparison with equation (5.4.17) that the two incremental-type constitutive equations (5.4.14) and (5.4.17) are equivalent provided the condition

$$\left(\frac{1}{2\hat{A}} - \frac{1}{\varphi_1}\right)\delta\mathbf{G}' = \mathbf{B}'\, \delta\alpha_1 + (\mathbf{B}^{-1})'\, \delta\alpha_{-1} \tag{5.4.21}$$

can be satisfied. Regarding equation (5.4.21) as a relation defining the parameter φ_1 in terms of m, it is evident that the response coefficients are now defined by the left-hand relation of equation (5.4.18) in the form

$$\beta_1 = \frac{3q}{(qI_{\mathbf{B}} + II_{\mathbf{B}})}\hat{A} = -q\beta_{-1} \qquad q = \frac{\mathrm{d}II_{\mathbf{B}}/\mathrm{d}p}{\mathrm{d}I_{\mathbf{B}}/\mathrm{d}p} \tag{5.4.22}$$

where p is any parameter which determines the sequence of events.

Equation (5.4.22) can be rearranged to give the relation

$$\hat{A} = \tfrac{1}{3}(\beta_1 I_{\mathbf{B}} - \beta_{-1} II_{\mathbf{B}}). \tag{5.4.23}$$

In the reference, that is the natural, undeformed state, $I_{\mathbf{B}} = II_{\mathbf{B}} = 3$ and $\hat{A}$ takes the limiting form

$$[\hat{A}]_{I_{\mathbf{B}} = II_{\mathbf{B}} = 3} = [\beta_1 - \beta_{-1}]_{I_{\mathbf{B}} = II_{\mathbf{B}} = 3}. \tag{5.4.24}$$

It is to be noted that, whatever form is chosen for the response coefficients, this form must satisfy the generalised Lode relation of equation (4.2.38).

5.5 DECOMPOSITION OF THE STRETCHING TENSOR

The compatibility condition of equation (5.1.14) gives the total velocity gradient tensor of equations (2.1.60) and (2.1.67) in the form (Lee 1969),

$$\mathbf{L}=(\dot{\mathbf{F}}^{e}\mathbf{F}^{p}+\mathbf{F}^{e}\dot{\mathbf{F}}^{p})(\mathbf{F}^{p})^{-1}(\mathbf{F}^{e})^{-1}=\mathbf{L}^{e}+\mathbf{F}^{e}\mathbf{L}^{p}(\mathbf{F}^{e})^{-1} \tag{5.5.1}$$

and hence the total stretching tensor,

$$\mathbf{D}=\mathbf{D}^{e}+(\mathbf{D}^{*})^{p} \tag{5.5.2}$$

where

$$\begin{aligned}(\mathbf{D}^{*})^{p}=&\tfrac{1}{2}\{\mathbf{F}^{e}\mathbf{W}^{p}(\mathbf{F}^{e})^{-1}+[\mathbf{F}^{e}\mathbf{W}^{p}(\mathbf{F}^{e})^{-1}]^{T}\}\\ &+\tfrac{1}{2}\{\mathbf{F}^{e}\mathbf{D}^{p}(\mathbf{F}^{e})^{-1}+[\mathbf{F}^{e}\mathbf{D}^{p}(\mathbf{F}^{e})^{-1}]^{T}\}\end{aligned} \tag{5.5.3}$$

and where $\mathbf{D}^{p}$ is the plastic component of the stretching tensor in the intermediate configuration. It is now possible, at least in principle, to enter the form for $\mathbf{D}$ given by equation (5.5.2) into equation (5.4.9) and separate the resulting relation into a constitutive equation for the elastic part and a constitutive equation for the plastic part (see for example Lee 1966, 1969, Lee and Lui 1967, Lee and McMeeking 1980, Lubarda and Lee 1981). In practice, a unique decomposition is not possible because there are terms which cannot be identified specifically as either contributing solely to the elastic behaviour, or solely to the plastic deformation.

The alternative approach is simply to assume that the total strain can be decomposed into the sum of elastic and plastic strains (see for example Green 1956, Green and Naghdi 1965, Hill 1958, Hill and Hutchinson 1975, Thomas 1955). This approach is taken to imply a relation of the form

$$\mathbf{D}=\tilde{\mathbf{D}}^{e}+\tilde{\mathbf{D}}^{p} \tag{5.5.4}$$

where, for convenience $\tilde{\mathbf{D}}^{e}$ and $\tilde{\mathbf{D}}^{p}$ are generally referred to as the elastic and plastic parts, respectively, of the total stretching tensor, having noted, however, that in general the $\tilde{\mathbf{D}}^{e}$ and $\tilde{\mathbf{D}}^{p}$ need not satisfy rate of deformation compatibility conditions. If the elastic deformation is small in an appropriate sense then, as Lee (1969) indicates, the compatibility condition of equation (5.1.14) is equivalent to equation (5.5.4). An example of this approach is the study by Spencer and Ferrier (1973) of some solutions for a class of plastic–elastic solids. For an incompressible material, Spencer and Ferrier take

$$2G_0\tilde{\mathbf{D}}^{e}=(\mathbf{T}')^{r}+\alpha[(\mathbf{D}'\mathbf{T}'+\mathbf{T}'\mathbf{D}')-\tfrac{2}{3}\operatorname{tr}(\mathbf{D}'\mathbf{T}')]=\frac{\mathscr{D}}{\mathscr{D}t}\mathbf{T}' \tag{5.5.5}$$

$$\tilde{\mathbf{D}}^{p}=\mathbf{T}'\lambda \tag{5.5.6}$$

where α is a dimensionless constant and λ is generally a variable with units of (time stress)$^{-1}$. Entering the form for the $\mathbf{D}^{e}$ and $\mathbf{D}^{p}$ given by equations (5.5.5) and (5.5.6) into equation (5.5.4) gives the total stretching tensor,

$$\mathbf{D}=\frac{1}{2G_0}\frac{\mathscr{D}}{\mathscr{D}t}\mathbf{T}'+\lambda\mathbf{T}' \tag{5.5.7}$$

which, with $2G_0 \equiv \tilde{\varphi}_1$, $\lambda \equiv \dot{m}$, $\alpha \equiv -\tilde{\psi}_5$, is just equation (5.3.33).

REFERENCES

Bridgman P W 1923 *Proc. Am. Acad. Arts Sci.* **58** 163–242

Cotter B and Rivlin R S 1955 *Quart. Appl. Math.* **13** 177–82

Crossland B 1954 *Inst. Mech. Eng. Proc.* **168** 935–46

Drucker D C 1951 *Proc. 1st US Nat. Congr. Appl. Mech.* (New York: ASME) pp 487–91

—— 1959 *J. Appl. Mech.* **26** 101–6

—— 1960 *Proc. 2nd Symp. Naval Structural Mechanics* ed. E H Lee and P S Symonds (Oxford: Pergamon) pp 170–84

Freund L B 1970 *Int. J. Solids Struct.* **6** 1193–209

Green A E 1956 *Proc. R. Soc.* A **234** 46–59

Green A E and Naghdi P M 1965 *Arch. Ration. Mech. Anal.* **18** 251–81

Hahn H T 1974 *Int. J. Solids and Struct.* **10** 111–21

Hahn H T and Jaunzemis W 1971 *Int. J. Eng. Sci.* **11** 1065–78

Havner K S 1982 *The Theory of Finite Plastic Deformation of Crystalline Solids in Mechanics of Solids* ed. H G Hopkins and M J Sewell (Oxford: Pergamon) pp 265–302

Hill R 1958 *J. Mech. Phys. Solids* **6** 236–49

—— 1968 *J. Mech. Phys. Solids* **16** 315–22

Hill R and Hutchinson J W 1975 *J. Mech. Phys. Solids* **23** 239–64

Jaumann G 1911 *Sitzungsber. Akad. Wiss. Wien* (IIa) **120** 385–530

Lee EH 1966 *Elastic–Plastic Waves of One-Dimensional Strain Proc. 5th US Natl. Congr. Appl. Mech.* (New York: ASME) pp 405–20

—— 1969 *J. Appl. Mech.* **36** 1–6

—— 1981 *Int. J. Solids Struct.* **17** 859–72

Lee E H and Germain P 1974 *Elastic–plastic Theory at Finite Strain in Problems of Plasticity* ed. A Sawczuk (Leyden: Noordhoff) pp 117–30

Lee E H and Lui D T 1967 *J. Appl. Phys.* **38** 19–27

Lee E H and McMeeking R M 1980 *Int. J. Solids Struct.* **16** 715–21

Lianis G and Ford H 1957 *J. Mech. Phys. Solids* **5** 215–22

Lubarda V A and Lee E H 1981 *J. Appl. Mechs.* **48** 35–40

Naghdi P M 1960 *Stress–Strain Relations in Plasticity and Thermoplasticity* in *Plasticity* ed. E H Lee and P Symonds (Oxford: Pergamon) pp 121–69

Noll W 1958 *Arch. Ration. Mech. Anal.* **2** 197–226

Prager W 1938 *Proc. 5th Int. Congr. Appl. Mech. Cambridge, MA* (New York: Wiley) pp 234–7

—— 1942 *Duke Math. J.* **9** 228–33

Reiner M 1945 *Am. J. Math.* **67** 350–62
Rivlin R S 1948 *Proc. R. Soc.* A **193** 260–81
—— 1949 *Proc. Camb. Phil. Soc.* **45** 88–91
Spencer A J M and Ferrier J E 1973 *Some Solutions for a Class of Plastic–Elastic Solids* in *Int. Symp. on Foundations of Plasticity* vol. 1 ed. A Sawczuk (Leyden: Noordhoff) pp 9–24
Thomas T Y 1955 *Proc. Natl Acad. Sci. USA* **41** 720–6
Truesdell C 1955a *J. Ration. Mech. Anal.* **4** 83–133
—— 1955b *J. Ration. Mech. Anal.* **4** 1019–20

GENERAL REFERENCES

Billington E W and Tate A 1981 *The Physics of Deformation and Flow* (New York: McGraw-Hill)
Bridgman P W 1952 *Studies in Large Plastic Flow and Fracture* (New York: McGraw-Hill)
Drucker D C 1967 *Introduction to the Mechanics of Deformable Solids* (New York: McGraw-Hill)
Freudenthal A M and Geiringer H 1958 *The Intrinsic Continuum* in *Handbuch der Physik* vol. IV ed. S Flügge (Berlin: Springer) pp 229–443
Fung Y C 1965 *Foundations of Solid Mechanics* (Englewood Cliffs, NJ: Prentice-Hall)
Hill R 1950 *Plasticity* (Oxford: Clarendon)
Hunter S C 1983 *Mechanics of Continuous Media* 2nd edn (Chichester: Ellis Horwood)
Kachanov L M 1971 *Foundations of the Theory of Plasticity* (Amsterdam: North Holland)
Luhbahn J D and Felgar R P 1961 *Plasticity and Creep of Metals* (New York: Wiley)
Nadai A 1950 *Theory of Flow and Fracture of Solids* (New York: McGraw-Hill)
Prager W 1959 *An Introduction to Plasticity* (Reading, MA: Addison-Wesley)
—— 1961 *Introduction to Mechanics of Continua* (Boston: Ginn)
Truesdell C and Noll W 1965 *The Non-linear Field Theories of Mechanics* in *Handbuch der Physik* vol. III/3 ed. S Flügge (Berlin: Springer)

6

Universal Solutions For a Class of Deformable Solids

6.1 HOMOGENEOUS ISOTROPIC BODIES: I, SIMPLE EXTENSION AND SIMPLE SHEAR

6.1.1 Static Universal Solutions

Throughout this section, $\mathscr{B}$ will denote a homogeneous body which is initially elastic, and $\boldsymbol{\kappa}$ will denote a homogeneous reference configuration of $\mathscr{B}$. Relative to $\boldsymbol{\kappa}$, the constitutive equation for a class of general elastic–plastic solids takes the form

$$\mathbf{G}(\mathbf{T}) = \mathbf{g}(\mathbf{F}) \tag{6.1.1}$$

where the loading function $\mathbf{G}$ is a tensor-valued function of the stress tensor $\mathbf{T}$, and where the response function $\mathbf{g}$ is independent of the body point. The value of the response function gives the loading function in any configuration $\boldsymbol{\psi}$ whose deformation gradient relative to $\boldsymbol{\kappa}$ is $\mathbf{F}$; thus

$$\mathbf{F} = \boldsymbol{\psi}_* \circ \boldsymbol{\kappa}_*^{-1} \tag{6.1.2}$$

and

$$\mathbf{G}(\mathbf{T}(X)) = \mathbf{g}(\boldsymbol{\psi}_{*X} \circ \boldsymbol{\kappa}_{*X}^{-1}) \qquad \forall\, \boldsymbol{x} \in \boldsymbol{\psi}(\mathscr{B}) \tag{6.1.3}$$

where $\boldsymbol{x}$ denotes the position of the body-point X in the configuration $\boldsymbol{\psi}$

$$\boldsymbol{x} = \boldsymbol{\psi}(X) \qquad X = \Psi(\boldsymbol{x}). \tag{6.1.4}$$

In the reference configuration $\boldsymbol{\kappa}$ the body-point X occupies the position $\boldsymbol{X}$, where

$$\boldsymbol{X} = \boldsymbol{\kappa}(X) = \boldsymbol{\kappa}(\Psi(\boldsymbol{x})). \tag{6.1.5}$$

Thus, the deformation $\boldsymbol{\kappa}$ to $\boldsymbol{\psi}$ maps the point $\boldsymbol{X} \in \boldsymbol{\kappa}(\mathscr{B})$ into the point $\boldsymbol{x} \in \boldsymbol{\psi}(\mathscr{B})$ (see §§2.1.1 and 2.1.3).

The simplest type of deformation is a *homogeneous* deformation, defined by the condition that the deformation gradient $\mathbf{F}$ be a constant tensor field: thus

$$\mathbf{F}(x)=\mathbf{F} \qquad \forall \boldsymbol{x}\in \psi(\mathscr{B}). \tag{6.1.6}$$

It follows that a homogeneous deformation can be given explicitly in the form

$$\boldsymbol{x}=\mathbf{F}(\boldsymbol{X}-\boldsymbol{O})+\boldsymbol{c} \tag{6.1.7}$$

where $\boldsymbol{O}$ denotes the reference origin and $\boldsymbol{c}$ denotes an arbitrary fixed place. A homogeneous deformation may be regarded as a linear transformation $\mathbf{F}$ followed by a rigid translation. In a homogeneous deformation every pair of non-coincident straight lines in $\boldsymbol{\kappa}(\mathscr{B})$ is mapped onto a pair of non-coincident straight lines in $\psi(\mathscr{B})$.

Since a homogeneous deformation always maps one homogeneous configuration onto another, that is all measures of deformation and rotation are constant throughout the body, the loading function $\mathbf{G}$ required to effect such a deformation is a constant tensor, and hence

$$\operatorname{div}\mathbf{G}=\boldsymbol{0}. \tag{6.1.8}$$

Entering the form for $\mathbf{G}$ given by equation (4.2.16) into equation (6.1.8) gives for symmetric stress $\mathbf{T}$

$$\operatorname{div}\mathbf{G}=\mathbf{I}\operatorname{grad}c_0+\mathbf{T}\operatorname{grad}c_1+c_1\operatorname{div}\mathbf{T}+\mathbf{T}^2\operatorname{grad}c_2+c_2\operatorname{div}\mathbf{T}^2=\boldsymbol{0}. \tag{6.1.9}$$

Use has been made of the identity

$$\operatorname{div}(c\mathbf{A})=\mathbf{A}^{\mathrm{T}}\operatorname{grad}c+c\operatorname{div}\mathbf{A}$$

having noted that $\mathbf{A}$ is an arbitrary second order tensor. Equation (6.1.9) can be rearranged using equation (4.2.17) into the form,

$$\frac{\partial c_0}{\partial I_a}\operatorname{grad}I_a+\frac{\partial c_1}{\partial I_a}\mathbf{T}\operatorname{grad}I_a+\frac{\partial c_2}{\partial I_a}\mathbf{T}^2\operatorname{grad}I_a+c_1\operatorname{div}\mathbf{T}+c_2\operatorname{div}\mathbf{T}^2=\boldsymbol{0}. \tag{6.1.10}$$

where the I_a denote the three principal invariants of $\mathbf{T}$, and hence the index a in equation (6.1.10) is to be summed from 1 to 3. Since the c_i $(i=0,1,2)$ are arbitrary functions, it follows from equation (6.1.10) that

$$\operatorname{grad}I_a=\boldsymbol{0} \qquad (a=1,2,3) \tag{6.1.11}$$

$$\operatorname{div}\mathbf{T}=\boldsymbol{0} \tag{6.1.12}$$

and

$$\operatorname{div}\mathbf{T}^2=\boldsymbol{0}. \tag{6.1.13}$$

Equation (6.1.12) is of particular physical significance since it implies that every configuration $\psi(\mathscr{B})$ of $\mathscr{B}$ which follows as a consequence of a homogeneous deformation from $\boldsymbol{\kappa}$ is a configuration of static equilibrium provided that the boundary $\partial_{\psi}(\mathscr{B})$ can be supported by the surface forces

$$t = \mathbf{T}^{\mathrm{T}} \boldsymbol{n}. \tag{6.1.14}$$

$\boldsymbol{n}$ denotes the outward unit normal to $\partial_{\psi}(\mathscr{B})$ (see §2.2.3).

Since the shape of the body and the particular response function **g** enter a specific loading programme only through the boundary condition (6.1.14), it follows that static equilibrium configurations are of considerable physical significance for the following reasons:

(i) forces of some description can, at least in principle, always be applied to a boundary;

(ii) the field equation of equilibrium is satisfied irrespective of the nature of the body;

(iii) if the boundary forces required for all homogeneous deformations are known, then the response function **g** is a determinable quantity.

With regard to the third property, that is the determination of **g**, equation (4.1.5) implies that information obtained from pure stretches should be sufficient.

6.1.2 Homogeneous Isotropic Solid Body

The constitutive equation (6.1.1) of $\mathscr{B}$ has the reduced form (cf equation (4.2.18)),

$$\mathbf{G}(\mathbf{T}) = \phi_0 \mathbf{I} + \phi_1 \mathbf{B} + \phi_2 \mathbf{B}^2 \tag{6.1.15}$$

which can be rearranged into the form

$$\mathbf{G}(\mathbf{T}) = \beta_0 \mathbf{I} + \beta_1 \mathbf{B} + \beta_{-1} \mathbf{B}^{-1} \tag{6.1.16}$$

where use has been made of the Cayley–Hamilton theorem and where

$$\beta_0 = \phi_0 - \phi_2 II_{\mathbf{B}} \qquad \beta_1 = \phi_1 + \phi_2 I_{\mathbf{B}} \qquad \beta_{-1} = \phi_2 III_{\mathbf{B}}. \tag{6.1.17}$$

Substituting for **G** in equation (6.1.8) from equation (6.1.16) gives

$$\frac{\partial \beta_0}{\partial I_a} \operatorname{grad} I_a + \frac{\partial \beta_1}{\partial I_a} \mathbf{B} \operatorname{grad} I_a + \frac{\partial \beta_{-1}}{\partial I_a} \mathbf{B}^{-1} \operatorname{grad} I_a + \beta_1 \operatorname{div} \mathbf{B} + \beta_{-1} \operatorname{div} \mathbf{B}^{-1} = \mathit{0} \tag{6.1.18}$$

where the I_a denote the three principal invariants $I_{\mathbf{B}}$, $II_{\mathbf{B}}$ and $III_{\mathbf{B}}$ of **B**, and hence the index a in equation (6.1.18) is to be summed from 1 to 3. Since the β_i $(i = 0, \pm 1)$ are arbitrary functions, it follows that

$$\operatorname{grad} I_a = \mathit{0} \qquad (a = 1, 2, 3) \tag{6.1.19}$$

$$\operatorname{div} \mathbf{B} = \mathit{0} \tag{6.1.20}$$

and

$$\operatorname{div} \mathbf{B}^{-1} = \mathbf{0}. \tag{6.1.21}$$

A consequence of the condition of integrability for equation (6.1.2) is that the curvature tensor R_{ijkl} (see §1.8), based on $\mathbf{B}^{-1}$ must satisfy the condition

$$R_{ijkl} = 0 \tag{6.1.22}$$

for all $\boldsymbol{x}$ in the domain of $\mathbf{F}$. This condition can be expressed in the form

$$\frac{1}{2}\left[\frac{\partial^2 (B^{-1})_{il}}{\partial x^j\,\partial x^k} + \frac{\partial^2 (B^{-1})_{kj}}{\partial x^i\,\partial x^l} - \frac{\partial^2 (B^{-1})_{ij}}{\partial x^l\,\partial x^k} - \frac{\partial^2 (B^{-1})_{kl}}{\partial x^i\,\partial x^j}\right] + B^{pq}(A_{kjp}A_{ilq} - A_{klp}A_{ijq}) = 0 \tag{6.1.23}$$

where

$$2A_{ilj} \equiv \frac{\partial (B^{-1})_{ij}}{\partial x^l} + \frac{\partial (B^{-1})_{lj}}{\partial x^i} + \frac{\partial (B^{-1})_{il}}{\partial x^j}. \tag{6.1.24}$$

Summing equation (6.1.23) with respect to the pairs of indices (i, l) and (j, k) and then using the conditions of equations (6.1.19), (6.1.20) and (6.1.21), gives

$$B^{pq} A_{ijp} A_{ijq} = 0 \tag{6.1.25}$$

which can be expressed in the form

$$(V^{pq} A_{ijp})^2 = 0 \tag{6.1.26}$$

where use has been made of equation (2.1.45), it being noted that $\mathbf{V}$ is the left stretch tensor. Since $\mathbf{V}$ is positive-definite (see §1.6.6), equation (6.1.26) implies that

$$A_{ijp} = 0. \tag{6.1.27}$$

But from equation (6.1.24), it follows that

$$\frac{\partial (B^{-1})_{jq}}{\partial x^i} = A_{ijq} + A_{qij} \tag{6.1.28}$$

Hence $\mathbf{B}$ must be a constant tensor field.

The condition of equation (6.1.27) implies that $\mathbf{F}$ is a constant tensor also (in accord with equation (6.1.6)), since from the transformation law for the Christoffel symbols A_{ijp}, equation (6.1.27) holds if and only if the coordinates (x^1, x^2, x^3) and (X^1, X^2, X^3) are related by an affine transformation (cf equation (6.1.7)).

6.1.3 Simple Extension: I, Deformation Response Function

Let the cartesian coordinates (X, Y, Z) be the referential coordinates of a body in the natural state and the cartesian coordinates (x, y, z) be the spatial coordinates in the deformed configuration. For simple extension with stretch in the Z-direction, the simple deformations are:

$$x=\lambda_1 X \qquad y=\lambda_2 Y \qquad z=\lambda_3 Z \tag{6.1.29}$$

where

$$\lambda_1=\lambda_2=\alpha\lambda_3 \tag{6.1.30}$$

and the λ_i $(i=1,2,3)$ are the principal stretches.

From equations (2.1.36), (2.1.45) and (6.1.29), the matrices of $\mathbf{F}$, $\mathbf{B}$ and $\mathbf{B}^{-1}$ are

$$[F_{i\alpha}]=\begin{bmatrix}\alpha\lambda_3 & 0 & 0\\ 0 & \alpha\lambda_3 & 0\\ 0 & 0 & \lambda_3\end{bmatrix} \tag{6.1.31}$$

$$[B_{ij}]=\begin{bmatrix}\alpha^2\lambda_3^2 & 0 & 0\\ 0 & \alpha^2\lambda_3^2 & 0\\ 0 & 0 & \lambda_3^2\end{bmatrix} \qquad [(B^{-1})_{ij}]=\begin{bmatrix}(\alpha\lambda_3)^{-2} & 0 & 0\\ 0 & (\alpha\lambda_3)^{-2} & 0\\ 0 & 0 & \lambda_3^{-2}\end{bmatrix} \tag{6.1.32}$$

The principal invariants are

$$I_{\mathbf{B}}=(2\alpha^2+1)\lambda_3^2 \qquad II_{\mathbf{B}}=(2+\alpha^2)/\alpha^2\lambda_3^2 \qquad III_{\mathbf{B}}=\alpha^4\lambda_3^6. \tag{6.1.33}$$

Substituting for the $b_i=\lambda_i^2$ in equation (4.1.27) from equation (6.1.30) gives $\kappa=-1$, which can be entered into equation (4.1.29) to give

$$\bar{\nu}=\kappa=-1. \tag{6.1.34}$$

The constitutive equation (6.1.1) can be used in the form of equation (4.2.40)

$$\mathbf{G}'(\mathbf{T}')=\mathbf{M}'(\mathbf{B}'). \tag{6.1.35}$$

From equations (4.2.40) and (6.1.32),

$$\begin{aligned} M'_{xx}&=\tfrac{1}{3}\beta_{-1}(1-\alpha^2)(q\lambda_3^2+1/\alpha^2\lambda_3^2)=M'_{yy}\\ M'_{zz}&=\tfrac{2}{3}(-\beta_{-1})(1-\alpha^2)(q\lambda_3^2+1/\alpha^2\lambda_3^2)=-2M'_{xx}\end{aligned} \tag{6.1.36}$$

where

$$q=-\beta_1/\beta_{-1}. \tag{6.1.37}$$

Equations (6.1.35) and (6.1.36) give

$$G'_{xx}=G'_{yy}=-\tfrac{1}{2}G'_{zz} \tag{6.1.38}$$

which when used with equation (4.2.3) gives

$$G'_{xx}-G'_{yy}=(\delta_1-\delta_2 T'_{zz})(T_{xx}-T_{yy})=0. \tag{6.1.39}$$

For the von Mises yield criterion (see equation (5.2.10)),

$$\delta_0=0, \qquad \delta_1=1, \qquad \delta_2=0 \qquad H=1, \qquad \Lambda=1, \qquad \mathbf{G}'=\mathbf{T}' \tag{6.1.40}$$

and it therefore follows that, if the present discussion is to include those materials which satisfy von Mises yield criterion, the term in δ_1 and δ_2 must be retained in equation (6.1.39). This condition implies that for simple extension,

$$T_{xx}=T_{yy}. \tag{6.1.41}$$

The principal stresses σ_i are the positive solutions for σ of the equation

$$\sigma^3-I_{\mathrm{T}}\sigma^2+II_{\mathrm{T}}\sigma-III_{\mathrm{T}}=0 \tag{6.1.42}$$

where I_{T}, II_{T} and III_{T} are the principal invariants of $\mathbf{T}$. For simple extension $T_{xy}=0$, $T_{yz}=0$, $T_{zx}=0$ and hence the characteristic equation (6.1.42) gives $\sigma_1=T_{xx}$, $\sigma_2=T_{yy}$, $\sigma_3=T_{zz}$, which together with equation (6.1.41) give $3\sigma_1'=T_{yy}-T_{zz}$ and $\sigma_3'-\sigma_2'=T_{zz}-T_{yy}$. Entering these values of σ_1' and $\sigma_3'-\sigma_2'$ into equation (4.1.26) gives $\mu=-1$, which can be entered into equation (4.2.34) to give $\bar{\mu}=\mu=-1$. These conditions on μ and $\bar{\mu}$ can be combined with equation (6.1.34) to give

$$\mu=\bar{\mu}=\bar{v}=\kappa=-1 \tag{6.1.43}$$

for simple extension, irrespective of which type of yield function the material satisfies. The condition of equation (6.1.43) is also independent of whether the material is compressible or incompressible.

From equations (4.2.27), (6.1.35) and (6.1.38) the following stress relation can be obtained

$$\begin{aligned} T_{zz}&=T_{yy}+\frac{1}{\Lambda}(G_{zz}'-G_{yy}') \\ &=T_{yy}+\frac{3}{2}\frac{1}{\Lambda}G_{zz}' \\ &=T_{yy}+(-\beta_{-1})\frac{(1-\alpha^2)}{\Lambda}\left(q\lambda_3^2+\frac{1}{\alpha^2\lambda_3^2}\right) \end{aligned} \tag{6.1.44}$$

where use has been made of equation (6.1.36), and where for simple extension,

$$\Lambda=\delta_1-\delta_2T_{xx}'=\delta_1-\delta_2\sigma_1'$$

is just equation (4.2.35).

The condition that the lateral faces $x=$constant, $y=$constant of the body are stress free is satisfied by setting $T_{xx}=T_{yy}=0$ which condition reduces equation (6.1.44) to

$$T_{zz}=(-\beta_{-1})\frac{(1-\alpha^2)}{\Lambda}\left(q\lambda_3^2+\frac{1}{\alpha^2\lambda_3^2}\right). \tag{6.1.45}$$

Equation (6.1.45) is a general stress relation for simple extension in so far as it does not distinguish between materials which are compressible and those which are incompressible. Thus, there are two cases to consider:

(i) ***Compressible material.*** There are no well defined ways in which the dependence of the response coefficients β_i $(i=0\ \pm 1)$ upon λ_3 can be evaluated for a compressible material. Similarly, the dependence of the parameter α upon λ_3 cannot be evaluated for a compressible material.

Uniform dilatation, defined by

$$\lambda_1=\lambda_2=\lambda_3=\lambda \tag{6.1.46}$$

is the special case of simple extension for a compressible material for which $\alpha=1$. For uniform dilatation, it follows from equations (6.1.41) and (6.1.44) that for $\alpha=1$, we must have

$$T_{xx}=T_{yy}=T_{zz} \tag{6.1.47}$$

and using the constitutive equation in the form of equation (4.2.25) it is evident that

$$\mathbf{G}=-P\mathbf{I} \qquad P=\kappa(\lambda) \tag{6.1.48}$$

where

$$\kappa(\lambda)=-\phi_0(3\lambda^2,3\lambda^{-2},\lambda^6)-\lambda^2\phi_1(3\lambda^2,3\lambda^{-2},\lambda^6)-\lambda^4\phi_2(3\lambda^2,3\lambda^{-2}\lambda^6) \tag{6.1.49}$$

use having been made of equations (4.2.15) and (6.1.33).

(ii) ***Incompressible material.*** For an incompressible material, $III_{\mathbf{B}}=1$, and equation (2.1.85) gives

$$\lambda_1\lambda_2\lambda_3=\det\mathbf{F}=1 \qquad (III_{\mathbf{B}}=1). \tag{6.1.50}$$

The condition of equation (6.1.50), when used with equation (6.1.30), gives $\alpha=\lambda_3^{-3/2}$, it being noted that

$$\lambda_1=\lambda_2=\lambda_3^{-1/2}. \tag{6.1.51}$$

The condition of equation (6.1.51) reduces the invariants of equation (6.1.33) to the form

$$I_{\mathbf{B}}=(2/\lambda_3)+\lambda_3^2 \qquad II_{\mathbf{B}}=2\lambda_3+1/\lambda_3^2 \qquad III_{\mathbf{B}}=1.$$

Entering the value $\alpha=\lambda_3^{-3/2}$ into equation (6.1.45) gives

$$T_{zz}=\left(\beta_1-\frac{1}{\lambda_3}\beta_{-1}\right)\frac{(\lambda_3^3-1)}{\lambda_3}=(-\beta_{-1})(q\lambda_3+1)\frac{(\lambda_3^3-1)}{\lambda_3^2} \tag{6.1.52}$$

where q is defined by equation (6.1.37).

It is a matter of observation that materials exist for which the response in simple tension and simple compression has equal but opposite effects. Hence, if the theory underlying the formulation of the constitutive equation of a simple elastic material is to have maximum generality, then it should be possible for the stress relation of equation (6.1.52) to satisfy this condition. For this condition to be satisfied, replacing λ_3 in equation (6.1.52) by $1/\lambda_3$ should simply change the sign of the response. Replacing λ_3 by $(1/\lambda_3)$ in equation (6.1.52) gives

$$T_{zz} = -\left(\frac{\beta_1}{\lambda_3} - \beta_{-1}\right)\frac{(\lambda_3^3 - 1)}{\lambda_3} = -(-\beta_{-1})(q + \lambda_3)\frac{(\lambda_3^3 - 1)}{\lambda_3^2}.$$

The required condition can, in principle, be satisfied by setting,

$$\begin{aligned} &\text{(i)}\quad \beta_1 = -\beta_{-1} && q = 1 \\ &\text{(ii)}\quad \beta_1 = \frac{3q\hat{A}(I_{\mathbf{B}}, II_{\mathbf{B}})}{2(qI_{\mathbf{B}} + II_{\mathbf{B}})} = -q\beta_{-1} && q = \frac{dII_{\mathbf{B}}}{dI_{\mathbf{B}}} = \frac{1}{\lambda_3}. \end{aligned} \qquad (6.1.53)$$

In practice, both these conditions are too restrictive.

The constitutive equation for an incompressible body has the reduced form (cf equation (4.2.25))

$$\mathbf{G}(\mathbf{T}) = -P\mathbf{I} + \beta_1\mathbf{B} + \beta_{-1}\mathbf{B}^{-1}.$$

Entering this form for $\mathbf{G}$ into equation (6.1.8) gives

$$\operatorname{div}(\beta_1\mathbf{B} + \beta_{-1}\mathbf{B}^{-1}) = \operatorname{grad} P$$

it being noted that P is related to the deformation only through this condition. Thus, given $\mathbf{B}$ it is evident that P can always be adjusted to satisfy some particular condition specified by the nature of the mode of deformation. For homogeneous strain fields, it is evident that P is a uniform constant. In the case of simple extension, the condition $T_{xx} = T_{yy} = 0$ gives

$$P = \beta_1\lambda_3^{-1} + \beta_{-1}\lambda_3$$

where use has been made of the above form of the constitutive equation.

This discussion of the simple extension of an incompressible, isotropic solid, when compared with the simple extension of a compressible isotropic material, illustrates the general observation that it is a simple problem to determine the response of an incompressible isotropic solid.

6.1.4 Simple Extension: II, Strain Response Function

The present discussion will be restricted to incompressible materials. For simple extension, the only non-zero component of stress has been shown in

§6.1.3 to be T_{zz}, which as a condition leads to the conclusion that

$$\mu = \bar{\mu} \equiv \bar{\nu} = \nu = -1$$

for all λ_3, use having been made of equation (4.2.47).

For an incompressible material,

$$[B_{ij}] = \begin{bmatrix} (\lambda_2\lambda_3)^{-2} & 0 & 0 \\ 0 & \lambda_2^2 & 0 \\ 0 & 0 & \lambda_3^2 \end{bmatrix} \qquad [(B^{-1})_{ij}] = \begin{bmatrix} \lambda_2^2\lambda_3^2 & 0 & 0 \\ 0 & \lambda_2^{-2} & 0 \\ 0 & 0 & \lambda_3^{-2} \end{bmatrix} \tag{6.1.54}$$

where use has been made of equation (6.1.50). The principal deviatoric strains

$$\varepsilon_i' = \tfrac{1}{4}[b_i' - (b^{-1})_i'] \qquad (i = 1, 2, 3)$$

can be inserted into equation (4.2.48) to give

$$\nu = -\frac{(2 + \lambda_2^2 + \lambda_3^2 + 2\lambda_2^2\lambda_3^2)(\lambda_2^2\lambda_3^2 - 1)}{(\lambda_2^2\lambda_3^2 + 1)(\lambda_3^2 - \lambda_2^2)}. \tag{6.1.55}$$

Here use has been made of equation (6.1.50), together with the condition $b_i = \lambda_i^2$. Setting $\nu = -1$ in equation (6.1.55) and solving for λ_3 gives

$$\lambda_1 = \lambda_2 = \lambda_3^{-1/2}$$

which is just equation (6.1.51). Using this condition and substituting into equation (4.2.20) from equation (6.1.54) gives

$$[\bar{E}_{ij}] = \begin{bmatrix} (1 - \lambda_3^2)/4\lambda_3 & 0 & 0 \\ 0 & (1 - \lambda_3^2)/4\lambda_3 & 0 \\ 0 & 0 & (\lambda_3^4 - 1)/4\lambda_3^2 \end{bmatrix}. \tag{6.1.56}$$

The principal deviatoric strains are

$$\bar{\varepsilon}_1' = \bar{\varepsilon}_2' = -\tfrac{1}{2}\bar{\varepsilon}_3' \qquad \bar{\varepsilon}_3' = (\lambda_3 + 1)(\lambda_3^3 - 1)/6\lambda_3^2. \tag{6.1.57}$$

Substituting in the constitutive equation (4.2.44) from equation (6.1.56) gives

$$G_{zz}' - G_{yy}' = \Lambda(T_{zz} - T_{yy}) = 3\bar{G}\Psi\bar{\varepsilon}_{zz} \tag{6.1.58}$$

where

$$\bar{\varepsilon}_{zz} = \frac{(\lambda_3 + 1)(\lambda_3^3 - 1)}{6\lambda_3^2} \qquad (\equiv \bar{\varepsilon}_3') \tag{6.1.59}$$

and where, from equations (4.2.35) and (4.3.13)

$$\Lambda = \delta_1 - \delta_2 T_{xx}' \qquad \Psi = \varphi_1 - \varphi_2\bar{\varepsilon}_1' \tag{6.1.60}$$

it being noted that equations (4.2.50) and (4.2.51) give

$$\varphi_1 = \frac{2}{3}\frac{(2\lambda_3^4+2\lambda_3^3+3\lambda_3^2+4\lambda_3+1)(\beta_1+\beta_{-1})}{(\lambda_3+1)^2(\lambda_3^2+1)\bar{G}} - 2\frac{\beta_{-1}}{\bar{G}} \tag{6.1.61}$$

$$\varphi_2 = \frac{8\lambda_3^2(\beta_1+\beta_{-1})}{(\lambda_3+1)^2(\lambda_3^2+1)\bar{G}}. \tag{6.1.62}$$

For simple extension, $T_{yy}=0$, and equation (6.1.58) reduces to,

$$T_{zz} = 3[\bar{G}\Psi/\Lambda]_{\nu=-1}\varepsilon_{zz}. \tag{6.1.63}$$

For the material response in compression and tension to have equal but opposite effects it is necessary for

$$\varepsilon_{zz}(\lambda_3) = -\varepsilon_{zz}(1/\lambda_3)$$

a condition which equation (6.1.59) satisfies, and hence ε_{zz} is an appropriate measure of simple tensile strain ε_T and simple compressive strain $\varepsilon_c = -\varepsilon_T$.

In the limit as $\lambda_3 \to 1$, equation (6.1.63) gives,

$$\lim_{\lambda_3\to 1}\frac{T_{zz}}{3\varepsilon_{zz}} = \lim_{\lambda_3\to 1}\frac{T_{zz}}{3\tilde{e}_{zz}} = \left[\frac{\bar{G}}{\Lambda}\right]_{\lambda_3=1} = \frac{G_0}{[\Lambda]_{\mu=-1}}$$

where $\tilde{e}_{zz} = \lambda_3 - 1$ and where G_0 is the ground state shear modulus. That

$$[\bar{G}]_{\lambda_3=1} = G_0$$

can be seen as follows. In the ground state, the material satisfies von Mises yield criterion (cf equation (6.1.40)), for which $\Lambda = 1$, and in the limit of infinitely small strains ($\lambda_3 \to 1$), the ratio $T_{zz}/(3\tilde{e}_{zz})$ must equal the classical shear modulus G_0.

Entering the condition $\mu = \bar{\mu} \equiv \bar{\nu} = \nu = -1$ into equations (3.1.12) and (4.3.9) gives $\omega = 1$, $\Omega = 1$, which as limiting conditions, when entered into equation (4.3.11) give

$$\varphi_1 = \left[N - 3\frac{dN}{d\Omega}\right]_{\Omega=-1} \qquad \varphi_2 = -\frac{3}{\bar{\varepsilon}_1'}\left[\frac{dN}{d\Omega}\right]_{\Omega=-1}$$

For the ground state, $N = 1$, and hence,

$$\varphi_1 = 1 \qquad \varphi_2 = 0$$

and from equation (6.1.62) for

$$\varphi_2 = 0 \qquad \beta_1 + \beta_{-1} = 0 \qquad (\lambda_3 \geqslant 1)$$

and thus the condition of equation (6.1.53(i)) restricts the stress relation of equation (6.1.52) to materials which are in the ground state, that is materials which satisfy only the von Mises type stress and strain intensity functions.

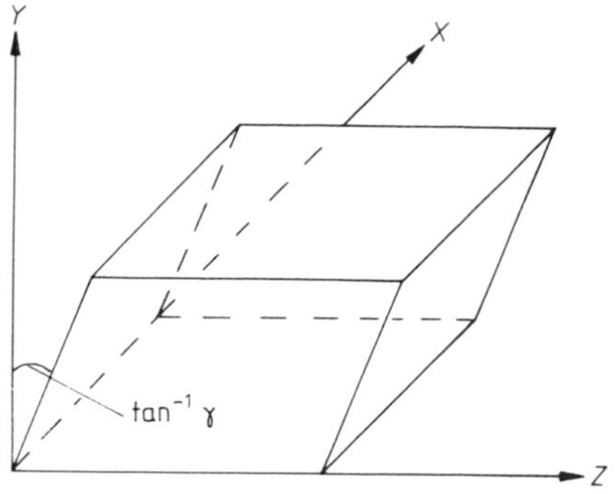

Figure 6.1 Simple shear.

6.1.5 Simple Shear: I, Deformation Response Function

Let the cartesian coordinates (X, Y, Z) be the referential coordinates of a body in the natural state and the cartesian coordinates (x, y, z) be the spatial coordinates in the deformed configuration.

Shown in figure 6.1 is the simple shear of a rectangular block aligned with the cartesian coordinate axes (X, Y, Z). A simple shear of amount γ is given by

$$x = X \qquad y = Y \qquad z = Z + \gamma Y. \tag{6.1.64}$$

From equations (2.1.36), (2.1.45) and (6.1.64), the matrices of $\mathbf{F}$, $\mathbf{B}$ and $\mathbf{B}^{-1}$ are:

$$[F_{i\alpha}] = \begin{bmatrix} 1 & 0 & 0 \\ 0 & 1 & 0 \\ 0 & \gamma & 1 \end{bmatrix} \tag{6.1.65}$$

$$[B_{ij}] = \begin{bmatrix} 1 & 0 & 0 \\ 0 & 1 & \gamma \\ 0 & \gamma & 1+\gamma^2 \end{bmatrix} \qquad [(B^{-1})_{ij}] = \begin{bmatrix} 1 & 0 & 0 \\ 0 & 1+\gamma^2 & -\gamma \\ 0 & -\gamma & 1 \end{bmatrix}. \tag{6.1.66}$$

The principal invariants are

$$I_{\mathbf{B}} = II_{\mathbf{B}} = 3 + \gamma^2 \qquad III_{\mathbf{B}} = 1 \tag{6.1.67}$$

from which it follows that, since $III_{\mathbf{B}} = 1$, simple shear is an isochoric deformation (see §2.1.6).

The left Cauchy–Green deformation tensor has the spectral representation

$$\mathbf{B} = \sum_{r=1}^{3} \lambda_r^2 \boldsymbol{q}_r \otimes \boldsymbol{q}_r \tag{6.1.68}$$

where the orthonormal triplets $\boldsymbol{q}_i$ specify the current stretch axes and where the λ_i are the principal stretches. In terms of the principal stretches, $III_{\mathbf{B}} = \lambda_1^2\lambda_2^2\lambda_3^2$ and hence from equations (6.1.67) and (2.1.85)

$$\lambda_1\lambda_2\lambda_3 = \det \mathbf{F} = 1. \tag{6.1.69}$$

The principal stretches are the positive solutions for λ of the equation

$$\lambda^6 - I_{\mathbf{B}}\lambda^4 + II_{\mathbf{B}}\lambda^2 - 1 = 0 \tag{6.1.70}$$

which can be rearranged, using the condition of equation (6.1.67) into the form

$$(\lambda^2 - 1)[\lambda^4 - (2 + \gamma^2)\lambda^2 + 1] = 0. \tag{6.1.71}$$

It follows from equations (6.1.69) and (6.1.71) that one of the principal stretches is unity, and the other two principal stretches are reciprocal to each other. Thus, it follows that equation (6.1.71) gives

$$\begin{aligned} &\lambda_1^2 = b_1 = 1 \qquad (\lambda_3 - \lambda_2 = \gamma) \\ &\lambda_2^2 = 1 + \tfrac{1}{2}\gamma^2 - \gamma(1 + \tfrac{1}{4}\gamma^2)^{1/2} = b_2 \\ &\lambda_3^2 = 1 + \tfrac{1}{2}\gamma^2 + \gamma(1 + \tfrac{1}{4}\gamma^2)^{1/2} = b_3 = \frac{1}{\lambda_2^3} \end{aligned} \tag{6.1.72}$$

from which it follows, since one of the principal stretches is unity, simple shear is an example of a deformation which is two-dimensional.

It is evident from equations (2.1.36) and (6.1.65) that for simple shear $\mathbf{F}$ can be expressed in the form

$$\mathbf{F} = \boldsymbol{e}_1 \otimes \boldsymbol{E}_1 + \boldsymbol{e}_2 \otimes \boldsymbol{E}_2 + \boldsymbol{e}_3 \otimes \boldsymbol{E}_3 + \gamma \boldsymbol{e}_3 \otimes \boldsymbol{E}_2. \tag{6.1.73}$$

Let the unit vectors $\boldsymbol{L}$, $\boldsymbol{M}$, $\boldsymbol{N}$ form an orthonormal set associated with the referential description and let the unit vectors $\boldsymbol{l}$, $\boldsymbol{m}$, $\boldsymbol{n}$ form an orthonormal set associated with the spatial description. In the context of these two sets of orthonormal vectors, equation (6.1.73) is taken to imply

$$\mathbf{F} = \boldsymbol{l} \otimes \boldsymbol{L} + \boldsymbol{m} \otimes \boldsymbol{M} + \boldsymbol{n} \otimes \boldsymbol{N} + \gamma \boldsymbol{n} \otimes \boldsymbol{M}. \tag{6.1.74}$$

From equations (2.1.45) and (6.1.74)

$$\mathbf{B} = \mathbf{F}\mathbf{F}^{\mathrm{T}} = \boldsymbol{l} \otimes \boldsymbol{l} + \boldsymbol{m} \otimes \boldsymbol{m} + (1 + \gamma^2)\boldsymbol{n} \otimes \boldsymbol{n} + \gamma(\boldsymbol{n} \otimes \boldsymbol{m} + \boldsymbol{m} \otimes \boldsymbol{n}) \tag{6.1.75}$$

$$\mathbf{B}^{-1} = \boldsymbol{l} \otimes \boldsymbol{l} + (1 + \gamma^2)\boldsymbol{m} \otimes \boldsymbol{m} + \boldsymbol{n} \otimes \boldsymbol{n} - \gamma(\boldsymbol{n} \otimes \boldsymbol{m} + \boldsymbol{m} \otimes \boldsymbol{n}) \tag{6.1.76}$$

and hence,

$$\begin{aligned} &\mathbf{B}\boldsymbol{l} = \boldsymbol{l} \\ &\mathbf{B}\boldsymbol{m} = \boldsymbol{m} + \gamma\boldsymbol{n} \\ &\mathbf{B}\boldsymbol{n} = \boldsymbol{n} + \gamma(\boldsymbol{m} + \gamma\boldsymbol{n}) = \boldsymbol{n} + \gamma\mathbf{B}\boldsymbol{m} \end{aligned} \tag{6.1.77}$$

The vector $\boldsymbol{n}$ is said to define the *direction of shear* and the planes orthogonal to $\boldsymbol{m}$ and $\boldsymbol{l}$ are called the *glide planes* and the *planes of shear* respectively. The deformation is two-dimensional with respect to the planes of shear.

Use of the conditions $\lambda_1 = 1, \lambda_2 = \lambda_3^{-1}$ gives the spectral representation of $\mathbf{B}$ in the form

$$\mathbf{B} = \boldsymbol{l} \otimes \boldsymbol{l} + \frac{1}{\lambda_3^2} \boldsymbol{q}_2 \otimes \boldsymbol{q}_2 + \lambda_3^2 \boldsymbol{q}_3 \otimes \boldsymbol{q}_3. \tag{6.1.78}$$

Since the orthogonal pair of vectors $\boldsymbol{q}_2$ and $\boldsymbol{q}_3$ are orthogonal to $\boldsymbol{l}$ they can be expressed in the form

$$\begin{aligned} \boldsymbol{q}_1 &= \boldsymbol{l} \\ \boldsymbol{q}_2 &= \cos\theta\, \boldsymbol{m} - \sin\theta\, \boldsymbol{n} \\ \boldsymbol{q}_3 &= \sin\theta\, \boldsymbol{m} + \cos\theta\, \boldsymbol{n}. \end{aligned} \tag{6.1.79}$$

Substituting for the $\boldsymbol{q}_i$ in equation (6.1.78) and comparing the resulting form for $\mathbf{B}$ with that of equation (6.1.75) gives

$$\begin{aligned} 1+\gamma^2 &= \lambda_3^2 \cos^2\theta + \lambda_3^{-2} \sin^2\theta \\ 1 &= \lambda_3^2 \sin^2\theta + \lambda_3^{-2} \cos^2\theta \\ \gamma &= (\lambda_3^2 - \lambda_3^{-2}) \cos\theta \sin\theta. \end{aligned} \tag{6.1.80}$$

Equation (6.1.80) gives

$$\begin{aligned} \cot\theta &= \tfrac{1}{2}\gamma + (1 + \tfrac{1}{4}\gamma^2)^{1/2} \\ &= (\lambda_3^2 - 1)/\gamma \end{aligned} \tag{6.1.81}$$

where use has also been made of equation (6.1.72). From equations (6.1.72) and (6.1.81)

$$\lambda_3^2 = 1 + \gamma \cot\theta = \lambda_2^{-2}. \tag{6.1.82}$$

For $\gamma = 0$, equation (6.1.79) reduces to

$$\boldsymbol{q}_1 = \boldsymbol{l} \qquad \boldsymbol{q}_2 = \frac{1}{\sqrt{2}} \boldsymbol{m} - \frac{1}{\sqrt{2}} \boldsymbol{n} \qquad \boldsymbol{q}_3 = \frac{1}{\sqrt{2}} \boldsymbol{m} + \frac{1}{\sqrt{2}} \boldsymbol{n} \tag{6.1.83}$$

while for $\gamma \to \infty$

$$\boldsymbol{q}_1 = \boldsymbol{l} \qquad \boldsymbol{q}_2 \to -\boldsymbol{n} \qquad \boldsymbol{q}_3 \to \boldsymbol{m}. \tag{6.1.84}$$

Thus, the principal axes $\boldsymbol{q}_2$ and $\boldsymbol{q}_3$ rotate about the axis associated with $\boldsymbol{l}$, that is the x-axis through an angle $\pi/4$ as γ ranges from 0 to ∞.

From equations (2.1.44) and (6.1.74)

$$\begin{aligned} \mathbf{C} = \mathbf{F}^{\mathrm{T}}\mathbf{F} = \boldsymbol{L} \otimes \boldsymbol{L} + (1+\gamma^2)\boldsymbol{M} \otimes \boldsymbol{M} + \boldsymbol{N} \otimes \boldsymbol{N} \\ + \gamma(\boldsymbol{M} \otimes \boldsymbol{N} + \boldsymbol{N} \otimes \boldsymbol{M}) \end{aligned} \tag{6.1.85}$$

and hence

$$\begin{aligned} \mathbf{C}\boldsymbol{L} &= \boldsymbol{L} \\ \mathbf{C}\boldsymbol{M} &= \boldsymbol{M}+\gamma(\boldsymbol{N}+\gamma\boldsymbol{M})=\boldsymbol{M}+\gamma\mathbf{C}\boldsymbol{N} \\ \mathbf{C}\boldsymbol{N} &= \boldsymbol{N}+\gamma\boldsymbol{M}. \end{aligned} \tag{6.1.86}$$

The right Cauchy–Green deformation tensor has the spectral representation

$$\mathbf{C}=\sum_{r=1}^{3}\lambda_r^2\boldsymbol{p}_r\otimes\boldsymbol{p}_r \tag{6.1.87}$$

where the orthonormal triplet $\boldsymbol{p}_\alpha$ specifies the referential stretch axes. Entering the conditions $\lambda_1=1$, $\lambda_2=\lambda_3^{-1}$ into equation (6.1.87) gives

$$\mathbf{C}=\boldsymbol{L}\otimes\boldsymbol{L}+\lambda_3^{-2}\boldsymbol{p}_2\otimes\boldsymbol{p}_2+\lambda_3^2\boldsymbol{p}_3\otimes\boldsymbol{p}_3. \tag{6.1.88}$$

Since the orthogonal pair of vectors $\boldsymbol{p}_2$ and $\boldsymbol{p}_3$ are orthogonal to $\boldsymbol{L}$ they can be expressed in the form

$$\begin{aligned} \boldsymbol{p}_1 &= \boldsymbol{L} \\ \boldsymbol{p}_2 &= \cos\Theta\,\boldsymbol{M}-\sin\Theta\,\boldsymbol{N} \\ \boldsymbol{p}_3 &= \sin\Theta\,\boldsymbol{M}+\cos\Theta\,\boldsymbol{N}. \end{aligned} \tag{6.1.89}$$

Substituting for the $\boldsymbol{p}_\alpha$ in equation (6.1.88) and comparing the resulting form for $\mathbf{C}$ with that of equation (6.1.85) gives a set of relations which correspond to those of equation (6.1.80) and which, when solved, give

$$\cot\Theta=-\left[\tfrac{1}{2}\gamma+(1+\tfrac{1}{4}\gamma^2)^{1/2}\right].$$

Hence, by comparison with equation (6.1.81), it follows that

$$\Theta=\tfrac{1}{2}\pi-\theta. \tag{6.1.90}$$

The rotation tensor $\mathbf{R}$ is obtained by way of equations (1.6.149) and (1.7.51) in the form

$$\mathbf{R}=\boldsymbol{q}_r\otimes\boldsymbol{p}_r=\boldsymbol{l}\otimes\boldsymbol{L}+\sin 2\theta(\boldsymbol{m}\otimes\boldsymbol{M}+\boldsymbol{n}\otimes\boldsymbol{N})-\cos 2\theta(\boldsymbol{m}\otimes\boldsymbol{N}-\boldsymbol{n}\otimes\boldsymbol{M}) \tag{6.1.91}$$

which when compared with equation (1.7.51), shows that $\mathbf{R}$ represents a rotation of amount $(2\theta-\frac{1}{2}\pi)$ about the axis associated with $\boldsymbol{l}$, that is the x-axis.

The constitutive equation (6.1.1) will be used in the form of equation (4.2.40)

$$\mathbf{G}'(\mathbf{T}')=\mathbf{M}'(\mathbf{B}'). \tag{6.1.92}$$

Equation (6.1.92) gives

$$\begin{aligned} \mathbf{G}'=&-\tfrac{1}{3}(\beta_1+\beta_{-1})\gamma^2\boldsymbol{l}\otimes\boldsymbol{l}-\tfrac{1}{3}(\beta_1-2\beta_{-1})\gamma^2\boldsymbol{m}\otimes\boldsymbol{m} \\ &+\tfrac{1}{3}(2\beta_1-\beta_{-1})\gamma^2\boldsymbol{n}\otimes\boldsymbol{n}+(\beta_1-\beta_{-1})\gamma(\boldsymbol{n}\otimes\boldsymbol{m}+\boldsymbol{m}\otimes\boldsymbol{n}) \end{aligned} \tag{6.1.93}$$

where use has been made of equations (4.2.40), (6.1.75) and (6.1.76). From equation (6.1.93),

$$\boldsymbol{l}\cdot(\mathbf{G}'\boldsymbol{l}) = -\tfrac{1}{3}(\beta_1+\beta_{-1})\gamma^2 = \tfrac{1}{3}\beta_{-1}(q-1)\gamma^2 \tag{6.1.94}$$

$$\boldsymbol{m}\cdot(\mathbf{G}'\boldsymbol{m}) = -\tfrac{1}{3}(\beta_1-2\beta_{-1})\gamma^2 = \tfrac{1}{3}\beta_{-1}(q+2)\gamma^2 \tag{6.1.95}$$

$$\boldsymbol{n}\cdot(\mathbf{G}'\boldsymbol{n}) = \tfrac{1}{3}(2\beta_1-\beta_{-1})\gamma^2 = -\tfrac{1}{3}\beta_{-1}(2q+1)\gamma^2 \tag{6.1.96}$$

$$\boldsymbol{n}\cdot(\mathbf{G}'\boldsymbol{m}) = (\beta_1-\beta_{-1})\gamma = \boldsymbol{m}\cdot(\mathbf{G}'\boldsymbol{n}) = -\beta_{-1}(q+1)\gamma \tag{6.1.97}$$

$$\boldsymbol{n}\cdot(\mathbf{G}'\boldsymbol{l}) = \boldsymbol{l}\cdot(\mathbf{G}'\boldsymbol{n}) = 0 \qquad \boldsymbol{l}\cdot(\mathbf{G}'\boldsymbol{m}) = \boldsymbol{m}\cdot(\mathbf{G}'\boldsymbol{l}) = 0 \tag{6.1.98}$$

where q is defined by equation (6.1.37).

The components of $\mathbf{T}$ follow from equations (1.6.61) and (2.2.45). In terms of the orthogonal set of vectors $\{\boldsymbol{l}, \boldsymbol{m}, \boldsymbol{n}\}$, the normal components of stress can be expressed in the form

$$\sigma_{(n)} \equiv \sigma_{(s)} = \boldsymbol{s}\cdot(\mathbf{T}\boldsymbol{s}) \qquad (s = l, m, n) \tag{6.1.99}$$

and for a symmetric state of stress the off-diagonal components can be expressed in the form

$$\begin{aligned} \tau_{(l)} &= \boldsymbol{m}\cdot(\mathbf{T}\boldsymbol{n}) = \boldsymbol{n}\cdot(\mathbf{T}\boldsymbol{m}) \\ \tau_{(m)} &= \boldsymbol{l}\cdot(\mathbf{T}\boldsymbol{n}) = \boldsymbol{n}\cdot(\mathbf{T}\boldsymbol{l}) \\ \tau_{(n)} &= \boldsymbol{l}\cdot(\mathbf{T}\boldsymbol{m}) = \boldsymbol{m}\cdot(\mathbf{T}\boldsymbol{l}). \end{aligned} \tag{6.1.100}$$

It is to be noted that

$$\boldsymbol{l}\cdot(\mathbf{T}'\boldsymbol{l}) + \boldsymbol{m}\cdot(\mathbf{T}'\boldsymbol{m}) + \boldsymbol{n}\cdot(\mathbf{T}'\boldsymbol{n}) = 0$$

and, for example, that

$$\begin{aligned} \boldsymbol{n}\cdot(\mathbf{T}'^2\boldsymbol{m}) &\equiv [\boldsymbol{n}\cdot(\mathbf{T}'\boldsymbol{s})][\boldsymbol{s}\cdot(\mathbf{T}'\boldsymbol{m})] \qquad (s = l, m, n) \\ &= [\boldsymbol{n}\cdot(\mathbf{T}'\boldsymbol{l})][\boldsymbol{l}\cdot(\mathbf{T}'\boldsymbol{m})] - [\boldsymbol{l}\cdot(\mathbf{T}'\boldsymbol{l})][\boldsymbol{n}\cdot(\mathbf{T}'\boldsymbol{m})] \\ &= \tau_{(m)}\tau_{(n)} - \tau_{(l)}[\boldsymbol{l}\cdot(\mathbf{T}'\boldsymbol{l})]. \end{aligned} \tag{6.1.101}$$

With this as the basis, and provided that δ_1 and δ_2 are not both zero, it follows from equations (6.1.98) and (6.1.100), that

$$\tau_{(m)} = \boldsymbol{l}\cdot(\mathbf{T}\boldsymbol{n}) = 0 \qquad \tau_{(n)} = \boldsymbol{l}\cdot(\mathbf{T}\boldsymbol{m}) = 0. \tag{6.1.102}$$

The ratio of the difference between equations (6.1.96) and (6.1.95) to equation (6.1.97) defines the stress ratio

$$\xi = [\boldsymbol{n}\cdot(\mathbf{G}'\boldsymbol{n}) - \boldsymbol{m}\cdot(\mathbf{G}'\boldsymbol{m})]/\boldsymbol{n}\cdot(\mathbf{G}'\boldsymbol{m}) = \gamma$$

or in terms of the components of stress defined by equations (6.1.99) and (6.1.100),

$$\xi = \frac{\sigma_{(n)} - \sigma_{(m)}}{\tau_{(l)}} = \gamma \tag{6.1.103}$$

which, because it is independent of the material response coefficients β_0, β_1 and β_{-1}, is referred to as a *universal relation.* Equation (6.1.103) shows that the shear stress $\tau_{(l)}$ arises directly from the normal stress difference $\sigma_{(n)} - \sigma_{(m)}$, and since it is a universal relation it follows that the shear stress is produced in exactly the same way in every isotropic elastic–plastic solid.

It is evident from equation (6.1.103) that the normal stresses $\sigma_{(m)}$ and $\sigma_{(n)}$ are unequal, a property of the material which is referred to as the *Poynting effect* in the initial elastic range of deformation and as the *Swift effect* in the post-yield plastic range of deformation.

Substituting for the $b_i = \lambda_i^2$ in equation (4.1.27) from equation (6.1.72) gives

$$\kappa = -(\lambda_3^2 - 1)/(\lambda_3^2 + 1) = -\gamma/(4+\gamma^2)^{1/2} \tag{6.1.104}$$

which can be entered into equation (4.1.29) to give

$$\bar{\nu} = \kappa\left(1 - \frac{2}{(1+q)}\right) = \kappa\frac{(q-1)}{(q+1)} = \bar{\mu}. \tag{6.1.105}$$

q is defined by equation (6.1.37), it being noted that κ lies in the range $(-1 \leqslant \kappa \leqslant 0)$.

The deviators σ_i' of the principal stresses σ_i are the positive solutions for σ' of the equation

$$\sigma'^3 - J_2'\sigma' - J_3' = 0 \tag{6.1.106}$$

where J_2' and J_3' are defined by equation (3.1.4). Equation (6.1.106) gives

$$\sigma_i' = \frac{2}{\sqrt{3}}(J_2')^{1/2}\cos[\tfrac{1}{3}\cos^{-1}\sqrt{\omega} + 2(i-1)\pi/3] \qquad (i = 1, 2, 3) \tag{6.1.107}$$

where ω is defined by equation (3.1.12). From equation (6.1.107)

$$\begin{aligned} \sigma_1' &= \zeta\tau_{(l)} \\ \sigma_2' &= -\tfrac{1}{2}\zeta\tau_{(l)} - \tfrac{1}{2}(4+\xi^2)^{1/2}\tau_{(l)} \\ \sigma_3' &= -\tfrac{1}{2}\zeta\tau_{(l)} + \tfrac{1}{2}(4+\xi^2)^{1/2}\tau_{(l)} \end{aligned} \tag{6.1.108}$$

where the stress ratio

$$\zeta = \frac{2\sigma_{(l)} - \sigma_{(m)} - \sigma_{(n)}}{3\tau_{(l)}} = \frac{\mu}{\bar{\mu}}\frac{\boldsymbol{l}\cdot(\mathbf{G}'\boldsymbol{l})}{\boldsymbol{n}\cdot(\mathbf{G}'\boldsymbol{m})}. \tag{6.1.109}$$

Also, using equations (4.1.26), (6.1.108) and (6.1.103)

$$\mu = \frac{3\sigma_1'}{\sigma_3' - \sigma_2'} = \frac{3\zeta}{(4+\xi^2)^{1/2}} = \frac{3\zeta}{(4+\gamma^2)^{1/2}} \tag{6.1.110}$$

having noted that $\bar{\mu}$ is defined by equation (4.2.32).

The components of $\mathbf{G}'$ can be expressed in terms of ξ, ζ, μ and $\bar{\mu}$: thus

$$[G'_{ij}]=\Lambda\tau_{(l)}\begin{bmatrix}(\bar{\mu}/\mu)\zeta & 0 & 0\\ 0 & -\tfrac{1}{2}[\xi+(\bar{\mu}/\mu)\zeta] & 1\\ 0 & 1 & \tfrac{1}{2}[\xi-(\bar{\mu}/\mu)\zeta]\end{bmatrix} \tag{6.1.111}$$

where Λ is defined by equation (4.2.35). Similarly, expressing the normal components of $\mathbf{B}$ in the form

$$B_{(n)}\equiv B_{(s)}=\boldsymbol{s}\cdot(\mathbf{B}\boldsymbol{s}) \qquad (s=l,m,n)$$

and the off-diagonal components in the form

$$\mathscr{B}_{(l)}=\boldsymbol{m}\cdot(\mathbf{B}\boldsymbol{n}) \qquad \mathscr{B}_{(m)}=\boldsymbol{n}\cdot(\mathbf{B}\boldsymbol{l}) \qquad \mathscr{B}_{(n)}=\boldsymbol{l}\cdot(\mathbf{B}\boldsymbol{m})$$

allows the components of the response function to be expressed in the form

$$[M'_{ij}]=\Phi\mathscr{B}_{(l)}\begin{bmatrix}(\bar{\nu}/\kappa)\rho & 0 & 0\\ 0 & -\tfrac{1}{2}[\chi+(\bar{\nu}/\kappa)\rho] & 1\\ 0 & 1 & \tfrac{1}{2}[\chi-(\bar{\nu}/\kappa)\rho]\end{bmatrix} \tag{6.1.112}$$

where

$$\Phi=\rho_1-\rho_2B'_{(l)}=\beta_1-B_{(l)}\beta_{-1} \tag{6.1.113}$$

$$\chi=\frac{B_{(n)}-B_{(m)}}{\mathscr{B}_{(l)}}=\frac{\boldsymbol{n}\cdot(\mathbf{M}'\boldsymbol{n})-\boldsymbol{m}\cdot(\mathbf{M}'\boldsymbol{m})}{\boldsymbol{n}\cdot(\mathbf{M}'\boldsymbol{m})} \tag{6.1.114}$$

$$\rho=\frac{2B_{(l)}-B_{(m)}-B_{(n)}}{3\mathscr{B}_{(l)}}=\frac{\kappa}{\bar{\nu}}\frac{\boldsymbol{l}\cdot(\mathbf{M}'\boldsymbol{l})}{\boldsymbol{n}\cdot(\mathbf{M}'\boldsymbol{m})} \tag{6.1.115}$$

and where κ and $\bar{\nu}$ are defined by equations (6.1.104) and (6.1.105).

From equations (6.1.92), (6.1.111) and (6.1.112) the following stress relation can be obtained

$$\tau_{(l)}=(\beta_1-\beta_{-1})\gamma/\Lambda \tag{6.1.116}$$

where Λ is defined by equation (4.2.35).

The corresponding relations for the simple shear of a simple elastic material are obtained by noting that the simple elastic material satisfies only the von Mises yield criterion for which $\Lambda=1$, $\mu=\bar{\mu}$, $\mathbf{G}'=\mathbf{T}'$. For the simple elastic material it is found that the stress ratio ξ remains that of equation (6.1.103), the principal differences being that

$$\zeta=-\frac{1}{3}\frac{(q-1)}{(q+1)}\gamma \qquad \tau_{(l)}=(\beta_1-\beta_{-1})\gamma. \tag{6.1.117}$$

In the case of the simple elastic material, the problem reduces to determining a form for the response coefficients β_i $(i=0,\pm 1)$, whereas for a non-simple elastic, or non-simple elastic–plastic material there is the additional problem of determining the loading coefficients δ_1 and δ_2 by way of the stress intensity function h discussed in §4.2.1.

The above discussion of simple shear does not distinguish between materials which are compressible and those which are incompressible. Thus, there are two cases to consider.

(i) ***Compressible material.*** There are no well-defined ways in which the dependence of the response coefficients β_i $(i=0, \pm 1)$ upon γ can be evaluated for a compressible material.

(ii) ***Incompressible material.*** Since γ lies in the range $(0 \leqslant \gamma \leqslant \infty)$ it follows from equation (6.1.104) that κ lies in the range $(-1 \leqslant \kappa \leqslant 0)$. In general, if κ varies with γ, then so also will μ vary with γ, it being noted that μ must lie in the correspondng range $(-1 \leqslant \mu \leqslant 0)$. The conditions,

$$\beta_1 = -\beta_{-1} \qquad q=1 \tag{6.1.118}$$

when substituted into equation (6.1.105) give the generalised Lode relation,

$$\bar{\mu} \equiv \bar{v} = 0 \qquad \text{for } (-1 \leqslant \mu \leqslant 0) \tag{6.1.119}$$

and for all γ. The conditions of equation (6.1.119) are just a specific form of equation (3.2.8) which, in the context of equations (4.2.1) and (4.2.2), apply only to the Tresca type stress and strain intensity functions.

6.1.6 Simple Shear: II, Strain Response Function

Substituting in equation (4.2.20) from equation (6.1.66) gives

$$[\bar{E}_{ij}] = \begin{bmatrix} 0 & 0 & 0 \\ 0 & -\frac{1}{4}\gamma^2 & \frac{1}{2}\gamma \\ 0 & \frac{1}{2}\gamma & \frac{1}{4}\gamma^2 \end{bmatrix} = [\bar{E}'_{ij}] \tag{6.1.120}$$

and hence

$$I_{\bar{\mathrm{E}}} = 0 \qquad II_{\bar{\mathrm{E}}} = -\tfrac{1}{4}\gamma^2(1+\tfrac{1}{4}\gamma^2) \qquad III_{\bar{\mathrm{E}}} = 0. \tag{6.1.121}$$

The principle deviatoric strains,

$$\bar{\varepsilon}'_i = \tfrac{1}{4}[b'_i - (b^{-1})'_i] \qquad (i=1,2,3) \tag{6.1.122}$$

are obtained by substituting for the b_i from equation (6.1.72) to give

$$\bar{\varepsilon}'_1 = 0 \qquad \bar{\varepsilon}'_2 = -\bar{\varepsilon}'_3 \qquad \bar{\varepsilon}'_3 = (\lambda_3^4 - 1)/4\lambda_3^2. \tag{6.1.123}$$

Entering these values for the $\bar{\varepsilon}'_i$ into equation (4.2.48) gives $v=0$ and hence, from equation (4.2.38), $\bar{\mu} \equiv \bar{v} = v = 0$ for all γ. In the context of equations (4.2.34), (4.2.38) and (4.2.47) the condition $\bar{\mu}=0$ for all γ is taken to imply $\mu=0$ and hence, for simple shear, the generalised Lode relation reduces to the Lode relation

$$\mu = \bar{\mu} \equiv \bar{v} = v = 0 \tag{6.1.124}$$

of equation (4.6.4) for all γ.

The constitutive equation will be used in the form of equation (4.2.44); thus

$$\mathbf{G}'(\mathbf{T}') = 2\bar{G}(\varphi_0 \mathbf{I} + \varphi_1 \bar{\mathbf{E}}' + \varphi_2 \bar{\mathbf{E}}'^2). \tag{6.1.125}$$

The conditions of equation (6.1.67) reduce equation (4.2.52) to

$$\varphi_0 = -\frac{\gamma^2(\beta_1 + \beta_{-1})}{6\bar{G}} \qquad \varphi_1 = \frac{(\beta_1 - \beta_{-1})}{\bar{G}} \qquad \varphi_2 = \frac{4(\beta_1 + \beta_{-1})}{(4+\gamma^2)\bar{G}}. \tag{6.1.126}$$

Substituting for **B** and $\mathbf{B}^{-1}$ in equation (4.2.20) from equations (6.1.75) and (6.1.76) gives,

$$\bar{\mathbf{E}} = \tfrac{1}{4}\gamma[\gamma(\boldsymbol{n} \otimes \boldsymbol{n} - \boldsymbol{m} \otimes \boldsymbol{m}) + 2(\boldsymbol{n} \otimes \boldsymbol{m} + \boldsymbol{m} \otimes \boldsymbol{n})] \tag{6.1.127}$$

and therefore

$$\bar{\mathbf{E}}^2 = \tfrac{1}{16}\gamma^2(4+\gamma^2)(\boldsymbol{m} \otimes \boldsymbol{m} + \boldsymbol{n} \otimes \boldsymbol{n}). \tag{6.1.128}$$

Noting that for simple shear $\bar{\mathbf{E}} = \bar{\mathbf{E}}'$, the form for $\bar{\mathbf{E}}'$ and $\bar{\mathbf{E}}'^2$ can be entered into equation (6.1.125) from equations (6.1.127) and (6.1.128) to give,

$$\begin{aligned}\mathbf{G}' = 2G\{&\varphi_0 \boldsymbol{l} \otimes \boldsymbol{l} + [\varphi_0 - \tfrac{1}{4}\gamma^2\varphi_1 + \tfrac{1}{16}\gamma^2(4+\gamma^2)\varphi_2]\boldsymbol{m} \otimes \boldsymbol{m} \\ &+ [\varphi_0 + \tfrac{1}{4}\gamma^2\varphi_1 + \tfrac{1}{16}\gamma^2(4+\gamma^2)\varphi_2]\boldsymbol{n} \otimes \boldsymbol{n} \\ &+ \tfrac{1}{2}\gamma\varphi_1(\boldsymbol{n} \otimes \boldsymbol{m} + \boldsymbol{m} \otimes \boldsymbol{n})\}.\end{aligned} \tag{6.1.129}$$

From equation (6.1.129)

$$\boldsymbol{l}\cdot(\mathbf{G}'\boldsymbol{l}) = 2\bar{G}\varphi_0 = -\tfrac{1}{3}(\beta_1 + \beta_{-1})\gamma^2 \tag{6.1.130}$$

$$\boldsymbol{m}\cdot(\mathbf{G}'\boldsymbol{m}) = 2\bar{G}[\varphi_0 - \tfrac{1}{4}\gamma^2\varphi_1 + \tfrac{1}{16}\gamma^2(4+\gamma^2)\varphi_2] = -\tfrac{1}{3}(\beta_1 - 2\beta_{-1})\gamma^2 \tag{6.1.131}$$

$$\boldsymbol{n}\cdot(\mathbf{G}'\boldsymbol{n}) = 2\bar{G}[\varphi_0 + \tfrac{1}{4}\gamma^2\varphi_1 + \tfrac{1}{16}\gamma^2(4+\gamma^2)\varphi_2] = \tfrac{1}{3}(2\beta_1 - \beta_{-1}) \tag{6.1.132}$$

$$\boldsymbol{n}\cdot(\mathbf{G}'\boldsymbol{m}) = \bar{G}\varphi_1\gamma = (\beta_1 - \beta_{-1})\gamma = \boldsymbol{m}\cdot(\mathbf{G}'\boldsymbol{n}) \tag{6.1.133}$$

$$\boldsymbol{n}\cdot(\mathbf{G}'\boldsymbol{l}) = \boldsymbol{l}\cdot(\mathbf{G}'\boldsymbol{n}) = 0 \qquad \boldsymbol{l}\cdot(\mathbf{G}'\boldsymbol{m}) = \boldsymbol{m}\cdot(\mathbf{G}'\boldsymbol{l}) = 0 \tag{6.1.134}$$

which are just equations (6.1.94) to (6.1.98).

From equations (6.1.100), (6.1.101), (6.1.102), (6.1.125) and (6.1.126),

$$\boldsymbol{n}\cdot(\mathbf{G}'\boldsymbol{m}) = [\delta_1 - \delta_2 \boldsymbol{l}\cdot(\mathbf{T}'\boldsymbol{l})]\tau_{(l)} = \bar{G}\varphi_1\gamma$$

which can be rearranged to give

$$\tau_{(l)} = [\bar{G}\varphi_1/\Lambda]_{v=0}\gamma \tag{6.1.135}$$

where

$$\Lambda = \delta_1 - \delta_2 \boldsymbol{l}\cdot(\mathbf{T}'\boldsymbol{l}).$$

For those materials which satisfy the von Mises type stress and strain intensity functions, $\varphi_1 = 1$, $\Lambda = 1$, and equation (6.1.135) reduces to,

$$\tau_{(l)} = [\bar{G}]_{v=0}\gamma.$$

6.2 HOMOGENEOUS ISOTROPIC BODIES: II, PURE SHEAR

6.2.1 Deformation Response Function

The present discussion is concerned with the mechanical response of a perfectly elastic material, regarded as incompressible with respect to volume. The material is assumed to be isotropic relative to its undeformed and unstressed, i.e. its natural, configuration. Let the cartesian coordinates (X^1, X^2, X^3) be the referential coordinates of a body in the natural state and the cartesian coordinates (x^1, x^2, x^3) be the spatial coordinates in the deformed configuration.

Assume a homogeneous deformation such that,

$$x^1 = \lambda_1 X^1 \qquad x^2 = \lambda_2 X^2 \qquad x^3 = \lambda_3 X^3 \tag{6.2.1}$$

the condition of incompressibility giving,

$$\lambda_1 \lambda_2 \lambda_3 = \det \mathbf{F} = 1. \tag{6.2.2}$$

Equation (6.2.2) implies that only two of the principal stretches are independently assignable; these will be taken to be λ_2 and λ_3. Entering this condition into equation (4.1.27) gives,

$$\kappa = -\frac{\lambda_2^2\lambda_3^2(\lambda_2^2 + \lambda_3^2) - 2}{\lambda_2^2\lambda_3^2(\lambda_3^2 - \lambda_2^2)} \tag{6.2.3}$$

use having been made of equation (6.2.2).

Consider a deformation for which,

$$\lambda_2 = 1 \qquad \lambda_3 = 1/\lambda_1 \geqslant 1. \tag{6.2.4}$$

From equations (2.1.36), (2.1.45), (6.2.1) and (6.2.4) the matrices of $\mathbf{F}$, $\mathbf{B}$ and $\mathbf{B}^{-1}$ are:

$$[F_{i\alpha}] = \begin{bmatrix} 1/\lambda_3 & 0 & 0 \\ 0 & 1 & 0 \\ 0 & 0 & \lambda_3 \end{bmatrix} \tag{6.2.5}$$

$$[B_{ij}] = \begin{bmatrix} 1/\lambda_3^2 & 0 & 0 \\ 0 & 1 & 0 \\ 0 & 0 & \lambda_3^2 \end{bmatrix} \qquad [(B^{-1})_{ij}] = \begin{bmatrix} \lambda_3^2 & 0 & 0 \\ 0 & 1 & 0 \\ 0 & 0 & 1/\lambda_3^2 \end{bmatrix}. \tag{6.2.6}$$

The principal invariants are

$$I_{\mathbf{B}} = II_{\mathbf{B}} = \lambda_3^2 + \lambda_3^{-2} + 1 \qquad III_{\mathbf{B}} = 1. \tag{6.2.7}$$

Entering the conditions of equation (6.2.4) into equation (6.2.3) gives

$$\kappa = -(1 + 2/\lambda_3^2) \qquad (-3 \leqslant \kappa \leqslant -1). \tag{6.2.8}$$

The relations of equations (6.2.4), (6.2.7) and (6.2.8) can be entered into equation (4.1.29) to give,

$$\bar{\nu} = -\left[1 + 2\frac{(\lambda_3^2 + q)}{(1 + q\lambda_3^2)}\right] \qquad [-3 \leqslant \bar{\nu} \leqslant -1 - (2/q_\infty)] \tag{6.2.9}$$

where

$$q = -\beta_1/\beta_{-1} \qquad [q]_{\lambda_3 = \infty} = q_\infty. \tag{6.2.10}$$

The only non-zero components of stress are taken to be,

$$T_{22} = \sigma_2 \qquad T_{33} = \sigma_3 \qquad (T_{11} = \sigma_1 = 0) \tag{6.2.11}$$

which for $T_{11} = 0$ follow from equation (6.1.42). Entering these conditions into equation (4.1.26) gives the Lode stress parameter,

$$\mu = -[1 + 2\sigma_2/(\sigma_3 - \sigma_2)]. \tag{6.2.12}$$

The form for μ given by equation (6.2.12) can be entered into equation (4.2.34) to give

$$\bar{\mu} = -\left[1 + 2\frac{\sigma_2}{(\sigma_3 - \sigma_2)}\left(1 - \frac{3\sigma_3}{3(\delta_1/\delta_2) + \sigma_2 + \sigma_3}\right)\right]. \tag{6.2.13}$$

In the undeformed state, $\lambda_3 = 1$ and equations (6.2.8) and (6.2.9) give

$$\lambda_3 = 1 \qquad \bar{\nu} = \kappa = -3. \tag{6.2.14}$$

The generalised Lode relation of equation (4.2.38) requires that $\bar{\mu} \equiv \bar{\nu}$ for all λ_3 and hence, must have

$$\bar{\mu} \equiv \bar{\nu} = \kappa = -3 \qquad \text{for } \lambda_3 = 1. \tag{6.2.15}$$

The constitutive equation (4.2.40) can be rearranged into the form,

$$(\tfrac{1}{2}\operatorname{tr}\mathbf{G}'^2)^{1/2} = (\tfrac{1}{2}\operatorname{tr}\mathbf{M}'^2)^{1/2} \tag{6.2.16}$$

where

$$(\tfrac{1}{2}\operatorname{tr}\mathbf{G}'^2)^{1/2} = (1/2\sqrt{3})\Lambda(3 + \bar{\mu}^2)^{1/2}(\sigma_3 - \sigma_2) \tag{6.2.17}$$

$$(\tfrac{1}{2}\operatorname{tr}\mathbf{M}'^2)^{1/2} = (1/2\sqrt{3})\Phi(3 + \bar{\nu}^2)^{1/2}(b_3 - b_2) \tag{6.2.18}$$

and where

$$\Lambda = \delta_1 - \delta_2\sigma_1' \qquad \Phi = \beta_1 - \beta_{-1}b_1. \tag{6.2.19}$$

The generalised Lode relation of equation (4.2.38) can be used to reduce equation (6.2.16) to the form

$$\tau_1=\tfrac{1}{2}(\sigma_3-\sigma_2)=2\,\frac{-\beta_{-1}}{\Lambda}(q\lambda_3^2+1)\frac{(\lambda_3^2-1)}{4\lambda_3^2}. \tag{6.2.20}$$

The constitutive equation can also be used in the form of equation (4.2.42)

$$g_i'=\delta_0+\delta_1\sigma_i'+\delta_2\sigma_i'^2=m_i'=\beta_1 b_i'+\beta_{-1}(b^{-1})_i' \tag{6.2.21}$$

it being noted that $b_i=\lambda_i^2$. From equations (6.2.4) and (6.2.21)

$$m_1'=\beta_{-1}[q(\lambda_3^2+2)+2\lambda_3^2+1](\lambda_3^2-1)/3\lambda_3^2 \tag{6.2.22}$$

$$m_2'=\beta_{-1}(q-1)(\lambda_3^2-1)^2/3\lambda_3^2 \tag{6.2.23}$$

$$m_3'=-\beta_{-1}[q(2\lambda_3^2+1)+\lambda_3^2+2](\lambda_3^2-1)/3\lambda_3^2 \tag{6.2.24}$$

where q is defined by equation (6.2.10). The difference between equations (6.2.24) and (6.2.22) can be rearranged using equation (6.2.21) to give,

$$\sigma_3=\frac{(-\beta_{-1})}{(\delta_1-\delta_2\sigma_2')}(q+1)\frac{(\lambda_3^4-1)}{\lambda_3^2} \tag{6.2.25}$$

and the difference between equations (6.2.23) and (6.2.22) can be rearranged using equation (6.2.21) to give

$$\sigma_2=\frac{(-\beta_{-1})}{(\delta_1-\delta_2\sigma_3')}(q+\lambda_3^2)\frac{(\lambda_3^2-1)}{\lambda_3^2}. \tag{6.2.26}$$

The ratio of equations (6.2.25) and (6.2.26) can be rearranged to give

$$\frac{\sigma_3}{2\sigma_2}=\left(1-\frac{\delta_2(\sigma_3-\sigma_2)}{(\delta_1-\delta_2\sigma_2')}\right)\left(1+\frac{(q-1)(\lambda_3^2-1)}{2(q+\lambda_3^2)}\right) \tag{6.2.27}$$

which can be entered into equation (6.2.12) to give,

$$\mu=-\left\{1+2\left[2\left(1-\frac{\delta_2(\sigma_3-\sigma_2)}{(\delta_1-\delta_2\sigma_2')}\right)\left(1+\frac{(q-1)(\lambda_3^2-1)}{2(q+\lambda_3^2)}\right)-1\right]^{-1}\right\}. \tag{6.2.28}$$

In the undeformed state, $\lambda_3=1, \sigma_2=0, \sigma_3=0$, and equation (6.2.28) reduces to $\mu=-3$, which as a limiting condition, when combined with equation (6.2.15) gives,

$$\mu=\bar{\mu}\equiv\bar{\nu}=\kappa=-3 \qquad \lim_{\lambda_3\to 1}\left(\frac{\sigma_3}{2\sigma_2}\right)=1 \qquad (\lambda_3=1) \tag{6.2.29}$$

for the undeformed state, use having been made of equation (6.2.27).

For the limiting condition $\lambda_3\to\infty$, equations (6.2.8) and (6.2.9) give

$$\kappa=-1 \qquad \bar{\nu}=-\left(1+\frac{2}{q_\infty}\right)\equiv\bar{\mu} \qquad \Gamma=1 \tag{6.2.30}$$

where Γ is defined by equation (4.3.2).

There are two cases to be considered. These are:

(i) ***Simple elastic solid.*** For this class of material

$$\delta_0=0 \qquad \delta_1=1 \qquad \delta_2=0 \qquad \mathbf{G}'=\mathbf{T}'$$

which, as conditions, reduce equations (6.2.27) and (6.2.28) to

$$\frac{\sigma_3}{2\sigma_2}=1+\frac{(q-1)(\lambda_3^2-1)}{2(q+\lambda_3^2)} \qquad \mu=-3\left(1-\frac{2}{3}\frac{(q-1)(\lambda_3^2-1)}{(1+q\lambda_3^2)}\right)\equiv\bar{\nu}. \quad (6.2.31)$$

From equations (4.1.28), (4.1.29), (4.3.2), (4.3.6) and (4.3.7)

$$\mu\equiv\bar{\nu}=\kappa\left(1+\frac{9(3+\kappa^2)(1-\kappa^2)(9-\kappa^2)(\mathrm{d}\ln K/\mathrm{d}\Gamma)}{(3+\kappa^2)^3-9\kappa^2(1-\kappa^2)(9-\kappa^2)(\mathrm{d}\ln K/\mathrm{d}\Gamma)}\right)$$

and since it is evident from §4.4 that the deformation intensity function m for a simple elastic solid must be restricted to one that is continuously differentiable, it follows from the above relation for μ that for $\kappa=-1$, we must have $\bar{\nu}\equiv\mu=-1$. Thus, for $\lambda_3\to\infty$, the following must be satisfied

$$\mu\equiv\bar{\nu}=\kappa=-1 \qquad [q]_{\lambda_3=\infty}=q_\infty=\infty \qquad \Gamma=1$$

where use has been made of equation (6.2.30).

In general, for the constitutive equation of a simple, incompressible elastic solid, the response coefficients

$$\beta_a=\beta_a(I_{\mathbf{B}}, II_{\mathbf{B}}) \qquad (a=\pm 1) \qquad (III_{\mathbf{B}}=1).$$

However, for pure shear, the condition of equation (6.2.7) implies,

$$\beta_a=\beta_a(I_{\mathbf{B}}) \qquad (a=\pm 1)$$

for all λ_3.

It is evident from equation (6.2.7) that replacing λ_3 by $(1/\lambda_3)$ leaves $I_{\mathbf{B}}$ unchanged, which implies that the response coefficients $\beta_a=\beta_a(I_{\mathbf{B}})$, $(a=\pm 1)$, must also remain unchanged when λ_3 is replaced by $1/\lambda_3$. From this it follows that q, defined by equation (6.2.10), also remains unchanged when λ_3 is replaced by $(1/\lambda_3)$. This conclusion regarding q when applied to equation (6.2.25), gives

$$\sigma_3(1/\lambda_3)=-\sigma_3(\lambda_3) \qquad (\delta_0=0, \quad \delta_1=1, \quad \delta_2=0)$$

from which it follows that, for pure shear, replacing λ_3 by $(1/\lambda_3)$ in equation (6.2.25) simply reverses the sign of σ_3, leaving its magnitude unchanged.

Replacing λ_3 in equation (6.2.26) by $(1/\lambda_3)$ gives,

$$\sigma_2(1/\lambda_3)=-(-\beta_{-1})(q\lambda_3^2+1)(\lambda_3^2-1)/\lambda_3^2 \qquad (\delta_0=0, \quad \delta_1=1, \quad \delta_2=0).$$

In principle, it is possible for a class of isotropic, incompressible solids to exist which satisfy the conditions

$$\sigma_2(1/\lambda_3) = -\sigma_2(\lambda_3) \qquad \sigma_3(1/\lambda_3) = -\sigma_3(\lambda_3).$$

It is evident that for a class of isotropic, incompressible elastic solids to satisfy these conditions, they must have $q=1$ for all λ_3.

The condition $q=1$ for all λ_3 reduces equation (6.2.31) to the form,

$$\sigma_3 = 2\sigma_2 \qquad \mu = -3 \qquad q=1$$

for all λ_3.

The conditions $q=1$, $\mu=-3$ for all λ_3 are irreconcilable with the limiting conditions of equation (6.2.32).

It is of interest to note that the condition of equation (6.2.7), when entered into equation (5.4.22) gives

$$\beta_1 = \frac{3\lambda_3^2}{2(\lambda_3^4+\lambda_3^2+1)}\hat{A}(I_{\mathrm{B}}) = -\beta_{-1} \qquad q=1$$

for all λ_3.

(ii) ***Generalised loading function.*** From equations (4.2.4), (4.2.34) and (4.2.35)

$$\bar{\mu} = \mu\left(1 + \frac{9(3+\mu^2)(1-\mu^2)(9-\mu^2)(\mathrm{d}\ln H/\mathrm{d}\omega)}{(3+\mu^2)^3 - 9\mu^2(1-\mu^2)(9-\mu^2)(\mathrm{d}\ln H/\mathrm{d}\omega)}\right)$$

and from equations (4.1.29), (4.3.2), (4.3.6) and (4.3.7)

$$\bar{\nu} = \kappa\left(1 + \frac{9(3+\kappa^2)(1-\kappa^2)(9-\kappa^2)(\mathrm{d}\ln K/\mathrm{d}\Gamma)}{(3+\kappa^2)^3 - 9\kappa^2(1-\kappa^2)(9-\kappa^2)(\mathrm{d}\ln K/\mathrm{d}\Gamma)}\right).$$

For the continuously differentiable forms of stress intensity and strain intensity functions it follows that the limiting condition $\kappa=-1$ requires that

$$\mu = \bar{\mu} \equiv \bar{\nu} = \kappa = -1 \qquad [q]_{\lambda_3=\infty} = q_\infty = \infty \qquad \Gamma = 1 \qquad \omega = 1 \qquad (6.2.33)$$

where use has been made of equation (6.2.30). For any piecewise continuous stress intensity and strain intensity functions the associated vertices occur at,

$$\mu^2 = \kappa^2 = \begin{cases} 0, 9 \\ 1, \infty \end{cases} \qquad \omega = \Gamma = \begin{cases} 0 \\ 1 \end{cases}.$$

Hence, in the transition through a given vertex, both $\bar{\mu}$ and $\bar{\nu}$ must pass discontinuously through the above values of μ and κ appropriate to the given vertex. Thus, even for the piecewise continuous forms of stress intensity and strain intensity functions, equation (6.2.33) has at some stage to be satisfied.

It follows from equation (6.2.12) that reversing the sign of both σ_2 and σ_3 leaves the sign of μ unchanged. From equations (4.2.4) and (3.1.12),

$$\delta_2\sigma_1' = 6\mu^2(\mu^2-9)(3+\mu^2)^{-1}(\mathrm{d}H/\mathrm{d}\omega)$$

and since $H=H(\omega)$ and $\omega=\omega(\mu^2)$ it follows that the sign of $\delta_2\sigma_1'$ remains unchanged when the sign of both σ_2 and σ_3 is reversed. Similarly, since

$$\delta_2\sigma_2'=-\delta_1\sigma_1'(2\sigma_2-\sigma_3)/(\sigma_2+\sigma_3)$$

$$\delta_2\sigma_3'=-\delta_1\sigma_1'(2\sigma_3-\sigma_2)/(\sigma_2+\sigma_3)$$

it follows that the sign of both $\delta_2\sigma_2'$ and $\delta_2\sigma_3'$ remain unchanged when the signs of both σ_2 and σ_3 are reversed. In the same way, it follows from equation (4.2.4) that the sign of δ_1 remains unchanged when the signs of both σ_2 and σ_3 are reversed. These observations, together with the above discussion of the simple elastic solid lead to the conclusion that for the generalised loading function $\mathbf{G}'$, equation (6.2.25) gives

$$\sigma_3(1/\lambda_3)=-\sigma_3(\lambda_3)$$

from which it follows that, for the pure shear mode of deformation of a material which satisfies the constitutive equation (6.2.16), replacing λ_3 by $(1/\lambda_3)$ in equation (6.2.25) simply reverses the sign of σ_3, leaving its magnitude unchanged.

Replacing λ_3 in equation (6.2.26) by $(1/\lambda_3)$ gives

$$\sigma_2(1/\lambda_3)=-\frac{(-\beta_{-1})}{(\delta_1-\delta_2\sigma_3')}(q\lambda_3^2+1)\frac{(\lambda_3^2-1)}{\lambda_3^2}. \qquad (6.2.34)$$

The remaining discussion now parallels that above for the simple elastic solid, it being concluded that for the generalised loading function $\mathbf{G}'$ the conditions

$$\sigma_2(1/\lambda_3)=-\sigma_2(\lambda_3) \qquad \sigma_3(1/\lambda_3)=-\sigma_3(\lambda_3)$$

can only be satisfied if $q=1$ for all λ_3.

The condition $q=1$ for all λ_3 reduces equation (6.2.9) to $\bar{v}=-3$ for all λ_3, and hence the generalised Lode relation gives $\bar{\mu}(\equiv\bar{v})=-3$ for all λ_3. These conditions imply,

$$\mu=\bar{\mu}\equiv\bar{v}=-3 \qquad (6.2.35)$$

for all λ_3, a condition which is not in accord with the limiting condition of equation (6.2.33).

6.2.2 Strain Response Function

It is evident from equations (2.1.36) and (6.2.5) that for pure shear $\mathbf{F}$ can be expressed in the form

$$\mathbf{F}=\frac{1}{\lambda_3}\boldsymbol{e}_1\otimes\boldsymbol{E}_1+\boldsymbol{e}_2\otimes\boldsymbol{E}_2+\lambda_3\boldsymbol{e}_3\otimes\boldsymbol{E}_3. \qquad (6.2.36)$$

Let the unit vectors $\boldsymbol{L}$, $\boldsymbol{M}$, $\boldsymbol{N}$ form an orthonormal set associated with the referential description and let the unit vectors $\boldsymbol{l}$, $\boldsymbol{m}$, $\boldsymbol{n}$ form an orthonormal set associated with the spatial description. In the context of these two sets of orthonormal vectors, equation (6.2.36) is taken to imply,

$$\mathbf{F}=\lambda_3^{-1}\boldsymbol{l}\otimes\boldsymbol{L}+\boldsymbol{m}\otimes\boldsymbol{M}+\lambda_3\boldsymbol{n}\otimes\boldsymbol{N}. \tag{6.2.37}$$

From equations (2.1.45) and (6.2.37)

$$\mathbf{B}=\mathbf{F}\mathbf{F}^{\mathrm{T}}=\lambda_3^{-2}\boldsymbol{l}\otimes\boldsymbol{l}+\boldsymbol{m}\otimes\boldsymbol{m}+\lambda_3^2\boldsymbol{n}\otimes\boldsymbol{n} \tag{6.2.38}$$

$$\mathbf{B}^{-1}=\lambda_3^2\boldsymbol{l}\otimes\boldsymbol{l}+\boldsymbol{m}\otimes\boldsymbol{m}+\lambda_3^{-2}\boldsymbol{n}\otimes\boldsymbol{n}. \tag{6.2.39}$$

Let the principal axes of $\mathbf{B}$ at x be defined by the orthonormal basis $q=\{\boldsymbol{q}_1,\boldsymbol{q}_2,\boldsymbol{q}_3\}$, it being noted that $b_i=\lambda_i^2$, where the b_i are the proper numbers of $\mathbf{B}$. With this as basis, $\mathbf{B}$ has the spectral representation,

$$\mathbf{B}=\frac{1}{\lambda_3^2}\boldsymbol{q}_1\otimes\boldsymbol{q}_1+\boldsymbol{q}_2\otimes\boldsymbol{q}_2+\lambda_3^2\boldsymbol{q}_3\otimes\boldsymbol{q}_3. \tag{6.2.40}$$

Since the orthogonal pair of vectors $\boldsymbol{q}_1$ and $\boldsymbol{q}_3$ are orthogonal to $\boldsymbol{m}$ they can be expressed in the form,

$$\boldsymbol{q}_1=\cos\theta\boldsymbol{l}-\sin\theta\boldsymbol{n} \qquad \boldsymbol{q}_2=\boldsymbol{m} \qquad \boldsymbol{q}_3=\sin\theta\boldsymbol{l}+\cos\theta\boldsymbol{n}. \tag{6.2.41}$$

Substituting for the $\boldsymbol{q}_i$ in equation (6.2.40) and comparing the resulting form for $\mathbf{B}$ with that of equation (6.2.38) gives,

$$\frac{1}{\lambda_3^2}=\frac{1}{\lambda_3^2}\cos^2\theta+\lambda_3^2\sin^2\theta \qquad \lambda_3^2=\lambda_3^2\cos^2\theta+\frac{1}{\lambda_3^2}\sin^2\theta. \tag{6.2.42}$$

From equations (2.1.44) and (6.2.37)

$$\mathbf{C}=\mathbf{F}^{\mathrm{T}}\mathbf{F}=\lambda_3^{-2}\boldsymbol{L}\otimes\boldsymbol{L}+\boldsymbol{M}\otimes\boldsymbol{M}+\lambda_3^2\boldsymbol{N}\otimes\boldsymbol{N}. \tag{6.2.43}$$

Let the principal axes of $\mathbf{C}$ at X be defined by the orthonormal basis $p=\{\boldsymbol{p}_1,\boldsymbol{p}_2,\boldsymbol{p}_3\}$, it being noted that $c_i=\lambda_i^2$, where the c_i are the proper numbers of $\mathbf{C}$. With this as the basis, $\mathbf{C}$ has the spectral representation,

$$\mathbf{C}=\lambda_3^{-2}\boldsymbol{p}_1\otimes\boldsymbol{p}_1+\boldsymbol{p}_2\otimes\boldsymbol{p}_2+\lambda_3^2\boldsymbol{p}_3\otimes\boldsymbol{p}_3. \tag{6.2.44}$$

Since the orthogonal pair of vectors $\boldsymbol{p}_1$ and $\boldsymbol{p}_3$ are orthogonal to $\boldsymbol{M}$ they can be expressed in the form,

$$\boldsymbol{p}_1=\cos\Theta\boldsymbol{L}-\sin\Theta\boldsymbol{N} \qquad \boldsymbol{p}_2=\boldsymbol{M} \qquad \boldsymbol{p}_3=\sin\Theta\boldsymbol{L}+\cos\boldsymbol{\Theta}\boldsymbol{N}. \tag{6.2.45}$$

Substituting for the $\boldsymbol{p}_i$ in equation (6.2.44) and comparing the resulting form for $\mathbf{C}$ with that of equation (6.2.43) gives

$$\frac{1}{\lambda_3^2}=\frac{1}{\lambda_3^2}\cos^2\Theta+\lambda_3^2\sin^2\Theta \qquad \lambda_3^2=\lambda_3^2\cos^2\Theta+\frac{1}{\lambda_3^2}\sin^2\Theta. \tag{6.2.46}$$

Comparison of equations (6.2.42) and (6.2.46) leads to the conclusion that,

$$\theta=\Theta \qquad (\lambda_3\geqslant 1) \tag{6.2.47}$$

for all λ_3. The rotation tensor $\mathbf{R}$ is obtained by way of equations (1.6.149) and (1.7.51) in the form,

$$\mathbf{R}=\sum_{r=1}^{3} \boldsymbol{q}_r \otimes \boldsymbol{p}_r = \boldsymbol{l}\otimes \boldsymbol{L}+\boldsymbol{m}\otimes \boldsymbol{M}+\boldsymbol{n}\otimes \boldsymbol{N} \tag{6.2.48}$$

from which it follows that the rotation is zero, a conclusion which follows directly from equation (6.2.44), and also from the relation

$$\mathbf{B}=\mathbf{R}\mathbf{C}\mathbf{R}^{\mathrm{T}} \tag{6.2.49}$$

since for pure shear it is evident that $\mathbf{B}\equiv\mathbf{C}$, as follows from equations (6.2.5) and (2.1.44).

It is evident that a given simple shear can be resolved into a pure shear followed by or preceded by a rotation about an axis normal to its plane.

The constitutive equation will be used in the form of equation (4.2.44)

$$\mathbf{G}'(\mathbf{T}')=\delta_0\mathbf{I}+\delta_1\mathbf{T}'+\delta_2\mathbf{T}'^2=2\bar{G}(\varphi_0\mathbf{I}+\varphi_1\bar{\mathbf{E}}'+\varphi_2\bar{\mathbf{E}}'^2). \tag{6.2.50}$$

From equations (4.2.20), (6.2.38) and (6.2.39),

$$\bar{\mathbf{E}}'=-\gamma_1(\boldsymbol{l}\otimes\boldsymbol{l}-\boldsymbol{n}\otimes\boldsymbol{n}) \tag{6.2.51}$$

where

$$\gamma_1=(\lambda_3^4-1)/4\lambda_3^2. \tag{6.2.52}$$

Entering the form for $\bar{\mathbf{E}}'$ given by equation (6.2.51) into equation (6.2.50) gives

$$\mathbf{G}'(\mathbf{T}')=-2\gamma_1\bar{G}(\varphi_1+\gamma_1\varphi_2)\boldsymbol{l}\otimes\boldsymbol{l}-4\gamma_1^2\bar{G}\boldsymbol{m}\otimes\boldsymbol{m}+2\gamma_1\bar{G}(\varphi_1-\gamma_1\varphi_2)\boldsymbol{n}\otimes\boldsymbol{n} \tag{6.2.53}$$

where use has been made of equation (4.2.45) to give $\varphi_0=-2\gamma_1^2\varphi_2$. From equation (6.2.53),

$$\operatorname{tr}\mathbf{G}'=-6\gamma_1^2\bar{G}\varphi_2=0 \tag{6.2.54}$$

and since $\gamma_1\geqslant 0$, $\bar{G}>0$, which requires that $\varphi_2=0$ for all λ_3 and this condition reduces equation (6.2.53) to

$$\mathbf{G}'(\mathbf{T}')=-2\gamma_1\bar{G}\varphi_1(\boldsymbol{l}\otimes\boldsymbol{l}-\boldsymbol{n}\otimes\boldsymbol{n}). \tag{6.2.55}$$

Equation (6.2.55) gives

$$\boldsymbol{l}\cdot(\mathbf{G}'\boldsymbol{l})=-2\gamma_1\bar{G}\varphi_1 \tag{6.2.56}$$

$$\boldsymbol{m}\cdot(\mathbf{G}'\boldsymbol{m})=0 \tag{6.2.57}$$

$$\boldsymbol{n}\cdot(\mathbf{G}'\boldsymbol{n})=2\gamma_1\bar{G}\varphi_1. \tag{6.2.58}$$

In terms of the orthogonal set of vectors $\{\boldsymbol{l},\boldsymbol{m},\boldsymbol{n}\}$, the normal components of stress can be expressed in the form,

$$\begin{gathered}\sigma_{(l)}=\boldsymbol{l}\cdot(\mathbf{T}\boldsymbol{l})=\sigma_1=0 \qquad \sigma_{(m)}=\boldsymbol{m}\cdot(\mathbf{T}\boldsymbol{m})=\sigma_2\\ \sigma_{(n)}=\boldsymbol{n}\cdot(\mathbf{T}\boldsymbol{n})=\sigma_3.\end{gathered} \tag{6.2.59}$$

The difference between equations (6.2.58) and (6.2.57) gives

$$\boldsymbol{n}\cdot(\mathbf{G}'\boldsymbol{n})-\boldsymbol{m}\cdot(\mathbf{G}'\boldsymbol{m})=\Lambda(\sigma_3-\sigma_2)=2\bar{G}\gamma_1\varphi_1 \tag{6.2.60}$$

where

$$\Lambda=\delta_1-\delta_2\sigma_1' \tag{6.2.61}$$

is just equation (4.2.35).

Entering the conditions $\lambda_1=1/\lambda_3$, $\lambda_2=1$ into the relation

$$\bar{\varepsilon}_i'=[b_i'-(b^{-1})_i']/4 \qquad (i=1,2,3) \tag{6.2.62}$$

gives

$$\bar{\varepsilon}_1'=(\lambda_3^4-1)/4\lambda_3^2=-\bar{\varepsilon}_3' \qquad \bar{\varepsilon}_2'=0 \tag{6.2.63}$$

which when entered into equations (4.2.48) and (4.2.52) gives

$$\nu=-3 \qquad \varphi_1=\frac{(\beta_1-\beta_{-1})}{\bar{G}} \qquad \varphi_2=\frac{4(\beta_1+\beta_{-1})\lambda_3^2}{(1+\lambda_3^2)^2\bar{G}}. \tag{6.2.64}$$

From equation (6.2.54), since $\gamma_1\geqslant 0$, $\bar{G}>0$, $\varphi_2=0$, and hence equations (6.2.64), (4.3.9) and (4.3.11) give

$$\nu=-3 \qquad \Omega=0 \qquad \varphi_1=2\beta_1/\bar{G}=1 \qquad \varphi_2=0 \qquad \beta_1=-\beta_{-1} \tag{6.2.65}$$

for all λ_3. The conditions of equation (6.2.65) reduce equation (6.2.60) to

$$\tau_1=\tfrac{1}{2}(\sigma_3-\sigma_2)=\left[\frac{\bar{G}}{\Lambda}\right]_{\nu=-3}\frac{(\lambda_3^4-1)}{4\lambda_3^2}=2\frac{(-\beta_{-1})}{\Lambda}\gamma_1 \tag{6.2.66}$$

which is to be compared with equation (6.2.20).

The conditions of equation (6.2.65) when entered into equation (4.2.47), give

$$\bar{\mu}\equiv\bar{\nu}=\nu=-3 \tag{6.2.67}$$

for all λ_3, where use has been made of the generalised Lode relation of equation (4.2.38). Setting $\varphi_1=1$ in equation (6.2.55) gives

$$\mathbf{G}'=\delta_0\mathbf{I}+\delta_1\mathbf{T}'+\delta_2\mathbf{T}'^2=2\bar{G}\bar{\mathbf{E}}' \tag{6.2.68}$$

use having been made of equations (4.2.3) and (6.2.51). Entering the spectral representations of the symmetric tensors $\mathbf{G}'$, $\mathbf{T}'$ and $\bar{\mathbf{E}}'$ into equation (6.2.68) gives

$$g_i'=\delta_0+\delta_1\sigma_i'+\delta_2\sigma_i'^2=2\bar{G}\bar{\varepsilon}_i' \qquad (i=1,2,3) \tag{6.2.69}$$

which is just a reduced form of equation (4.2.46), corresponding to the conditions of equation (6.2.65). Equation (6.2.69) can be used to obtain a form for the ratio $\delta_1/(\delta_2\sigma_1')$ which can then be entered into equation (4.2.34) with $\bar{\mu}=-3$, the solution for μ simply being equation (6.2.12). For a continuously differentiable stress intensity function, as for example that characterised by equation (3.3.6), it follows from equation (3.3.5) that the condition $\bar{\mu}=-3$ for all λ_3 requires that,

$$\mu=\bar{\mu}\equiv\bar{\nu}=\nu=-3 \quad \text{(continuously differentiable)} \tag{6.2.70}$$

for all λ_3, where use has been made of equation (6.2.67). In the case of a twelve-sided, piecewise continuous stress intensity function, it follows from equations (3.1.18) and (3.2.30) that

$$\bar{\mu}\equiv\bar{\mu}_n=\begin{cases}\bar{\mu}_1 \\ \bar{\mu}_2\end{cases} \quad \text{for} \begin{cases}(-3\leqslant\mu\leqslant-1) \\ (-3\geqslant\mu\geqslant-\infty)\end{cases} \tag{6.2.71}$$

and hence μ can only vary for this type of stress intensity function when either $\bar{\mu}\equiv\bar{\mu}_1=$constant or when $\bar{\mu}\equiv\bar{\mu}_2=$constant. For a piecewise continuous stress intensity function, $\bar{\mu}=-3$ corresponds to the vertex at which the linear sections characterised by $\bar{\mu}\equiv\bar{\mu}_1=$constant and $\bar{\mu}\equiv\bar{\mu}_2=$constant meet, and hence the particular value $\bar{\mu}=-3$ lies between $\bar{\mu}_1$ and $\bar{\mu}_2$. Hence, it follows from equation (6.2.71) that for $\bar{\mu}=-3$, we must have,

$$\mu=\bar{\mu}\equiv\bar{\nu}=\nu=-3 \quad (\bar{\mu}_0=\text{constant}) \tag{6.2.72}$$

for all λ_3, where $\bar{\mu}_0$ is defined by equation (3.2.31). Thus, for pure shear, we must have

$$\mu=\bar{\mu}\equiv\bar{\nu}=\nu=-3 \quad \text{(pure shear)} \tag{6.2.73}$$

for all λ_3, and all forms of stress intensity function h defined by equation (4.2.1) (considered in terms of the initial yield condition discussed in Chapter 3). Setting $\mu=-3$ in equation (6.2.12) gives

$$\sigma_3=2\sigma_2 \quad (\mu=-3,\ \sigma_1=0) \tag{6.2.74}$$

for all λ_3. The condition of equation (6.2.74) is independent of the loading coefficients δ_i $(i=0,1,2)$ and is therefore a *universal relation* (cf equation (6.1.103)).

Use of the strain response function has two advantages over the use of the deformation response function. These are

(i) the form of the stress–strain relationship does not impose any restriction on the form of the stress and strain intensity functions;

(ii) the identification of a universal relation which can be directly subjected to the test of experiment.

6.3 COMBINED STRESSING OF A SHEET OF INCOMPRESSIBLE MATERIAL

6.3.1 The Stress–Strain Relations

Let the cartesian coordinates (X^1, X^2, X^3) be the referential coordinates of a body in the natural state and the cartesian coordinates (x^1, x^2, x^3) be the spatial coordinates in the deformed configuration. The discussion will be restricted to an incompressible sheet of thickness $2h_0$ aligned with the cartesian coordinate system (X^1, X^2, X^3). The material, regarded as incompressible with respect to volume, is assumed to be isotropic relative to its undeformed and unstressed, i.e. its natural, configuration. Consider the deformation to be produced by means of edge tractions only, so that the major surfaces $X^1 = \pm h_0$ are free from applied stress.

Assume a homogeneous deformation such that

$$x^1 = \lambda_1 X^1 \qquad x^2 = \lambda_2 X^2 \qquad x^3 = \lambda_3 X^3 \tag{6.3.1}$$

the condition of incompressibility giving,

$$\lambda_1 \lambda_2 \lambda_3 = \det \mathbf{F} = 1 \tag{6.3.2}$$

where the λ_i $(i = 1, 2, 3)$ are the principal stretches. Equation (6.3.2) implies that only two of the principal stretches are independently assignable. Since the major surfaces $X^1 = \pm h_0$ are free from applied stress, the two independently assignable principal stretches will be taken as λ_2 and λ_3.

From equations (2.1.36), (2.1.45) and (6.3.1), the matrices of $\mathbf{F}$, $\mathbf{B}$ and $\mathbf{B}^{-1}$ are,

$$[F_{i\alpha}] = \begin{bmatrix} (\lambda_2\lambda_3)^{-1} & 0 & 0 \\ 0 & \lambda_2 & 0 \\ 0 & 0 & \lambda_3 \end{bmatrix}$$

$$[B_{ij}] = \begin{bmatrix} (\lambda_2\lambda_3)^{-2} & 0 & 0 \\ 0 & \lambda_2^2 & 0 \\ 0 & 0 & \lambda_3^2 \end{bmatrix} \qquad [(B^{-1})_{ij}] = \begin{bmatrix} \lambda_2^2\lambda_3^2 & 0 & 0 \\ 0 & \lambda_2^{-2} & 0 \\ 0 & 0 & \lambda_3^{-2} \end{bmatrix}. \tag{6.3.3}$$

The principal invariants are

$$\begin{aligned} I_{\mathbf{B}} &= (\lambda_2^4\lambda_3^2 + \lambda_2^2\lambda_3^4 + 1)/(\lambda_2^2\lambda_3^2) \\ II_{\mathbf{B}} &= (\lambda_2^2 + \lambda_3^2 + \lambda_2^4\lambda_3^4)/(\lambda_2^2\lambda_3^2) \qquad III_{\mathbf{B}} = 1. \end{aligned} \tag{6.3.4}$$

Substituting for the $b_i = \lambda_i^2$ in equation (6.1.122) gives

$$\begin{aligned} \bar{\varepsilon}_1' &= (2 + \lambda_2^2 + \lambda_3^2 + 2\lambda_2^2\lambda_3^2)(1 - \lambda_2^2\lambda_3^2)/12\lambda_2^2\lambda_3^2 \\ \bar{\varepsilon}_2' &= (1 + 2\lambda_3^2 + 2\lambda_2^2\lambda_3^2 + \lambda_2^2\lambda_3^4)(\lambda_2^2 - 1)/12\lambda_2^2\lambda_3^2 \\ \bar{\varepsilon}_3' &= (1 + 2\lambda_2^2 + 2\lambda_2^2\lambda_3^2 + \lambda_2^4\lambda_3^2)(\lambda_3^2 - 1)/12\lambda_2^2\lambda_3^2 \end{aligned} \tag{6.3.5}$$

where use has been made of equation (6.3.2). Entering the values of $\bar{\varepsilon}_i'$ into equation (4.2.48) from equation (6.3.5) gives

$$\nu = -\frac{(2 + \lambda_2^2 + \lambda_3^2 + 2\lambda_2^2\lambda_3^2)(\lambda_2^2\lambda_3^2 - 1)}{(\lambda_2^2\lambda_3^2 + 1)(\lambda_3^2 - \lambda_2^2)}. \tag{6.3.6}$$

For an incompressible material, the φ_i $(i=1,2)$ of equation (4.2.52) reduce to

$$\varphi_1=\frac{2}{3}\frac{[\beta_1(3+2I_{\mathbf{B}}+II_{\mathbf{B}})-\beta_{-1}(3+I_{\mathbf{B}}+2II_{\mathbf{B}})]}{(2+I_{\mathbf{B}}+II_{\mathbf{B}})\bar{G}} \tag{6.3.7}$$

$$\varphi_2=\frac{8(\beta_1+\beta_{-1})}{(2+I_{\mathbf{B}}+II_{\mathbf{B}})\bar{G}}. \tag{6.3.8}$$

Since the major surfaces, $X^1=\pm h_0$, are taken to be free from applied stress, the only non-zero components of stress are the normal components: thus in terms of the principal stresses (cf equation (2.2.40)),

$$\sigma_1\equiv T_{11}=0 \qquad \sigma_2\equiv T_{22} \qquad \sigma_3\equiv T_{33}\geqslant 0. \tag{6.3.9}$$

For $\sigma_1=0$, Lode's (1926) stress parameter μ can be expressed in the form

$$\mu=-(\sigma_2+\sigma_3)/(\sigma_3-\sigma_2) \tag{6.3.10}$$

where use has been made of equation (4.1.26).

The constitutive equation (6.1.1) will be used in the form of equation (4.2.46)

$$g_i'=\delta_0+\delta_1\sigma_i'+\delta_2\sigma_i'^2=m_i'=2\bar{G}(\varphi_0+\varphi_1\bar{\varepsilon}_i'^2+\varphi_2\bar{\varepsilon}_i'^2)=2\bar{G}\bar{m}_i' \quad (i=1,2,3). \tag{6.3.11}$$

Equation (6.3.11) gives

$$g_3'-g_2'=(\sigma_3'-\sigma_2')(\delta_1-\delta_2\sigma_1')=2\bar{G}\Psi(\bar{\varepsilon}_3'-\bar{\varepsilon}_2') \tag{6.3.12}$$

which can be rearranged into the form

$$\tau_1=2\left[\bar{G}\frac{\Psi}{\Lambda}\right]_\nu\gamma_1 \tag{6.3.13}$$

where

$$\tau_1=\tfrac{1}{2}(\sigma_3-\sigma_2) \qquad \gamma_1=\tfrac{1}{2}(\bar{\varepsilon}_3'-\bar{\varepsilon}_2')=\frac{(\lambda_2^2\lambda_3^2+1)(\lambda_3^2-\lambda_2^2)}{8\lambda_2^2\lambda_3^2} \tag{6.3.14}$$

and where Λ and Ψ are defined by equations (4.2.35) and (4.3.13) respectively.

The stress–strain relationship can also be obtained from equation (6.3.11),

$$\tfrac{1}{2}(\sigma_2+\sigma_3)=3\left[\bar{G}\frac{\Psi}{\Lambda}\right]_\nu\left(\frac{\mu}{\nu}\right)\frac{(2+\lambda_2^2+\lambda_3^2+2\lambda_2^2\lambda_3^2)(\lambda_2^2\lambda_3^2-1)}{12\lambda_2^2\lambda_3^2} \tag{6.3.15}$$

which finds application in the equibiaxial extension of a thin sheet.

6.3.2 Effective Stress, Reduced Strain, and the Generalised Shear Modulus

The constitutive equation (6.1.1) can also be used in the form of equation (4.2.44)

$$\mathbf{G}'(\mathbf{T}')=2\bar{G}(\varphi_0\mathbf{I}+\varphi_1\bar{\mathbf{E}}'+\varphi_2\bar{\mathbf{E}}'^2)=2\bar{G}\bar{\mathbf{M}}'(\bar{\mathbf{E}}'). \tag{6.3.16}$$

A question of particular interest, in relation to the assumption of isotropy, is whether it is possible to obtain a single, that is a universal stress–strain relation. From the constitutive equation (6.3.16) there can be obtained the scalar relation,

$$(\tfrac{1}{2}\operatorname{tr}\mathbf{G}'^2)^{1/2}=2\bar{G}(\tfrac{1}{2}\operatorname{tr}\bar{\mathbf{M}}'^2)^{1/2}$$

where

$$(\tfrac{1}{2}\operatorname{tr}\mathbf{G}'^2)^{1/2}=(3+\bar{\mu}^2)^{1/2}\Lambda\tau_1/\sqrt{3} \qquad \tau_1=\tfrac{1}{2}(\sigma_3-\sigma_2) \tag{6.3.17}$$

and where

$$(\tfrac{1}{2}\operatorname{tr}\bar{\mathbf{M}}'^2)^{1/2}=(3+\bar{\nu}^2)^{1/2}\Psi\gamma_1/\sqrt{3} \qquad \gamma_1=\tfrac{1}{2}(\bar{\varepsilon}_3-\bar{\varepsilon}_2) \tag{6.3.18}$$

it being noted that Λ and Ψ are defined by equations (4.2.35) and (4.3.13) respectively.

The formulation of an effective stress–effective strain relation must satisfy the essential condition that at initial yield,

$$f=J_2'H(\omega)-k^2=0 \tag{6.3.19}$$

which is just equation (3.1.11).

The condition of equation (6.3.19) implies that the stress–strain relation must be of the form

$$\sigma''=3\frac{[N^{1/2}]_{\nu=-1}}{[H^{1/2}]_{\mu=-1}}G\varepsilon'' \qquad G=\bar{G}\left[\frac{\Psi(3+\mu\bar{\mu})}{\Lambda(3+\nu\bar{\nu})}\right]^{1/2} \tag{6.3.20}$$

where use has been made of the generalised Lode relation of equation (4.2.38) to eliminate the terms involving $\bar{\mu}$ and $\bar{\nu}$, and where the *effective stress*

$$\sigma''=\frac{(3J_2'H)^{1/2}}{[H^{1/2}]_{\mu=-1}}=\frac{[H(3+\mu^2)]^{1/2}}{[H^{1/2}]_{\mu=-1}}\tau_1 \tag{6.3.21}$$

the *effective strain*

$$\varepsilon''=\frac{(\tfrac{4}{3}K_2'N)^{1/2}}{[N^{1/2}]_{\nu=-1}}=\frac{2}{3}\frac{[N(3+\nu^2)]^{1/2}}{[N^{1/2}]_{\nu=-1}}\gamma_1 \tag{6.3.22}$$

and where τ_1 and γ_1 are defined by equations (6.3.17) and (6.3.18) respectively, use having been made of the relations,

$$\Lambda=H\frac{(3+\mu^2)}{(3+\mu\bar{\mu})} \qquad \Psi=N\frac{(3+\nu^2)}{(3+\nu\bar{\nu})}$$

which follow from equations (3.1.17) and (4.3.12).

The remaining discussion is restricted to those materials for which

$$\lim_{\varepsilon''\to 0} G=G_0$$

where G_0 is the ground state, that is the classical, shear modulus. With this as basis, equation (6.3.20) can be rearranged into the linear relation,

$$\bar{\sigma}'' = 3G_0\bar{e}'' \tag{6.3.23}$$

where the *reduced stress*

$$\bar{\sigma}'' = (3J_2')^{1/2} = (3+\mu^2)^{1/2}\tau_1 \qquad \tau_1 = \tfrac{1}{2}(\sigma_3 - \sigma_2) \tag{6.3.24}$$

and where the *reduced strain*

$$\begin{aligned}\bar{e}'' &= \frac{\bar{G}}{G_0}\left[\frac{\Psi(3+\mu\bar{\mu})}{H\Lambda(3+\nu\bar{\nu})}\right]^{1/2}\varepsilon'' \\ &= \frac{2}{3}\frac{\bar{G}}{G_0}\left[\frac{N\Psi(3+\mu\bar{\mu})}{H\Lambda(3+\nu\bar{\nu})}\right]^{1/2}(3+\nu^2)^{1/2}\gamma_1 \qquad \gamma_1 = \tfrac{1}{2}(\bar{\varepsilon}_3' - \bar{\varepsilon}_2').\end{aligned} \tag{6.3.25}$$

6.3.3 Biaxial Extension

It is evident from the discussion given in §§3.3 and 4.3 that for all forms of continuously differentiable stress intensity and strain intensity functions

$$\mu^2 = \bar{\mu}^2 \equiv \bar{\nu}^2 = \nu^2 = \begin{cases} 0, 9 \\ 1, \infty \end{cases} \qquad \text{for } \omega = \Omega = \begin{cases} 0 \\ 1 \end{cases} \tag{6.3.26}$$

where ω and Ω are defined by equations (3.1.12) and (4.3.9) respectively and where use has been made of the generalised Lode relation of equation (4.2.38) (see for example equations (3.3.5) and (4.3.16)). For any piecewise continuous stress intensity and strain intensity functions the associated vertices occur at

$$\mu^2 = \nu^2 = \begin{cases} 0, 9 \\ 1, \infty \end{cases} \qquad \text{for } \omega = \Omega = \begin{cases} 0 \\ 1 \end{cases}.$$

Hence, in the transition through a given vertex, both $\bar{\mu}$ and $\bar{\nu}$ must pass discontinuously through the above values of $\mu \equiv \nu$ appropriate to the given vertex. Thus, even for the piecewise continuous forms of stress intensity and strain intensity functions, equation (6.3.26) has at some stage to be satisfied. It is evident from equation (6.3.26) that the values of μ corresponding to $\omega \in [0, 1]$ and the values of ν for $\Omega \in [0, 1]$, cover the entire range of μ and ν. These considerations are taken to imply that particular physical significance is to be attached to the values of $\nu = 0, -1, -3, -\infty$ and hence to the particular conditions

$$\mu = \bar{\mu} \equiv \bar{\nu} = \nu = \begin{cases} 0, 3 \\ -1, -\infty \end{cases} \qquad \text{for } \omega = \Omega = \begin{cases} 0 \\ 1 \end{cases} \tag{6.3.27}$$

where use has been made of equations (6.3.26).

(i) $\nu = 0$. Solving equation (6.3.6) for this value of ν gives

$$\lambda_1 = 1 \qquad \lambda_2 = \lambda_3^{-1} \tag{6.3.28}$$

where use has been made of the equation (6.3.2). Entering the condition of equation (6.3.28) into equations (6.3.4), (6.3.5), (6.3.7) and (6.3.8) gives

$$I_{\mathbf{B}}=II_{\mathbf{B}}=\lambda_3^2+\lambda_3^{-2}+1$$
$$\Omega=0 \qquad \varphi_0=0 \qquad \varphi_1=1 \qquad \varphi_2=0 \qquad \Psi=N=1 \tag{6.3.29}$$

$$\bar{\varepsilon}_1'=0 \qquad \bar{\varepsilon}_3'=-\bar{\varepsilon}_2'=\gamma_1 \qquad \gamma_1=(\lambda_3^4-1)/4\lambda_3^2 \tag{6.3.30}$$

for all $\lambda_3 \geqslant 1$, where use has been made of equations (4.3.8), (4.3.9) and (4.3.11). The conditions of equation (6.3.29), when entered into equation (4.2.47), give $\bar{v}=0$ and hence, $\bar{\mu}\equiv\bar{v}=v=0$, where use has been made of the generalised Lode relation of equation (4.2.38). Thus, for equation (6.3.27) to be satisfied (i.e. for $\mu=0$), equation (6.3.10) gives,

$$\sigma_3=-\sigma_2 \qquad (\mu=0,\ \sigma_1=0) \tag{6.3.31}$$

from which it follows that the simple mode of deformation for which $v=0$ is an example of combined tension ($\sigma_3 \geqslant 0$, $\lambda_3 \geqslant 1$) and compression ($\sigma_2 \leqslant 0$, $\lambda_2 \leqslant 1$). The conditions $\mu=0$, $\bar{\mu}=0$ require that

$$\delta_0=0, \qquad \delta_1=1, \qquad \delta_2=0, \qquad H=\Lambda=1, \qquad \omega=0. \tag{6.3.32}$$

The conditions of equations (6.3.29) to (6.3.32) reduce equations (6.3.21), (6.3.22), (6.3.24) and (6.3.25) to

$$\sigma''=\frac{\sqrt{3}}{[H^{1/2}]_{\mu=-1}}\sigma_3 \qquad \varepsilon''=\frac{2}{\sqrt{3}[N^{1/2}]_{v=-1}}\gamma_1$$
$$\bar{\sigma}''=\sqrt{3}\,\tau_1=\sqrt{3}\sigma_3 \qquad \bar{e}''=\frac{2}{\sqrt{3}}\frac{[\bar{G}]_{v=0}}{G_0}\gamma_1 \tag{6.3.33}$$

which can be entered into equations (6.3.20) and (6.3.23) to give

$$\tau_1=2[\bar{G}]_{v=0}\gamma_1 \qquad (\tau_1=\sigma_3). \tag{6.3.34}$$

For infinitely small strains, equation (6.3.34) gives,

$$\lim_{\lambda_3\to 1}\frac{\tau_1}{2\gamma_1}=[\bar{G}]_{\lambda_3=1}=G_0. \tag{6.3.35}$$

Since the proper numbers c_i of $\mathbf{C}$ are identical to the proper numbers b_i of $\mathbf{B}$, it follows from the identity $b_i=\lambda_i^2=c_i$ that for this simple mode of deformation, **B** and **C** are indistinguishable and hence from equation (6.2.49) it is concluded that the rotation is zero. Thus, this simple mode of deformation is an example of *pure shear*.

(ii) $v=-1$. Solving equation (6.3.6) for $v=-1$ gives

$$\lambda_1=\lambda_2=\lambda_3^{-1/2} \tag{6.3.36}$$

where use has been made of equation (6.3.2), it being noted that equation

(6.3.36) is just equation (6.1.51) which for an incompressible material is a characteristic of simple extension. Entering the condition of equation (6.3.36) into equations (6.3.4), (6.3.5), (6.3.7) and (6.3.8) gives for all λ_3,

$$I_{\mathbf{B}}=\frac{2}{\lambda_3}+\lambda_3^2 \qquad II_{\mathbf{B}}=2\lambda_3+\frac{1}{\lambda_3^2} \qquad \Omega=1$$

$$\varphi_1=2\frac{[\beta_1(2\lambda_3^3+3\lambda_3+1)-\beta_{-1}(\lambda_3^3+3\lambda_3^2+2)]}{3\bar{G}(\lambda_3+1)(\lambda_3^2+1)} \tag{6.3.37}$$

$$\varphi_2=\frac{8}{\bar{G}}\frac{(\beta_1+\beta_{-1})\lambda_3^2}{(\lambda_3+1)^2(\lambda_3^2+1)} \qquad \Psi=\frac{2(\beta_1\lambda_3-\beta_{-1})}{\bar{G}(\lambda_3+1)}=N$$

$$\bar{\varepsilon}_1'=\bar{\varepsilon}_2'=-\tfrac{1}{2}\bar{\varepsilon}_3' \qquad \bar{\varepsilon}_3'=(\lambda_3+1)(\lambda_3^3-1)/6\lambda_3^2 \tag{6.3.38}$$

where use has been made of equations (4.3.8), (4.3.9) and (4.3.11). The conditions of equation (6.3.37) when entered into equation (4.2.47), give $\bar{\nu}=-1$, and hence, $\bar{\mu}\equiv\bar{\nu}=\nu=-1$, where use has been made of the generalised Lode relation of equation (4.2.38). Thus, for equation (6.3.27) to be satisfied (i.e. for $\mu=-1$), equation (6.3.10) gives

$$\sigma_2=0 \qquad \sigma_3\geqslant 0 \qquad (\mu=-1, \sigma_1=0) \tag{6.3.39}$$

from which it follows that the simple mode of deformation for which equation (6.3.27) is satisfied, with $\mu=-1$, is an example of simple extension. The conditions $\mu=-1$, $\bar{\mu}=-1$, require that

$$\omega=1 \qquad \Lambda=\delta_1+\tfrac{1}{3}\delta_2\sigma_3=H. \tag{6.3.40}$$

The conditions of equations (6.3.37) to (6.3.40) reduce equations (6.3.21), (6.3.22), (6.3.24) and (6.3.25) to

$$\sigma''=\sigma_3 \qquad \varepsilon''=\bar{\varepsilon}_3'=(\lambda_3+1)(\lambda_3^3-1)/6\lambda_3^2$$

$$\bar{\sigma}''=\sigma_3 \qquad \bar{e}''=\left[\frac{\bar{G}}{G_0}\frac{\Psi}{\Lambda}\right]_{\mu=\nu=-1}\bar{\varepsilon}_3' \tag{6.3.41}$$

which can be entered into equations (6.3.20) and (6.3.23) to give

$$\sigma_3=3\left[\bar{G}\frac{\Psi}{\Lambda}\right]_{\mu=-1}\bar{\varepsilon}_3'=\bar{E}\bar{\varepsilon}_3' \tag{6.3.42}$$

where

$$\bar{E}=3\left[\bar{G}\frac{\Psi}{\Lambda}\right]_{\mu=\nu=-1}. \tag{6.3.43}$$

For infinitely small strains, equation (6.3.42) gives,

$$\lim_{\lambda_3\to 1}\frac{\sigma_3}{\bar{\varepsilon}_3'}=3\left[\bar{G}\frac{\Psi}{\Lambda}\right]_{\lambda_3=1}=[\bar{E}]_{\lambda_3=1}=3G_0\left[\frac{\Psi}{\Lambda}\right]_{\lambda_3=1}=E_0 \tag{6.3.44}$$

having noted that for the ground state ($\Psi/\Lambda=1$), and for the above condition, equation (6.3.44) is just the stress–strain relation of classical infinitesimal elasticity for simple tension. The associated condition $E_0=3G_0$ is taken to imply that the $\bar{E}$ defined by equation (6.3.43) can be regarded as a generalised form of Young's modulus. Equation (6.3.42) implies that the universal stress–strain relation of equation (6.3.20) has been normalised to the stress–strain relation in simple tension (or simple compression). It is to be noted that $\bar{\varepsilon}_3'$ is identical with the $\bar{\varepsilon}_{zz}$ defined by equation (6.1.59). Since $\bar{\varepsilon}_3'(\lambda_3)=-\bar{\varepsilon}_3'(1/\lambda_3)$, the material's response in simple tension and simple compression has equal but opposite effects.

(iii) $v=-3$. Solving equation (6.3.6) for this value of v gives

$$\lambda_2=1 \qquad \lambda_3=\lambda_1^{-1} \tag{6.3.45}$$

where use has been made of equation (6.3.2). Entering the condition of equation (6.3.45) into equations (6.3.4), (6.3.5), (6.3.7) and (6.3.8) gives

$$\begin{gathered} I_{\mathbf{B}}=II_{\mathbf{B}}=\lambda_3^2+\lambda_3^{-2}+1 \\ \Omega=0 \qquad \varphi_1=1 \qquad \varphi_2=0 \qquad \Psi=N=1 \end{gathered} \tag{6.3.46}$$

$$\bar{\varepsilon}_1'=-\bar{\varepsilon}_3' \qquad \bar{\varepsilon}_2'=0 \qquad \bar{\varepsilon}_3'=(\lambda_3^4-1)/4\lambda_3^2=2\gamma_1 \tag{6.3.47}$$

where use has been made of equations (4.3.8), (4.3.9) and (4.3.11). The conditions of equation (6.3.46), when entered into equation (4.2.47), give $\bar{v}=-3$ and hence, $\bar{\mu}\equiv\bar{v}=v=0$, where use has been made of the generalised Lode relation of equation (4.2.38). Thus, for equation (6.3.27) to be satisfied (i.e. for $\mu=-3$), equation (6.3.10) gives

$$\sigma_3=2\sigma_2 \qquad (\mu=-3,\ \sigma_1=0) \tag{6.3.48}$$

for all λ_3. The condition of equation (6.3.48) is independent of both the loading coefficients δ_i ($i=0,1,2$) and the material response coefficients φ_i ($i=0,1,2$) and is therefore a *universal relation* (equation (6.3.48) is just equation (6.2.74)). It is evident that this simple mode of deformation is the pure shear mode discussed in §6.2. The conditions, $\mu=-3$, $\bar{\mu}=-3$, require that

$$\delta_0=0 \qquad \delta_1=1 \qquad \delta_2=0 \qquad H=1 \qquad \Lambda=1 \qquad \omega=0. \tag{6.3.49}$$

The conditions of equations (6.3.46) to (6.3.49) reduce equations (6.3.21), (6.3.22), (6.3.24) and (6.3.25) to

$$\begin{aligned} \sigma''&=\frac{\sqrt{3}}{2[H^{1/2}]_{\mu=-1}}\sigma_3 & \varepsilon''&=\frac{4}{\sqrt{3}[N^{1/2}]_{v=-3}}\gamma_1 \\ \bar{\sigma}''&=\tfrac{1}{2}\sqrt{3}\,\sigma_3 & \bar{e}''&=\frac{4}{\sqrt{3}}\frac{[\bar{G}]_{v=-3}}{G_0}\gamma_1 \end{aligned} \tag{6.3.50}$$

which can be entered into equations (6.3.20) and (6.3.23) to give

$$\tau_1 = 2[\bar{G}]_{\nu=-3}\gamma_1 \qquad (\tau_1 = \tfrac{1}{4}\sigma_3). \tag{6.3.51}$$

For infinitely small strains, equation (6.3.51) gives

$$\lim_{\lambda_3 \to 1} \frac{\tau_1}{2\gamma_1} = [\bar{G}]_{\lambda_3=1} = G_0 \tag{6.3.52}$$

where G_0 is the ground state, or classical shear modulus.

(iv) $\nu = -\infty$. Solving equation (6.3.6) for this value of ν gives,

$$\lambda_2 = \lambda_3 = \lambda_1^{-1/2} \tag{6.3.53}$$

where use has been made of equation (6.3.2). Entering the condition of equation (6.3.53) into equations (6.3.4), (6.3.5), (6.3.7) and (6.3.8) gives

$$I_{\mathbf{B}} = 2\lambda_3^2 + \frac{1}{\lambda_3^4} \qquad II_{\mathbf{B}} = \frac{2}{\lambda_3^2} + \lambda_3^4 \qquad \Omega = 1$$

$$\varphi_1 = 2\frac{[(2+3\lambda_3^4+\lambda_3^6)\beta_1 - (1+3\lambda_3^2+2\lambda_3^6)\beta_{-1}]}{3\bar{G}(\lambda_3^2+1)(\lambda_3^4+1)} \tag{6.3.54}$$

$$\varphi_2 = \frac{8}{\bar{G}}\frac{(\beta_1+\beta_{-1})\lambda_3^4}{(\lambda_3^2+1)^2(\lambda_3^4+1)} \qquad \Psi = \frac{2(\beta_1\lambda_3^4 - \beta_{-1})}{\bar{G}(\lambda_3^4+1)} = N$$

$$\bar{\varepsilon}_2' = \bar{\varepsilon}_3' = -\tfrac{1}{2}\bar{\varepsilon}_1' \qquad \bar{\varepsilon}_1' = -\frac{(\lambda_3^2+1)(\lambda_3^6-1)}{6\lambda_3^4} \tag{6.3.55}$$

where use has been made of equations (4.3.8), (4.3.9) and (4.3.11). The conditions of equation (6.3.54), when entered into equation (4.2.47), give $\bar{\nu} = -\infty$, and hence

$$\bar{\mu} \equiv \bar{\nu} = \nu = -\infty$$

where use has been made of the generalised Lode relation of equation (4.2.38). Thus, for equation (6.3.27) to be satisfied (i.e. for $\mu = -\infty$), equation (6.3.10) must give

$$\sigma_2 = \sigma_3 \qquad (\mu = -\infty, \sigma_1 = 0) \tag{6.3.56}$$

which is just the condition for equibiaxial extension, having noted that for this simple mode of deformation we must also have $\lambda_2 = \lambda_3$ which is just the condition of equation (6.3.53). The conditions, $\mu = -\infty$, $\bar{\mu} = -\infty$, require that

$$\omega = 1 \qquad \Lambda = \delta_1 + \tfrac{2}{3}\delta_2\sigma_3 = H. \tag{6.3.57}$$

For equibiaxial extension it is necessary to rearrange equations (6.3.21), (6.3.22), (6.3.24) and (6.3.25) into the form

$$\sigma'' = \frac{[(\sigma_3-\sigma_2)^2 + \sigma_3\sigma_2]^{1/2}H^{1/2}}{[H^{1/2}]_{\mu=-1}} \qquad \varepsilon'' = \frac{[(\bar{\varepsilon}_3' - \bar{\varepsilon}_2')^2 + 3(\bar{\varepsilon}_3' + \bar{\varepsilon}_2')^2]^{1/2}N^{1/2}}{\sqrt{3}[N^{1/2}]_{\nu=-1}} \tag{6.3.58}$$

$$\bar{\sigma}''=[(\sigma_3-\sigma_2)^2+\sigma_3\sigma_2]^{1/2} \qquad \bar{e}''=\frac{\bar{G}}{G_0}\left[\frac{\Psi(3+\mu\bar{\mu})}{H\Lambda(3+\nu\bar{\nu})}\right]^{1/2}\varepsilon''. \tag{6.3.59}$$

The conditions of equations (6.3.54) to (6.3.57) reduce equations (6.3.58) and (6.3.59) to

$$\sigma''=\frac{[H^{1/2}]_{\mu=-\infty}}{[H^{1/2}]_{\mu=-1}}\sigma_3 \qquad \varepsilon''=2\frac{[N^{1/2}]_{\nu=-\infty}}{[N^{1/2}]_{\nu=-1}}\bar{\varepsilon}_3'$$
$$\bar{\sigma}''=\sigma_3 \qquad \bar{e}''=2\left[\frac{\bar{G}\Psi}{G_0\Lambda}\right]_{\nu=-\infty}\bar{\varepsilon}_3' \tag{6.3.60}$$

which can be entered into equations (6.3.20) and (6.3.23) to give

$$\sigma_3=6\left[\bar{G}\frac{\Psi}{\Lambda}\right]_{\nu=-\infty}\varepsilon_{\mathrm{E}} \qquad \varepsilon_{\mathrm{E}}=\frac{(\lambda_3^2+1)(\lambda_3^6-1)}{12\lambda_3^4}\ (\equiv\bar{\varepsilon}_3'). \tag{6.3.61}$$

For the material response in equibiaxial compression and equibiaxial tension to have equal but opposite effect it is necessary that $\varepsilon_{\mathrm{E}}(\lambda_3)=-\varepsilon_{\mathrm{E}}(1/\lambda_3)$ a condition which equation (6.3.61) satisfies, and hence ε_{E} is an appropriate measure of simple equibiaxial tensile and compressive strains.

For the above simple modes of deformation the ratio $[3N^{1/2}/H^{1/2}]_{\mu=\nu=-1}$, appearing in equation (6.3.20), is a constant scaling factor. Hence, if the non-negative scalar factor of proportionality G, defined in equation (6.3.20), is solely a function of the effective strain ε'', then given a form for the stress and strain intensity functions, H and N, the four separate stress–strain relations obtained using equations (6.3.20), (6.3.33), (6.3.41), (6.3.50) and (6.3.60) should be indistinguishable, thus giving a single, universal, $(\sigma'', \varepsilon'')$ relation for an incompressible isotropic material.

As an example, consider the stress intensity function to be of the form of the modified Tresca type, for which equation (3.3.10) gives

$$H=\cos^2[\tfrac{1}{6}\cos^{-1}(1-2a\omega)] \qquad (0\leqslant a\leqslant 1) \tag{6.3.62}$$

where a is a constant characteristic of material properties.

The form for N is determined by whether

$$[\bar{\varepsilon}_3']_Y/[\gamma_1]_k>2/\sqrt{3}$$

for a given value of σ'', or whether

$$[\bar{\varepsilon}_3']_Y/[\gamma_1]_k<2/\sqrt{3}.$$

As an example, consider

$$N=\tfrac{4}{3}\cos^2[\tfrac{1}{6}\cos^{-1}(2a\Omega-1)] \qquad \text{for} \quad [\bar{\varepsilon}_3']_Y/[\gamma_1]_k<2/\sqrt{3} \tag{6.3.63}$$

where $\bar{\varepsilon}_3'$ is defined by equation (6.3.38) and γ_1 is defined by equation (6.3.30).

Entering the forms for H and N given by equations (6.3.62) and (6.3.63) into equations (6.3.33), (6.3.41), (6.3.50) and (6.3.60) gives

(i) $v=0$:

$$p''=\frac{\sqrt{3}\sigma_2}{\cos[\frac{1}{6}\cos^{-1}(1-2a)]}=2\sqrt{3}\frac{\cos[\frac{1}{6}\cos^{-1}(2a-1)]}{\cos[\frac{1}{6}\cos^{-1}(1-2a)]}G \times\frac{\gamma_1}{\cos[\frac{1}{6}\cos^{-1}(2a-1)]}. \tag{6.3.64}$$

(ii) $v=-1$:

$$\sigma''=\sigma_3=2\sqrt{3}\frac{\cos[\frac{1}{6}\cos^{-1}(2a-1)]}{\cos[\frac{1}{6}\cos^{-1}(1-2a)]}G\bar{\varepsilon}_3'. \tag{6.3.65}$$

(iii) $v=-3$:

$$\sigma''=\frac{\sqrt{3}\,\sigma_3}{2\cos[\frac{1}{6}\cos^{-1}(1-2a)]}=2\sqrt{3}\frac{\cos[\frac{1}{6}\cos^{-1}(2a-1)]}{\cos[\frac{1}{6}\cos^{-1}(1-2a)]}G \times\frac{2\gamma_1}{\cos[\frac{1}{6}\cos^{-1}(2a-1)]}. \tag{6.3.66}$$

(iv) $v=-\infty$:

$$\sigma''=\sigma_3=2\sqrt{3}\frac{\cos[\frac{1}{6}\cos^{-1}(2a-1)]}{\cos[\frac{1}{6}\cos^{-1}(1-2a)]}G2\bar{\varepsilon}_3' \tag{6.3.67}$$

where use has been made of equation (6.3.20).

6.4 NONHOMOGENEOUS DEFORMATIONS

The constitutive equation (6.1.1) for an incompressible body $\mathscr{B}$ has the reduced form

$$\mathbf{G}(\mathbf{T})=-P\mathbf{I}+\beta_1\mathbf{B}+\beta_{-1}\mathbf{B}^{-1} \tag{6.4.1}$$

where the response coefficients

$$\beta_a=\beta_a(I_{\mathbf{B}}, II_{\mathbf{B}}) \qquad (a=\pm 1) \tag{6.4.2}$$

are functions only of

$$I_{\mathbf{B}}=II_{\mathbf{B}^{-1}} \qquad II_{\mathbf{B}}=I_{\mathbf{B}^{-1}} \tag{6.4.3}$$

since for an incompressible material $III_{\mathbf{B}}=1$. Entering the form for $\mathbf{G}$ given by equation (6.4.1) into equation (6.1.8) gives

$$\operatorname{div}(\beta_1\mathbf{B}+\beta_{-1}\mathbf{B}^{-1})=\operatorname{grad} P \tag{6.4.4}$$

it being noted that P is related to the deformation only through this condition.

Thus, given **B** it is evident that P can always be adjusted to satisfy some particular condition specified by the nature of the mode of deformation. In order for a scalar field P satisfying equation (6.4.4) to exist, it is necessary that

$$\operatorname{curl}\operatorname{div}(\beta_1\mathbf{B}+\beta_{-1}\mathbf{B}^{-1})=\boldsymbol{0}. \tag{6.4.5}$$

Setting grad $P=\boldsymbol{s}$ in equation (6.4.4) gives

$$\boldsymbol{s}=\operatorname{div}(\beta_1\mathbf{B}+\beta_{-1}\mathbf{B}^{-1}). \tag{6.4.6}$$

Using equation (1.9.9), the gradient of $\boldsymbol{s}$ can be expressed in the form

$$\begin{aligned} s_{i,p}=&b^1_{ip}\frac{\partial^2\beta_1}{\partial I^2_{\mathbf{B}}}+b^2_{ip}\frac{\partial^2\beta_1}{\partial I_{\mathbf{B}}\,\partial II_{\mathbf{B}}}+b^3_{ip}\frac{\partial^2\beta_1}{\partial II^2_{\mathbf{B}}}+b^4_{ip}\frac{\partial\beta_1}{\partial I_{\mathbf{B}}}\\ &+b^5_{ip}\frac{\partial\beta_1}{\partial II_{\mathbf{B}}}+b^6_{ip}\beta_1+b^7_{ip}\frac{\partial^2\beta_{-1}}{\partial I^2_{\mathbf{B}}}+b^8_{ip}\frac{\partial^2\beta_{-1}}{\partial I_{\mathbf{B}}\,\partial II_{\mathbf{B}}}\\ &+b^9_{ip}\frac{\partial^2\beta_{-1}}{\partial II^2_{\mathbf{B}}}+b^{10}_{ip}\frac{\partial\beta_{-1}}{\partial I_{\mathbf{B}}}+b^{11}_{ip}\frac{\partial\beta_{-1}}{\partial II_{\mathbf{B}}}+b^{12}_{ip}\beta_{-1}. \end{aligned} \tag{6.4.7}$$

Here

$$\begin{aligned} b^1_{ip}&=I_{\mathbf{B},q}I_{\mathbf{B},p}B^q{}_i\\ b^2_{ip}&=(I_{\mathbf{B},q}II_{\mathbf{B},p}+III_{\mathbf{B},q}I_{\mathbf{B},p})B^q{}_i\\ b^3_{ip}&=II_{\mathbf{B},q}II_{\mathbf{B},p}B^q{}_i\\ b^4_{ip}&=I_{\mathbf{B},p}B^q{}_{i,q}+(I_{\mathbf{B},q}B^q{}_i)_{,p}\\ b^5_{ip}&=II_{\mathbf{B},p}B^q{}_{i,q}+(II_{\mathbf{B},q}B^q{}_i)_{,p}\\ b^6_{ip}&=B^q{}_{i,qp} \end{aligned} \tag{6.4.8}$$

and where $b^7_{ip},\ldots,b^{12}_{ip}$ are the same as $b^1_{ip},\ldots,b^6_{ip}$ respectively, except that **B** is replaced by $\mathbf{B}^{-1}$. For the deformation tensor **B** to correspond to a universal solution, the condition of equation (6.4.5) must hold for all choices of β_1 and β_{-1}. Hence from equations (6.4.5) to (6.4.8) the governing equations for the field **B** are

$$b^\Delta_{ip}=b^\Delta_{pi}\qquad \Delta=1,\ldots,12 \tag{6.4.9}$$

having noted that **B** must also satisfy the constraint

$$\det\mathbf{B}=1 \tag{6.4.10}$$

together with the compatibility condition of equation (6.1.22).

The governing equations (6.4.9), (6.4.10) and (6.1.22) form a highly over-determined system for the deformation tensor field **B**. Following the extensive studies of Rivlin (1947, 1948a,b, 1949a,b) and Rivlin and Saunders (1951), Ericksen (1954) established the existence of four distinct groups of inhomogeneous deformations, and has concluded that these, together with the isochoric homogeneous deformations, and the possible existence of two very restricted groups, constitute all possible solutions. One of the possible, restricted class of solutions has been shown by Morris and Shiau (1970) to be empty. With regard to the remaining restricted class of solutions, see for example Klingbeil and Shield (1966) and Syngh and Pipkin (1965). However, for certain classes of deformations, it has been proved that the known groups of solutions are complete (see for example, Fosdick and Schuler 1969, Kafadar 1972, Müller 1970).

The present discussion of nonhomogeneous deformations will be restricted to static, universal solutions for incompressible bodies in the context of a cylindrical coordinate system.

6.5 EXTENSION, INFLATION AND TORSION OF AN INCOMPRESSIBLE CIRCULAR CYLINDRICAL TUBE

6.5.1 The Stress Ratios

A classical problem of nonlinear elasticity theory is the combined extension and torsion of an incompressible solid rod or cylindrical tube. This is a typical case of a nonhomogeneous deformation.

Consider a right-circular tube or solid rod of incompressible material. Let (R, Θ, Z) be the cylindrical referential coordinates in the initial state of a particle that is located in the deformed configuration by the cylindrical spatial coordinates (r, θ, z).

For a material body $\mathscr{B}$ which is permanently at rest the acceleration is zero and the equation (2.2.81) of motion reduces to the equilibrium equation

$$\operatorname{div} \mathbf{T} + \rho \boldsymbol{b} = \boldsymbol{0}$$

which in the absence of body forces ($\boldsymbol{b} = \boldsymbol{0}$), reduces to

$$\operatorname{div} \mathbf{T} = \boldsymbol{0}. \tag{6.5.1}$$

The equations of equilibrium in terms of the physical components $T\langle ij\rangle$ (cf §1.7.4) can be expressed in the form

$$\begin{aligned}
\frac{\partial T\langle rr\rangle}{\partial r} + \frac{1}{r}\frac{\partial T\langle r\theta\rangle}{\partial \theta} + \frac{\partial T\langle rz\rangle}{\partial z} + \frac{1}{r}(T\langle rr\rangle - T\langle \theta\theta\rangle) &= 0 \\
\frac{\partial T\langle r\theta\rangle}{\partial r} + \frac{1}{r}\frac{\partial T\langle \theta\theta\rangle}{\partial \theta} + \frac{\partial T\langle \theta z\rangle}{\partial z} + \frac{2}{r}T\langle r\theta\rangle &= 0 \\
\frac{\partial T\langle rz\rangle}{\partial r} + \frac{1}{r}\frac{\partial T\langle \theta z\rangle}{\partial \theta} + \frac{\partial T\langle zz\rangle}{\partial z} + \frac{1}{r}T\langle rz\rangle &= 0.
\end{aligned} \tag{6.5.2}$$

The non-zero, physical components of stress associated with a tube which has its principal axis aligned with the Z axis are

$$[T\langle ij\rangle]=\begin{bmatrix}\sigma_{rr} & 0 & 0\\ 0 & \sigma_{\theta\theta} & \tau_{\theta z}\\ 0 & \tau_{z\theta} & \sigma_{zz}\end{bmatrix} \qquad (\tau_{\theta z}=\tau_{z\theta}). \tag{6.5.3}$$

The deviators σ_i' $(i=1,2,3)$ of the principal stresses σ_i are the positive solutions for σ' of the equation

$$\sigma'^3-J_2'\sigma'-J_3'=0 \tag{6.5.4}$$

where the invariants J_2' and J_3' are defined by equation (3.1.4). Equation (6.5.4) gives for $\sigma_3>\sigma_1>\sigma_2$ the relation

$$\sigma_i'=\frac{2}{\sqrt{3}}(J_2')^{1/2}\cos\left(\tfrac{1}{3}\cos^{-1}\sqrt{\omega}+\frac{2(i-1)\pi}{3}\right) \qquad (i=1,2,3) \tag{6.5.5}$$

where ω is defined by equation (3.1.12). From equation (6.5.5)

$$\sigma_1'=\zeta\tau_{\theta z} \tag{6.5.6}$$

$$\sigma_2'=-\tfrac{1}{2}\zeta\tau_{\theta z}-\tfrac{1}{2}(4+\xi^2)^{1/2}\tau_{\theta z} \tag{6.5.7}$$

$$\sigma_3'=-\tfrac{1}{2}\zeta\tau_{\theta z}+\tfrac{1}{2}(4+\xi^2)^{1/2}\tau_{\theta z} \tag{6.5.8}$$

where the stress ratios ξ and ζ are defined by the relations

$$\xi=\frac{\sigma_{zz}-\sigma_{\theta\theta}}{\tau_{\theta z}}=\frac{G'\langle zz\rangle-G'\langle\theta\theta\rangle}{G'\langle\theta z\rangle} \tag{6.5.9}$$

$$\zeta=\frac{2\sigma_{rr}-\sigma_{\theta\theta}-\sigma_{zz}}{3\tau_{\theta z}}=\frac{\mu\, G'\langle rr\rangle}{\bar{\mu}\, G'\langle\theta z\rangle} \tag{6.5.10}$$

and where, entering the values of the σ_i' $(i=1,2,3)$ given by equations (6.5.6), (6.5.7) and (6.5.8) into equation (4.1.26), Lode's (1926) stress parameter

$$\mu=\frac{3\sigma_1'}{\sigma_3'-\sigma_2'}=\frac{3\zeta}{(4+\xi^2)^{1/2}}. \tag{6.5.11}$$

With regard to equations (6.5.9) and (6.5.10), the matrix of components of the loading function $\mathbf{G}'$ follows from equations (4.2.3) and (6.5.3)

$$[G'\langle ij\rangle]=\Lambda\tau_{\theta z}\begin{bmatrix}\dfrac{\bar{\mu}}{\mu}\zeta & 0 & 0\\ 0 & -\tfrac{1}{2}\left(\xi+\dfrac{\bar{\mu}}{\mu}\zeta\right) & 1\\ 0 & 1 & \tfrac{1}{2}\left(\xi-\dfrac{\bar{\mu}}{\mu}\zeta\right)\end{bmatrix} \tag{6.5.12}$$

where Λ is defined by equation (4.2.35) and where $\bar{\mu}$ is defined by equation (4.2.32) (see also equations (3.1.16) and (4.2.34)).

6.5.2 The Strain Ratios

The simple deformations to be considered are

$$r=(\eta/F)^{1/2}R \qquad \theta=\Theta+DZ \qquad z=FZ \tag{6.5.13}$$

where

$$\eta=1+\frac{K}{R^2}=\frac{r^2}{R^2}F=\frac{v}{V} \tag{6.5.14}$$

is the ratio of the current volume v to the undeformed volume V of a cylinder of radius R, and where

$$F=l/L \tag{6.5.15}$$

is the ratio of the current length l to the undeformed length L of a tube. In equation (6.5.13), the constants D, F and K have values such that $(K+R^2)/F>0$ when R is in the interval $(R_1 \geqslant R \geqslant R_2)$, where R_1 and R_2 are the outer and inner radii respectively of an undeformed tube.

The matrix of components of the deformation gradient $\mathbf{F}$ relative to (R, Θ, Z) and (r, θ, z) follows from equation (6.5.13)

$$[F^i{}_\alpha]=\begin{bmatrix} \dfrac{R}{rF} & 0 & 0 \\ 0 & 1 & D \\ 0 & 0 & F \end{bmatrix}. \tag{6.5.16}$$

Using equation (2.1.45), the contravariant components of $\mathbf{B}$ in the cylindrical coordinate system (r, θ, z) are given by,

$$B^{ij}=F^i{}_\alpha F^j{}_\beta \delta^{\alpha\beta}. \tag{6.5.17}$$

Thus,

$$[B^{ij}]=\begin{bmatrix} \dfrac{1}{\eta F} & 0 & 0 \\ 0 & \dfrac{\eta}{r^2F}(1+R^2D^2) & DF \\ 0 & DF & F^2 \end{bmatrix}. \tag{6.5.18}$$

The covariant components of $\mathbf{B}^{-1}$ are related to the contravariant components of $\mathbf{B}$ by

$$(B^{-1})_{ip}B^{pj}=\delta^j{}_i \tag{6.5.19}$$

and hence

$$[(B^{-1})_{ij}]=\begin{bmatrix} \eta F & 0 & 0 \\ 0 & \dfrac{r^2F}{\eta} & -\dfrac{r^2}{\eta}D \\ 0 & -\dfrac{r^2}{\eta}D & \dfrac{(1+R^2D^2)}{F^2} \end{bmatrix}. \tag{6.5.20}$$

From equations (6.5.18) and (6.5.20), the principal invariants of **B** are

$$I_{\mathbf{B}}=B^{pq}g_{pq}=\frac{1}{\eta F}+\frac{\eta}{F}+F^2+\frac{\eta}{F}R^2D^2 \tag{6.5.21}$$

$$II_B=I_{\mathbf{B}^{-1}}=(B^{-1})_{pq}g^{pq}=\eta F+\frac{F}{\eta}+\frac{1}{F^2}+\frac{R^2}{F^2}D^2 \tag{6.5.22}$$

where g_{pq} and g^{pq} are the components of the metric tensor in the cylindrical coordinate system (r, θ, z). The physical components of **B** and $\mathbf{B}^{-1}$ in terms of (r, θ, z) are given by

$$B\langle pq\rangle=\sqrt{g_{pp}g_{qq}}\,B^{pq} \qquad \text{(no sum)} \tag{6.5.23}$$

$$B^{-1}\langle pq\rangle=\sqrt{g^{pp}g^{qq}}\,(B^{-1})_{pq} \qquad \text{(no sum)}. \tag{6.5.24}$$

Thus

$$[B\langle ij\rangle]=\begin{bmatrix} \dfrac{1}{\eta F} & 0 & 0 \\ 0 & \dfrac{\eta}{F}(1+R^2D^2) & rDF \\ 0 & rDF & F^2 \end{bmatrix}$$

$$[B^{-1}\langle ij\rangle]=\begin{bmatrix} \eta F & 0 & 0 \\ 0 & \dfrac{F}{\eta} & -\dfrac{rD}{\eta} \\ 0 & -\dfrac{rD}{\eta} & \dfrac{(1+R^2D^2)}{F^2} \end{bmatrix}. \tag{6.5.25}$$

From equations (4.2.20) and (6.5.25)

$$[\bar{E}\langle ij\rangle]=\begin{bmatrix} \dfrac{1-\eta^2F^2}{4\eta F} & 0 & 0 \\ 0 & \dfrac{(\eta^2-F^2)}{4\eta F}+\tfrac{1}{4}r^2D^2 & \dfrac{(\eta F+1)}{4\eta}rD \\ 0 & \dfrac{(\eta F+1)}{4\eta}rD & \dfrac{(F^4-1)}{4F^2}-\dfrac{r^2D^2}{4\eta F} \end{bmatrix}. \tag{6.5.26}$$

The components of the strain response function follow from equations (4.2.44) and (6.5.26)

$$[\bar{M}'\langle ij\rangle]=\Psi\bar{E}\langle\theta z\rangle\begin{bmatrix}\frac{\bar{v}}{v}\rho & 0 & 0\\ 0 & -\frac{1}{2}\left(\chi+\frac{\bar{v}}{v}\rho\right) & 1\\ 0 & 1 & \frac{1}{2}\left(\chi-\frac{\bar{v}}{v}\rho\right)\end{bmatrix} \qquad (6.5.27)$$

where the strain ratios χ and ρ are defined by the relations

$$\chi=\frac{\bar{E}\langle zz\rangle-\bar{E}\langle\theta\theta\rangle}{\bar{E}\langle\theta z\rangle}=\frac{\bar{M}\langle zz\rangle-\bar{M}'\langle\theta\theta\rangle}{\bar{M}'\langle\theta z\rangle}=\frac{G'\langle zz\rangle-G'\langle\theta\theta\rangle}{G'\langle\theta z\rangle}\equiv\xi$$

$$=\frac{(F+1)(F^3-1)}{(\eta F+1)DF^2}\frac{\eta}{r}-\frac{(\eta-1)(F^2+\eta)}{(\eta F+1)DF}\frac{1}{r}-\frac{D}{F}r\equiv\xi$$

$$=\frac{(F^3-1)}{DF^2}\frac{1}{r}-\frac{(\eta-1)}{DF^2}\frac{1}{\mathrm{r}}-\frac{D}{F}r\equiv\xi \qquad (6.5.28)$$

$$\rho=\frac{\bar{E}'\langle rr\rangle}{\bar{E}\langle\theta z\rangle}=\frac{2(\bar{E}\langle rr\rangle-\bar{E}\langle\theta\theta\rangle)}{3\bar{E}\langle\theta z\rangle}-\frac{1}{3}\chi=\frac{v}{\mu}\zeta$$

$$=-\frac{1}{3}\chi-\frac{2\eta F}{3(\eta F+1)}\left(\frac{(\eta^2-1)(F^2+1)}{\eta DF^2}\frac{1}{r}+\frac{D}{F}r\right) \qquad (6.5.29)$$

and where, from equations (4.2.48) and (4.3.13)

$$\Psi=\varphi_1-\varphi_2\bar{\varepsilon}_1'=\varphi_1-\varphi_2\bar{E}'\langle rr\rangle \qquad (6.5.30)$$

$$v=\frac{3\bar{\varepsilon}_1'}{\bar{\varepsilon}_3'-\bar{\varepsilon}_2'}=\frac{3\rho}{(4+\chi^2)^{1/2}} \qquad (6.5.31)$$

having noted that the deviators $\bar{\varepsilon}_i'$ $(i=1,2,3)$ of the proper numbers $\bar{\varepsilon}_i$ of $\bar{\mathbf{E}}$ are the positive solutions for $\bar{\varepsilon}'$ of the equation

$$\bar{\varepsilon}'^3-K_2'\bar{\varepsilon}'-K_3'=0 \qquad (6.5.32)$$

where K_2', K_3' are defined by equation (4.2.28). For $\bar{\varepsilon}_3>\bar{\varepsilon}_1>\bar{\varepsilon}_2$, equation (6.5.32) gives

$$\bar{\varepsilon}_i'=\frac{2}{\sqrt{3}}(K_2')^{1/2}\cos[\tfrac{1}{3}\cos^{-1}\sqrt{\Omega}+\tfrac{2}{3}(i-1)\pi] \qquad (i=1,2,3) \qquad (6.5.33)$$

where Ω is defined by equation (4.3.9). From equation (6.5.33)

$$\begin{aligned}\bar{\varepsilon}_1'&=\rho\bar{E}\langle\theta z\rangle\\ \bar{\varepsilon}_2'&=-\tfrac{1}{2}\rho\bar{E}\langle\theta z\rangle-\tfrac{1}{2}(4+\chi^2)^{1/2}\bar{E}\langle\theta z\rangle\\ \bar{\varepsilon}_3'&=-\tfrac{1}{2}\rho\bar{E}\langle\theta z\rangle+\tfrac{1}{2}(4+\chi^2)^{1/2}\bar{E}\langle\theta z\rangle\end{aligned} \qquad (6.5.34)$$

which when entered into equation (4.2.48) give equation (6.5.31).

6.5.3 The Stress–Strain Relations

The constitutive equation (6.1.1) will be used in the form of equation (4.2.44)

$$\mathbf{G}'(\mathbf{T}')=\delta_0\mathbf{I}+\delta_1\mathbf{T}'+\delta_2\mathbf{T}'^2=\mathbf{M}'(\bar{\mathbf{E}}')=2\bar{G}(\varphi_0\mathbf{I}+\varphi_1\bar{\mathbf{E}}'+\varphi_2\bar{\mathbf{E}}'^2)=2\bar{G}\bar{\mathbf{M}}'(\bar{\mathbf{E}}'). \tag{6.5.35}$$

In the context of equation (6.5.35), it is evident from §6.4 that the physical components of the loading function $\mathbf{G}'$ with respect to a cylindrical coordinate system (r, θ, z) are functions of r alone. This implies, by way of equation (4.2.18) that the physical components of the stress with respect to a cylindrical coordinate system (r, θ, z) are also functions of r alone, and hence the equations (6.5.2) of equilibrium reduce to,

$$\begin{aligned}
&\frac{\partial\sigma_{rr}}{\partial r}+\frac{1}{r}(\sigma_{rr}-\sigma_{\theta\theta})=0\\
&\frac{\partial}{\partial r}(r^2\sigma_{r\theta})-r\frac{\partial P}{\partial\theta}=0\\
&\frac{\partial}{\partial r}(r\sigma_{rz})-r\frac{\partial P}{\partial z}=0
\end{aligned} \tag{6.5.36}$$

where use has been made of the notation of equation (6.5.3).

The restricted stress system of equation (6.5.3) further reduces the equilibrium of equation (6.5.36) to

$$\frac{\partial\sigma_{rr}}{\partial r}+\frac{1}{r}(\sigma_{rr}-\sigma_{\theta\theta})=0 \tag{6.5.37}$$

$$\partial P/\partial\theta=0 \qquad \partial P/\partial z=0 \tag{6.5.38}$$

having noted that all the quantities σ_{rr}, $\sigma_{\theta\theta}$, σ_{zz}, $\tau_{\theta z}$ and P are functions of r only.

Using equation (6.5.37), the stress relations can be obtained from equation (4.2.27)

$$\sigma_{rr}=\int_{r_2}^{r}[(\sigma_{\theta\theta}-\sigma_{rr})/r]\,\mathrm{d}r+[\sigma_{rr}]_{r=r_2} \tag{6.5.39}$$

$$\sigma_{\theta\theta}=\sigma_{rr}+\frac{1}{\Lambda}[G'\langle\theta\theta\rangle-G'\langle rr\rangle+\tfrac{3}{2}(1-\mu/\bar{\mu})G'\langle rr\rangle] \tag{6.5.40}$$

$$\sigma_{zz}=\sigma_{rr}+\frac{1}{\Lambda}[G'\langle zz\rangle-G'\langle rr\rangle+\tfrac{3}{2}(1-\mu/\bar{\mu})G'\langle rr\rangle] \tag{6.5.41}$$

$$\tau_{\theta z}=\frac{1}{\Lambda}G'\langle\theta z\rangle \tag{6.5.42}$$

$$\tau_{rz}=0 \qquad \tau_{r\theta}=0 \tag{6.5.43}$$

where, from equations (4.2.35) and (6.5.6)

$$\Lambda=\delta_1-\delta_2\sigma'_{rr}=H\left(1-9\frac{(1-\mu^2)}{(9-\mu^2)}\frac{\omega}{H}\frac{\mathrm{d}H}{\mathrm{d}\omega}\right) \tag{6.5.44}$$

and where μ and $\bar{\mu}$ are defined by equations (6.5.11) and (4.2.32).

Entering the form for the components of

$$\mathbf{G}'=\mathbf{M}'=2\bar{G}\bar{\mathbf{M}}' \tag{6.5.45}$$

obtained by use of equations (6.5.45) and (6.5.27) into equations (6.5.40), (6.5.41) and (6.5.42) gives

$$\sigma_{\theta\theta}=\sigma_{rr}-\frac{3}{2}\left[\bar{G}\frac{\Psi}{\Lambda}\right]\bar{\varepsilon}_{zz}+\left[\bar{G}\frac{\Psi}{\Lambda}\right]\left(\frac{(\eta-1)(F^2+\eta)}{4\eta F}+\frac{(\eta F+1)r^2D^2}{4\eta F}\right)$$
$$-3\left[\bar{G}\frac{\Psi}{\Lambda}\right]\frac{\mu}{v}\rho\bar{E}\langle\theta z\rangle \tag{6.5.46}$$

$$\sigma_{zz}=\sigma_{rr}+\frac{3}{2}\left[\bar{G}\frac{\Psi}{\Lambda}\right]\bar{\varepsilon}_{zz}-\left[\bar{G}\frac{\Psi}{\Lambda}\right]\left(\frac{(\eta-1)(F^2+\eta)}{4\eta F}+\frac{(\eta F+1)r^2D^2}{4\eta F}\right)$$
$$-3\left[\bar{G}\frac{\Psi}{\Lambda}\right]\frac{\mu}{v}\rho\bar{E}\langle\theta z\rangle \tag{6.5.47}$$

$$\tau_{\theta z}=2\left[\bar{G}\frac{\Psi}{\Lambda}\right]\frac{(\eta F+1)}{4\eta}rD=\left[\bar{G}\frac{\Psi}{\Lambda}\right]\gamma_{\theta z} \tag{6.5.48}$$

where

$$\bar{\varepsilon}_{zz}=\frac{(F+1)(F^3-1)}{6F^2} \qquad \gamma_{\theta z}=\frac{(\eta F+1)}{2\eta}rD. \tag{6.5.49}$$

The resultant torque

$$\Gamma=2\pi\int_{r_2}^{r_1}r^2\tau_{\theta z}\,\mathrm{d}r$$
$$=\pi D\int_{r_2}^{r_1}\left(\bar{G}\frac{\Psi}{\Lambda}\right)\frac{(\eta F+1)}{\eta}r^3\,\mathrm{d}r \tag{6.5.50}$$

where r_1 and r_2 are the outer and inner radii respectively of the deformed tube.

6.5.4 Extension and Inflation of a Thin-Walled Tube

If the outer surface of the tube is free from surface traction, the inflating pressure P_i on the inner surface can be obtained from equation (6.5.39) in the form,

$$P_i = -[\sigma_{rr}]_{r=r_2} = \int_{r_2}^{r_1} (\sigma_{\theta\theta} - \sigma_{rr}) \frac{dr}{r} \tag{6.5.51}$$

where the stress difference $\sigma_{\theta\theta} - \sigma_{rr}$ is given by equation (6.5.46).

Of particular interest is the simultaneous extension and inflation of a thin-walled cylindrical tube for the particular conditions

$$D=0 \qquad \tau_{\theta z}=0 \tag{6.5.52}$$

which reduce equations (6.5.46) and (6.5.47) to

$$\sigma_{\theta\theta} - \sigma_{rr} = \left[\bar{G}\frac{\Psi}{\Lambda}\right]\left[\left(\frac{\mu}{\nu}-1\right)\frac{(F^4-1)}{4F^2} + \left(1+\frac{\mu}{\nu}\right)\frac{(\eta^2-F^2)}{4\eta F} + 2\frac{\mu}{\nu}\frac{(\eta^2F^2-1)}{4\eta F}\right] \tag{6.5.53}$$

$$\sigma_{zz} - \sigma_{rr} = \left[\bar{G}\frac{\Psi}{\Lambda}\right]\left[\left(\frac{\mu}{\nu}+1\right)\frac{(F^4-1)}{4F^2} - \left(1-\frac{\mu}{\nu}\right)\frac{(\eta^2-F^2)}{4\eta F} + 2\frac{\mu}{\nu}\frac{(\eta^2F^2-1)}{4\eta F}\right]. \tag{6.5.54}$$

The conditions of equations (6.5.52) reduce equation (6.5.11) to

$$\mu = -1 - 2\frac{(\sigma_{\theta\theta} - \sigma_{rr})}{(\sigma_{zz} - \sigma_{\theta\theta})}. \tag{6.5.55}$$

Substituting for $\sigma_{zz} - \sigma_{\theta\theta}$ and $\sigma_{\theta\theta} - \sigma_{rr}$ in equation (6.5.55) using equations (6.5.53) and (6.5.54) gives

$$\nu = -1 - 2\frac{(\eta^2 - F^2) + (\eta^2F^2 - 1)}{\eta(F^4-1) - F(\eta^2 - F^2)} F. \tag{6.5.56}$$

The inflating pressure can be adjusted so that at some particular radius $r=r_a$, the condition $[r/R]_{r=r_a}=1$ can be maintained for all F. This condition reduces equation (6.5.14) to

$$\eta = F \qquad (D=0,\ r_a/R_a=1).$$

Entering the condition $\eta=F$ into equation (6.5.56) gives $\nu=-3$ and hence from equation (6.3.27),

$$\mu = \bar{\mu} \equiv \bar{\nu} = \nu = -3 \qquad (D=0,\ r_a/R_a=1).$$

for all F.

For a large diameter tube with a sufficiently thin wall, the condition $[r/R]_{r=r_a}=1$ for all F implies that $(r/R \simeq 1)$ for all r in the interval $(r_2 \leqslant r \leqslant r_1)$. The approximation $(r/R \simeq 1)$ for all F reduces equation (6.5.14) to

$$\eta \simeq F \qquad (D=0,\ r/R \simeq 1) \tag{6.5.57}$$

for all F. Entering the conditions $(\eta = F)$, $(\mu/\nu = 1)$ into equations (6.5.53) and

(6.5.54) gives

$$\sigma_{\theta\theta}-\sigma_{rr}=2\left[\bar{G}\frac{\Psi}{\Lambda}\right]_{v=-3}\frac{(F^4-1)}{4F^2}\qquad \sigma_{zz}-\sigma_{rr}=4\left[\bar{G}\frac{\Psi}{\Lambda}\right]_{v=-3}\frac{(F^4-1)}{4F^2}\qquad (r/R\simeq 1)\quad (6.5.58)$$

from which it follows that

$$\sigma_{zz}-\sigma_{rr}=2(\sigma_{\theta\theta}-\sigma_{rr})\qquad (r/R\simeq 1)\qquad (6.5.59)$$

for all F.

At $r=r_a$, $[\sigma_{rr}]_{r=r_a}=0$ and equation (6.5.59) reduces for all F to the *universal relation*

$$\sigma_3=2\sigma_2\qquad (\sigma_3\equiv\sigma_{zz},\sigma_2\equiv\sigma_{\theta\theta},r=r_a)\qquad (6.5.60)$$

which is just equation (6.2.74), and therefore it can be concluded that simultaneous inflation and extension of a thin-walled tube, subject to the condition $(r_a/R_a=1)$ is just a pure shear mode of deformation characterised by $v=-3$ (see §6.2).

For the condition that σ_{zz} is independent of r, the following relation can be obtained from equation (6.5.59)

$$\sigma_{zz}=2\left(\frac{R_a}{d}P_i\right)+\left(\frac{r_2}{d}\int_{r_2}^{r_1}\sigma_{rr}\frac{dr}{r}-2\frac{(R_a-r_2)}{d}P_i\right)\qquad (6.5.61)$$

where

$$d=r_1-r_2\qquad \ln(r_1/r_2)\simeq(r_1-r_2)/r_2.$$

6.5.5 Resultant Longitudinal Force on Plane End of Tube

The resultant force acting on the ends of the cylinder is

$$N_z=2\pi\int_{r_2}^{r_1}r\sigma_{zz}\,dr\qquad (6.5.62)$$

where σ_{zz} is given by equation (6.5.41). There are two cases to consider which depend upon whether or not the inner surface is free of traction.

(i) ***Inner surface free of traction.*** The condition that the cylinders $R=R_1$ and $R=R_2$ are deformed into cylinders free of traction follows from equation (6.5.39)

$$\int_{r_2}^{r_1}[(\sigma_{\theta\theta}-\sigma_{rr})/r]\,dr=0.\qquad (6.5.63)$$

This can be used to express the longitudinal force N_z in the form

$$N_z=2\pi\int_{r_2}^{r_1}r\sigma_{zz}\,dr=\tfrac{3}{2}\pi\int_{r_2}^{r_1}\tau_{\theta z}(\xi-\zeta)r\,dr.\qquad (6.5.64)$$

Substituting for ξ and ζ in equation (6.5.64) from equations (6.5.28) and (6.5.29) gives

$$N_z = \tfrac{3}{2}\pi \int_{r_2}^{r_1} \tau_{\theta z}\left(\chi - \frac{\mu}{\nu}\rho\right) r\,\mathrm{d}r \tag{6.5.65}$$

the strain ratios χ and ρ being given by equations (6.5.28) and (6.5.29).

(ii) ***Stress acting on inner surface.*** Equation (6.5.39) can be used to express the resultant normal force N_z in the form

$$N_z = \tfrac{1}{2}\pi \int_{r_2}^{r_1} \tau_{\theta z}\left[\left(3 - \frac{r_2^2}{r^2}\right)\xi - 3\left(1 + \frac{r_2^2}{r^2}\right)\zeta\right] r\,\mathrm{d}r \tag{6.5.66}$$

where use has been made of equation (6.5.62). Substituting for ξ and ζ in equation (6.5.66) from equations (6.5.28) and (6.5.29) gives

$$N_z = \tfrac{1}{2}\pi \int_{r_2}^{r_1} \tau_{\theta z}\left[\left(3 - \frac{r_2^2}{r^2}\right)\chi - 3\frac{\mu}{\nu}\left(1 + \frac{r_2^2}{r^2}\right)\rho\right] r\,\mathrm{d}r \tag{6.5.67}$$

the strain ratios χ and ρ being given by equations (6.5.28) and (6.5.29).

6.6 EXTENSION AND TORSION OF AN INCOMPRESSIBLE SOLID CIRCULAR CYLINDER

6.6.1 Stress–Strain Relations

For a solid rod

$$R_2 = r_2 = 0 \qquad \eta = 1 \tag{6.6.1}$$

for all D and F, where the notation is that of §6.5. These limiting conditions reduce equations (6.5.28) and (6.5.29) to

$$\chi = \frac{(F^3 - 1)}{DF^2}\frac{1}{r} - \frac{D}{F} r \equiv \xi \tag{6.6.2}$$

$$\rho = -\frac{(F^3 - 1)}{3DF^2}\frac{1}{r} + \frac{1}{3}\frac{D}{F} r - \frac{2D}{3(F+1)} r. \tag{6.6.3}$$

The limiting conditions of equation (6.6.1) reduce the stress–strain relations of equations (6.5.46), (6.5.47) and (6.5.48) to

$$\sigma_{\theta\theta} = \sigma_{rr} - \left[\bar{G}\frac{\Psi}{\Lambda}\right]\left[\frac{3}{2}\left(1 - \frac{\mu}{\nu}\right)\bar{\varepsilon}_{zz} - \frac{(F+1)r^2D^2}{4F}\left(1 + \frac{\mu}{\nu}\frac{(F-1)}{(F+1)}\right)\right] \tag{6.6.4}$$

$$\sigma_{zz} = \sigma_{rr} + \left[\bar{G}\frac{\Psi}{\Lambda}\right]\left[\frac{3}{2}\left(1 + \frac{\mu}{\nu}\right)\bar{\varepsilon}_{zz} - \frac{(F+1)r^2D^2}{4F}\left(1 - \frac{\mu}{\nu}\frac{(F-1)}{(F+1)}\right)\right] \tag{6.6.5}$$

$$\tau_{\theta z}=\left[\bar{G}\frac{\Psi}{\Lambda}\right]\gamma_{\theta z} \qquad \gamma_{\theta z}=\tfrac{1}{2}(F+1)rD. \tag{6.6.6}$$

Using equation (6.6.4), the condition of equilibrium in the radial direction, i.e. equation (6.5.37), gives

$$\sigma_{rr}=-\frac{(F+1)}{4F}D^2\int_r^{r_1}\left(1+\frac{\mu}{\nu}(2F-1)\right)\left[\bar{G}\frac{\Psi}{\Lambda}\right]r\,\mathrm{d}r$$
$$+\tfrac{3}{2}\bar{\varepsilon}_{zz}\int_r^{r_1}\left[\bar{G}\frac{\Psi}{\Lambda}\right]\left(1-\frac{\mu}{\nu}\right)\frac{\mathrm{d}r}{r}. \tag{6.6.7}$$

6.6.2 Resultant Longitudinal Force on Plane End of Solid Rod

The identity

$$\int_r^{r_1} r\sigma_{rr}\,\mathrm{d}r=\tfrac{1}{2}r_1^2[\sigma_{rr}]_{r=r_1}-\tfrac{1}{2}r^2\sigma_{rr}-\frac{1}{2}\int_r^{r_1} r^2\left(\frac{\mathrm{d}\sigma_{rr}}{\mathrm{d}r}\right)\mathrm{d}r \tag{6.6.8}$$

can be rearranged using equation (6.5.37) to give

$$\int_r^{r_1} r(\sigma_{rr}+\sigma_{\theta\theta})\,\mathrm{d}r=r_1^2[\sigma_{rr}]_{r=r_1}-r^2\sigma_{rr}. \tag{6.6.9}$$

The right-circular solid rod can be deformed by simple torsion and remain a right-circular solid rod, the deformation being achieved with zero applied traction to the outer curved surface at $r=r_1$, such that $[\sigma_{rr}]_{r=r_1}=0$ for all D. This condition reduces equation (6.6.9) to,

$$\int_r^{r_1} r(\sigma_{rr}+\sigma_{\theta\theta})\,\mathrm{d}r=-r^2\sigma_{rr} \tag{6.6.10}$$

which applies to the annular region bounded by the outer curved surface at $r=r_1$ and the surface at some arbitrary r. The condition of equilibrium in the radial direction must still apply in this annular region, and hence it follows from equations (6.5.37) and (6.6.10) that at any arbitrary r, we must have,

$$\sigma_{rr}+\sigma_{\theta\theta}=0 \tag{6.6.11}$$

for all values of D.

Entering the condition of equation (6.6.11) into equation (6.6.4) gives,

$$\sigma_{rr}=\frac{1}{2}\left[\bar{G}\frac{\Psi}{\Lambda}\right]\left[\frac{3}{2}\left(1-\frac{\mu}{\nu}\right)\bar{\varepsilon}_{zz}-\frac{(F+1)}{4F}\left(1+\frac{\mu}{\nu}\frac{(F-1)}{(F+1)}\right)r^2D^2\right] \tag{6.6.12}$$

which can, in turn, be entered into equation (6.6.5) to give

$$\sigma_{zz}=\frac{1}{4}\left[\bar{G}\frac{\Psi}{\Lambda}\right]\left[3\left(3+\frac{\mu}{\nu}\right)\bar{\varepsilon}_{zz}-\frac{(F+1)}{2F}\left(3-\frac{\mu}{\nu}\frac{(F-1)}{(F+1)}\right)r^2D^2\right]. \tag{6.6.13}$$

Entering the form for σ_{zz} given by equation (6.6.13) into equation (6.5.62) gives the resultant longitudinal force on the plane end of a solid rod in the form

$$N_z = \tfrac{3}{2}\pi\bar{\varepsilon}_{zz} \int_0^{r_1} \left(\bar{G}\frac{\Psi}{\Lambda}\right)\left(3+\frac{\mu}{\nu}\right) r\,\mathrm{d}r - \pi\frac{(F+1)}{4F} D^2 \int_0^{r_1} \left(\bar{G}\frac{\Psi}{\Lambda}\right)\left(3-\frac{\mu(F-1)}{\nu(F+1)}\right) r^3\,\mathrm{d}r. \tag{6.6.14}$$

For the von Mises type stress and strain intensity functions

$$\mu=\nu \qquad \Psi=1 \qquad \Lambda=1 \tag{6.6.15}$$

for all D and F. Entering these limiting conditions into equation (6.6.14) gives

$$N_z = 6\pi\bar{\varepsilon}_{zz} \int_0^{r_1} \bar{G} r\,\mathrm{d}r - \pi\frac{(F+2)}{2F} D^2 \int_0^{r_1} \bar{G} r^3\,\mathrm{d}r \tag{6.6.16}$$

and for infinitely small strains $\bar{G} \simeq G_0 = \text{constant}$, $F \to 1$ and equation (6.6.16) reduces to,

$$\frac{N_z}{\pi R_1^2} = 3G_0\bar{\varepsilon}_{zz} - \tfrac{3}{8}G_0 R_1^2 D^2 \tag{6.6.17}$$

where G_0 is the classical shear modulus.

6.7 STRESS–STRAIN RELATIONS FOR SIMPLE EXTENSION, SIMPLE TORSION AND PURE SHEAR

The present discussion is concerned with the formulation of a universal stress–strain relation, using the constitutive equation (6.1.1) in the form of equation (4.2.44)

$$\mathbf{G}'(\mathbf{T}') = 2\bar{G}(\varphi_0\mathbf{I} + \varphi_1\bar{\mathbf{E}}' + \varphi_2\bar{\mathbf{E}}'^2) = 2\bar{G}\bar{\mathbf{M}}'(\bar{\mathbf{E}}'). \tag{6.7.1}$$

The universal stress–strain relation will be normalised to the stress–strain relation in simple tension (or simple compression); thus, using the notation of §6.5 and 6.6

$$\sigma'' = 3\frac{[N^{1/2}]_{\nu=-1}}{[H^{1/2}]_{\mu=-1}} G\varepsilon'' \qquad G = \bar{G}\left(\frac{\Psi(3+\mu\bar{\mu})}{\Lambda(3+\nu\bar{\nu})}\right)^{1/2} \tag{6.7.2}$$

where use has been made of the generalised Lode relation of equation (4.2.38) to eliminate the terms involving $\bar{\mu}$ and $\bar{\nu}$, and where the *effective stress* is

$$\sigma'' = \frac{(3J_2'H)^{1/2}}{[H^{1/2}]_{\mu=-1}} = \frac{[H(3+\mu^2)(4+\xi^2)]^{1/2}}{2[H^{1/2}]_{\mu=-1}} \tau_{\theta z} \tag{6.7.3}$$

and the *effective strain*

$$\varepsilon''=\frac{(\frac{4}{3}K_2'N)^{1/2}}{[N^{1/2}]_{v=-1}}=\frac{[N(3+v^2)(4+\chi^2)]^{1/2}}{3[N^{1/2}]_{v=-1}}\bar{E}\langle\theta z\rangle \tag{6.7.4}$$

having noted that Λ and Ψ are defined by equations (4.2.35) and (4.3.13) respectively.

The remaining discussion is restricted to those materials for which

$$\lim_{\varepsilon''\to 0} G=G_0 \tag{6.7.5}$$

where G_0 is the classical, that is the ground state shear modulus. With this as the basis, equation (6.7.2) can be rearranged into the linear relation,

$$\bar{\sigma}''=3G_0\bar{e}'' \tag{6.7.6}$$

where the *reduced stress* is

$$\bar{\sigma}''=[(3+\mu^2)(1+\tfrac{1}{4}\xi^2)]^{1/2}\tau_{\theta z} \tag{6.7.7}$$

and where the *reduced strain*

$$\bar{e}''=\frac{1}{3}\frac{\bar{G}}{G_0}\left(\frac{\Psi N(3+\mu\bar{\mu})}{\Lambda H(3+v\bar{v})}\right)^{1/2}[(3+v^2)(1+\tfrac{1}{4}\chi^2)]^{1/2}\bar{E}\langle\theta z\rangle. \tag{6.7.8}$$

(i) ***Simple extension of a solid rod.*** For this simple mode of deformation

$$D=0,\qquad \eta=1,\qquad \tau_{\theta z}=0,\qquad \sigma_{\theta\theta}=0,\qquad \sigma_{rr}=0 \tag{6.7.9}$$

and the only non-vanishing component of stress is σ_{zz}. This mode of deformation has been discussed in §6.1.4 where it was shown that

$$\begin{aligned}&\mu=\bar{\mu}\equiv\bar{v}=v=-1 \qquad \omega=1,\qquad \Lambda=H \qquad \Omega=1 \qquad \Psi=N\\ &\bar{\varepsilon}_1'=\bar{\varepsilon}_2'=-\tfrac{1}{2}\bar{\varepsilon}_3' \qquad \bar{\varepsilon}_3'=[(F+1)(F^3-1)/6F^2]\ (\equiv\bar{\varepsilon}_{zz})\end{aligned} \tag{6.7.10}$$

for all F. The conditions of equations (6.7.9) and (6.7.10) reduce equations (6.7.3), (6.7.4), (6.7.7) and (6.7.8) to

$$\begin{aligned}&\sigma''=\sigma_{zz} \qquad \varepsilon''=\bar{\varepsilon}_{zz}\\ &\bar{\sigma}''=\sigma_{zz} \qquad \bar{e}''=\left[\frac{\bar{G}\Psi}{G_0\Lambda}\right]_{\mu=-1}\bar{\varepsilon}_{zz}\end{aligned} \tag{6.7.11}$$

which can be entered into equations (6.7.2) and (6.7.6) to give

$$\sigma_{zz}=3\left[\bar{G}\frac{\Psi}{\Lambda}\right]_{\mu=v=-1}\bar{\varepsilon}_{zz}. \tag{6.7.12}$$

(ii) ***Pure shear.*** For pure shear

$$D=0 \qquad \tau_{\theta z}=0. \tag{6.7.13}$$

This mode of deformation has been discussed in §6.5.4 where it was shown that

$$\mu=\bar{\mu}\equiv\bar{\nu}=\nu=-3 \qquad \sigma_{zz}=2\sigma_{\theta\theta}-\sigma_{rr} \tag{6.7.14}$$

which as conditions give

$$\begin{aligned} &\omega=0 \qquad \Lambda=H=1 \qquad \Omega=0 \qquad \Psi=N=1 \\ &\bar{\varepsilon}_2'=0 \qquad \bar{\varepsilon}_1'=-\bar{\varepsilon}_3'=-2\gamma_1 \qquad \gamma_1=[(F^4-1)/8F^2] \end{aligned} \tag{6.7.15}$$

for all F, having made use of equations (6.5.26) and (6.5.34). The conditions of equations (6.7.13), (6.7.14) and (6.7.15) reduce equations (6.7.3), (6.7.4), (6.7.7) and (6.7.8) to

$$\begin{aligned} &\sigma''=\frac{\sqrt{3}\,(\sigma_{zz}-\sigma_{\theta\theta})}{[H^{1/2}]_{\mu=-1}} \qquad \varepsilon''=\frac{2}{\sqrt{3}\,[N^{1/2}]_{\nu=-1}}\gamma_1 \\ &\bar{\sigma}''=\sqrt{3}\,(\sigma_{zz}-\sigma_{\theta\theta}) \qquad \bar{e}''=\frac{2}{\sqrt{3}}\frac{\bar{G}}{G_0}\gamma_1 \end{aligned} \tag{6.7.16}$$

which can be entered into equations (6.7.2) and (6.7.6) to give

$$\tfrac{1}{2}(\sigma_{zz}-\sigma_{\theta\theta})=[\bar{G}]_{\mu=\nu=-3}\gamma_1. \tag{6.7.17}$$

(iii) ***Simple torsion.*** For this mode of deformation, the effective stress and effective strain are given by equations (6.7.3) and (6.7.4) in the form

$$\sigma''=\frac{[H(3+\mu^2)(4+\chi^2)]^{1/2}}{2[H^{1/2}]_{\mu=-1}}\tau_{\theta z} \tag{6.7.18}$$

$$\varepsilon''=\frac{[N(3+\nu^2)(4+\chi^2)]^{1/2}}{6[N^{1/2}]_{\nu=-1}}\gamma_{\theta z} \tag{6.7.19}$$

where $\gamma_{\theta z}(\equiv 2\bar{E}\langle\theta z\rangle)$ is defined by equation (6.5.49) and where use has been made of the identity of equation (6.5.28). Equations (6.7.7) and (6.7.8) take the form

$$\bar{\sigma}''=[(3+\mu^2)(1+\tfrac{1}{4}\chi^2)]^{1/2}\tau_{\theta z} \tag{6.7.20}$$

$$\bar{\varepsilon}''=\frac{1}{6}\frac{\bar{G}}{G_0}\left(\frac{\Psi N(3+\mu\bar{\mu})}{\Lambda H(3+\nu\bar{\nu})}\right)^{1/2}[(3+\nu^2)(1+\tfrac{1}{4}\chi^2)]^{1/2}\gamma_{\theta z}. \tag{6.7.21}$$

Given a form for the stress intensity and strain intensity functions, the three separate stress–strain relations obtained using equations (6.7.2), (6.7.11), (6.7.16), (6.7.18) and (6.7.19) should be indistinguishable, thus giving a single, universal, $(\sigma'', \varepsilon'')$ relation for an incompressible isotropic material.

As an example, consider the stress intensity function to be of the modified Tresca type, for which it has been shown in §3.3.2 that

$$H=\cos^2[\tfrac{1}{6}\cos^{-1}(1-2a\omega)] \qquad (0\leqslant a\leqslant 1) \tag{6.7.22}$$

where a is a constant characteristic of material properties.

The form for N is determined by whether $[\bar{\varepsilon}_{zz}]_Y/[\gamma_{\theta z}]_k > 1/\sqrt{3}$ for a given value of σ'', or whether $[\bar{\varepsilon}_{zz}]_Y/[\gamma_{\theta z}]_k < 1/\sqrt{3}$. As an example, consider

$$N=\tfrac{4}{3}\cos^2[\tfrac{1}{6}\cos^{-1}(2a\Omega-1)] \quad \text{for} \quad [\bar{\varepsilon}_{zz}]_Y/[\gamma_{\theta z}]_k<1/\sqrt{3}. \tag{6.7.23}$$

Entering the forms for H and N given by equations (6.7.22) and (6.7.23) into equations (6.7.11), (6.7.16), (6.7.18) and (6.7.19) gives

(*a*) *Simple extension:*

$$\sigma''=\sigma_{zz}=2\sqrt{3}\,\frac{\cos[\frac{1}{6}\cos^{-1}(2a-1)]}{\cos[\frac{1}{6}\cos^{-1}(1-2a)]}\,G\varepsilon_{zz}. \tag{6.7.24}$$

(*b*) *Simple torsion:*

$$\begin{aligned}\sigma''&=\frac{\sqrt{3}\,[H(1+\frac{1}{3}\mu^2)(1+\frac{1}{4}\chi^2)]^{1/2}}{\cos[\frac{1}{6}\cos^{-1}(1-2a)]}\,\tau_{\theta z}\\&=2\sqrt{3}\,\frac{\cos[\frac{1}{6}\cos^{-1}(2a-1)]}{\cos[\frac{1}{6}\cos^{-1}(1-2a)]}\,G\,\frac{[N(1+\frac{1}{3}\nu^2)(1+\frac{1}{4}\chi^2)]^{1/2}}{2\cos[\frac{1}{6}\cos^{-1}(2a-1)]}\,\gamma_{\theta z}.\end{aligned} \tag{6.7.25}$$

(*c*) *Pure shear:*

$$\begin{aligned}\sigma''&=\frac{\sqrt{3}}{2\cos[\frac{1}{6}\cos^{-1}(1-2a)]}\,\sigma_{zz}\\&=2\sqrt{3}\,\frac{\cos[\frac{1}{6}\cos^{-1}(2a-1)]}{\cos[\frac{1}{6}\cos^{-1}(1-2a)]}\,G\,\frac{2\gamma_1}{\cos[\frac{1}{6}\cos^{-1}(2a-1)]}\end{aligned} \tag{6.7.26}$$

use having been made of equation (6.7.2).

6.8 BENDING OF A BLOCK INTO AN ANNULAR WEDGE

Consider a rectangular block of incompressible material for which the initial, that is the reference configuration $\boldsymbol{\kappa}$ is described in the cartesian coordinate system (X, Y, Z). The loaded, that is the spatial configuration ψ is described in the cylindrical coordinate system (r, θ, z).

The simple deformations to be considered are:

$$r=(2AX)^{1/2} \qquad \theta=BY \qquad z=Z. \tag{6.8.1}$$

For these deformations to be isochoric, it is necessary and sufficient that

$$AB=1. \tag{6.8.2}$$

The matrix of components of the deformation gradient $\mathbf{F}$ relative to

(X, Y, Z) and (r, θ, z) follows from equation (6.8.1)

$$[F^i{}_\alpha] = \begin{bmatrix} A/r & 0 & 0 \\ 0 & B & 0 \\ 0 & 0 & 1 \end{bmatrix}. \qquad (6.8.3)$$

Using equation (2.1.45), the contravariant components of **B** in the cylindrical coordinate system (r, θ, z) are given by

$$B^{ij} = F^i_\alpha F^j{}_\beta \delta^{\alpha\beta}. \qquad (6.8.4)$$

Thus

$$[B^{ij}] = \begin{bmatrix} A^2/r^2 & 0 & 0 \\ 0 & B^2 & 0 \\ 0 & 0 & 1 \end{bmatrix}. \qquad (6.8.5)$$

The covariant components of $\mathbf{B}^{-1}$ are related to the contravariant components of **B** by

$$(B^{-1})_{ip} B^{pj} = \delta^j{}_i \qquad (6.8.6)$$

and hence

$$[(B^{-1})_{ij}] = \begin{bmatrix} r^2/A^2 & 0 & 0 \\ 0 & A^2 & 0 \\ 0 & 0 & 1 \end{bmatrix}. \qquad (6.8.7)$$

From equations (6.8.5) and (6.8.7), the principal invariants of **B** are

$$I_{\mathbf{B}} = B^{pq} g_{pq} = \frac{A^2}{r^2} + \frac{r^2}{A^2} + 1 \qquad (6.8.8)$$

$$II_{\mathbf{B}} = I_{\mathbf{B}^{-1}} = (B^{-1})_{pq} g^{pq} = \frac{r^2}{A^2} + \frac{A^2}{r^2} + 1 = I_{\mathbf{B}} \qquad (6.8.9)$$

where g_{pq} and g^{pq} are the components of the metric tensor in the cylindrical coordinate system (r, θ, z). The physical components of **B** and $\mathbf{B}^{-1}$ in (r, θ, z) are given by

$$B\langle pq\rangle = (g_{pp} g_{qq})^{1/2} B^{pq} \qquad \text{(no sum)} \qquad (6.8.10)$$

$$B^{-1}\langle pq\rangle = (g^{pp} g^{qq})^{1/2} (B^{-1})_{pq} \qquad \text{(no sum).} \qquad (6.8.11)$$

Thus

$$[B\langle ij\rangle] = \begin{bmatrix} A^2/r^2 & 0 & 0 \\ 0 & r^2/A^2 & 0 \\ 0 & 0 & 1 \end{bmatrix}$$

$$[(B^{-1})\langle ij\rangle]=\begin{bmatrix} r^2/A^2 & 0 & 0 \\ 0 & A^2/r^2 & 0 \\ 0 & 0 & 1 \end{bmatrix}. \tag{6.8.12}$$

From equations (4.2.20) and (6.8.12)

$$[\bar{E}'\langle ij\rangle]=\begin{bmatrix} \dfrac{(A^4-r^4)}{4r^2A^2} & 0 & 0 \\ 0 & -\dfrac{(A^4-r^4)}{4r^2A^2} & 0 \\ 0 & 0 & 0 \end{bmatrix} \tag{6.8.13}$$

where the prime denotes a deviator.

The constitutive equation (6.1.1) will be used in the form of equation (4.2.44)

$$\mathbf{G}'(\mathbf{T}')=\delta_0\mathbf{I}+\delta_1\mathbf{T}'+\delta_2\mathbf{T}'^2=\mathbf{M}'(\bar{\mathbf{E}}')=2\bar{G}(\varphi_0\mathbf{I}+\varphi_1\bar{\mathbf{E}}'+\varphi_2\bar{\mathbf{E}}'^2)=2\bar{G}\bar{\mathbf{M}}'(\bar{\mathbf{E}}'). \tag{6.8.14}$$

In the context of equation (6.8.14), it is evident from §6.4 that, provided P is a function of r alone, the physical components of the loading function $\mathbf{G}'$ with respect to the cylindrical coordinate system (r, θ, z) are functions of r alone. This implies, by way of equation (6.8.14) that the physical components of the stress with respect to a cylindrical coordinate system (r, θ, z) are also functions of r alone, and hence for $\tau_{rz}=0$, $\tau_{\theta r}=0$, the equations (6.5.2) of equilibrium reduce to

$$\frac{\partial\sigma_{rr}}{\partial r}+\frac{1}{r}(\sigma_{rr}-\sigma_{\theta\theta})=0 \tag{6.8.15}$$

$$\frac{\partial P}{\partial\theta}=0 \qquad \frac{\partial P}{\partial z}=0 \tag{6.8.16}$$

it being noted that all the quantities σ_{rr}, $\sigma_{\theta\theta}$, σ_{zz}, $\tau_{\theta z}$ and P are functions of r only.

Using equation (6.8.15), the following stress relations can be obtained from equation (6.8.14)

$$\sigma_{rr}=\int[(\sigma_{\theta\theta}-\sigma_{rr})/r]\,\mathrm{d}r \tag{6.8.17}$$

$$\sigma_{\theta\theta}=\sigma_{rr}+\frac{1}{\Lambda}\left[G'\langle\theta\theta\rangle-G'\langle rr\rangle+\tfrac{3}{2}\left(1-\frac{\mu}{\bar{\mu}}\right)G'\langle rr\rangle\right] \quad 6.8.18)$$

$$\sigma_{zz}=\sigma_{rr}+\frac{1}{\Lambda}\left[G'\langle zz\rangle-G'\langle rr\rangle+\tfrac{3}{2}\left(1+\frac{\mu}{\bar{\mu}}\right)G'\langle rr\rangle\right] \tag{6.8.19}$$

$$\tau_{\theta z}=\frac{1}{\Lambda}G'\langle\theta z\rangle \qquad \tau_{rz}=0 \qquad \tau_{zr}=0 \tag{6.8.20}$$

where Λ is defined by equation (4.2.35), it being noted that $\sigma_1'=\sigma_{rr}'$. Equations (6.8.17) to (6.8.20) can be rearranged into the form

$$\sigma_{rr}=\int\left(\bar{G}\frac{\Psi}{\Lambda}\right)\left(1+3\frac{\mu}{\nu}\right)\frac{(r^4-A^4)}{r^2A^2}\frac{\mathrm{d}r}{r} \tag{6.8.21}$$

$$\sigma_{\theta\theta}=\sigma_{rr}-\left(\bar{G}\frac{\Psi}{\Lambda}\right)\left(1+3\frac{\mu}{\nu}\right)\frac{(A^4-r^4)}{r^2A^2} \tag{6.8.22}$$

$$\sigma_{zz}=\sigma_{rr}+\left(\bar{G}\frac{\Psi}{\Lambda}\right)\left(1-3\frac{\mu}{\bar{\mu}}\right)\frac{\bar{\nu}}{\nu}\frac{(A^4-r^4)}{r^2A^2} \tag{6.8.23}$$

$$\tau_{\theta z}=0 \tag{6.8.24}$$

where use has been made of the generalised Lode relation of equation (4.2.38).

Let the deformation carry the block into an annular wedge bounded by the cylinders $r=r_1=(2AX_1)^{1/2}$, $r=r_2=(2AX_2)^{1/2}$ and the planes $\theta=\pm\theta_0=\pm Y_0/A$, $z=\pm z_0=\pm Z_0$. For the particular case that σ_{rr} is zero at $r=r_1$, that is $[\sigma_{rr}]_{r=r_1}=0$, then σ_{rr} can be expressed by way of equation (6.8.21) in the form

$$\sigma_{rr}=\int_{r_1}^{r}\left(\bar{G}\frac{\Psi}{\Lambda}\right)\left(1+3\frac{\mu}{\nu}\right)\left(\frac{1}{A^4}-\frac{1}{r^4}\right)r\,\mathrm{d}r. \tag{6.8.25}$$

The condition that both surfaces are free of traction, that is $[\sigma_{rr}]_{r=r_1}=0$, $[\sigma_{rr}]_{r=r_2}=0$, can be obtained from equation (6.8.25) in the form

$$\int_{r_1}^{r_2}\left(\bar{G}\frac{\Psi}{\Lambda}\right)\left(1+3\frac{\mu}{\nu}\right)\left(\frac{1}{A^4}\right)r\,\mathrm{d}r=\int_{r_1}^{r_2}\left(\bar{G}\frac{\Psi}{\Lambda}\right)\left(1+3\frac{\mu}{\nu}\right)\frac{\mathrm{d}r}{r^3} \tag{6.8.26}$$

which, in principle, gives a relation from which A can be evaluated. This special case of pure bending specifically for a simple elastic material was first discussed by Rivlin (1949a,b).

REFERENCES

Ericksen J L 1954 *Z. Angew. (Math.) Phys.* **5** 466–89
Fosdick R L and Schuler K W 1969 *Int. J. Eng. Sci.* **7** 217–33
Kafadar C B 1972 *Arch. Ration. Mech. Anal.* **47** 15–27
Klingbeil W W and Shield R T 1966 *Z. Angew. (Math.) Phys.* **17** 489–511
Morris A W and Shiau J F 1970 *Arch. Ration. Mech. Anal.* **36** 135–60
Müller W C 1970 *Z. Angew. (Math.) Phys.* **21** 633–6
Rivlin R S 1947 *J. Appl. Phys.* **18** 444–9
—— 1948a *Phil. Trans. R. Soc.* A **240** 459–90
—— 1948b *Phil. Trans. R. Soc.* A **241** 379–97

—— 1949a *Proc. R. Soc.* A **195** 463–73
—— 1949b *Phil. Trans. R. Soc.* A **242** 173–95
Rivlin R S and Saunders D W 1951 *Phil. Trans. R. Soc.* A **243** 251–88
Syngh M and Pipkin A C 1965 *Z. Angew. (Math.) Phys.* **16** 706–9

GENERAL REFERENCES

Chadwick P 1976 *Continuum Mechanics* (London: Allen and Unwin)
Green A E and Adkins J E 1960 *Large Elastic Deformations* (Oxford: Clarendon)
Truesdell C and Noll W 1965 *The Non-Linear Field Theories of Mechanics* in *Handbuch der Physik* Vol. III/3 ed. S Flügge (Berlin: Springer)
Wang C-C and Truesdell C 1973 *Introduction to Rational Elasticity* (Leyden: Noordhoff)

7

Material Response in the Spatial Description

7.1 METALS: I, CHARACTERISTIC BEHAVIOUR

7.1.1 Initial Yield Condition

Implicit in the formulation of classical plasticity theory is the assumption that the idealised, perfectly plastic material satisfies von Mises (1913) yield criterion. In the context of classical plasticity theory, Lode (1926) established

$$\mu = \nu \qquad \text{(Lode's relation)}$$

for all μ. Observation that a material satisfies Lode's relation has been widely taken as evidence that the material satisfies the von Mises yield criterion at initial yield. Deviations from the condition $\mu = \nu$ have been correctly interpreted as evidence that a material does not satisfy the von Mises yield condition (Drucker 1949, Freudenthal and Gou 1969, Prager 1945). However, materials which do satisfy the condition $\mu = \nu$ do not necessarily have to satisfy the von Mises yield criterion.

For a non-simple material which satisfies the constitutive equation (4.2.44), and which also satisfies the generalised, isotropic yield criterion formulated by equation (3.1.11), the generalised Lode relation of equation (4.2.38), i.e. the fundamental identity

$$\bar{\mu} \equiv \bar{\nu} \qquad \text{(generalised Lode relation)}$$

must be satisfied for all μ and all ν. In general, a non-simple material will not satisfy the Lode relation $\mu = \nu$. However, exceptions do exist. Elastic–plastic materials have been identified, which when extensively work-hardened, satisfy the Lode relation $\mu = \nu$, but do not satisfy von Mises yield criterion. In this connection, neither the formulation of the constitutive equation (4.2.44) for a non-simple material, nor the formulation of the generalised yield criterion, exclude the possibility that a non-simple material satisfies both the generalised

Lode relation $\bar{\mu} \equiv \bar{v}$, and the Lode relation $\mu = v$. These observations are in accord with the fundamental condition that the only way in which the initial yield function can be determined is by establishing the form of the $(\mu, \bar{\mu})$ relation in the context of the discussion given in §§3.2, 3.3 and 3.4. The condition $\mu = \bar{\mu}$ for all μ is the only condition which establishes that a material satisfies von Mises yield criterion (see §3.3.1).

Thus, with regard to the problem of establishing the particular type of initial yield condition which a material satisfies, the (μ, v) relation is of no relevance whatsoever.

There are two different types of experiment for determining the form of the $(\mu, \bar{\mu})$ relation. One type of experiment is concerned with the plastic response of thin-walled tubes, as for example those of Lode (1926) and Taylor and Quinney (1931). The other type of experiment deals with the axial loading of a notched strip using the method proposed by Hill (1952, 1953).

(i) ***Combined stressing of a thin-walled tube.*** It is the loading of thin-walled tubes in axial tension and simple torsion in the way used by Taylor and Quinney (1931) which is of interest at present. There are two possible approaches to this type of experiment. They can be used to evaluate the form of the yield function without having recourse to any particular constitutive equation. This approach has two disadvantages. It requires direct measurement of changes in volume of the bore of the tubes, a measurement which remains the most difficult to make within an acceptable degree of accuracy. The second limitation is that it requires direct observation of the yield stress in simple tension and in simple torsion. There is no method by which both yield values can be independently measured to an acceptable degree of accuracy. In the measurements of Taylor and Quinney (1931), the onset of plastic distortion on reloading in torsion can only be evaluated by empirical methods. Assuming the form of the constitutive equation for the purely elastic behaviour to be known, an alternative method has been developed (Billington 1985) for determining the initial yield function from combined stress measurements of the type described by Taylor and Quinney (1931). This approach has the advantage that it does not require an independent measurement of the yield stress in simple torsion.

Let the thin-walled tube be given an initial extension by means of a directly applied load W, which can be interpreted as a maximum tensile stress σ'' in simple uniaxial, that is pure tension. Partial removal of the load leaves a fraction, mW $(0 \leqslant m \leqslant 1)$ which can be interpreted as a tensile stress $\sigma_{zz} = m\sigma''$. A stress which is effectively equivalent to the maximum tensile stress σ'' can be generated by holding this reduced load constant and applying a gradually increasing torque, thus giving rise to a shear stress τ_m at the onset of further irrecoverable extension.

The method developed by Taylor and Quinney (1931) is concerned only with the purely elastic deformation of the test metals. For metals, the purely elastic deformations are sufficiently small for the constitutive equation to be approximated to that of equation (4.5.16). Equation (4.5.16) can be expressed in the form,

$$\mathbf{G}'(\mathbf{T}') = 2G_0\tilde{\mathbf{e}}'$$

where a prime denotes a deviator. Using the notation of §6.5, the simple deformations of a thin-walled tube are

$$r = (\eta/F)^{1/2}R \qquad \theta = \Theta + DZ \qquad z = FZ$$

which is just equation (6.5.13) having noted that

$$\eta = (r/R)^2 F = v/V$$

where v and V are the deformed and undeformed volumes, respectively, of a cylinder of radius R and where $F = l/L$ is the ratio of the current length to the undeformed length L of a tube. For infinitely small strains, $D \to 0$, $F \to 1$, $\eta \to 1$, and the components of $\tilde{\mathbf{e}}'$ follow from equation (6.5.26) in the form

$$[\tilde{e}'\langle ij\rangle] = \begin{bmatrix} -\frac{1}{2}e_{zz} - \frac{1}{2}(\eta - 1) & 0 & 0 \\ 0 & -\frac{1}{2}e_{zz} + \frac{1}{2}(\eta - 1) & \frac{1}{2}\gamma_{\theta z} \\ 0 & \frac{1}{2}\gamma_{\theta z} & e_{zz} \end{bmatrix}$$

which can be used with the constitutive equation to give a set of stress–strain relations which can in turn be combined to give

$$\lim_{D\to 0} \xi = 3\frac{e_{zz}}{\gamma_{\theta z}} - \frac{\eta - 1}{\gamma_{\theta z}} = \frac{\sigma_{zz} - \sigma_{\theta\theta}}{\tau_{\theta z}} \tag{7.1.1}$$

$$\lim_{D\to 0} \zeta = -\frac{\mu}{\bar{\mu}}\left(\frac{e_{zz}}{\gamma_{\theta z}} + \frac{\eta - 1}{\gamma_{\theta z}}\right) = \frac{\sigma'_{rr}}{\tau_{\theta z}}. \tag{7.1.2}$$

These can be combined to give

$$\frac{e_{zz}}{\gamma_{\theta z}} = \frac{1}{12}\xi\left[3 - \bar{\mu}\left(1 + \frac{4}{\xi^2}\right)^{1/2}\right] \tag{7.1.3}$$

where

$$e_{zz} = F - 1 (\equiv \tilde{e}_{zz}) \qquad \gamma_{\theta z} = rD/F. \tag{7.1.4}$$

Use has also been made of equation (6.5.11), having noted that for the stress ratios ξ and ζ

$$\sigma_{zz} = m\sigma'' \qquad (0 \leqslant m \leqslant 1). \tag{7.1.5}$$

Equation (7.1.3) is independent of the change in volume $[(\eta - 1)V]_{r=r_2}$ of the inner bore of the tube.

The conditions at initial yield are obtained from equation (4.2.1) by setting $a=k^2$, $[h]_{a=k^2}=f$ to give

$$f=J_2'H(\omega)-k^2=0 \tag{7.1.6}$$

which is just equation (3.1.11), k being the maximum shear stress at yielding in a state of pure shear. Three types of yield function are of interest. These are given by equation (7.1.6) with H defined by equations (3.3.6), (3.2.34) and (3.4.1). Thus

(*a*) *Continuously differentiable yield condition.* From §3.3.1,

$$H=(1-c\omega)(\equiv\hat{H}_{(2)}) \qquad \bar{\mu}=\mu\left[1-\frac{9c(3+\mu^2)(1-\mu^2)(9-\mu^2)}{(3+\mu^2)^3-8c\mu^4(9-\mu^2)}\right] \quad (c>0) \tag{7.1.7}$$

where c is a constant characteristic of material properties. Entering the H of equation (7.1.7) into equation (7.1.6) and rearranging gives

$$\xi=\frac{2\sqrt{3}\,mb_{(1)}(1-c\omega)^{1/2}}{\{4-(3+b_{(2)}^2)b_{(1)}^2m^2-c[4-(3+b_{(2)}^2)m^2b_{(1)}^2\omega]\}^{1/2}} \tag{7.1.8}$$

which can be entered into equation (7.1.3) to give

$$\frac{e_{zz}}{\gamma_{\theta z}}=\frac{1}{2\sqrt{3}}\frac{mb_{(1)}(1-c\omega)^{1/2}[3+b_{(2)}(1-c\varkappa(\mu))]}{\{4-(3+b_{(2)}^2)m^2b_{(1)}^2-c[4-(3+b_{(2)}^2)m^2b_{(1)}^2\omega]\}^{1/2}}. \tag{7.1.9}$$

Here

$$b_{(1)}=1-\frac{\sigma_{\theta\theta}}{m\sigma''} \qquad b_{(2)}=1-2\frac{(\sigma_{rr}-\sigma_{\theta\theta})}{(m\sigma''-\sigma_{\theta\theta})} \tag{7.1.10}$$

and, from equation (7.1.7)

$$\varkappa(\mu)=\frac{9(3+\mu^2)(1-\mu^2)(9-\mu^2)}{(3+\mu^2)^3-8c\mu^4(9-\mu^2)}. \tag{7.1.11}$$

The von Mises yield criterion is obtained by setting $C=0$ thus giving $H=1$ in accord with equation (3.3.1). For the infinitely small strains up to and including the initial yield of many metals, it is customary to assume that $\sigma_{rr}\simeq 0$, $\sigma_{\theta\theta}\simeq 0$. These limiting conditions reduce equations (7.1.8) and (7.1.9) for the von Mises yield criterion to the form,

$$\xi=\frac{\sqrt{3}\,m}{(1-m^2)^{1/2}} \tag{7.1.12}$$

$$\frac{e_{zz}}{\gamma_{\theta z}}=\frac{1}{\sqrt{3}}\frac{m}{(1-m^2)^{1/2}} \tag{7.1.13}$$

(*b*) *Piecewise continuous yield condition.* From §3.2.4

$$H = \frac{(3+\bar{\mu}_n\mu)^2}{(3+\mu^2)}\left[\frac{(3+\mu^2)}{(3+\bar{\mu}_n\mu)^2}\right]_{\omega=0} \quad (\equiv \hat{H}_{(1)}) \qquad (n=0,1,2) \quad \bar{\mu}=\text{constant} \tag{7.1.14}$$

Substituting the H of equation (7.1.14) into equation (7.1.6) and rearranging gives

$$\xi = \frac{6mb_{(1)}}{\{[3+|\bar{\mu}_0|(1-mb_{(1)}b_{(2)})]^2 - 9m^2b_{(1)}^2\}^{1/2}} \tag{7.1.15}$$

which can then be entered into equation (7.1.3) to give

$$\frac{e_{zz}}{\gamma_{\theta z}} = \frac{1}{6}\frac{\{9mb_{(1)} + |\bar{\mu}_0|[3+|\bar{\mu}_0|(1-mb_{(1)}b_{(2)})]\}}{\{[3+|\bar{\mu}_0|(1-mb_{(1)}b_{(2)})]^2 - 9m^2b_{(1)}^2\}^{1/2}}. \tag{7.1.16}$$

Equations (7.1.1) and (7.1.2) can be combined to give

$$2\frac{(\sigma_{rr}-\sigma_{\theta\theta})}{\tau_{\theta z}} = 3\left(1-\frac{\mu}{\bar{\mu}_0}\right)\frac{e_{zz}}{\gamma_{\theta z}} - \left(1+3\frac{\mu}{\bar{\mu}_0}\right)\frac{(\eta-1)}{\gamma_{\theta z}} = 3\zeta + \xi. \tag{7.1.17}$$

For sufficiently small strains and for $m \to 1$, $\sigma_{rr} \to 0$, $\sigma_{\theta\theta} \to 0$, and the stress ratios $(\sigma_{rr}-\sigma_{\theta\theta})/\tau_{\theta z}$ and $\sigma_{\theta\theta}/m\sigma''$ can therefore both be neglected, so giving

$$b_{(1)} = 1, \qquad b_{(2)} = 1 \qquad \xi = -3\zeta \qquad (\sigma_{rr}=0,\ \sigma_{\theta\theta}=0) \tag{7.1.18}$$

These conditions reduce equations (7.1.15) and (7.1.16) to

$$\xi = \frac{6m}{\{[3+|\bar{\mu}_0|(1-m)]^2 - 9m^2\}^{1/2}} \qquad (\sigma_{rr}=0,\ \sigma_{\theta\theta}=0) \tag{7.1.19}$$

$$\frac{e_{zz}}{\gamma_{\theta z}} = \frac{1}{6}\frac{\{9m + |\bar{\mu}_0|[3+|\bar{\mu}_0|(1-m)]\}}{\{[3+|\bar{\mu}_0|(1-m)]^2 - 9m^2\}^{1/2}}. \tag{7.1.20}$$

As m is increased from zero, the effect of the applied torque is rapidly attenuated, so giving way to the effect of the increasing applied tensile stress. There will therefore be a particular value of $m = m^* \ll 1$, above which equation (7.1.20) will be expected to apply.

The limiting condition $m=0$ corresponds to simple torsion for which $F \geqslant 1$. Setting $m=0$ reduces equation (7.1.16) to

$$\frac{e_{zz}}{\gamma_{\theta z}} = \frac{1}{6}\frac{\{|\bar{\mu}_0|[3+|\bar{\mu}_0|(\sigma''-\sigma_{\theta\theta}-2\sigma_{rr})/\sigma''] - 9(\sigma_{\theta\theta}/\sigma'')\}}{\{[3+|\bar{\mu}_0|(\sigma''-\sigma_{\theta\theta}-2\sigma_{rr})/\sigma'']^2 - 9(\sigma_{\theta\theta}^2/\sigma''^2)\}^{1/2}} \tag{7.1.21}$$

and for $\sigma_{\theta\theta}=0$ at $\gamma_{\theta z}=0$, equation (7.1.21) gives

$$[e_{zz}/\gamma_{\theta z}]_{\gamma_{\theta z}=0} = \tfrac{1}{6}|\bar{\mu}_0| \qquad (m=0,\ \sigma_{\theta\theta}=0). \tag{7.1.22}$$

Thus, for a material to satisfy a twelve-sided, piecewise continuous yield condition, for which $|\bar{\mu}_0| = \text{constant}$ for all μ, the material must respond to simple torsion in such a way as to give $e_{zz}/\gamma_{\theta z}=$constant at $\gamma_{\theta z}=0$, a limiting condition for which there is no direct experimental evidence (see, however, Rose and Stüwe 1968).

(*c*) *Composite yield function.* From §3.4

$$H=\phi\hat{H}_{(1)}+(\hat{H}_{(2)}-\phi\hat{H}_{(1)})U \tag{7.1.23}$$

where

$$\phi=\frac{3}{3+\mu^{*2}} \qquad (\mu^*\equiv\bar{\mu}_0) \qquad U=\begin{cases}1, & (0\leqslant\mu^2\leqslant\mu^{*2})\\ 0, & (\mu^{*2}\leqslant\mu^2\leqslant 1)\end{cases}. \tag{7.1.24}$$

A material of particular interest is one which satisfies the yield condition obtained using equations (7.1.6) and (7.1.23), the transition from one type of yield condition to the other occurring in such a way that the ϕ of equation (7.1.24) is to be identified with the value of $\mu=\bar{\mu}_0$ at which $m=m^*\ll 1$, above which equation (7.1.20) will be expected to apply. For this type of composite yield condition, equation (7.1.9) applies for $(0\leqslant m\leqslant m^*)$. In this way the composite yield function satisfies the essential condition that for $m=0$, $e_{zz}/\gamma_{\theta z}=0$ at $\gamma_{\theta z}=0$.

The programme of combined stressing described by Taylor and Quinney (1931) is just that considered above. Central to the combined stress measurements of Taylor and Quinney is the quantity

$$\frac{2}{\xi}=\frac{2\tau_{\theta z}}{\sigma_{zz}-\sigma_{\theta\theta}} \qquad (\equiv\tan 2\theta \text{ in their notation}). \tag{7.1.25}$$

It is therefore assumed that Taylor and Quinney will have determined this quantity as accurately as their measurements permit, having noted that ξ is defined by equation (6.5.9). The values of $2/\xi$ have been recalculated and, with the exception of $m=0.90$ for the copper tubes, found to be in good accord with the values given by Taylor and Quinney in their Table 1. Taking their value of $2/\xi$, recalculation gives $m=0.886$, which to the degree of accuracy of the original analysis is taken to imply the value 0.89 which is used in the present discussion.

In the context of the present equation (7.1.5), the maximum tensile stress σ'' is the initial yield stress Y in simple uniaxial tension. Values of Y $(\equiv\sigma'')$ and σ_{zz} are given in table 7.1 for the copper tubes tested by Taylor and Quinney (1931).

The effect of applying the load W is to extend the tube from its initial length L to a length l_0, and hence from equation (6.5.15), $F_0=l_0/L$ $(\equiv 1+e_0$ in the notation of Taylor and Quinney).

Partial removal of the load leaves a fraction mW which for a given tube can be interpreted as the tensile stress $\sigma_{zz}=mY$ defined by the present equation (7.1.5) with σ'' set equal to Y.

For any given value of m, the reloading in torsion gives to a good

Table 7.1 Combined stress measurements using the copper tubes of Taylor and Quinney (1931).

m	0.0252	0.280	0.515	0.65	0.70	0.80	0.89
Y (MPa)	217.1	222.9	216.5	216.5	180.3	218.0	222.9
σ_{zz} (MPa)	5.47	62.4	111.5	140.7	126.2	174.4	198.4
$e_{zz}/\gamma_{\theta z}$†	0.0155	0.166	0.340	0.468	0.530	0.717	1.033
$e_{zz}/\gamma_{\theta z}$‡	0.0146	0.168	0.340	0.468	0.532	0.714	1.034
$-\bar{\mu}$	0.0232	0.2416	0.3049	0.3018	0.2900	0.3280	0.2988
$-\mu$	0.0219	0.2444	0.4908	0.6279	0.6803	0.7829	0.8804

† Taylor and Quinney (1931).
‡ From equations (7.1.13) and (7.1.20).

approximation a constant value for the quantity

$$\psi^* = \frac{LF_0D}{\delta l} \qquad \left(\equiv \frac{\chi}{\delta l} \text{ in their notation}\right) \tag{7.1.26}$$

where δl is the additional extension of the tube resulting from application of an increasing torque. This gives rise to an axial strain,

$$e_{zz} = \frac{(1+e_0)^{1/2}}{\psi^* R_{m0}} \gamma_{\theta z} \tag{7.1.27}$$

where R_{m0} is the mean radius of the tube before the first stretching, and where $\gamma_{\theta z}$ is the shear strain. Using standard data retrieval techniques, the actual values of the total angle of twist, LF_0D and the extension of the tube δl have been obtained from figure 7a of Taylor and Quinney (1931). The $(LF_0D, \delta l)$ results have been analysed, assuming the linear relation of equation (7.1.26) but using only the measured values of LF_0D and δl. The values of ψ^* obtained in this way have been used with equation (7.1.27) to give the value of the ratio $e_{zz}/\gamma_{\theta z}$ shown in table 7.1. The variation of these values of $e_{zz}/\gamma_{\theta z}$ with m is shown as the circles in figure 7.1. Using the same approach, values of the ratio $e_{zz}/\gamma_{\theta z}$ for the aluminium tubes have been obtained from figure 7b of Taylor and Quinney (1931), and are given in table 7.2. The variation of these values of $e_{zz}/\gamma_{\theta z}$ with m is also shown as triangles in figure 7.1, the experimental $(e_{zz}/\gamma_{\theta z}, m)$ curve being displaced by 0.2 units parallel to the ordinate.

Using the value $\bar{\mu}_0 = -0.305$ for the range $(m^* \leqslant m \leqslant 1)$, the full line $(e_{zz}/\gamma_{\theta z}, m)$ curve shown in figure 7.1 has been calculated for the copper tubes from equation (7.1.20) and is seen to be indistinguishable from the experimental $(e_{zz}/\gamma_{\theta z}, m)$ curve. The values of $e_{zz}/\gamma_{\theta z}$ calculated in this way for $m \geqslant m^* = 0.45$ are given in table 7.1 for direct comparison with the values of the ratio $e_{zz}/\gamma_{\theta z}$ determined for the measurements of Taylor and Quinney (1931). The calculated extension of the $(e_{zz}/\gamma_{\theta z}, m)$ curve for $(0 \leqslant m \leqslant m^*)$ is shown as the broken curve in figure 7.1. The full $(e_{zz}/\gamma_{\theta z}, m)$ curve shown in figure 7.1 for

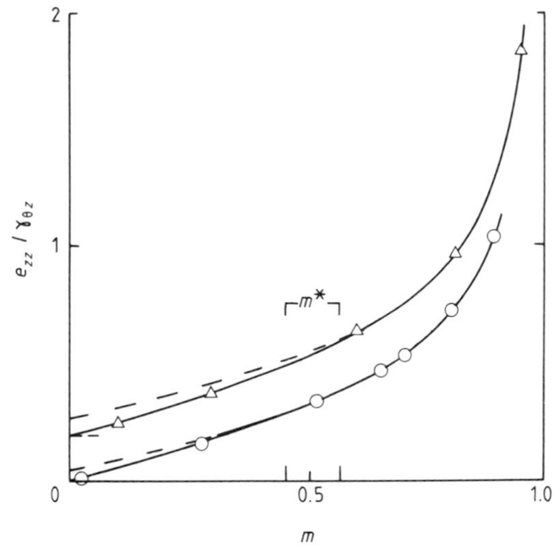

Figure 7.1 Variation of $e_{zz}/\gamma_{\theta z}$ with m for the combined stress measurements of Taylor and Quinney (1931) using their results for copper (○) and aluminium (△) tubes.

the aluminium tubes has been calculated from equation (7.1.20) using $\bar{\mu}_0 = -0.454$ and is seen to be indistinguishable from the experimental $(e_{zz}/\gamma_{\theta z}, m)$ curve for $m \geqslant m^* = 0.563$. The calculated values of $e_{zz}/\gamma_{\theta z}$ for $m \geqslant m^*$ are given in table 7.2 for direct comparison with the values of the ratio $e_{zz}/\gamma_{\theta z}$ determined from experiment. The calculated extension of the $(e_{zz}/\gamma_{\theta z}, m)$ curve is shown as the broken curve in figure 7.1. It is concluded that the combined stress measurements of Taylor and Quinney (1931) can be accurately correlated for m in the range $(m^* \leqslant m \leqslant 1)$ by the twelve-sided, linear, piecewise continuous yield condition defined by equation (7.1.6) with H given by equation (7.1.14).

Table 7.2 Combined stress measurements using the aluminium tubes of Taylor and Quinney (1931).

m	0.10	0.30	0.60	0.81	0.95
Y (MPa)	93.8	93.3	94.6	94.6	93.8
σ_{zz} (MPa)	9.38	28.0	56.8	76.7	89.1
†$e_{zz}/\gamma_{\theta z}$	0.050	0.166	0.435	0.761	1.641
‡$e_{zz}/\gamma_{\theta z}$	0.058	0.182	0.435	0.761	1.642
$-\bar{\mu}$	0.0748	0.2416	0.4540	0.4540	0.4498
$-\mu$	0.0867	0.2634	0.5657	0.7875	0.9429

† Taylor and Quinney (1931)
‡ From equations (7.1.13) and (7.1.20)

The full $(e_{zz}/\gamma_{\theta z}, m)$ curves shown in figure 7.1 have been calculated for the range $(0 \leqslant m \leqslant m^*)$ from equation (7.1.13) for the limiting conditions $\sigma_{rr}=0$, $\sigma_{\theta\theta}=0$, and are seen to be indistinguishable from the experimental $(e_{zz}/\gamma_{\theta z}, m)$ curves for both the copper and aluminium tubes. The values of $e_{zz}/\gamma_{\theta z}$ calculated in this way for $m \leqslant m^*$ are given in tables 7.1 and 7.2 for direct comparison with the values of the ratio $e_{zz}/\gamma_{\theta z}$ determined from the measurements of Taylor and Quinney (1931). It is concluded that the combined stress measurements of Taylor and Quinney (1931) can be correlated for m in the range $(0 \leqslant m \leqslant m^*)$ by the von Mises yield criterion defined by equation (7.1.6) with $H=1$ for all μ.

It is of particular interest to note that this reanalysis of the combined-stress measurements of Taylor and Quinney (1931) does not in any way depend on the change in volume, $[(\eta-1)V]_{r=r_2}$, of the inner bore of the tubular specimen; furthermore, it does not require a measurement of the shear stress τ_m corresponding to onset of plastic distortion.

Values of $\bar{\mu}$ for m in the range $(m^* \leqslant m \leqslant 1)$ have been calculated from equation (7.1.20) using the experimental values of the ratio $e_{zz}/\gamma_{\theta z}$ determined by Taylor and Quinney (1931) and given in the present tables 7.1 and 7.2. Using these values of $\bar{\mu}$ $(\equiv \bar{\mu}_0)$, corresponding values of ξ have been calculated from equation (7.1.19), these values of ξ being substituted into equation (6.5.11) to give values of μ for the limiting condition $\xi=-3\zeta$ of equation (7.1.18). These values of $\bar{\mu}$ $(\equiv \bar{\mu}_0)$ and μ are given in tables 7.1 and 7.2 and are shown as the $(\mu, \bar{\mu})$ curves in figure 7.2.

For m in the range $(0 \leqslant m \leqslant m^*)$, the values of ξ have been calculated from the relation of equation (7.1.12) and then substituted into equation (6.5.11) to give values of μ for the limiting condition $\xi=-3\zeta$ of equation (7.1.18). Corresponding values of $\bar{\mu}=\mu$ have been evaluated by way of equations

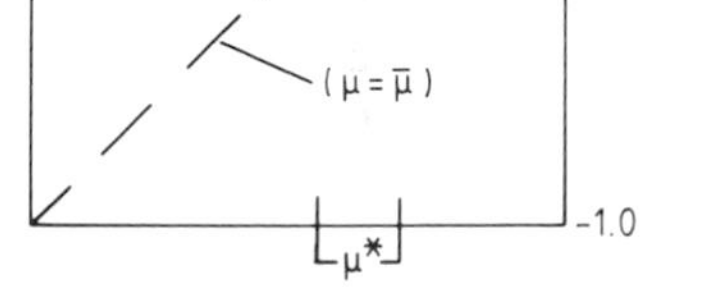

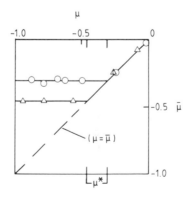

Figure 7.2 Values of μ and $\bar{\mu}$ for the combined stress measurements of Taylor and Quinney (1931) using their results for copper (○) and aluminium (△) tubes.

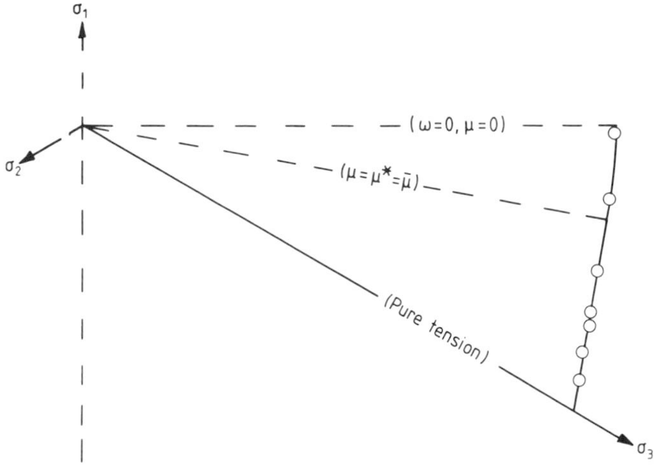

Figure 7.3 Yield surface in the principal stress space showing the yield curve in the π-plane for the copper tubes (Taylor and Quinney 1931).

(6.5.11) and (7.1.18) and the relation of equation (7.1.12), using the experimental values of the ratio $e_{zz}/\gamma_{\theta z}$ determined by Taylor and Quinney (1931) and given in tables 7.1 and 7.2. These values of $\mu=\bar{\mu}$ for $m \leqslant m^*$ are given in tables 7.1 and 7.2 and are shown as the $(\mu, \bar{\mu})$ curves in figure 7.2.

The form for the $(\mu, \bar{\mu})$ relation shown in figure 7.2 is taken to imply that the material of the copper and aluminium tubes satisfy a composite yield function defined by equations (7.1.6) and (7.1.23). The yield surface in the principal stress space is shown in figure 7.3, that is the yield curve in the deviatoric plane (see §3.1.2 and figure 3.2).

(ii) ***Notched strip.*** In this method (Hill 1952, 1953), a thin uniform rectangular strip subjected to axial load is deformed plastically under conditions of plane stress. The distortion is restricted to a narrow straight zone by notching the strip unsymmetrically, as shown in figure 7.4. If the thickness of the strip is sufficiently small, use of notches which are deep and sharp restricts the plastic deformation to a narrow straight neck joining the roots of the notches. Deformation is produced by the application of an axial tensile load F, assumed to be uniformly distributed along the length l of the neck. The load F produces a relative displacement of the two halves of the strip, the direction of this relative displacement being inclined at an angle ψ to the direction of the neck, as indicated in figure 7.4.

By considering the normal and shearing stress, in the context of equations (2.2.45) to (2.2.49), it follows that (Hill 1953)

$$\frac{F \sin \theta}{h_0 l} = (\sigma_2 + \sigma_3) + (\sigma_3 - \sigma_2) \sin \psi \qquad (7.1.28)$$

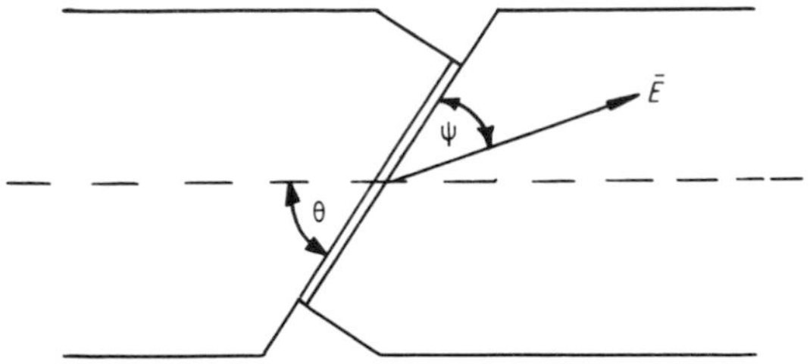

Figure 7.4 Typical notched-strip specimen.

$$\frac{F\cos\theta}{h_0 l}=(\sigma_3-\sigma_2)\cos\psi \tag{7.1.29}$$

where use has been made of the notation of §6.3, having assumed that the major surfaces $X^1=\pm h_0$ are free from applied stress so that $\sigma_1=0$ for all states of loading. Equations (7.1.28) and (7.1.29) can be solved to give

$$\sigma_2=\frac{F}{2h_0 l\cos\psi}[\sin(\theta-\psi)-\cos\theta]. \tag{7.1.30}$$

$$\sigma_3=\frac{F}{2h_0 l\cos\psi}[\sin(\theta-\psi)+\cos\theta]. \tag{7.1.31}$$

Entering the form for σ_2 and σ_3 given by equations (7.1.30) and (7.1.31) into equation (4.1.26) gives,

$$\mu=-\sin(\theta-\psi)/\cos\theta. \tag{7.1.32}$$

From equations (3.1.15), (7.1.30) and (7.1.31)

$$u=\frac{1}{\sqrt{6}h_0 l}\left(F\frac{\sin(\theta-\psi)}{\cos\psi}\right) \qquad v=\frac{1}{\sqrt{6}\,h_0 l}\left(\sqrt{3}\,F\frac{\cos\theta}{\cos\psi}\right) \tag{7.1.33}$$

having noted that

$$\mu=\sqrt{3}\,u/v \qquad \bar{\mu}=-\sqrt{3}\,\mathrm{d}v/\mathrm{d}u \tag{7.1.34}$$

where use has been made of equation (3.1.14).

The strain tensor in the neck has the principal deviatoric components

$$\begin{aligned}\bar{\varepsilon}_1'&=-E\sin\psi\\ \bar{\varepsilon}_2'&=\tfrac{1}{2}E\sin\psi-\tfrac{1}{2}E\\ \bar{\varepsilon}_3'&=\tfrac{1}{2}E\sin\psi+\tfrac{1}{2}E\end{aligned} \tag{7.1.35}$$

which can be entered into equation (4.2.48) to give

$$\nu=-3\sin\psi \tag{7.1.36}$$

having made use of the incompressibility condition to obtain equation (7.1.35).

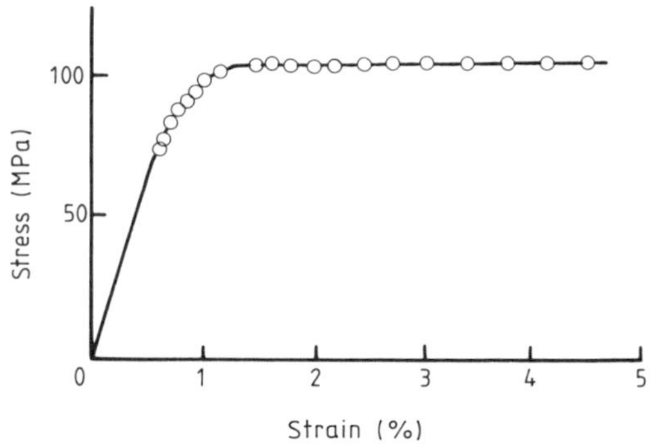

Figure 7.5 Typical tensile stress–strain curve for a strip of aluminium rolled at the temperature of liquid air (Lianis and Ford 1957).

The notched strip technique proposed by Hill (1953) can be used only for metals having a very small rate of hardening.

In the context of the above technique, the conditions at initial yield have been studied by Lianis and Ford (1957) using notched strips of commercially pure aluminium rolled at the temperature of liquid air. A typical stress–strain curve is shown in figure 7.5 and is seen to satisfy the required condition of a very small rate of hardening. With regard to the analysis of Hill (1952, 1953) and the studies of Lianis and Ford (1957), use was made of Lode's variables in a form which can be obtained by replacing the $\sigma_1', \sigma_2', \sigma_3'$ in equation (4.1.26) by σ_2', σ_3', σ_1' respectively, and the $\bar{\varepsilon}_1'$, $\bar{\varepsilon}_2'$, $\bar{\varepsilon}_3'$ in equation (4.2.48) by $\bar{\varepsilon}_2'$, $\bar{\varepsilon}_3'$, $\bar{\varepsilon}_1'$ respectively. Using a technique developed by Hundy and Green (1954) for measuring the angle ψ, Lianis and Ford (1957) obtained values of u, v, μ and ν. The variation of these values of u with the corresponding values of v are shown in figure 7.6. The measurements of Hundy and Green (1954) were restricted to determining μ and ν for specimens of copper, zinc and stainless steel. Both the measurements of Hundy and Green (1954) and the measurements of Lianis and Ford (1957) are in accord with the relation $\mu = \nu$ which when used with equations (7.1.32) and (7.1.36) gives

$$\tan\theta = 4\tan\psi \qquad (\mu = \nu) \tag{7.1.37}$$

for all μ.

In §3.3.1, it has been shown that for a material which satisfies von Mises (1913) yield criterion, the condition $\mu = \bar{\mu}$ for all μ requires that

$$u^2 + v^2 = 2k^2 \qquad (\mu = \bar{\mu}). \tag{7.1.38}$$

Here k is the maximum shear stress at yielding in a state of pure shear and u and v are defined in general by equation (3.1.15) and in a specific form for a notched strip by equation (7.1.33). Values of u and v have been obtained from

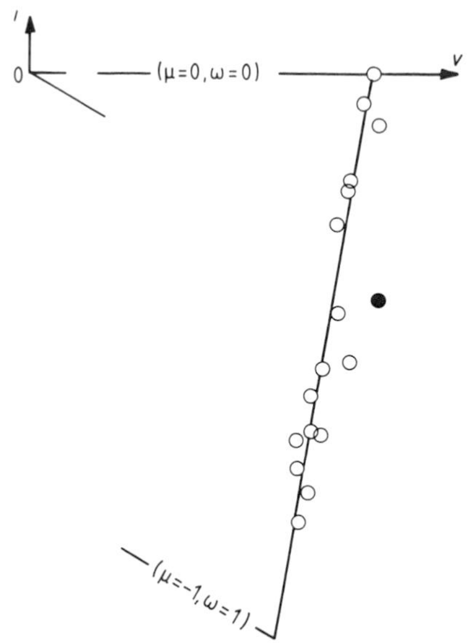

Figure 7.6 Variation of u with v for notched strip specimens of aluminium rolled at the temperature of liquid air (Lianis and Ford 1957).

figure 7 of Lianis and Ford (1957) (see figure 7.6), using standard data retrieval techniques. Using these values of u and v, it has been found that u^2 varies with v^2 according to the relation

$$u^2 + Av^2 = B^2 \tag{7.1.39}$$

where $A = 0.638$, and $B = 76.2$ MPa, the constants A and B being evaluated using a standard linear regression technique, the associated correlation coefficient $r = 0.870$.

For a material which satisfies a twelve-sided, linear, piecewise continuous yield condition it follows from equations (3.1.11), (3.1.15), (3.1.23), (3.2.34) and (7.1.33) that

$$u = \sqrt{3}\,(v_0 - v)/\bar{\mu}_0 \qquad (v_0 = \sqrt{2}\,k). \tag{7.1.40}$$

Using values of u and v corresponding to the open circles in figure 7.6, it has been found that u varies with v according to equation (7.1.40) with

$$\bar{\mu}_0 = -0.290 \qquad v_0 = 93.24 \text{ MPa} \qquad (r = 0.924).$$

Values of $\bar{\mu}_0$ have been calculated from equation (7.1.40) using the experimental values of u, v and v_0. The variation of these values of $\bar{\mu}_0$ with μ is shown in figure 7.7. With the exception of the result corresponding to the full circle in figure 7.6, it is evident from figure 7.7 that the measurements of Lianis and Ford (1957) imply that the material satisfies a twelve-sided, piecewise linear yield condition.

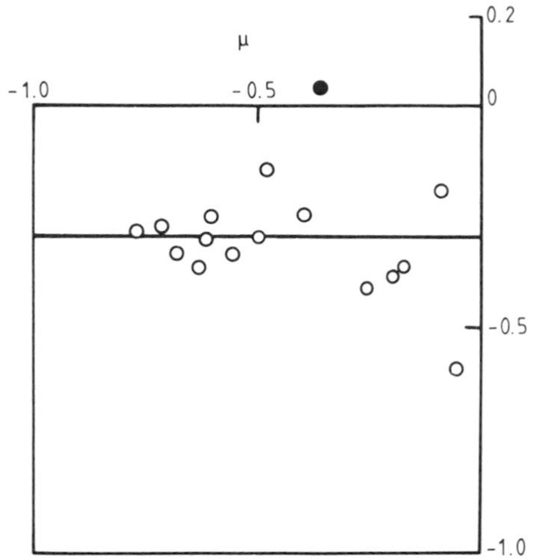

Figure 7.7 Variation of μ with $\bar{\mu}_0$ for notched strip specimens of aluminium rolled at the temperature of liquid air (Lianis and Ford 1957).

Both techniques are subject to one major disadvantage. Determination of the $(\mu, \bar{\mu})$ relation from experiment requires the use of several specimens which, in principle, have to be identical in all their properties. In practice, it is almost impossible to rigorously satisfy this condition. However, the method developed by Taylor and Quinney (1931) has one major advantage. The initial yield stress in simple tension can be determined to a high degree of accuracy. Using experimental techniques only recently available, the initial deformation in tension can be adjusted to give, the same value of the initial yield stress in simple tension to a high degree of accuracy for all test specimens.

7.1.2 Poynting Effect

For solids, the mechanical response of a rod or tube twisted in simple torsion $(F > 1)$ is characterised by a non-linear normal stress effect. This is the Poynting effect which relates to the observation by Poynting (1909, 1912) that the length of various steel, copper and brass wires increased when twisted in the elastic range and that the elongation was proportional to the square of the twist. The measurements by Foux (1964) confirm these observations. Lengthening of rubber filaments in simple torsion was also observed by Poynting (1913).

That the Poynting effect is predicted by a non-simple elastic material has been shown in the context of §6.1.5. However, technically useful studies of the Poynting effect cannot be obtained from the simple shear mode of deformation in the form discussed in §6.1.5. It is more usual to study the Poynting effect by way of the simple torsion of a solid rod. In this context, it is the measurements of Foux (1964) which are of particular interest.

For a solid rod subject to the combined effect of an axial tensile strain and simple torsion, it has been shown in §6.7 that the axial strain $\bar{\varepsilon}_{zz}$ varies with F according to the relation

$$\bar{\varepsilon}_{zz} = \frac{(F+1)(F^3-1)}{6F^2} \quad (\equiv \bar{\varepsilon}_3')$$

which is just equation (6.7.10). The Poynting effect is characterised by an axial elongation $e_{(\mathrm{P})}$ for which $F \to 1$, and hence

$$\lim_{F \to 1} \bar{\varepsilon}_{zz} = F - 1 \equiv e_{(\mathrm{P})} \tag{7.1.41}$$

which is observed for the particular condition that the resultant longitudinal force on the plane end of a solid rod is zero. For the purely elastic deformation of most metals, $D \to 0$, $F \to 1$, $\bar{G} \to G_0$, and therefore we can set $N_z = 0$ in equation (6.6.14) which gives

$$\bar{\varepsilon}_{zz} = D^2 \frac{\displaystyle\int_0^{r_1} \left(G_0 \frac{\Psi}{\Lambda} \right) r^3 \, \mathrm{d}r}{\displaystyle\int_0^{r_1} \left(G_0 \frac{\Psi}{\Lambda} \right) \left(3 + \frac{\mu}{\nu} \right) r \, \mathrm{d}r} = e_{(\mathrm{P})}. \tag{7.1.42}$$

For the particular condition that the ratios $(G_0 \Psi/\Lambda)$ and μ/ν can each be regarded as being independent of r, equation (7.1.42) reduces to,

$$e_{(\mathrm{P})} = \frac{1}{2(3+\mu/\nu)} R_1^2 D^2. \tag{7.1.43}$$

Thus, in the absence of any dependence upon r, and noting that for the von Mises type stress and strain intensity functions

$$\mu = \nu \qquad \Psi = 1 \qquad \Lambda = 1 \tag{7.1.44}$$

for all D and F, the ratio

$$b = e_{(\mathrm{P})}/\gamma_{\theta z}^2 = \tfrac{1}{8} \qquad \text{(solid rod)} \tag{7.1.45}$$

where the shear strain,

$$\gamma_{\theta z} = R_1 D \qquad (r_1 \simeq R_1 \text{ as } D \to 0). \tag{7.1.46}$$

The results of the experimental studies of Poynting (1909, 1912) and of Foux (1964) are in good accord with the prediction that in simple torsion the associated axial extension $e_{(\mathrm{P})}$ is proportional to $\gamma_{\theta z}^2$ in the elastic range of deformation. However, there remains the question of the extent to which the results of these experimental studies are in quantitative agreement with the predictions of nonlinear constitutive theory, in particular the value of the ratio $b = e_{(P)}/\gamma_{\theta z}^2$.

For present purposes, use will be made of equation (6.6.17) expressed in the form

$$e_{(P)} = e_0 + b\gamma_{\theta z}^2 \qquad (b = \tfrac{1}{8}) \tag{7.1.47}$$

where retention of the term

$$e_0 = N_z/(3\pi R_1^2 G_0)$$

recognises that, although the weight of the apparatus may have negligible effect in pure tension, the question of whether it changes the torsional modulus in the simple torsion mode of deformation cannot be resolved. Foux (1964) analysed his extensive test series on hard steel wires in simple torsion assuming that the axial elongation could be expressed as a polynomial expansion in $\gamma_{\theta z}$. The present approach differs from that of Foux, in so far as $\gamma_{\theta z}$ is simply replaced by $\gamma_{\theta z} \pm \gamma_0$ where γ_0 is a constant for all $\gamma_{\theta z}$. Replacing $\gamma_{\theta z}$ in equation (7.1.47) by $\gamma_{\theta z} \pm \gamma_0$ and rearranging gives

$$e_{(P)} = a_0 + a\gamma_{\theta z} + b\gamma_{\theta z}^2 \tag{7.1.48}$$

where

$$a_0 = e_0 + b\gamma_0^2 \qquad a = \pm 2b\gamma_0.$$

The presence of the term in a_0 necessitates a slightly different interpretation of the measurements given by Foux (1964) in his table 2. For present purposes, the positive and negative cycles have been independently analysed to give two values of b. The variation of the average of these two values of b with $(R_1/L)^2$ is shown in figure 7.8 for wire numbers 4, 6, 7 and 9, in the as-drawn state and after heat treatment. Wire numbers 2, 5 and 8 are not shown because they are from different sources. Wire number 10 is omitted because the effect of reversing the direction of shear of the as-drawn specimen is to reverse the direction of the change in length of the wire. This is taken to imply that for this radius of wire, the material properties are markedly different, and hence there is the possibility that heat treatment will not necessarily produce the required change in material properties. With regard to the omission of wire numbers 1 and 3, it is to be noted that, although equation (7.1.47) is independent of material properties, differences in material properties may affect the proposed empirical correlation of b with $(R_1 L)^2$. It is evident from figure 7.8 that the values of b are dependent upon the cross-section of the wires. There are two possible explanations for such a dependence upon r. This type of measurement is subject to an error arising from an apparent strengthening by the material nearer to the longitudinal axis of the solid rod specimen which is stressed at a lower level. A second contributing factor is the linear dimension of the grains which for some materials may approach that of the diameter of the

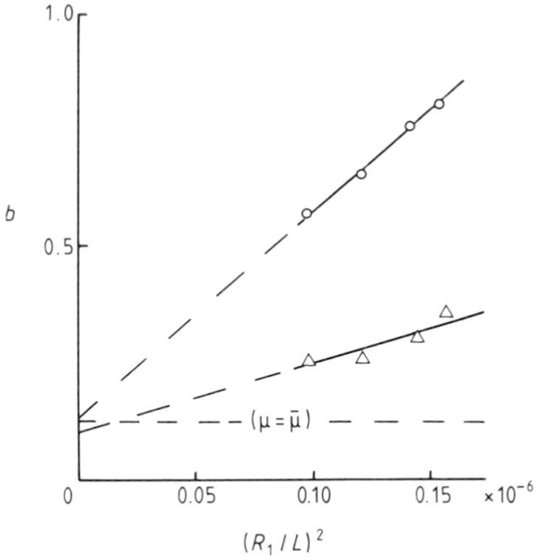

Figure 7.8 Variation of b with $(R_1/L)^2$ for wire numbers 4, 6, 7 and 9 in the as-drawn state (○) and after heat treatment (△) (Billington 1985).

wire. Because of these limitations, and having regard to the extent of the extrapolation, these results cannot give a conclusive value of b. Subject to these limitations, the $[b,(R_1/L)^2]$ correlation can be represented by the linear relation

$$b = b_0 + c(R_1/L)^2$$

where

$$b_0 = \begin{cases} 0.104 \\ 0.069 \end{cases} \quad \text{for} \begin{cases} \text{as-drawn state} \\ \text{after heat treatment} \end{cases}$$

and where c is a constant characteristic of material properties.

The way in which the total, static axial strain $e_{(PS)}$ varies with the shear strain $\gamma_{\theta z}$ is shown in figure 7.9 for a thin-walled tubular specimen of aluminium provided with a close fitting plug. Although the plug is close fitting, there is a small gap between the outer surface of the plug and the inner surface of the tube which is packed with an extreme pressure lubricant prior to the test. The lubricant does not restrict the initial free movement of the inner surface of the tube in a radial direction. Hence the inner surface of the thin-walled tubular specimen can be regarded as being free of traction, and the material should therefore respond according to equation (6.5.65) with $N_z = 0$, for an initial, limited range of $\gamma_{\theta z}$. For sufficiently small strains, $D \to 0$, $F \to 1$, $\eta \to 1$, $e_{(PS)} \to e_{(P)}$ and with $N_z = 0$, equation (6.5.65) reduces to

$$4e_{(P)} \int_{r_2}^{r_1} \left(\frac{\tau_{\theta z}}{r}\right) r \, dr = D^2 \int_{r_2}^{r_1} \left(\frac{\tau_{\theta z}}{r}\right) r^3 \, dr \tag{7.1.49}$$

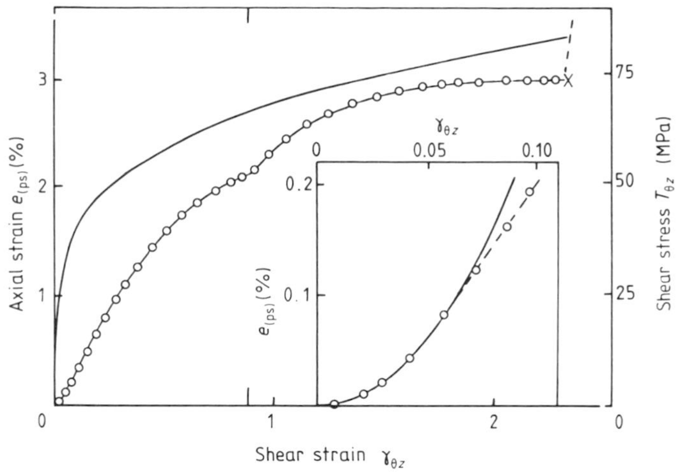

Figure 7.9 Variation of the static axial strain $e_{(PS)}$ with the static shear strain $\gamma_{\theta z}$ for a thin-walled tubular specimen of aluminium provided with a close fitting plug. The associated static stress–strain $(\tau_{\theta z}, \gamma_{\theta z})$ curve in simple torsion is also shown. X denotes the fracture point (Billington 1985).

and for the particular condition that the ratio $\tau_{\theta z}/r$ is regarded as being independent of r, equation (7.1.49) gives

$$e_{(P)} = \frac{1}{4}\left[\frac{2(r_1^2 + r_2^2)}{(r_1 + r_2)^2}\right]\gamma_{\theta z}^2 \qquad (7.1.50)$$

where, for a thin-walled tube

$$\gamma_{\theta z} = \tfrac{1}{2}(r_1 + r_2)D. \qquad (7.1.51)$$

For a sufficiently thin-walled tube, equation (7.1.50) gives

$$e_{(P)} \simeq \tfrac{1}{4}\gamma_{\theta z}^2 \qquad \text{(thin-walled tube).} \qquad (7.1.52)$$

Shown in the inset in figure 7.9 is the much expanded initial region of the $(e_{(PS)}(\equiv e_{(P)}), \gamma_{\theta z})$ curve shown in figure 7.9. The full curve has been calculated from equation (7.1.52) and is seen to provide a good correlation to the experimental $(e_{(PS)}(\equiv e_{(P)}), \gamma_{\theta z})$ curve over an initial limited range of $\gamma_{\theta z}$ (Billington 1985).

7.1.3 Poynting–Swift Effect

Swift (1947) observed that the mechanical response of a permanently twisted rod or tube is characterised by a nonlinear normal stress effect which is the direct counterpart to the Poynting effect in the elastic range. The Swift effect relates to the observation by Swift that a permanently twisted rod elongates in the direction of the principal axis about which the rod is being twisted, and this permanent elongation increases with increasing shear strain. It was observed by Swift that after a reversal of the torque there is an initial, small, shortening of the specimen, but continued twisting produces, however, continued axial extension. This means that the axial extension from twisting in one direction appears to continue for large reversed shear strains. An accumulation of such lengthenings in plastic torsion has also been observed by Freudenthal and Ronay (1966) and by Ronay (1966, 1967, 1968).

That the Poynting–Swift effect is predicted by a non-simple elastic–plastic material has been shown in the context of §6.1.5. However, technically useful studies of the Poynting–Swift effect cannot be obtained from the simple shear mode of deformation in the form discussed in §6.1.5. It is more useful to study the Poynting–Swift effect by way of the simple torsion of a thin-walled tube.

The $(e_{(PS)}, \gamma_{\theta z})$ curve shown in figure 7.9 is typically characteristic of the behaviour associated with the Poynting–Swift effect. Measurements of the Poynting–Swift effect in the form of the characteristic $(e_{(PS)}, \gamma_{\theta z})$ curve enable values of μ and $\bar{\mu}$ to be evaluated for any given value of $\gamma_{\theta z}$.

It is evident from the discussion given in §6.5.4, that studies of the Poynting–Swift effect using thin-walled tubes require the dependence of η upon $\gamma_{\theta z}$ to be known, having noted that η is defined by equation (6.5.14). The use of thin-walled tubes provided with a close-fitting plug for the purpose of evaluating η has been described by Billington (1977b). Except for a small initial range of $\gamma_{\theta z}$, the inner wall of the tube is effectively in contact with the outer surface of the plug. For such specimens $r_2 \simeq R_2$ for almost all D, and hence from equation (6.5.14)

$$\eta \simeq 1 + \frac{R_2^2}{R^2}(F-1) = 1 + \frac{R_2^2}{R^2} e_{(PS)} \tag{7.1.53}$$

and for a sufficiently thin-walled tube

$$\eta \simeq F \qquad (r_2 \simeq r_1). \tag{7.1.54}$$

For sufficiently thin-walled tubes it will be further assumed that the ratios $\tau_{\theta z}/r$ and μ/v can be regarded as being independent of r. These limiting conditions together with the condition $N_z = 0$ and the condition of equation (7.1.54) reduce equation (6.5.67) to a form which can be rearranged to give the ratio,

$$\frac{\mu}{v} = \frac{2(F+1)\left(\dfrac{r_2^2 \ln(r_1^2/r_2^2)}{(r_1^2 - r_2^2)} - 3\right) e_{(PS)} + (3r_1^2 + r_2^2) D^2}{(F+1)\left[6\left(1 + \dfrac{r_2^2 \ln(r_1^2/r_2^2)}{(r_1^2 - r_2^2)}\right) e_{(PS)} + \dfrac{(r_1^2 + 3r_2^2)}{(F^2+1)} e_{(PS)} D^2\right]} \tag{7.1.55}$$

and for a sufficiently thin-walled tube, equation (7.1.55) reduces to,

$$\frac{\mu}{v}=\frac{1-(F+1)(e_{(PS)}/\gamma_{\theta z}^2)}{(F+1)[3+\gamma_{\theta z}^2/(F^2+1)](e_{(PS)}/\gamma_{\theta z}^2)} \qquad (r_1 \simeq r_2) \qquad (7.1.56)$$

where

$$e_{(PS)}=F-1 \qquad \gamma_{\theta z}=(r_1+r_2)D/(2F). \qquad (7.1.57)$$

Values of μ/v can be evaluated from equation (7.1.56) using corresponding values of $e_{(PS)}$ and $\gamma_{\theta z}$ as given by the $(e_{(PS)}, \gamma_{\theta z})$ curve characteristic of the Poynting–Swift effect.

Values of v can be evaluated from equation (6.5.31) expressed in the form

$$v=-\frac{F(F+1)[3+\gamma_{\theta z}^2/(F^2+1)](e_{(PS)}/\gamma_{\theta z}^2)}{\{4F^6+[(F+1)(e_{(PS)}/\gamma_{\theta z}^2)-F^2]^2\gamma_{\theta z}^2\}^{1/2}}\gamma_{\theta z} \qquad (7.1.58)$$

from which it follows that when $\gamma_{\theta z}=0$, $v=0$, where use has been made of the fact that $(e_{(PS)}/\gamma_{\theta z}^2=\text{constant})$ for $\gamma_{\theta z}=0$. These values of v can be combined with the values of μ/v evaluated from equation (7.1.56) to give values of μ.

From equation (6.5.12) the following can be obtained

$$\bar{\mu}=\frac{\mu}{\zeta}\frac{G'\langle rr\rangle}{G'\langle\theta z\rangle}=\frac{3}{(4+\xi^2)^{1/2}}\frac{M'\langle rr\rangle}{M'\langle\theta z\rangle} \qquad (7.1.59)$$

where use has been made of equation (6.5.11). The identify of equation (6.5.28) can be used to express equation (7.1.59) in the form

$$\bar{\mu}=\frac{3}{(4+\chi^2)^{1/2}}\frac{M'\langle rr\rangle}{M'\langle\theta z\rangle}=\frac{3}{(4+\chi^2)^{1/2}}\frac{\bar{M}'\langle rr\rangle}{\bar{M}'\langle\theta z\rangle} \qquad (7.1.60)$$

which can be rearranged into the form

$$\bar{\mu}=\frac{3}{(4+\chi^2)^{1/2}}\left[\rho+\frac{\varphi_2}{\Psi}\left(\frac{2\bar{E}'^2\langle rr\rangle-\frac{1}{3}\operatorname{tr}\bar{\mathbf{E}}'^2}{\bar{\mathbf{E}}\langle\theta z\rangle}\right)\right]\equiv\bar{v}. \qquad (7.1.61)$$

Alternatively, the constitutive equation can be used in the form of equation (4.2.40), which gives equation (7.1.60) in the form

$$\bar{\mu}=\frac{\gamma_{\theta z}}{F^4(4+\chi^2)^{1/2}}\left((F^3+F^2+2F+2)(e_{(PS)}/\gamma_{\theta z}^2)+\frac{[(qF^2-1)F^4+2(F^4-1)(F+1)(e_{(PS)}/\gamma_{\theta z}^2)]}{(qF^2+1)}\right) \qquad (7.1.62)$$

where

$$q=-\beta_1/\beta_{-1}. \qquad (7.1.63)$$

Values of $\bar{\mu}$ can only be determined from experiment if the form of either the φ_i $(i=1,2)$ or the form of the β_i $(i=\pm 1)$ are known. Using the form for the β_i

($i = \pm 1$) given by equation (5.4.22), values of q have been determined from equation (5.4.22). Use has been made of the relations

$$I_{\mathbf{B}} \simeq 3 + \frac{(F+1)^2}{F^2} e_{(\mathrm{PS})}^2 + F^2 \gamma_{\theta z}^2$$

$$II_{\mathbf{B}} \simeq 3 + \frac{(F+1)^2}{F^2} e_{(\mathrm{PS})}^2 + \gamma_{\theta z}^2 \tag{7.1.64}$$

which were obtained from equations (6.5.21), (6.5.22) and (7.1.54). These values of q, together with the corresponding values of F, $e_{(\mathrm{PS})}$, $\gamma_{\theta z}$ and χ, can be entered into equation (7.1.62) to give values of $\bar{\mu}$.

It is to be noted that in this approach to the evaluation of μ and $\bar{\mu}$, all non-zero components of stress are effectively, though indirectly, accounted for in terms of the stress-ratios ξ and ζ.

Initial studies of the Poynting–Swift effect were based on the use of thin-walled tubular specimens of the same wall thickness for all materials. The wall thickness was made as small as possible, consistent with uniformity of material properties across the width of the tube (Billington 1977b). However, it was observed that very thin-walled tubular specimens of low strength materials develop a fold in the direction of the helix associated with the shear strain $\gamma_{\theta z}$. The $(e_{(\mathrm{PS})}, \gamma_{\theta z})$ curve for many of these materials passes through a maximum. This maximum in the $(e_{(\mathrm{PS})}, \gamma_{\theta z})$ curve is not observed when use is made of thick-walled tubes. In this connection, it has been shown (Billington 1978), that there is a range of wall thickness over which the thin-walled tube approximation in the form of equation (7.1.54), can be applied. Thus, studies of the Poynting–Swift effect therefore have to be restricted to specimens having a minimum wall thickness which is determined by the condition that the specimen must not form a fold. Comparatively high strength materials, as for example iron (Billington 1977b, figure 3), do not form a fold, their minimum wall thickness being determined solely by the need for uniformity of material properties across the wall.

Values of μ and $\bar{\mu}$, evaluated in the way described above, are shown in figures 7.10 and 7.11 as the $(\mu, \bar{\mu})$ curves for specimens of aluminium, copper and iron, use having been made of the $(e_{(\mathrm{PS})}, \gamma_{\theta z})$ curve characteristic of the Poynting–Swift effect for these materials (Billington 1978). Over an initial range of μ, these $(\mu, \bar{\mu})$ curves are seen to exhibit a marked formal similarity to the curves obtained from the measurements of Taylor and Quinney (1931), shown in figure 7.2.

It is evident from figure 7.10 that for copper in the fully annealed state there is a range ($-0.25 \leqslant \mu \leqslant -0.05$) over which $\bar{\mu}$ can be regarded as being a constant, a condition which is in accord with the assumption of a twelve-sided, piecewise continuous yield condition. Thus, over the limited range ($-0.25 \leqslant \mu \leqslant 0$), the yield surface simply expands without change in shape in accord with the concepts underlying classical isotropic hardening. For $\mu \leqslant -0.25$, both μ

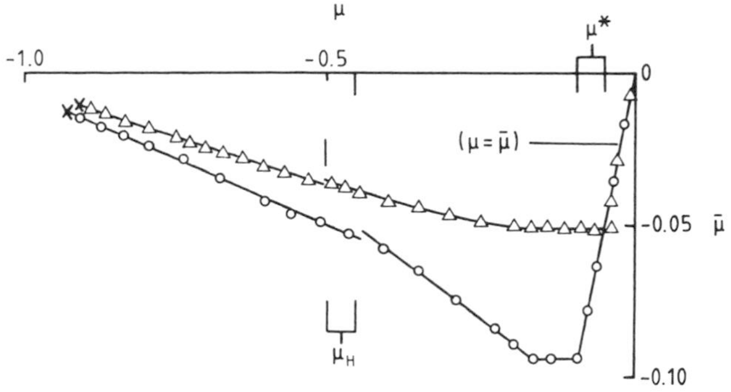

Figure 7.10 Values of the parameters μ and $\bar{\mu}$ for thin-walled specimens of Al, ○, and fully annealed Cu, △, determined from measurements in simple torsion (Billington 1985).

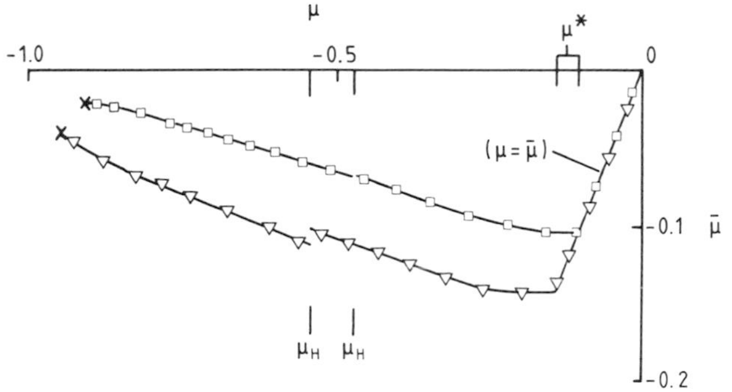

Figure 7.11 Values of the parameters μ and $\bar{\mu}$ for thin-walled specimens of Cu, □, and Fe, ▽, both in the as-received state, determined from measurements in simple torsion (Billington 1985).

and $\bar{\mu}$ vary with the parameter a which has been introduced into the stress intensity function of equation (4.2.1) to take account of the work-hardening properties of the material. Thus, for $\mu \leqslant -0.25$, each value of $\bar{\mu}$ corresponds to a different state of hardness of the material. However, having regard to the combined stress measurements of Taylor and Quinney (1931), in particular figure 7.2, it is tentatively concluded that for $\mu \leqslant -0.25$, each value of $\bar{\mu}$ corresponds to a different twelve-sided, piecewise continuous yield condition. The same interpretation is considered to apply to the other three $(\mu, \bar{\mu})$ curves shown in figures 7.10 and 7.11. The $(\mu, \bar{\mu})$ results for copper obtained from the measurements of the Poynting–Swift effect, are compared in figure 7.12 with the results obtained from the combined stress measurements of Taylor and Quinney (1931).

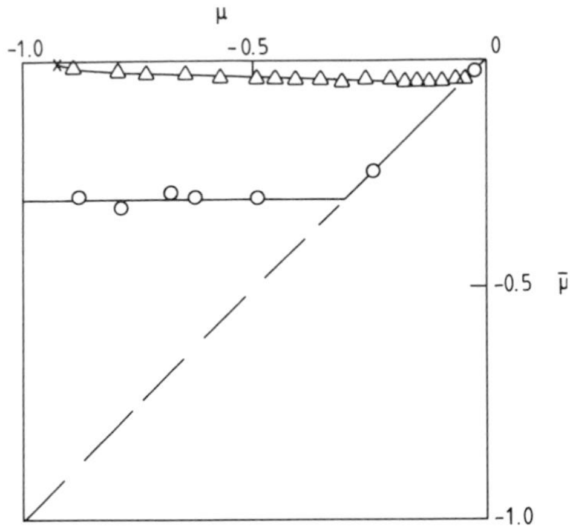

Figure 7.12 Values of the parameters μ and $\bar{\mu}$ for copper obtained from measurements of the Poynting–Swift effect (△) compared with the results from the combined stress measurements of Taylor and Quinney (1931) (○).

7.1.4 Stress–Strain Relations

The present discussion is concerned solely with the stress–strain relations considered in §6.7.

An example of a stress–strain curve which approximates the behaviour characteristic of an elastic–perfectly plastic material is shown in figure 7.5. The full curve has been calculated using equation (6.3.20) with G taken to be of the form

$$G = G_0\left\{1 - \frac{(\varepsilon'' - \varepsilon_0'')}{\varepsilon''(\exp[-\phi\lambda(\varepsilon'' - \varepsilon_0'')] + 1)}\left(1 - \frac{\tanh[\lambda(\varepsilon'' - \varepsilon_0'')]}{\lambda(\varepsilon'' - \varepsilon_0'')}\right)\right\}$$

subject to the conditions

$$\sigma_m'' = 3G_0\varepsilon_m^* \qquad \varepsilon_m^* = \varepsilon_0'' + \frac{1}{\lambda} \qquad (1 \leqslant \lambda \leqslant \infty) \qquad (\phi \gg 1)$$

where σ_m'' ($=104.4$ MPa) is the maximum stress, and where $G_0 = 4.24$ GPa, $\varepsilon_m^* = 0.821\%$, $\lambda = 239.2$, $\phi = 10$. Setting $\lambda = \infty$ reduces equation (6.3.20) to the constitutive equation describing the classical idealisation of an elastic–perfectly plastic material which is characterised by a piecewise continuous stress–strain curve with the initial yield occurring at the vertex.

The comparison of the stress–strain curves, characteristic of the response of a material tested in compression and in tension, in terms of the Cauchy stress σ_{zz} and the natural strain $\varepsilon_{zz} = \ln F$, was first suggested by Ludwik (1909) and shown to give a single stress–strain curve by Ludwik and Scheu (1925). Using the notation of the present §6.7 and the results for F given in table 2 of Ludwik

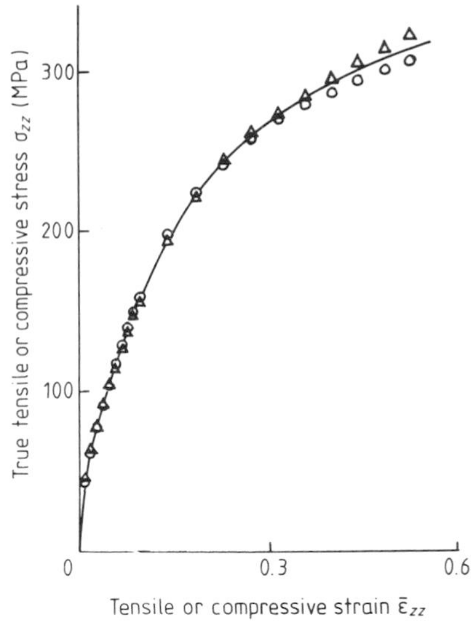

Figure 7.13 Comparison of the true stress–uniaxial tensile strain curves for annealed copper tested in tension (△) and compression (○) (Ludwik and Scheu 1925).

and Scheu (1925), values of $\bar{\varepsilon}_{zz}$ have been calculated from equation (6.7.10). The variation of $\bar{\varepsilon}_{zz}$ with the values of σ_{zz} given by Ludwik and Scheu (1925) is shown in figure 7.13 for annealed copper.

For simple compression, characterised by $\nu = -1$, that is case (i) of §6.7, equation (6.7.11) gives

$$\sigma'' = \sigma_{zz} \qquad \varepsilon'' = \bar{\varepsilon}_{zz} = (F+1)(F^3-1)/6F^2 \tag{7.1.65}$$

for a solid rod. The variation of $\sigma'' = \sigma_{zz}$ with $\varepsilon'' \equiv \bar{\varepsilon}_{zz}$ is shown in figure 7.14(○) for aluminium tested in simple compression. This sample of aluminium is from the same source as that used to obtain the results shown in figure 7.10.

The results for aluminium shown in figure 7.10 are taken to imply that to a first approximation, the test specimen of aluminium satisfies a Tresca type stress intensity and strain intensity function.

The shear stress–shear strain $(\tau_{\theta z}, \gamma_{\theta z})$ curve has been evaluated from measurements in simple torsion using a thin-walled tubular specimen of aluminium. Although the material exhibits the Poynting–Swift effect, the axial elongation $e_{zz} = F-1$, characteristic of the Poynting–Swift effect, is sufficiently small for F to be approximated to unity. Hence $\eta \sim 1$ and equation (6.5.28) gives $\chi \simeq -\gamma_{\theta z} = -rD$. Entering this limiting condition into equation (6.7.25) gives

$$\sigma'' = 2(1+\tfrac{1}{4}\gamma_{\theta z}^2)^{1/2}\tau_{\theta z} = 4G\tfrac{1}{2}(1+\tfrac{1}{4}\gamma_{\theta z}^2)^{1/2}\gamma_{\theta z} = 4G\varepsilon'' \tag{7.1.66}$$

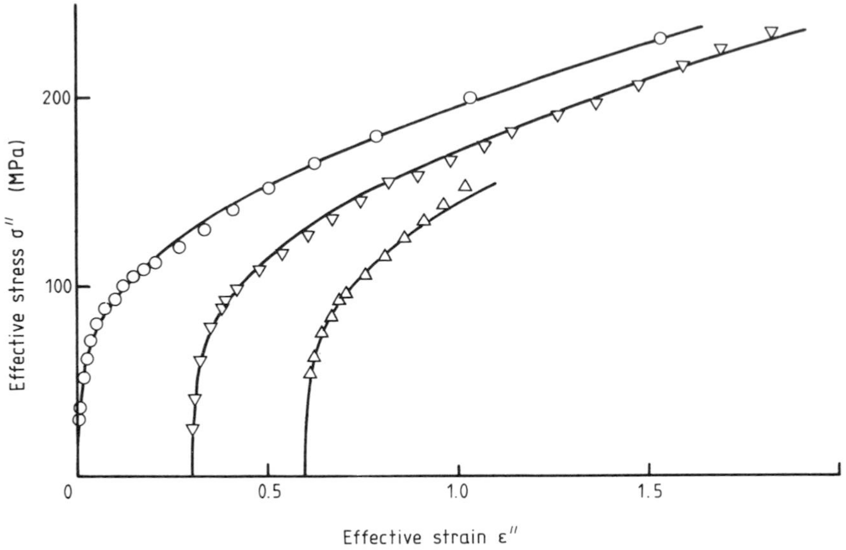

Figure 7.14 Variation of σ'' with ε'' for aluminium: ○, compression; △, pure shear; ▽, simple torsion.

where use has been made of equation (3.2.11). Values of σ'' and ε'' have been evaluated from equation (7.1.66) using corresponding values of $\tau_{\theta z}$ and $\gamma_{\theta z}$ obtained from the experimental $(\tau_{\theta z}, \gamma_{\theta z})$ curve. The variation of these values of σ'' with ε'' is shown in figure 7.14(▽), the origin for this curbe being shifted by 0.3 units parallel to the abscissa.

Pure shear can be approximated by a thin-walled tube fitted with a plug, the interface between the inner surface of the tube and that of the plug being lubricated with an extreme pressure lubricant (Billington 1977a). The presence of the close fitting plug maintains $r_2 \simeq R_2$ for all F, and hence

$$\eta_{\mathrm{m}} \simeq 1 + \frac{R_2^2}{R_{\mathrm{m}}^2}(F-1) \qquad R_{\mathrm{m}} = \tfrac{1}{2}(R_1 + R_2) \tag{7.1.67}$$

where R_{m} denotes the mean R and use has been made of equation (7.1.53). The difference between equations (6.5.54) and (6.5.53) gives,

$$[(\sigma_{zz} - \sigma_{\theta\theta})]_{r=r_{\mathrm{m}}} = 3\bar{G}\frac{\bar{\Psi}}{\Lambda}\bar{\gamma}_1 \qquad r_{\mathrm{m}} = \tfrac{1}{2}(r_1 + r_2) \tag{7.1.68}$$

where

$$\bar{\gamma}_1 = \bar{\varepsilon}_{zz} - \frac{(\eta_{\mathrm{m}} - 1)(F^2 + \eta_{\mathrm{m}})}{6\eta_{\mathrm{m}}F} \tag{7.1.69}$$

and where η_m is to be evaluated from equation (7.1.67). It is now assumed, to a first approximation, that for an infinitely thin-walled tube the conditions of pure shear can be applied to equation (7.1.68) to give

$$\sigma_{zz}=6\bar{G}\frac{\Psi}{\Lambda}\bar{\gamma}_1 \qquad (r_a\simeq r_{\rm m}) \tag{7.1.70}$$

where $r_{\rm m}$ denotes the mean r and use has been made of equation (6.7.14) subject to the conditions

$$r_a\simeq r_{\rm m} \qquad r_a/R_a=1 \qquad [\sigma_{rr}]_{r=r_{\rm m}}=0. \tag{7.1.71}$$

Equation (7.1.70) is to be compared with equation (6.7.17) subject to the condition $\sigma_{zz}=2\sigma_{\theta\theta}$ at $r=r_a\simeq r_{\rm m}$. The effective stress and effective strain are given by equation (6.7.26) with $a=1$ and $2\gamma_1$ replaced by $3\bar{\gamma}_1/2$ thus

$$\sigma''=\sigma_{zz}=4G\tfrac{3}{2}\bar{\gamma}_1=4G\varepsilon''. \tag{7.1.72}$$

The variation of the (σ'',ε'') curve obtained from this approximation to pure shear is shown in figure 7.14($\triangle$), the origin for this curve being shifted by 0.6 units parallel to the abscissa.

In figure 7.14, the three full curves are identical, and hence it is concluded that the three individual stress–strain relations for these three simple modes of deformation can be combined using equation (6.7.2) to give a single (universal) effective stress–effective strain curve. All the full curves shown in figure 7.14 have been calculated from equation (6.7.2) using,

$$G=G_0\left[1+\phi_1\left(\frac{\varepsilon''}{1+\phi_2\varepsilon''}\right)^n\right]^{-1} \tag{7.1.73}$$

with $G_0=72$ GPa, $\phi_1=1625$, $\phi_2=0.145$ and $n=0.72$.

The variation of $\bar{\sigma}''$ with $\bar{e}''$ is shown in figure 7.15 for simple compression, pure shear and simple torsion using equations (7.1.73), (6.7.11), (6.7.16), (6.7.20) and (6.7.21). The origin for the $(\bar{\sigma}'', \bar{e}'')$ curves for simple torsion and pure shear have been displaced parallel to the ordinate by 40 MPa and 80 MPa respectively. The full lines are identical and represent the linear relation of equation (6.7.6). The experimental $(\bar{\sigma}'', \bar{e}'')$ curves are seen to be in good accord with the linear relation of equation (6.7.6).

7.1.5 Assumption of Isotropy and Submacroscopic Mechanisms of Deformation

The individual stress–strain relations, obtained from measurements in simple compression, pure shear and simple torsion, compound together to give a single effective stress–effective strain curve over a wide range of strain. This observation is taken as evidence in support of the assumption that the materials can be regarded as being isotropic.

It would at present appear unlikely that the behaviour characteristic of the

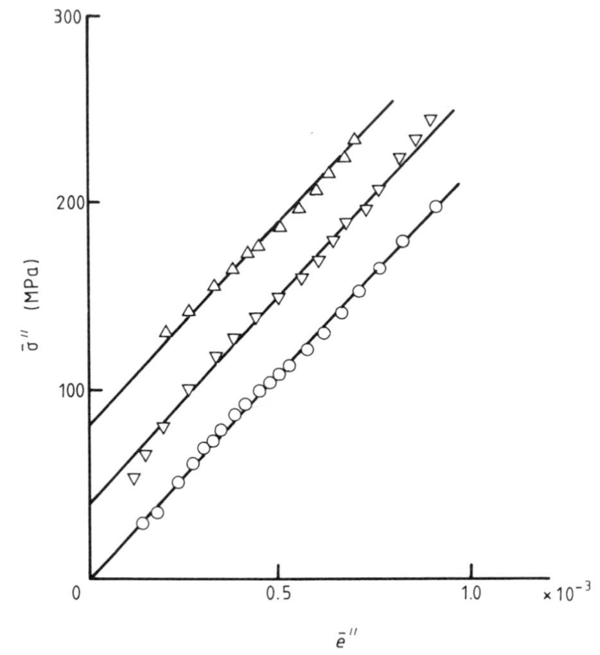

Figure 7.15 Variation of $\bar{\sigma}''$ with $\bar{e}''$ for aluminium ○, compression; △, pure shear; ▽, simple torsion.

Poynting–Swift effect can be completely accounted for in terms of the theories of crystal dislocation mechanics.

Although there is insufficient experimental evidence to resolve the question of the extent to which a metal must be considered as being compressible, the studies of Foux (1964) can be taken to imply that the Poynthing effect can be largely accounted for by a nonlinear theory of elasticity which assumes the material to be incompressible. In this connection, it is concluded by Freudenthal and Ronay (1966), Swift (1947) and Billington (1977b) from their studies of the Swift effect that, within the limits of experimental accuracy, the permanent axial elongation of a twisted rod or tube takes place without measurable change in volume. This conclusion led Swift (1947) to suggest that a polycrystalline body should be regarded as a two-component system consisting of a low concentration of relatively hard solid particles dispersed in a soft solid matrix. Axial elongation of a permanently twisted rod or tube arises, in part, from rotation of the relatively hard solid particles. This rotation of the shape, elongated by the distortion of the surrounding medium, will increase its dimension in the axial direction and proportionately reduce its transverse dimension.

The assumption that the bulk of the material deforms at the microstructural level according to the theories of crystal dislocation mechanics raises the following question: how can the use of a total deformation type constitutive equation for an isotropic, homogeneous, incompressible strain-hardening elastic–plastic solid with large deformation be reconciled with the microscopic mechanism of deformation which, since it is directional, requires that plastic deformation on a microscopic scale be regarded as physically anisotropic.

The most interesting comment regarding the above question is that made by Freudenthal, in the discussion to Freudenthal and Ronay (1968), who suggests from a consideration of sub-grain formation that in terms of its macroscopic response, *the material essentially produces the isotropy which it needs.* Thus, as noted by Freudenthal, effects usually attributed to the assumed anisotropy of the deformation are to be regarded as having their origin in the Swift effect. The observation that universal stress–strain curves can be obtained over an extended range of effective strain is in general accord with the above suggestion made by Freudenthal. The suggestion by Swift (1947) that a polycrystalline body should be regarded as a two-component system consisting of a low concentration of relatively hard solid particles dispersed in a soft solid matrix is of interest regarding the question of to what extent can a material undergoing plastic deformation be regarded as isotropic. A two-component system of this type will be characterised by differential grain motion with which there will be associated rotation of the relatively hard solid particles. This rotation of a particle, elongated by the distortion of the surrounding medium will to some extent retard the progressive development of anisotropy during the plastic deformation. In particular, the rotational motion will inhibit the development of any extensive, continuous regions involving a preferred direction of flow. Questions relating to isotropy and anisotropy in the context of the incremental theory of plasticity have been considered by Rees (1982).

7.2 METALS: II, APPLICATIONS

7.2.1 High Rates of Strain

There is a growing body of experimental evidence for the existence of a wide range of materials which exhibit a difference between the high strain rate stress levels and the corresponding effectively static stress levels for the same value of the strain; this behaviour has been reviewed by Campbell (1973) (see also Billington and Tate 1981). It is of interest to consider an elastic–plastic solid for which the uniaxial stress–strain curve, when tested at near zero strain rate, is of the form shown schematically in figure 5.2(*b*); that is the material exhibits the mechanical response characteristic of an elastic–perfectly plastic material when tested at near zero strain rate. Let the material be subjected to a near instantaneous compressive stress loading pulse, the magnitude of which exceeds the initial static yield stress. It is evident from the discussion given in

Chapter 5 that the stress should pass through a maximum, in excess of the initial yield, the stress decreasing with decreasing strain rate until it attains the value corresponding to the initial static yield stress. The unloading from this stress level would be purely elastic. Thus, the observed stress enhancement is in general accord with the constitutive equations considered in §§5.1 to 5.4. There are, however, materials whose mechanical response cannot be described in detail using the constitutive equations discussed in §§5.1 to 5.4.

The remaining discussion is concerned with materials which cannot be described in the general context of Chapter 5. For a given strain, and dynamic loading below the high energy level of shock loading, the stress for these materials initially increases with increasing strain rate, the stress approaching a saturation value such that for a limited range of high strain rates the individual dynamic stress–strain curves, one for each value of the loading pulse, tend to compound together. Using the notation of §§6.5 and 6.7(i), the way in which the true, i.e. the Cauchy, compressive stress σ_{zz} varies with the uniaxial compressive strain

$$\bar{\varepsilon}_{zz} = (F+1)(F^3-1)/6F^2$$

where F is defined by equation (6.5.15), is shown for aluminium in figure 7.16. Similar behaviour has been obtained for measurements which approximate closely to simple tension (using the technique described by Lindholm and Yeakley 1968). The variation of the stress σ_{zz} with the strain $\bar{\varepsilon}_{zz}$ is shown in figure 7.17 for specimens of aluminium from the same stock sample as that used for figure 7.16.

Using the technique described by Nicholas and Campbell (1972), measurements in simple torsion on aluminium specimens from the same stock

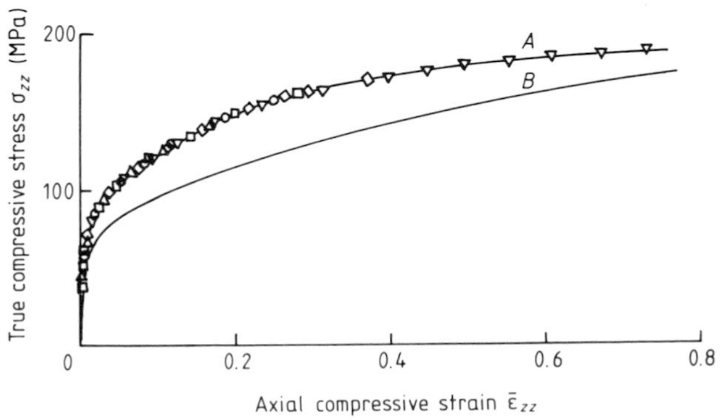

Figure 7.16 Dynamic (A) and static (B) stress–strain curves for aluminium in compression. Compressive strain loading pulse maximum $= \varepsilon_{I_{max}} \times 10^{-3}$: □, 0.875; △, 1.250; ◇, 1.317; ○, 1.846; ▽, 2.192 (Billington 1977a).

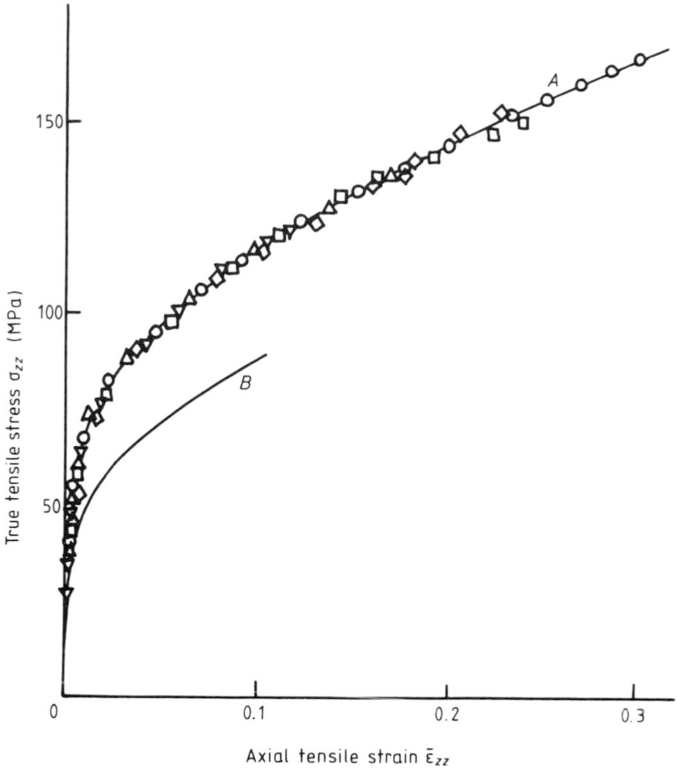

Figure 7.17 Dynamic (*A*) and static (*B*) stress–strain curves for aluminium in tension. Maximum strain rate $= \dot{e}_{max} \times 10^3\ s^{-1}$: ▽, 0.69; △, 0.83; □, 0.93; ◇, 1.27; ○, 1.69 (Billington 1977a).

sample as that used for figures 7.16 and 7.17 have been obtained and are shown in figure 7.18. Also shown in figures 7.16, 7.17 and 7.18 are the corresponding near static stress–strain curves (curve *B*).

It is evident from figures 7.16, 7.17 and 7.18 that the stress–strain response at high strain rates and near zero strain rates is formally very similar for these three simple modes of deformation. There is, however, one significant difference in behaviour between the dynamic response in torsion and the dynamic response in both tension and compression. For compression and tension, the dynamic stress levels for a given value of the strain are considerably larger than the corresponding effectively static stress levels. In marked contrast, only a comparatively small difference has been detected between the corresponding dynamic and static stress levels in torsion over a range of strain rates from $\sim 10^{-3}$ to $\sim 10^3\ s^{-1}$, as can be seen from figure 7.18.

With regard to figure 7.18, it is of interest to note the distinction between pure torsion and simple torsion. For present purposes, pure torsion is taken to

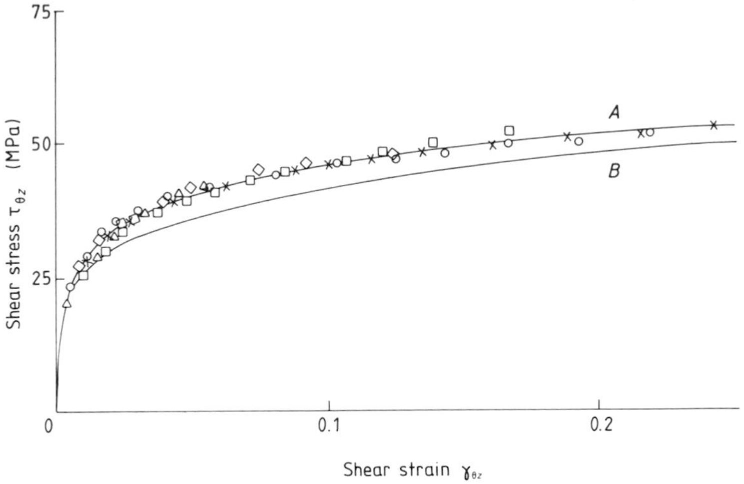

Figure 7.18 Dynamic and static stress–strain curves for aluminium in torsion. Maximum shear strain rate $= \dot{\gamma}_{max} \times 10^3\ s^{-1}$: ◇, 0.40; ○, 0.89; △, 0.95; □, 1.10. *A*, dynamic stress–strain curve; *B*, static stress–strain curve for simple torsion characterised by $F \geqslant 1$ for all $\gamma_{\theta z} \geqslant 0$; the crosses denote values of $\tau_{\theta z}$ for static measurements in pure torsion characterised by $F=1$ for all $\gamma_{\theta z}$ (Billington 1977a).

be characterised by the condition $F=1$ for all values of the total shear strain, $\gamma_{\theta z}$, i.e. a material deformed in pure torsion does not exhibit the behaviour characteristic of the Poynting–Swift effect in so far as $e_{(PS)} = F - 1 = 0$ for all $\gamma_{\theta z}$. Simple torsion is characterised by $F > 1$ for all $\gamma_{\theta z} > 0$ and is the mode of deformation most suited to the observation of both Poynting and Swift effects.

For those materials which exhibit the Poynting–Swift effect, the pure torsion mode of deformation, characterised by $F=1$ for all $\gamma_{\theta z}$, requires the application of a normal compressive force N_z, as can be seen from equation (6.6.17). Attenuation of the Poynting–Swift effect by applying an appropriate compressive force N_z gives rise to a shear stress in pure torsion ($F=1$ for all $\gamma_{\theta z}$), which is significantly higher than the shear stress in simple torsion ($F > 1$ for all $\gamma_{\theta z} > 0$) for the same $\gamma_{\theta z}$.

The values of the shear stress obtained from effectively static measurements in pure torsion ($F=1$ for all $\gamma_{\theta z}$), are shown by the cross symbols in figure 7.18, and are seen to be in good agreement with the high strain rate stress levels, for the same value of the shear strain. These results are typical for low strength materials.

For higher strength materials, such as iron, the presently observed shear stress levels obtained from effectively static measurements in pure torsion observed at present fall markedly below the corresponding high strain rate shear stress levels. However, for higher strength materials it has been observed in all cases that as the test machine has been made increasingly rigid in the axial direction, the magnitude of the static shear stress enhancement increases for the same given value of the shear strain. Thus, until the problem of how to obtain a sufficiently rigid test machine in the axial direction has been resolved, an acceptable comparison of the shear stress levels obtained from effectively static measurements in pure torsion with the corresponding high strain rate stress levels cannot be made.

From these observations it can be concluded that the Poynting–Swift effect is a source of weakness in a solid for the range of strengths presently available, in so far as its suppression (i.e. $F=1$ for all D), enhances the stress.

For various low strength polycrystalline metals which do not exhibit a well defined yield stress, it has been shown that the stress level in pure static torsion closely approaches the corresponding stress level in simple dynamic torsion for the same value of the shear strain. This observation, together with the conclusion that the Poynting–Swift effect is a source of weakness in a solid, gave rise to the tentative prediction that a material tested in dynamic torsion may not exhibit the Poynting–Swift effect. This prediction has been confirmed by Pukas (1977) who observed that, within the limits of experimental accuracy, the mechanical response of a material to high rates of strain is characterised by the absence of a detectable Poynting–Swift effect.

Swift's (1947) assumption of two or more minority components dispersed in a soft solid matrix is of interest in relation to the response at high rates of strain. Assume the Swift effect arises from some form of submacroscopic mechanism arising from the micromechanics of one of the minority components. The observation by Pukas (1977) that the mechanical response to high rates of strain is characterised by the absence of a dynamic Swift effect can now be accounted for by the further assumption that the submacroscopic mechanism which gives rise to the near static Swift effect is suppressed at high rates of strain. For example, the rotational motion of comparatively large units of subcontinua will be characterised by a time constant which is relatively long compared with the duration of the high strain rate loading pulse. Hence, for small strain rates, the rotational motion can follow the rate of deformation. As the strain rate is increased, there is a progressive phasing out of the rotational motion until at high strain rates the Swift effect arising from rigid body rotation has been attenuated to zero.

It has been shown (Billington 1977c), that the dynamic stress–strain curves for the three simple modes of deformation, that is tension, compression and torsion, compound together to give a single universal stress–strain curve over an initial, limited range of effective strain. Thus, if the static characteristic shear stress–shear strain curve for pure torsion could be shown to an acceptable degree of accuracy to approximate closely to the dynamic characteristic shear stress–shear strain curve, then the dynamic stress–strain curves in compression, tension and torsion could be obtained directly from the near static measurements in the three simple modes of deformation using the concept of a universal stress–strain curve. This approach, although only giving an estimate of the mechanical response at high rates of strain, could be technically useful for high strength materials whose properties at high rates of strain cannot be determined by direct measurement.

It is to be emphasised that, even if it can be shown that the static characteristic shear stress–shear strain curve for pure torsion approximates closely to the dynamic characteristic shear stress–shear strain curve, this cannot be taken as evidence that the same submacroscopic mechanisms of deformation are necessarily effective in both cases.

With §7.1.4 as the basis, the stress–strain curve shown in figure 7.16 for aluminium tested in simple compression at high rates of strain is reproduced as just the full ($\sigma''=\sigma_{zz}$, $\varepsilon''=\bar{\varepsilon}_{zz}$) curve in figure 7.19. The stress–strain curve shown in figure 7.17 for aluminium tested in simple tension at high rates of strain is also reproduced in figure 7.19 using full circles. In the case of the torsion measurements shown in figure 7.18, $F=1$, $\eta=1$ for all $\gamma_{\theta z}$ and hence

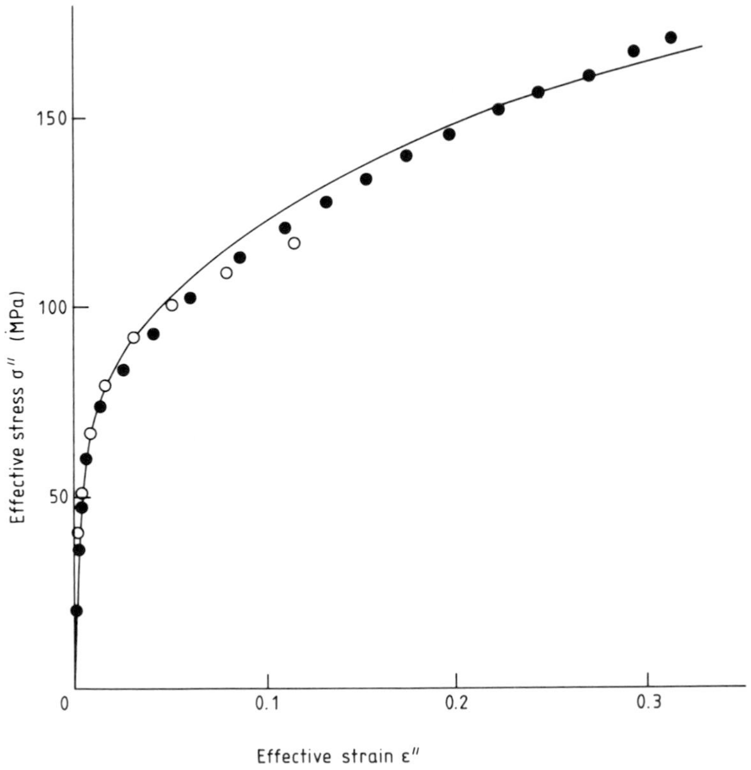

Figure 7.19 Variation of σ'' with ε'' for aluminium. Full curve, compression; ●, tension; ○, simple torsion.

equation (7.1.67) reduces to

$$\sigma'' = (2/\sqrt{3})[(3+\mu^2)(1+\tfrac{1}{4}\gamma_{\theta z}^2)]^{1/2}\tau_{\theta z}$$
$$\varepsilon'' = (1/2\sqrt{3})[(3+v^2)(1+\tfrac{1}{4}\gamma_{\theta z}^2)]^{1/2}\gamma_{\theta z}. \quad (7.2.1)$$

The variation of σ'' with ε'' for the torsion measurements at high rates of strain are shown as the open circles in figure 7.19. It is evident from figure 7.19 that, to within the limits of experimental accuracy, a single, effective stress–effective strain curve, has been obtained for deformation at high rates of strain.

7.2.2 The Effect of the Yield Surface Vortex on the Failure of Ductile Metals

In the general field of fracture mechanics, emphasis has been placed on Mode I cracking under tension, to the almost total exclusion of Modes II and III, such that without any qualification almost all fracture situations are assumed to be Mode I. However, plastic deformation in simple tension is effectively a deformation in shear, and hence the mechanical response in simple tension must be determined to some extent by the Poynting and Swift effects. Thus it follows that if crack propagation in simple torsion is in any way determined by the mechanisms underlying the Poynting and Swift effects, then crack propagation in specimens deformed plastically in simple tension will be too. There is a growing body of experimental evidence (for example, Billington (1977), Freudenthal and Ronay (1966), L'Hermite and Dawance (1947), Ronay (1966, 1967, 1968) and Swift (1947)), which can be taken to imply the possibility that the Swift effect when considered at the submacroscopic level may have a dominant effect on the observed mechanical response of a material deformed in the plastic range of deformation.

Since the Poynting–Swift effect is usually studied by way of the simple torsion of a thin-walled tube, the question arises of whether any useful information regarding the failure of ductile metals can be obtained from this type of measurement. It is this aspect of the failure of ductile metals which is of present interest.

The characteristic, shear stress–shear strain, that is the $(\tau_{\theta z}, \gamma_{\theta z})$ curve for a thin-walled tubular specimen of aluminium is shown in figure 7.9 together with the $(e_{(PS)}, \gamma_{\theta z})$ curve characteristic of the behaviour of the associated Poynting–Swift effect. The variation of the parameter $\bar{\mu}$ with Lode's stress parameter μ is shown in figure 7.10, where use has been made of the $(e_{(PS)}, \gamma_{\theta z})$ curve shown in figure 7.9 to evaluate μ and $\bar{\mu}$. Also shown in figure 7.10 is the $(\mu, \bar{\mu})$ curve for fully annealed copper, and the $(\mu, \bar{\mu})$ curves for copper and iron are shown in figure 7.11, both in the as-received state. It is evident from figures 7.10 and 7.11 that failure in simple torsion occurs for values of μ which approximate closely to $\mu = -1$. In this connection it should be noted that the values of μ and $\bar{\mu}$ shown in figures 7.10 and 7.11 have been evaluated at the mean current radius

$r=r_{\mathrm{m}}=(r_1+r_2)/2$ where r_1 and r_2 are the current values of the outer and inner radii respectively, of a thin-walled tube.

For the conditions of the Poynting–Swift effect,

$$3\zeta=\frac{2\sigma_{rr}-\sigma_{\theta\theta}}{\tau_{\theta z}}=2\frac{\sigma_{rr}}{\tau_{\theta z}}+\xi \tag{7.2.2}$$

where the stress ratios ξ and ζ are defined by equations (6.5.9) and (6.5.10) respectively. Entering the form for ζ given by equation (7.2.2) into equation (6.5.11) gives

$$\mu=\frac{2(\sigma_{rr}/\tau_{\theta z})}{(4+\xi^2)^{1/2}}+\frac{\xi}{(4+\xi^2)^{1/2}}. \tag{7.2.3}$$

For zero applied traction on the curved surface at $r=r_1$, such that $[\sigma_{rr}]_{r=r_1}=0$, equation (7.2.3) gives

$$[\mu]_{r=r_1}=\left[\frac{\xi}{(4+\xi^2)^{1/2}}\right]_{r=r_1}\equiv\left[\frac{\chi}{(4+\chi^2)^{1/2}}\right]_{r=r_1} \tag{7.2.4}$$

which can be evaluated from observations of the Poynting–Swift effect, use being made of the identity $\xi\equiv\chi$ of equation (6.5.28). At $r=r_2$, equation (7.2.3) gives

$$[\mu]_{r=r_2}=\left[\frac{\chi}{(4+\chi^2)^{1/2}}\left(1-2\frac{\sigma_{rr}}{\sigma_{\theta\theta}}\right)\right]_{r=r_2}. \tag{7.2.5}$$

Use of equations (7.2.4) and (7.2.5) leads to the conclusion that failure occurs to a good approximation at $\mu=-1$, $r=1.014r_2$ for the aluminium specimen shown in figure 7.10 and at $r=r_2$ for the copper specimens shown in figures 7.10 and 7.11. For the iron specimen shown in figure 7.11, $\mu=-1$ at $r=1.056r_2$. This although significantly larger than for other metals, is still sufficiently close to the inner radius of the tube for it to be tentatively concluded that the total deformation type constitutive equation (4.2.44) predicts that failure of ductile materials occurs in the immediate vicinity of the inner surface of the tubular specimens.

The observation that failure occurs at $r\simeq r_2$ when $\mu=-1$ could be taken to imply that at $\mu=-1$, one of the associated components of stress has attained the fracture stress for failure in the simple torsion mode of deformation, thus providing a physically acceptable basis for the observed failure of a ductile material. However, that this does not apply to the observed failure of ductile materials tested in simple torsion can be seen as follows.

It has been observed (Billington 1978) that for the metals of interest at present, fracture occurs, to a good approximation, in a transverse plane. This observation can be taken to imply that failure in simple torsion arises as a result of the thin-walled tubular specimen being subjected to some form of plastic deformation in the axial direction. However, since there is no directly applied tensile axial stress, the observation that fracture occurs approximately in the transverse plane is difficult to reconcile with the critical value of any one of the non-zero components of stress, as would appear to be implied by the observation that fracture always occurs at $\mu = -1$.

The observed failure in a transverse plane must be accompanied by axial elongation. In this connection, although there is no applied axial tensile stress, there is the axial tensile strain $e_{(PS)}$, characteristic of the Poynting–Swift effect.

For the twelve-sided linear, piecewise continuous yield condition discussed in §3.2.3, there is a discontinuous change at $\mu = -1$ from $\bar{\mu}_0$ to $\bar{\mu}_1$, the relation between $\bar{\mu}_0$ and $\bar{\mu}_1$ being obtained from equation (3.2.30) by setting $n = 1$ to give

$$|\bar{\mu}_1| = 4/(1 + |\bar{\mu}_0|) - 1 \qquad (\mu = -1). \tag{7.2.6}$$

It is the discontinuous change from $\bar{\mu}_0$ to $\bar{\mu}_1$ at $\mu = -1$ which is of interest in relation to the failure of a ductile metal at $\mu = -1$.

For the $(\mu, \bar{\mu})$ curve shown in figure 7.10, it follows from equation (7.2.6) that the ratio $\bar{\mu}_n/\mu$ at $\mu = -1$, and for $n = 0, 1$, would in the absence of failure have to increase discontinuously from 0.01 to 2.96 as $\bar{\mu}_n$ changes from $\bar{\mu}_0$ to $\bar{\mu}_1$. The value of $\bar{\mu} = \bar{\mu}_1$ at $\mu = -1$ and the value of $\gamma_{\theta z}$ at $\mu = -1$ can be substituted into equation (7.1.62) to give the predicted value which F, and hence $e_{(PS)}$, would have in the absence of failure. In this way the $(e_{(PS)}, \gamma_{\theta z})$ curve associated with the $(\mu, \bar{\mu})$ curve for aluminium shown in figure 7.10 predicts an axial elongation (Billington 1984)

$$[e_{(PS)}]_{\bar{\mu}=\bar{\mu}_1} = 79.3[e_{(PS)}]_{\bar{\mu}=\bar{\mu}_0} \qquad (\mu = -1) \tag{7.2.7}$$

which the tube cannot achieve. The magnitude of the ratio of the axial elongation $e_{(PS)}$ at $\bar{\mu} = \bar{\mu}_1$ to its value at $\bar{\mu} = \bar{\mu}_0$ given in equation (7.2.7) is typical of that obtained for various ductile metals in the polycrystalline state. Failure is thus to be associated with the incipient stage of a sudden and rapid increase in the axial elongation characteristic of the Poynting–Swift effect, the fracture criterion relating specifically to the discontinuous change in the parameter $\bar{\mu}_n$ $(n = 0, 1)$ which occurs at $\mu = -1$.

7.2.3 The Effect of Work-Hardening on the Failure of Ductile Metals

Ductile metals considered at the subcontinuum level exhibit a wide range of material properties, and in this context it would therefore seem unlikely that all ductile metals will fail when μ attains the value $\mu = -1$. That this is so can be seen by considering the Poynting–Swift effect for a specimen of an aluminium alloy.

The experimental $(e_{(PS)}, \gamma_{\theta z})$ curve characteristic of the Poynting–Swift effect for an aluminium alloy specimen is shown in figure 7.20 (triangles), the origin for this curve being displaced by 0.8 units parallel to the abscissa. Also shown

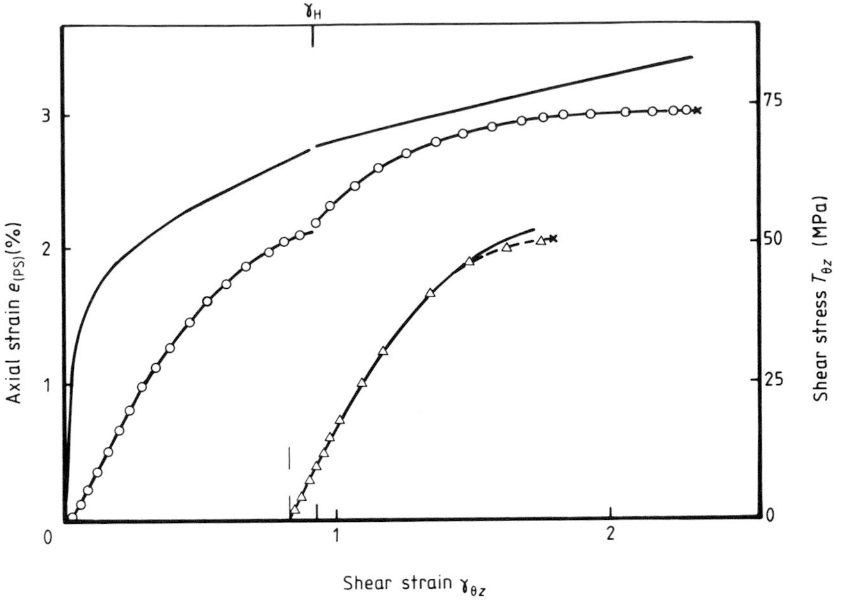

Figure 7.20 ○, variation of the static axial strain, $e_{(PS)}$, with the static shear strain $\gamma_{\theta z}$ for a thin-walled specimen of aluminium; △, a thin-walled specimen of an aluminium alloy, both specimens being provided with a close fitting plug. × denotes fracture. Also shown is the static stress–strain ($\tau_{\theta z}$, $\gamma_{\theta z}$) curve for aluminium in simple torsion.

displaced by 0.8 units parallel to the abscissa is the full curve for the specimen of aluminium which is seen to be almost indistinguishable from the experimental ($e_{(PS)}$, $\gamma_{\theta z}$) curve for the aluminium alloy. Details of the composition of the test materials are given in table 7.3. It is evident from figure 7.20 that within the limits of experimental accuracy, the ($e_{(PS)}$, $\gamma_{\theta z}$) curves for the specimen of aluminium and the specimen of aluminium alloy are indistinguishable, the only significant difference being that the aluminium alloy fails at a value of $\gamma_{\theta z}$ which is very much smaller than the value of $\gamma_{\theta z}$ at which the specimen of aluminium fails. This observation, together with the fact that the aluminium content of the two specimens is seen from table 7.3 to differ

Table 7.3 Chemical and spectrographic examination of specimens.

	Al	Cu	Fe	Mn	Si	Mg	Ti
Al	99.26	0.02	0.43	0.01	0.02	0.05	0.015
Al alloy	97.52	0.08	0.27	0.52	1.09	0.47	0.05

by only 1.74% is taken to imply that the Poynting–Swift effect is associated primarily with the properties of the aluminium.

A characteristic feature of both the stress–strain ($\tau_{\theta z}$, $\gamma_{\theta z}$) curve and the ($e_{(PS)}$, $\gamma_{\theta z}$) curve for the specimen of aluminium is the presence of a small discontinuity in these curves. This is identified in figure 7.20 as occurring at the particular value $\gamma_{\theta z}=\gamma_H$ of the shear strain. The discontinuity occurring at $\gamma_{\theta z}=\gamma_H$ gives a small, but detectable discontinuity in the ($\mu, \bar{\mu}$) curve in figure 7.10 occurring at the particular value $\mu=\mu_H$ which corresponds to the particular value γ_H of the shear strain. For present purposes, the discontinuity shown occurring at γ_H will be considered to give rise to a discontinuous change in $\bar{\mu}_0$ from $\bar{\mu}_0=\bar{\mu}_E$ to $\bar{\mu}_0=\bar{\mu}_A$ ($\bar{\mu}_A<\bar{\mu}_E$) for a given value of $\mu=\mu_H$ In the case of an incompressible material, the hydrostatic component of stress has no effect on the criterion of yield and it is therefore convenient to project the yield surface onto the deviatoric, or π-plane which is normal to the hydrostatic stress axis, in the way described in §3.1.2. The interpretation of the (μ, $\bar{\mu}$) curves shown in figures 7.10 and 7.11 has been based on the assumptions that for μ less than some particular value μ^*, both μ and $\bar{\mu}$ ($\equiv\bar{\mu}_0$) vary with the parameter a appearing in equation (4.2.1) and that each value of $\bar{\mu}$ ($\equiv\bar{\mu}_0$) corresponds to a different twelve-sided, piecewise continuous yield criterion. Thus, for the given state of hardness corresponding to $\mu=\mu_H$, the projection of the yield surface onto the deviatoric plane gives a twelve-sided, piecewise continuous yield curve characterised by $\bar{\mu}_0=\bar{\mu}_E$, and a second twelve-sided, piecewise continuous curve characterised by $\bar{\mu}_0=\bar{\mu}_A$. These two twelve-sided yield curves will intersect in the deviatoric plane, thus effectively giving rise to a vertex on the yield surface at $\mu=\mu_H$ corresponding to a particular state of hardness. If the discontinuous change in $\bar{\mu}_0$ from $\bar{\mu}_E$ to $\bar{\mu}_A$ is sufficiently large, then failure will be expected to occur at $\mu=\mu_H$ in the same manner as that proposed in §7.2.2 for failure at $\mu=-1$. For the aluminium shown in figure 7.20 the specimen has survived the discontinuous change at $\gamma_{\theta z}=\gamma_H$, from which it is concluded that the discontinuous change in $\bar{\mu}$ is too small to produce failure for this particular specimen of aluminium. In particular, all high purity metals in the polycrystalline state have been found to exhibit to varying degrees a similar discontinuity at $\gamma_{\theta z}=\gamma_H$, and to have survived the associated discontinuous change in $\bar{\mu}$ at the corresponding value of $\mu=\mu_H$, failure occurring in all cases in the manner discussed in §7.2.2, i.e. at $\mu=-1$.

With regard to the aluminium alloy, failure occurs at a value of $\gamma_{\theta z}$ which is seen from figure 7.20 to approximate closely to the value of $\gamma_{\theta z}=\gamma_H$ for the specimen of aluminium. Thus, for the aluminium alloy, failure may possibly have occurred due to the onset of a discontinuous change in the ($e_{(PS)}$, $\gamma_{\theta z}$) curve and hence in both $\bar{\mu}$ and $e_{(PS)}$, the latter exceeding the elongation which the specimen can produce discontinuously and survive without fracture occurring. The very act of failure precludes any possibility of independently confirming the existence of a vertex on the yield surface and its contribution to fracture. However, the above comparison with the corresponding behaviour of the bulk component in a high state of purity is taken as evidence for the conclusion that the proposed conditions at failure can, at least in principle, exist in the case of the aluminium alloy. It is of particular interest to note that for the specimen of aluminium, the discontinuity in the $(\tau_{\theta z}, \gamma_{\theta z})$ curve corresponds to a discontinuous decrease in the rate of hardening for $\gamma_{\theta z} \geqslant \gamma_{\mathrm{H}}$. Thus, the existence of a vertex on the yield surface is associated with a discontinuous change in the rate of hardening, a tentative conclusion which is taken to imply that, in the absence of failure, the aluminium alloy would exhibit a large decrease in the rate of hardening at $\gamma_{\theta z} = \gamma_{\mathrm{H}}$.

7.2.4 Some Submacroscopic Considerations

For a polycrystalline material, the initial stages of deformation are characterised by an essentially elastic response with only small amounts of plastic flow within individual grains. In these early stages of deformation, the microplasticity is prevented from propagating across the specimen by the grain boundaries or by obstacles within the grains. This blocked slip gives rise to a stress concentration which in turn results in the formation of microcracks in a brittle solid. In a ductile material there will be many easily activated Frank–Read sources within the neighbourhood of the stress concentration, and these will emit dislocations in such numbers that they will tend to neutralise this concentration and to propagate slip into the next nearest unit of subcontinua. These mechanisms will tend to inhibit the generation of microcracks in ductile materials. However, the rotational motion of comparatively large units of sub-continua in the manner postulated by Swift (1947) will be expected to generate voids which can be regarded as contributing to the generation of microcracks, and hence, offsetting to some extent the above mechanism of crystal dislocation mechanics. In principle, generation of voids by rotation of large units of subcontinua and hence the generation and growth of microcracks, could be a first order effect. Fracture arising from the growth and eventual merging of holes from microscopic inclusions within materials has been observed in ductile metals by Puttick (1959), Rogers (1960), Tipper (1949) and by Beachem (1963) and Pelloux (1963) using electron micrographs of replicas. In this connection, the motion of a cylindrical void in an infinite plane viscous body has been studied by Berg (1962). This work has been extended by McClintock *et al* (1966) to described fracture by the growth of cylindrical holes under various loadings, including a study of ductile fracture by hole growth in shear bands (see also, Budiansky *et al* (1982), Drucker (1974)).

In the case of a polycrystalline metal in a high state of purity, the differential grain motion proposed by Swift (1947) is between subcontinuum units of the same material. Hence in any given direction, differences in hardness, although significant in relation to the Poynting–Swift effect, are relatively small when considered in relation to macroscopic deformation. Hence the overall distortion in the immediate region of subcontinuum units undergoing differential motion will be expected to be such that the generation and growth of voids will be comparatively small with increasing shear strain, the subcontinuum units of structure remaining close packed. This continues until a stage is reached at which further deformation would, in the absence of failure be accompanied by a large decrease in the rate of hardening associated with $\mu = -1$, and the predicted discontinuous, large increase in $e_{(PS)}$ which is tentatively attributed to a large rigid body rotation at the subcontinuum level, thus producing a large expansion of one or more voids, which in turn leads to failure.

Failure of the aluminium alloy is attributed to a greater difference in hardness, in any given direction, between the bulk of the material and the minority component of relatively hard solid particles. With regard to figure 7.5 the difference between the full curve for the aluminium specimen and the open symbol denoting the specimen of aluminium alloy for $\gamma_{\theta z}$ in the region of $\gamma_{\theta z} = \gamma_H$ is attributed to this greater difference in hardness at the micromechanical level. This small difference between the $(e_{(PS)}, \gamma_{\theta z})$ curves in the region of $\gamma_{\theta z} = \gamma_H$ is taken to imply that, for the aluminium alloy, there is a reduction in the close packing giving rise to voids which are significantly larger than in pure aluminium. Failure of the aluminium alloy is attributed to the sudden growth of these slightly larger voids due to rigid body rotation associated with a discontinuous increase in $e_{(PS)}$, the latter arising from the discontinuous behaviour at a vertex on the yield surface.

7.3 POLYMERIC SOLIDS

Any physically defensible test of the constitutive equation for a non-simple elastic solid must be based on a quantitative prediction regarding the mechanical properties of the test material. Thus, insertion of a property of the material, such as incompressibility, into the kinematics of the system must lead to a specific prediction regarding the dynamics of the system. It is to be noted that simple tension, characterised by $\mu = -1$, $\nu = -1$ for all $\lambda_3 \geqslant 1$, does not provide a prediction regarding the dynamics of the system. In contrast, equations (6.3.31), (6.3.48) and (6.3.56) give the predictions

$$\sigma_3 = \begin{cases} -\sigma_2 \\ 2\sigma_2 \\ \sigma_2 \end{cases} \quad \text{for } \nu = \begin{cases} 0 \\ -3 \\ -\infty \end{cases} \tag{7.3.1}$$

for all $\lambda_3 \geqslant 1$. The simple mode of deformation for which $\nu = 0$ cannot be obtained in practice using a thin sheet of material in the way discussed in §6.3. The prediction of most interest is that for $v = -3$: that is $\sigma_3 = 2\sigma_2$.

There would appear to be only two sets of measurements available in a form suitable for comparison with the proposed constitutive equation (4.2.44), used in the form of the effective stress–effective strain relation of equation (6.3.20). These are the experimental studies of Jones and Treloar (1975) and James *et al* (1975). In the case of the measurements of Jones and Treloar, the numerical values of σ_2, σ_3, λ_2 and λ_3 have been obtained from table 2 of Haines and Wilson (1979). It is to be noted that in all the available studies of rubber-like solids, the present σ_3 can be identified with the results given in terms of σ_1. None of the available measurements of σ_2 and σ_3 for $v = -3$, $\lambda_2 = 1$ for all $\lambda_3 \geqslant 1$ satisfy the prediction $\sigma_3 = 2\sigma_2$. This is a rather unexpected observation in so far as the development of the appropriate class of simple materials has been directed almost exclusively to rubber-like solids, thus recognising that this type of material displays, to a very marked extent, the principal characteristic features of nonlinear elasticity.

REFERENCES

Beachem C D 1963 *Trans. Quart. ASM* **56** 318–26

Berg C A 1962 *Proc. 4th US Natl Congr. Appl. Mech.* **2** 885–92

Billington E W 1977a *J. Phys. D: Appl. Phys.* **10** 519–31

—— 1977b *J. Phys. D: Appl. Phys.* **10** 533–52

—— 1977c *J. Phys. D: Appl. Phys.* **10** 553–69

—— 1978 *J. Phys. D: Appl. Phys.* **11** 459–71

—— 1984 *Eng. Fracture Mech.* **19** 777–92

—— 1985 *Int. J. Solids Struct.* **21** 355–71

Budiansky B, Hutchinson J W and Slutsky S 1982 *Void Growth and Collapse in Viscous Solids* in *Mechanics of Solids* ed: H G Hopkins and M J Sewell (Oxford: Pergamon) pp 13–45

Campbell J D 1973 *Mat. Sci. Eng.* **12** 3–21

Drucker D C 1949 *J. Appl. Mech.* **16** 349–57

—— 1974 *On Plastic Analysis of the Microstructure of Metallic Alloys in Problems of Plasticity* (ed. A Sawczuk) pp 25–44 (Leyden: Noordhoff)

Foux A 1964 *Second Order Effects* in *Elasticity, Plasticity and Fluid Dynamics* ed. M Reiner and D Abir (Oxford: Pergamon) pp 228–51

Freudenthal A M and Gou P F 1969 *Acta Mechan.* **8** 34–52

Ferudenthal A M and Ronay M 1966 *Proc. R. Soc.* A **292** 14–50

Haines D W and Wilson W D 1979 *J. Mech. Phys. Solids* **27** 345–60

Hill R 1952 *J. Mech. Phys. Solids* **1** 19–30

—— 1953 *J. Mech. Phys. Solids* **1** 271–6

Hundy B B and Green A P 1954 *J. Mech. Phys. Solids* **3** 16–21

James A G, Green A and Simpson G M 1975 *J. Appl. Polmer Sci.* **19** 2033–58

Jones D F and Treloar L R G 1975 *J. Phys. D: Appl. Phys.* **8** 1285–1304

Lianis G and Ford H 1957 *J. Mech. Phys. Solids* **5** 215–22
L'Hermite R and Dawance G 1947 *Circul. Inst. Tech. Batim. Series G* No 16 p 258
Lindholm U S and Yeakley L M 1968 *Expt. Mech.* **8** 1–9
Lode W 1926 *Z. Phys.* **36** 913–39
Ludwik P 1909 *Elemente der Technologischen Mechanik* (Berlin: Springer)
Ludwik P and Scheu A 1925 *Stahl. Eisen* **45** 373–81
McClintock F A, Kaplan S M and Berg C A 1966 *Int. J. Fracture Mech.* **2** 614–27
Nicholas T and Campbell J D 1972 *Expt. Mech.* **12** 441–7
Pelloux R M N 1963 *The Analysis of Fracture Surfaces by Electron Microscopy* Solid State Physics Lab, (Boeing Scientific) Seattle, Washington
Poynting J H 1909 *Proc. R. Soc.* A **82** 546–59
—— 1912 *Proc. R. Soc.* A **86** 534–61
—— 1913 *India Rubber J.* **46** 616
Prager W 1945 *J. Appl. Phys.* **16** 837–40
Pukas S R 1976 *PhD Thesis* Thames Polytechnic, London
Puttick K E 1959 *Phil. Mag.* **4** 964–69
Rees D W A 1982 *Proc. R. Soc.* A **383** 333–57
Rogers H C 1960 *Trans. Metall. Soc. AIME* **218** 498–506
Ronay M 1966 *J. Inst. Metall.* **94** 392–4
—— 1967 *Int. J. Solids Struct.* **3** 167–75
—— 1968 *Int. J. Solids Struct.* **4** 509–15
Rose W and Stüwe H P 1968 *Zeit. Metallkde* **59** 396–9
Swift H W 1947 *Engineering* **163** 253–7
Taylor G I and Quinney H 1931 *Phil. Trans. R. Soc.* A **230** 323–62
Tipper C F 1949 *Metallurgia* **39** 133–57
von Mises R 1913 *Gottinger Nachrichten Math. Phys.* K L 582–92

GENERAL REFERENCES

Billington E W and Tate A 1981 *The Physics of Deformation and Flow* (New York: McGraw-Hill)
Drucker D C 1967 *Introduction to the Mechanics of Deformable Solids* (New York: McGraw-Hill)
Lubahn J D and Felgar R P 1961 *Plasticity and Creep of Metals* (New York: Wiley)
Nadai A 1950 *Theory of Flow and Fracture of Solids* (New York: McGraw-Hill)
Prager W 1959 *An Introduction to Plasticity* (Reading Mass: Addison-Wesley)

8

Purely Elastic Materials in the Referential Description

8.1 GENERALISED REFERENTIAL STRESS TENSOR

8.1.1 Some General Considerations

In the solution of particular problems, the analysis can on occasion be simplified by transforming the stress relation from the current configuration back to the reference configuration. This approach gives rise to the problem of determining an appropriate representation of the stress. The classical approach is by way of the first Piola–Kirchhoff stress tensor, or its transpose which is referred to as the nominal stress tensor (see for example Toupin 1964). However, the nominal stress tensor, which involves both the current configuration of a material body and a reference configuration, is not in general a symmetric tensor. Because it is not symmetric it is difficult to incorporate into the constitutive relation between stress and strain. In this context, a more useful form is the second Piola–Kirchhoff stress tensor which is symmetric and is concerned solely with the reference configuration. However, no direct physical interpretation can be given for the second Piola–Kirchhoff stress tensor in terms of the contact force acting on a surface segment. The properties of the Piola–Kirchhoff stress tensors have been discussed in §2.2.3 (see also Truesdell and Noll 1965). Thus there is the problem of defining a measure of stress which will provide a representation that, while transformed back to the reference configuration, is still a symmetric tensor and for which a physically acceptable interpretation can be given in terms of the contact force acting on a surface segment.

The constitutive equation of a simple elastic material has the form (see §4.1.1)

$$\mathbf{T}=\mathbf{M}(\mathbf{B})=a_0\mathbf{I}+a_1\mathbf{B}+a_2\mathbf{B}^2 \tag{8.1.1}$$

with response coefficients

$$a_k = a_k(I_{\mathbf{B}}, II_{\mathbf{B}}, III_{\mathbf{B}}) \qquad (k=0,1,2) \tag{8.1.2}$$

if and only if the material is isotropic. $\mathbf{T}$ is the (symmetric) Cauchy stress, and $\mathbf{B}=\mathbf{F}\mathbf{F}^{\mathrm{T}}$ is the positive definite, symmetric left Cauchy–Green deformation tensor discussed in §2.1.3. Since $\mathbf{B}$ is invertible, equation (8.1.1) can be rearranged into the form

$$\mathbf{T}=\mathbf{M}(\mathbf{B})=\beta_0\mathbf{I}+\beta_1\mathbf{B}+\beta_{-1}\mathbf{B}^{-1} \tag{8.1.3}$$

where the response coefficients

$$\beta_k=\beta_k(I_{\mathbf{B}}, II_{\mathbf{B}}, III_{\mathbf{B}}) \qquad (k=0,\pm 1) \tag{8.1.4}$$

are related to the response coefficients a_k $(k=0,1,2)$ by equations (4.1.13) and (4.1.14). The symmetric response function $\mathbf{M}$ of equation (8.1.3) remains that of equation (8.1.1), because equation (8.1.3) has been obtained simply by using the Cayley–Hamilton theorem.

From equation (8.1.3),

$$\mathbf{T}^2=\mathbf{M}^2(\mathbf{B})=\overset{(2)}{\beta}_0\mathbf{I}+\overset{(2)}{\beta}_1\mathbf{B}+\overset{(2)}{\beta}_{-1}\mathbf{B}^{-1} \tag{8.1.5}$$

where use has been made of the Cayley–Hamilton theorem to give

$$\begin{aligned}\overset{(2)}{\beta}_0&=\beta_0^2+2\beta_1\beta_{-1}-\beta_1^2 II_{\mathbf{B}}-\beta_{-1}^2 I_{\mathbf{B}}/III_{\mathbf{B}}\\ \overset{(2)}{\beta}_1&=2\beta_0\beta_1+\beta_1^2 I_{\mathbf{B}}+\beta_{-1}^2/III_{\mathbf{B}}\\ \overset{(2)}{\beta}_{-1}&=2\beta_0\beta_{-1}+\beta_1^2 III_{\mathbf{B}}+\beta_{-1}^2 II_{\mathbf{B}}/III_{\mathbf{B}}.\end{aligned} \tag{8.1.6}$$

Repeated application of the Cayley–Hamilton theorem gives

$$\mathbf{T}^n=\mathbf{M}^n(\mathbf{B})=\overset{(n)}{\beta}_0\mathbf{I}+\overset{(n)}{\beta}_1\mathbf{B}+\overset{(n)}{\beta}_{-1}\mathbf{B}^{-1} \qquad (n\neq 0). \tag{8.1.7}$$

The existence of a constitutive equation of the form

$$\overset{*}{\mathbf{T}}=\mathbf{K}(\mathbf{C})=\overset{*}{\alpha}_0\mathbf{I}+\overset{*}{\alpha}_1\mathbf{C}+\overset{*}{\alpha}_{-1}\mathbf{C}^{-1} \tag{8.1.8}$$

is assumed. $\overset{*}{\mathbf{T}}$ is a measure of the referential stress which has yet to be defined, $\mathbf{K}$ is a symmetric response function which is a tensor-valued function of the positive definite, symmetric, right Cauchy–Green tensor $\mathbf{C}$ discussed in §2.1.3 and the response coefficients

$$\overset{*}{\alpha}_k=\overset{*}{\alpha}_k(I_{\mathrm{C}}, II_{\mathrm{C}}, III_{\mathrm{C}}) \qquad (k=0,1,2) \tag{8.1.9}$$

are functions of the principal invariants $I_{\mathrm{C}}(=I_{\mathbf{B}})$, $II_{\mathrm{C}}(=II_{\mathbf{B}})$ and

$III_C (= III_{\mathbf{B}})$ of $\mathbf{C}$. Whereas the Cauchy stress is concerned solely with the current configuration B_t of the material body $\mathscr{B}$, the stress $\overset{*}{\mathbf{T}}$ is concerned exclusively with the reference configuration B_r. From equation (8.1.8)

$$\overset{*}{\mathbf{T}}{}^2 = \mathbf{K}^2(\mathbf{C}) = \overset{(2)}{\alpha}_0 \mathbf{I} + \overset{(2)}{\alpha}_1 \mathbf{C} + \overset{(2)}{\alpha}_{-1} \mathbf{C}^{-1} \tag{8.1.10}$$

where use has been made of the Cayley–Hamilton theorem to give

$$\begin{aligned}
\overset{(2)}{\alpha}_0 &= \overset{*}{\alpha}{}_0^2 + 2\overset{*}{\alpha}_1 \overset{*}{\alpha}_{-1} - \overset{*}{\alpha}{}_1^2 II_C - \overset{*}{\alpha}{}_{-1}^2 I_C / III_C \\
\overset{(2)}{\alpha}_1 &= 2\overset{*}{\alpha}_0 \overset{*}{\alpha}_1 + \overset{*}{\alpha}{}_1^2 I_C + \overset{*}{\alpha}{}_{-1}^2 / III_C \\
\overset{(2)}{\alpha}_{-1} &= 2\overset{*}{\alpha}_0 \overset{*}{\alpha}_{-1} + \overset{*}{\alpha}{}_1^2 III_C + \overset{*}{\alpha}{}_{-1}^2 II_C / III_C
\end{aligned} \tag{8.1.11}$$

which are to be compared with equation (8.1.6). Repeated application of the Cayley–Hamilton theorem gives

$$\overset{*}{\mathbf{T}}{}^n = \mathbf{K}^n(\mathbf{C}) = \overset{(n)}{\alpha}_0 \mathbf{I} + \overset{(n)}{\alpha}_1 \mathbf{C} + \overset{(n)}{\alpha}_{-1} \mathbf{C}^{-1} \qquad (n \neq 0) \tag{8.1.12}$$

which is to be compared with equation (8.1.7).

With regard to equations (8.1.7) and (8.1.12) it should be noted that

$$\overset{*}{\mathbf{T}}{}^n = J^n \mathbf{F}^{-1} \mathbf{T}^n (\mathbf{F}^{-1})^{\mathrm{T}} = \mathbf{K}^n(\mathbf{C}) = \overset{(n)}{\alpha}_0 \mathbf{I} + \overset{(n)}{\alpha}_1 \mathbf{C} + \overset{(n)}{\alpha}_{-1} \mathbf{C}^{-1} \qquad (n \neq 0) \tag{8.1.13}$$

where

$$\begin{aligned}
\overset{(n)}{\alpha}_0 &= J^n \overset{(n)}{\beta}_1 - J^{n-2} I_C \overset{(n)}{\beta}_{-1} \\
\overset{(n)}{\alpha}_1 &= J^{n-2} \overset{(n)}{\beta}_{-1} \qquad (n \neq 0) \\
\overset{(n)}{\alpha}_{-1} &= J^n \overset{(n)}{\beta}_0 + J^{n-2} II_C \overset{(n)}{\beta}_{-1}
\end{aligned} \tag{8.1.14}$$

and where J is an as-yet undefined scalar factor of proportionality.

As a definition of the referential stress tensor $\overset{*}{\mathbf{T}}$, equations (8.1.13) is restricted in so far as it has been formulated within the context of the particular constitutive equations (8.1.7) and (8.1.12). Thus the question arises of whether it can be stated as a definition of a generalised referential stress tensor.

8.1.2 Generalised Referential Stress Tensor

Consider the relation (Billington 1985a)

$$\overset{*}{\mathbf{T}}^n = J^n \mathbf{F}^{-1} \mathbf{T}^n (\mathbf{F}^{-1})^{\mathrm{T}} \qquad (n \neq 0) \tag{8.1.15}$$

where the symmetric Cauchy stress tensor $\mathbf{T}$ is concerned solely with the current configuration B_t of a material body $\mathscr{B}$, and where the referential stress tensor $\overset{*}{\mathbf{T}}$ is concerned exclusively with the reference configuration B_r of $\mathscr{B}$.

For $n=1$, equation (8.1.15) reduces to

$$\overset{*}{\mathbf{T}} \equiv \tilde{\mathbf{T}} = J\mathbf{F}^{-1}\mathbf{T}(\mathbf{F}^{-1})^{\mathrm{T}}$$

which, in the context of equation (2.2.54), identifies $\overset{*}{\mathbf{T}} \equiv \tilde{\mathbf{T}}$ with the second Piola–Kirchhoff stress tensor, subject to the condition

$$J = \det \mathbf{F} > 0$$

which is just equation (2.1.41).

From §2.2.3 the Cauchy stress tensor $\mathbf{T}$ and the stress vector $\boldsymbol{t}$ at x have the spectral representations,

$$\mathbf{T} = \sum_{r=1}^{3} \sigma_r \boldsymbol{q}_r \otimes \boldsymbol{q}_r \qquad \boldsymbol{t} = \sum_{r=1}^{3} \sigma_r n_r \boldsymbol{q}_r \tag{8.1.16}$$

where $\boldsymbol{n}$ is the unit normal vector of equation (2.2.29). Similarly, from §2.1.3, or directly from equation (4.1.20), $\mathbf{B}$ has the spectral representation

$$\mathbf{B} = \sum_{r=1}^{3} b_r \boldsymbol{q}_r \otimes \boldsymbol{q}_r \tag{8.1.17}$$

where the $b_i = \lambda_i^2$ are the proper numbers of $\mathbf{B}$. The common principal axes of $\mathbf{T}$ and $\mathbf{B}$ at x is defined by the orthonormal basis $\{\boldsymbol{q}_1, \boldsymbol{q}_2, \boldsymbol{q}_3\}$, and the λ_i are the principal stretches. The referential stress tensor $\overset{*}{\mathbf{T}}$ at X has the representation

$$\overset{*}{\mathbf{T}} = \sum_{r=1}^{3} \overset{*}{t}_r \boldsymbol{p}_r \otimes \boldsymbol{p}_r \tag{8.1.18}$$

where the $\overset{*}{t}_i$ are the principal referential stresses at X. From §2.1.3, the right Cauchy–Green deformation tensor $\mathbf{C}$ at X has the representation

$$\mathbf{C} = \sum_{r=1}^{3} c_r \boldsymbol{p}_r \otimes \boldsymbol{p}_r \tag{8.1.19}$$

where the $c_i = \lambda_i^2$ are the proper numbers of $\mathbf{C}$, the common principal axes of $\overset{*}{\mathbf{T}}$ and $\mathbf{C}$ at X being defined by the orthonormal basis $\{\boldsymbol{p}_1, \boldsymbol{p}_2, \boldsymbol{p}\ \}$.

Use of the right polar decomposition $\mathbf{F} = \mathbf{RU}$ of equation (2.1.42) enables equation (8.1.15) to be rearranged into the form

$$\mathbf{T}^n = J^{-n}\mathbf{R}\mathbf{U}\overset{*}{\mathbf{T}}{}^n\mathbf{U}\mathbf{R}^{\mathrm{T}} \qquad (n \neq 0) \tag{8.1.20}$$

The rotation tensor

$$\mathbf{R} = \sum_{r=1}^{3} \boldsymbol{q}_r \otimes \boldsymbol{p}_r \tag{8.1.21}$$

follows from equations (1.6.149) and (1.7.51) and

$$\mathbf{U} = \sum_{r=1}^{3} \lambda_r \boldsymbol{p}_r \otimes \boldsymbol{p}_r \tag{8.1.22}$$

is the right stretch tensor (see §2.1.3).

Entering the spectral representations of the symmetric tensors $\mathbf{T}$, $\overset{*}{\mathbf{T}}$, and $\mathbf{U}$ into equation (8.1.20) leads to the conclusion that

$$\overset{*}{t}_r/\sigma_r = J/\lambda_r^{2/n} > 0 \qquad (n \neq 0). \tag{8.1.23}$$

Since, from equation (2.1.41), $J > 0$ and also $\lambda_r > 0$, the $\overset{*}{t}_r$ always have the same sign as the σ_r $(r = 1, 2, 3)$.

8.2 SOME PHYSICAL CONSIDERATIONS

8.2.1 Contact Force on a Surface Element

Consider a material surface element with area $\mathrm{d}a$ and outward unit normal $\boldsymbol{n}$ in the present configuration B_t of a body $\mathscr{B}$. Let the force on the element of area $\mathrm{d}a$ be $\mathrm{d}\boldsymbol{F}$. This is related to $\mathbf{T}$ and $\overset{*}{\mathbf{T}}$ by the relation

$$\mathrm{d}\boldsymbol{F} = \mathbf{T}^{\mathrm{T}}\boldsymbol{n}\,\mathrm{d}a = \mathbf{R}\mathbf{U}^{(2-n)/n}\overset{*}{\mathbf{T}}{}^{\mathrm{T}}\boldsymbol{N}\,\mathrm{d}A \qquad (n \neq 0) \tag{8.2.1}$$

where the material surface element has area $\mathrm{d}A$ and outward unit normal $\boldsymbol{N}$ in the reference configuration B_r. Use has been made of the relation (see §2.2.2)

$$\boldsymbol{n}\,\mathrm{d}a = J(\mathbf{F}^{-1})^{\mathrm{T}}\boldsymbol{N}\,\mathrm{d}A. \tag{8.2.2}$$

There are two cases of particular interest. These are:

(i) ***First referential stress tensor.*** As already noted, setting $n = 1$ in equation (8.1.15) identifies $\overset{*}{\mathbf{T}} \equiv \tilde{\mathbf{T}}$ with the second Piola–Kirchhoff stress tensor. For $n = 1$, equation (8.2.1) reduces to

$$\mathbf{T}^{\mathrm{T}}\boldsymbol{n}\,\mathrm{d}a = \mathbf{F}\tilde{\mathbf{T}}^{\mathrm{T}}\boldsymbol{N}\,\mathrm{d}A \qquad (\overset{*}{\mathbf{T}} \equiv \tilde{\mathbf{T}}) \tag{8.2.3}$$

from which it is evident that no direct interpretation can be given for the first referential stress tensor $\tilde{\mathbf{T}}$ in terms of the contact force acting on a surface segment.

(ii) ***Second referential stress tensor.*** Setting $n=2$ and $\overset{*}{\mathbf{T}} \equiv \hat{\mathbf{T}}$ in equation (8.1.15) gives

$$\hat{\mathbf{T}}^2 = J^2\mathbf{F}^{-1}\mathbf{T}^2(\mathbf{F}^{-1})^{\mathrm{T}} \qquad (\hat{\mathbf{T}} \equiv \overset{*}{\mathbf{T}})$$

or, alternatively, since $\mathbf{U}$ commutes with $\hat{\mathbf{T}}$, and using the right polar decomposition $\mathbf{F}=\mathbf{RU}$, equation (8.1.20) gives, for $n=2$

$$\hat{\mathbf{T}} = J\mathbf{F}^{-1}\mathbf{TR} = J\mathbf{U}^{1/2}\mathbf{F}^{-1}\mathbf{TRU}^{-1/2} = J\mathbf{U}^{1/2}\mathbf{F}^{-1}\mathbf{T}(\mathbf{F}^{-1})^{\mathrm{T}}\mathbf{U}^{1/2}. \tag{8.2.4}$$

For $n=2$, equation (8.2.1) reduces to

$$\mathbf{T}^{\mathrm{T}}\boldsymbol{n}\,\mathrm{d}a = \mathbf{R}\hat{\mathbf{T}}^{\mathrm{T}}\boldsymbol{N}\,\mathrm{d}A \tag{8.2.5}$$

which can be expressed in the form

$$\mathbf{T}^{\mathrm{T}}\boldsymbol{n}\,\mathrm{d}a = (\mathbf{T}^0)^{\mathrm{T}}\boldsymbol{N}\,\mathrm{d}A$$

where $\mathbf{T}^0$ is the nominal stress tensor of equation (2.2.52). Thus, the interpretation of the second referential stress tensor $\hat{\mathbf{T}}$ differs from that of the nominal stress $\mathbf{T}^0$ by the rigid rotation $\mathbf{R}$.

The following relations can be obtained from equation (8.2.2)

$$(\mathrm{d}a/\mathrm{d}A)^2 = J^2\boldsymbol{N}\cdot(\mathbf{C}^{-1}\boldsymbol{N}) = J^2[\boldsymbol{n}\cdot(\mathbf{B}\boldsymbol{n})]^{-1} \tag{8.2.6}$$

which are just the relations of equation (2.2.16). Entering the spectral representations of the symmetric tensors $\mathbf{T}$, $\hat{\mathbf{T}}$ and either $\mathbf{B}$ or $\mathbf{C}$, into equations (8.2.5) and (8.2.6), leads by way of equation (8.2.5) to the relation,

$$\hat{t}_r = J\sigma_r/\lambda_r \tag{8.2.7}$$

which follows directly from equation (8.1.23) with $n=2$ and $\overset{*}{t}_r \equiv \hat{t}_r$. The Cartesian components of the second referential stress tensor have been introduced by Biot (1965) (see also Atluri 1984, Bufler 1983, Koiter 1976).

8.2.2 Field Equations

The field equation (2.2.56) form of the balance of linear momentum can be expressed in terms of the referential description using equation (2.2.52) in the form

$$\rho_{\mathrm{r}}\boldsymbol{a} = \mathrm{Div}\,\mathbf{T}^0 + \rho_{\mathrm{r}}\boldsymbol{b} \tag{8.2.8}$$

where use has been made of the relation,

$$\mathrm{Div}\,\mathbf{T}^0 = J\,\mathrm{div}\,\mathbf{T} \tag{8.2.9}$$

which can be obtained by use of equations (1.6.88), (1.9.21), (1.6.72), (2.1.36) and

(2.1.41). The stress power per unit volume in B_t, i.e. tr(**TD**), can be expressed in the form,

$$\mathrm{tr}(\mathbf{TD}) = J^{-1}\,\mathrm{tr}(\hat{\mathbf{T}}\dot{\mathbf{U}}). \tag{8.2.10}$$

This can be entered into equations (2.2.94) and (2.2.99) to give the field equation form of the balance of thermal energy in the referential description as

$$\rho_r\dot{\varepsilon} = \mathrm{tr}(\hat{\mathbf{T}}\dot{\mathbf{U}}) + (\rho_r/\rho)\,\mathrm{div}\,\boldsymbol{q} + \rho_r r \tag{8.2.11}$$

and

$$\rho_r\gamma\theta \equiv \rho_r\theta\dot{\eta} - \rho_r\dot{\varepsilon} + \mathrm{tr}(\hat{\mathbf{T}}\dot{\mathbf{U}}) + (\rho_r/\rho\theta)\boldsymbol{q}\cdot\mathrm{grad}\,\theta \geqslant 0. \tag{8.2.12}$$

8.3 CONSTITUTIVE EQUATIONS

8.3.1 Simple Elastic Materials

Although $\overset{*}{\mathbf{T}}$ can, in principle, have many different forms, it is evident from the discussion in §8.2, that only the second referential stress tensor has physical significance. For this reason, the present discussion will be restricted to the formulation of constitutive equations in terms of $\hat{\mathbf{T}}$ and **C**.

Entering the constitutive equation (8.1.3) into equation (8.2.4) gives

$$\hat{\mathbf{T}} = J(\beta_0\mathbf{U}^{-1} + \beta_1\mathbf{U} + \beta_{-1}\mathbf{U}^{-3})$$

a stress relation which is in general difficult to use in solutions of special problems. However, $\hat{\mathbf{T}}$ has the alternative representation

$$\hat{\mathbf{T}} = \hat{\alpha}_0\mathbf{I} + \hat{\alpha}_1\mathbf{C} + \hat{\alpha}_{-1}\mathbf{C}^{-1} \tag{8.3.1}$$

where, for $n=2$, equations (8.1.6), (8.1.11) and (8.1.14) give

$$\hat{\alpha}_0^2 + 2\hat{\alpha}_1\hat{\alpha}_{-1} - \hat{\alpha}_{-1}^2(I_{\mathbf{B}}/III_{\mathbf{B}}) - \hat{\alpha}_1^2 II_{\mathbf{B}}$$
$$= 2\beta_0(\beta_1 III_{\mathbf{B}} - \beta_{-1}I_{\mathbf{B}}) + \beta_{-1}^2[1 - (I_{\mathbf{B}}II_{\mathbf{B}}/III_{\mathbf{B}})]$$

$$2\hat{\alpha}_0\hat{\alpha}_1 + \hat{\alpha}_1^2 I_{\mathbf{B}} + (\hat{\alpha}_{-1}^2/III_{\mathbf{B}})$$
$$= 2\beta_0\beta_{-1} + \beta_1^2 III_{\mathbf{B}} + \beta_{-1}^2(II_{\mathbf{B}}/III_{\mathbf{B}})$$

$$2\hat{\alpha}_0\hat{\alpha}_{-1} + \hat{\alpha}_1^2 III_{\mathbf{B}} + \hat{\alpha}_{-1}^2(II_{\mathbf{B}}/III_{\mathbf{B}})$$
$$= \beta_0^2 III_{\mathbf{B}} + 2\beta_1\beta_{-1}III_{\mathbf{B}} + 2\beta_0\beta_{-1}II_{\mathbf{B}} + \beta_{-1}^2[(II_{\mathbf{B}}^2/III_{\mathbf{B}}) - I_{\mathbf{B}}].$$

Equation (8.3.1) can be rearranged using the Cayley–Hamilton theorem into the form

$$\hat{\mathbf{T}} = \mathbf{K}(\mathbf{C}) = \gamma_0\mathbf{I} + \gamma_1\mathbf{C} + \gamma_2\mathbf{C}^2 \tag{8.3.2}$$

where

$$\hat{\alpha}_0 = \gamma_0 - II_C\gamma_2 \qquad \hat{\alpha}_1 = \gamma_1 + I_C\gamma_2 \qquad \hat{\alpha}_{-1} = III_C\gamma_2$$

$$\gamma_0 = \hat{\alpha}_0 + \frac{II_C}{III_C}\hat{\alpha}_{-1} \qquad \gamma_1 = \hat{\alpha}_1 - \frac{I_C}{III_C}\hat{\alpha}_{-1} \qquad \gamma_2 = \frac{1}{III_C}\hat{\alpha}_{-1} \tag{8.3.3}$$

Entering the spectral representations of $\hat{\mathbf{T}}$ and $\mathbf{C}$ into equations (8.3.1) and (8.3.2) gives

$$\hat{t}_i = \gamma_0 + \gamma_1 c_i + \gamma_2 c_i^2 \tag{8.3.4}$$

or, alternatively

$$\begin{aligned}\hat{t}_i &= \hat{\alpha}_0 + \hat{\alpha}_1 c_i + \hat{\alpha}_{-1} c_i^{-1} \\ &= \hat{\alpha}_0 + \hat{\alpha}_1 \lambda_i^2 + \hat{\alpha}_{-1}\lambda_i^{-2}.\end{aligned} \tag{8.3.5}$$

The constitutive equations (8.3.1) and (8.3.2) can be expressed in the form

$$\begin{aligned}\hat{\mathbf{T}}' &= \hat{\rho}_0\mathbf{I} + \hat{\rho}_1\mathbf{C}' + \hat{\rho}_2\mathbf{C}'^2 \\ &= \hat{\beta}_1\mathbf{C}' + \hat{\beta}_{-1}(\mathbf{C}^{-1})' = \mathbf{K}'(\mathbf{C}')\end{aligned} \tag{8.3.6}$$

where a prime denotes a deviator and where

$$\hat{\rho}_0 = -\tfrac{1}{3}\hat{\rho}_2 \operatorname{tr}\mathbf{C}'^2 \qquad \hat{\rho}_1 = \gamma_1 + \tfrac{2}{3}I_C\gamma_2 \qquad \hat{\rho}_2 = \gamma_2$$

$$\hat{\rho}_1 = \hat{\beta}_1 - \tfrac{1}{3}\frac{I_C}{III_C}\hat{\beta}_{-1} \qquad \hat{\rho}_2 = \frac{1}{III_C}\hat{\beta}_{-1}. \tag{8.3.7}$$

Equations (8.1.18) and (8.1.19) can be used to express equation (8.3.6) in the form

$$\begin{aligned}\hat{t}'_i &= \hat{\rho}_0 + \hat{\rho}_1 c'_i + \hat{\rho}_2 c'^2_i \\ &= \hat{\beta}_1 c'_i + \hat{\beta}_{-1}(c^{-1})'_i = k'_i \qquad (i=1,2,3).\end{aligned} \tag{8.3.8}$$

The proper numbers $\hat{t}'_i$ $(i=1,2,3)$ of $\hat{\mathbf{T}}'$ are the deviators of the principal referential stresses $\hat{t}_i$ $(i=1,2,3)$ and similarly the proper numbers c'_i, $(c^{-1})'_i$ $(i=1,2,3)$ of $\mathbf{C}'$ and $(\mathbf{C}^{-1})'$ respectively, are the deviators of the proper numbers c_i and c_i^{-1} of $\mathbf{C}$ and $\mathbf{C}^{-1}$ respectively. The k'_i $(i=1,2,3)$ are the proper numbers of the response function $\mathbf{K}'(\mathbf{C}')$.

Setting

$$\hat{\mu} = \frac{3\hat{t}'_1}{\hat{t}'_3 - \hat{t}'_2} \qquad \tilde{v} = \frac{3k'_1}{k'_3 - k'_2} \qquad \hat{\kappa} = \frac{3c'_1}{c'_3 - c'_2} \tag{8.3.9}$$

and entering the values of $\hat{t}'_i$ given by equation (8.3.8) gives the identity

$$\begin{aligned}\hat{\mu} &= \hat{\kappa}\left(1 + \frac{3}{2}\frac{(1-\hat{\kappa}^2)}{\hat{\kappa}^2}\frac{1}{1-(\hat{\rho}_1/\hat{\rho}_2 c'_1)}\right) \equiv \tilde{v} \\ &= \hat{\kappa}\left(1 + \frac{(1-\hat{\kappa}^2)}{2\hat{\kappa}^2}\frac{(3c_1 - I_C)}{(c_1 + III_C\hat{q})}\right) \equiv \tilde{v}\end{aligned} \tag{8.3.10}$$

where

$$q = -\hat{\beta}_1/\hat{\beta}_{-1}. \tag{8.3.11}$$

The relation of equation (8.3.10) provides a possible way of determining whether a particular material satisfies the constitutive equation of a simple elastic material in the form of equation (8.3.6) using only information obtained from pure stretches.

8.3.2 Non-Simple Elastic Material

Since $\hat{\mathbf{T}}$ is a symmetric second order tensor, the conditions at initial yield can be obtained from Chapter 3 by replacing $\mathbf{T}$ everywhere by $\hat{\mathbf{T}}$. With this as basis it is assumed that there exists a stress intensity function

$$\hat{h} = \hat{J}_2\hat{H}(\hat{\omega}) - \hat{\varkappa} = 0 \tag{8.3.12}$$

where

$$\hat{\omega} = \frac{27\hat{J}_3'^2}{4\hat{J}_2'^2} = \hat{\mu}^2\frac{(\hat{\mu}^2-9)^2}{(3+\hat{\mu}^2)^3} \qquad (0 \leqslant \hat{\omega} \leqslant 1) \tag{8.3.13}$$

$$\hat{J}_2' = \tfrac{1}{2}\operatorname{tr}\hat{\mathbf{T}}'^2 \qquad \hat{J}_3' = \det\hat{\mathbf{T}}' \tag{8.3.14}$$

it being noted that $\hat{h}$ defines the stress systems that determine a given mode of deformation over the entire range of deformation.

The stress intensity function $\hat{h}$ is assumed to determine a loading function $\hat{\mathbf{G}}'$ by way of the relation

$$\hat{\mathbf{G}}' = (\partial\hat{h}/\partial\hat{\mathbf{T}}). \tag{8.3.15}$$

Substituting for $\hat{h}$ in equation (8.3.15) from equation (8.3.12) gives

$$\hat{\mathbf{G}}' = \hat{\delta}_0\mathbf{I} + \hat{\delta}_1\hat{\mathbf{T}}' + \hat{\delta}_2\hat{\mathbf{T}}'^2 \tag{8.3.16}$$

where the loading coefficients $\hat{\delta}_a$ $(a = 0, 1, 2)$ are given by

$$\hat{\delta}_0 = -\tfrac{2}{3}\hat{J}_2'\hat{\delta}_2 \qquad \hat{\delta}_1 = \hat{H}\left(1 - 3\frac{\hat{\omega}}{\hat{H}}\frac{d\hat{H}}{d\hat{\omega}}\right)$$
$$\hat{\delta}_2 = 2\hat{H}\left(\frac{\hat{J}_2'}{\hat{J}_3'}\right)\frac{\hat{\omega}}{\hat{H}}\frac{d\hat{H}}{d\hat{\omega}}. \tag{8.3.17}$$

With §4.2 as basis, the constitutive equation for a non-simple elastic material in the referential description can be expressed in the form

$$\hat{\mathbf{G}} = \hat{\mathfrak{b}}_0\mathbf{I} + \hat{\mathfrak{b}}_1\hat{\mathbf{T}} + \hat{\mathfrak{b}}_2\hat{\mathbf{T}}^2 = \mathbf{K}(\mathbf{C}) = \hat{\alpha}_0\mathbf{I} + \hat{\alpha}_1\mathbf{C} + \hat{\alpha}_2\mathbf{C}^2 \tag{8.3.18}$$

which is to be compared with equation (4.2.18). Using the Cayley–Hamilton theorem, the constitutive equation (8.3.18) can be rearranged into the alternative form

$$\hat{\mathbf{G}}=\hat{b}_0\mathbf{I}+\hat{b}_1\hat{\mathbf{T}}+\hat{b}_2\hat{\mathbf{T}}^2=\mathbf{K}(\hat{\mathbf{E}})=\hat{\varepsilon}_0\mathbf{I}+\hat{\varepsilon}_1\hat{\mathbf{E}}+\hat{\varepsilon}_2\hat{\mathbf{E}}^2 \tag{8.3.19}$$

where

$$\hat{\mathbf{E}}=\tfrac{1}{4}(\mathbf{C}-\mathbf{C}^{-1})=\tfrac{1}{2}[\mathbf{E}+\tfrac{1}{2}(\mathbf{I}-\mathbf{C}^{-1})] \tag{8.3.20}$$

and where

$$\hat{\varepsilon}_0=\hat{\alpha}_0+\frac{1}{4}\frac{II_C}{III_C}\hat{\varepsilon}_1+\frac{1}{16}\left(2+\frac{I_C}{III_C}-\frac{II_C^2}{III_C^2}\right)\hat{\varepsilon}_2 \tag{8.3.21}$$

$$\hat{\varepsilon}_1=\frac{1}{4}\frac{(II_C+III_C^2)}{III_C}\hat{\varepsilon}_2-4III_C\hat{\alpha}_2 \tag{8.3.22}$$

$$\hat{\varepsilon}_2=16\frac{III_C[\hat{\alpha}_1+\hat{\alpha}_2(I_C+III_C)]}{[1+II_C+(I_C+III_C)III_C]}. \tag{8.3.23}$$

In equation (8.3.20), $\mathbf{E}=(\mathbf{C}-\mathbf{I})/2$ is the classical Green–St Venant measure of strain.

Using a prime to denote a deviator, equations (8.3.18) and (8.3.19) can be expressed in the form

$$\hat{\mathbf{G}}'(\hat{\mathbf{T}}')=\hat{\delta}_0\mathbf{I}+\hat{\delta}_1\hat{\mathbf{T}}'+\hat{\delta}_2\hat{\mathbf{T}}'^2=\mathbf{K}'(\mathbf{C}')=\hat{\rho}_0\mathbf{I}+\hat{\rho}_1\mathbf{C}'+\hat{\rho}_2\mathbf{C}'^2=\hat{\beta}_1\mathbf{C}'+\hat{\beta}_{-1}(\mathbf{C}^{-1})' \tag{8.3.24}$$

$$\hat{\mathbf{G}}'(\hat{\mathbf{T}}')=\hat{\delta}_0\mathbf{I}+\hat{\delta}_1\hat{\mathbf{T}}'+\hat{\delta}_2\hat{\mathbf{T}}'^2=\mathbf{K}'(\hat{\mathbf{E}}')=2\hat{G}(\hat{\varphi}_0\mathbf{I}+\hat{\varphi}_1\hat{\mathbf{E}}'+\hat{\varphi}_2\hat{\mathbf{E}}'^2)=2\hat{G}\hat{\mathbf{K}}'(\hat{\mathbf{E}}) \tag{8.3.25}$$

Here

$$\hat{G}=\hat{G}(\hat{K}_2',\hat{K}_3') \tag{8.3.26}$$

is a non-negative scalar factor of proportionality and

$$\hat{K}_2'=\tfrac{1}{2}\operatorname{tr}\hat{\mathbf{E}}'^2 \qquad \hat{K}_3'=\det\hat{\mathbf{E}}'. \tag{8.3.27}$$

The response coefficients

$$\hat{\rho}_0=-\tfrac{1}{3}(\operatorname{tr}\mathbf{C}'^2)\hat{\rho}_2 \qquad \hat{\beta}_1=\hat{\alpha}_1+\hat{\alpha}_2I_C=\hat{\rho}_1+\tfrac{1}{3}I_C\hat{\rho}_2 \qquad \hat{\beta}_{-1}/III_C=\hat{\alpha}_2=\hat{\rho}_2 \tag{8.3.28}$$

$$\hat{\varphi}_0=-\tfrac{2}{3}\hat{K}_2'\hat{\varphi}_2 \qquad \hat{\varphi}_1=(\hat{\varepsilon}_1+\tfrac{2}{3}I_{\hat{E}}\hat{\varepsilon}_2)/2\hat{G} \qquad \hat{\varphi}_2=\hat{\varepsilon}_2/2\hat{G} \tag{8.3.29}$$

where use has been made of the Cayley–Hamilton theorem.

The symmetric tensors $\hat{\mathbf{G}}'$, $\mathbf{K}'$ and $\hat{\mathbf{K}}'$ at X have the spectral representations,

$$\hat{\mathbf{G}}'=\sum_{r=1}^{3}\hat{g}_r'\boldsymbol{p}_r\otimes\boldsymbol{p}_r \qquad \mathbf{K}'=\sum_{r=1}^{3}k_r'\boldsymbol{p}_r\otimes\boldsymbol{p}_r \qquad \hat{\mathbf{K}}'=\sum_{r=1}^{3}\hat{k}_r'\boldsymbol{p}_r\otimes\boldsymbol{p}_r \tag{8.3.30}$$

where the $\hat{g}_r'$, k_r', $\hat{k}_r'$ are the proper numbers of $\hat{\mathbf{G}}'$, $\mathbf{K}'$ and $\hat{\mathbf{K}}'$ respectively, the common principal axes of $\hat{\mathbf{G}}'$, $\mathbf{K}'$, $\hat{\mathbf{K}}'$, $\hat{\mathbf{T}}'$ and $\mathbf{C}'$ at X being defined by the orthonormal basis $\{\boldsymbol{p}_1, \boldsymbol{p}_2, \boldsymbol{p}_3\}$. Entering the spectral representations of $\hat{\mathbf{G}}'$, $\mathbf{K}'$, $\hat{\mathbf{K}}'$ and

$$\hat{\mathbf{E}}' = \sum_{r=1}^{3} \hat{\varepsilon}'_r \boldsymbol{p}_r \otimes \boldsymbol{p}_r \qquad \hat{\varepsilon}'_i = [c'_i - (c^{-1})'_i]/4 \qquad (i=1,2,3) \qquad (8.3.31)$$

into equation (8.3.25) gives

$$\hat{g}'_i = \hat{\delta}_0 + \hat{\delta}_1 \hat{t}'_i + \hat{\delta}_2 \hat{t}'^2_i = k'_i = 2\hat{G}(\hat{\varphi}_0 + \hat{\varphi}_1 \hat{\varepsilon}'_i + \hat{\varphi}_2 \hat{\varepsilon}'^2_i) = 2\hat{G}\hat{k}'_i \qquad (i=1,2,3) \qquad (8.3.32)$$

where use has been made of equation (8.1.18) with $\overset{*}{\mathbf{T}} \equiv \hat{\mathbf{T}}$ and $\overset{*}{t}_i \equiv \hat{t}_i$.

With the definition of Lode's (1926) stress parameter defined by equation (4.1.26) as basis, the following useful parameters can be defined

$$\tilde{\mu} = \frac{3\hat{g}'_1}{\hat{g}'_3 - \hat{g}'_2} \qquad \tilde{\nu} = \frac{3k'_1}{k'_3 - k'_2} = \frac{3\hat{k}'_1}{\hat{k}'_3 - \hat{k}'_2} \qquad \hat{\nu} = \frac{3\hat{\varepsilon}'_1}{\hat{\varepsilon}'_3 - \hat{\varepsilon}'_2}. \qquad (8.3.33)$$

Substituting for the $\hat{g}'_i$, k'_i, $\hat{k}'_i$, $\hat{\varepsilon}'_i$ in equation (8.3.31) from equation (8.3.32) gives the *referential form of the generalised Lode relation:* that is

$$\tilde{\mu} = \frac{3\hat{g}'_1}{\hat{g}'_3 - \hat{g}'_2} \equiv \frac{3k'_1}{k'_3 - k'_2} = \frac{3\hat{k}'_1}{\hat{k}'_3 - \hat{k}'_2} = \tilde{\nu}. \qquad (8.3.34)$$

Here

$$\tilde{\mu} = \hat{\mu}\left(1 + \frac{(1-\hat{\mu}^2)}{2\hat{\mu}^2} \frac{3}{1 - (\hat{\delta}_1/\hat{\delta}_2 \hat{t}'_1)}\right) \qquad (8.3.35)$$

and

$$\tilde{\nu} = \hat{\nu}\left(1 + \frac{(1-\hat{\nu}^2)}{2\hat{\nu}^2} \frac{3}{1 - (\hat{\varphi}_1/\hat{\varphi}_2 \hat{\varepsilon}'_1)}\right) \qquad (8.3.36)$$

having noted that $\hat{\mu}$ is defined by equation (8.3.9). The parameters $\hat{\mu}$ and $\hat{\nu}$ are the counterpart in the referential description to Lode's (1926) stress and strain parameters in the spatial description.

The relation between the response coefficients of equations (8.3.28) and (8.3.29), implied by the identity between the two forms of the response function $\mathbf{K}' = \mathbf{K}'(\mathbf{C}') \equiv \mathbf{K}'(\hat{\mathbf{E}}')$, can be expressed in the form

$$\hat{\varphi}_0 = -\tfrac{2}{3}\hat{K}'_2 \hat{\varphi}_2 \qquad \hat{\varphi}_2 = \frac{8III_C(\hat{\beta}_1 + \hat{\beta}_{-1})}{[1 + II_C + (I_C + III_C)III_C]\hat{G}}$$
$$\hat{\varphi}_1 = \frac{2}{3} \frac{[3(II_C + III_C^2) + 2(I_C - II_C)III_C]}{[1 + II_C + (I_C + III_C)III_C]\hat{G}} (\hat{\beta}_1 + \hat{\beta}_{-1}) - 2\frac{\hat{\beta}_{-1}}{\hat{G}}. \qquad (8.3.37)$$

For purposes of reference, $\mathbf{K}' = \mathbf{K}'(\mathbf{C}')$ will be referred to as the *referential deformation response function*, and $\mathbf{K}' = \mathbf{K}'(\hat{\mathbf{E}}')$ will be referred to as the *referential strain response function.*

8.4 THE RESPONSE COEFFICIENTS

8.4.1 Referential Deformation Response Function

By direct analogy with the formulation of the concept of a referential stress intensity function, it is assumed that there exists a *referential deformation intensity function*

$$\hat{m} = \hat{L}_2' \hat{S}(\hat{\Gamma}) - \hat{c} = 0 \qquad \hat{S}(0) = 1. \tag{8.4.1}$$

Here

$$\hat{\Gamma} = \frac{27}{4} \frac{\hat{L}_3'^2}{\hat{L}_2'^3} = \hat{\kappa}^2 \frac{(\hat{\kappa}^2 - 9)^2}{(3 + \hat{\kappa}^2)^3} \qquad (0 \leqslant \hat{\Gamma} \leqslant 1) \tag{8.4.2}$$

and

$$\hat{L}_2' = \tfrac{1}{2} \operatorname{tr} \mathbf{C}'^2 \qquad \hat{L}_3' = \det \mathbf{C}' \tag{8.4.3}$$

having noted that $\hat{c}$ is to be regarded as a constant for a given state of loading. One further assumption is made:

$$\mathbf{K}'(\mathbf{C}') = 2G^* \, \partial \hat{m} / \partial \mathbf{C} = \hat{\rho}_0 \mathbf{I} + \hat{\rho}_1 \mathbf{C}' + \hat{\rho}_2 \mathbf{C}'^2 \tag{8.4.4}$$

which should be compared with equation (8.3.15). Substituting for $\hat{m}$ in equation (8.4.4) from equation (8.4.1) and comparing with equation (8.3.24) gives the referential deformation response coefficients

$$\hat{\rho}_0 = -\tfrac{2}{3} \hat{L}_2' \hat{\rho}_2 \qquad \hat{\rho}_1 = 2G^* \hat{S}\left(1 - 3 \frac{\hat{\Gamma}}{\hat{S}} \frac{d\hat{S}}{d\hat{\Gamma}}\right)$$

$$\hat{\rho}_2 = -12G^* \frac{\hat{S}(3 + \hat{\kappa}^2)}{\hat{c}_1'(9 - \hat{\kappa}^2)} \frac{\hat{\Gamma}}{\hat{S}} \frac{d\hat{S}}{d\hat{\Gamma}} \tag{8.4.5}$$

to be compared with equation (8.3.17). Thus, given a form for $\hat{S}$, the referential deformation response coefficients $\hat{\rho}_a$ $(a = 0, 1, 2)$ can be evaluated from equation (8.4.5). Entering the form for the $\hat{\rho}_p$ $(p = 1, 2)$ given by equation (8.4.5) into the second part of equation (8.3.10) gives

$$\tilde{\nu} = \hat{\kappa}\left[1 - \frac{(3 + \hat{\kappa}^2)}{\hat{\kappa}^2}\left(1 - \frac{\hat{S}}{\hat{\psi}}\right)\right] \tag{8.4.6}$$

with

$$\hat{\psi} = (\hat{\rho}_1 - \hat{\rho}_2 \hat{c}_1') / 2G^* = \hat{S}\left(1 - \frac{(3 + \hat{\kappa}^2)}{6} \frac{\hat{\kappa}}{\hat{S}} \frac{d\hat{S}}{d\hat{\kappa}}\right) \tag{8.4.7}$$

where use has been made of equation (8.4.5).

8.4.2 Referential Strain Response Function

By direct analogy with the formulation of the concept of a referential stress intensity function, it is assumed that there exists a *referential strain intensity function*

$$\hat{g}=\hat{K}'_2\hat{N}(\hat{\Omega})-\hat{b}=0 \qquad \hat{N}(0)=1 \tag{8.4.8}$$

with

$$\hat{\Omega}=\frac{27}{4}\frac{\hat{K}'^2_3}{\hat{K}'^3_2}=\hat{v}^2\frac{(\hat{v}^2-9)^2}{(3+\hat{v}^2)^3} \qquad (0\leqslant\hat{\Omega}\leqslant 1) \tag{8.4.9}$$

having noted that b is to be regarded as a constant for a given state of loading. One further assumption is made:

$$\hat{\mathbf{K}}'=\partial\hat{g}/\partial\hat{\mathbf{E}} \tag{8.4.10}$$

which is to be compared with equation (8.3.15). Substituting for $\hat{g}$ in equation (8.4.10) from equation (8.4.8) and comparing with equation (8.3.25) gives the referential strain response coefficients

$$\hat{\varphi}_0=-\tfrac{2}{3}\hat{K}'_2\hat{\varphi}_2 \qquad \hat{\varphi}_1=\hat{N}\left(1-3\frac{\hat{\Omega}}{\hat{N}}\frac{\mathrm{d}\hat{N}}{\mathrm{d}\hat{\Omega}}\right) \qquad \hat{\varphi}_2=-6\frac{\hat{N}}{\tilde{\varepsilon}'_1}\frac{(3+\hat{v}^2)}{(9-\hat{v}^2)}\frac{\hat{\Omega}}{\hat{N}}\frac{\mathrm{d}\hat{N}}{\mathrm{d}\hat{\Omega}} \tag{8.4.11}$$

which can be compared with equation (8.3.17). Thus, given a form for $\hat{N}$, the referential strain response coefficients $\hat{\varphi}_a$ $(a=0,1,2)$ can be evaluated from equation (8.4.11). Entering the form for the $\hat{\varphi}_a$ $(a=1,2)$ given by equation (8.4.11) into equation (8.3.36) and rearranging gives

$$\tilde{v}=\hat{v}\left[1-\frac{(3+\hat{v}^2)}{\hat{v}^2}\left(1-\frac{\hat{N}}{\hat{\Psi}}\right)\right] \tag{8.4.12}$$

where

$$\hat{\Psi}=\hat{\varphi}_1-\hat{\varphi}_2\tilde{\varepsilon}'_1=\hat{N}\left(1-\frac{(3+\hat{v}^2)}{6}\frac{\hat{v}}{\hat{N}}\frac{\mathrm{d}\hat{N}}{\mathrm{d}\hat{v}}\right). \tag{8.4.13}$$

There are three forms $\hat{N}$ of interest at present. These are:

(i) ***Continuously differentiable.*** Taking

$$\hat{N}=1-\tilde{c}\hat{\Omega} \qquad (\tilde{c}=\text{constant}) \tag{8.4.14}$$

gives

$$\hat{\varphi}_0=-\tfrac{2}{3}\hat{K}'_2\hat{\varphi}_2 \qquad \hat{\varphi}_1=1+2\tilde{c}\hat{\Omega} \qquad \hat{\varphi}_2=6\frac{(3+\hat{v}^2)\tilde{c}\hat{\Omega}}{(9-\hat{v}^2)\,\tilde{\varepsilon}'_1}. \tag{8.4.15}$$

When this is entered into equation (8.4.12) it gives

$$\tilde{\nu}=\hat{\nu}\left(1-\frac{9\tilde{c}(3+\hat{\nu}^2)(1-\hat{\nu}^2)(9-\hat{\nu}^2)}{(3+\hat{\nu}^2)^3-8\tilde{c}\hat{\nu}^4(9-\hat{\nu}^2)}\right)\equiv\tilde{\mu} \qquad (8.4.16)$$

where use has also been made of equation (8.4.13).

(ii) ***Piecewise linear***. For a given state of loading the following relation can be obtained from equation (8.4.8)

$$\hat{b}=\tfrac{1}{12}(3+\hat{\nu}^2)(\hat{\varepsilon}_3'-\hat{\varepsilon}_2')^2\hat{N}(\hat{\Omega})$$

which can be differentiated with respect to $\hat{\varepsilon}_1'$ and then rearranged using equations (8.3.33) and (8.3.36) to give the condition

$$\tilde{\nu}=-\frac{\mathrm{d}}{\mathrm{d}\hat{\varepsilon}_1'}(\hat{\varepsilon}_3'-\hat{\varepsilon}_2'). \qquad (8.4.17)$$

An interesting form for the strain intensity function results from taking $\tilde{\nu}=$ constant in equation (8.4.17), integration and rearrangement giving

$$\tfrac{4}{3}\hat{K}_2'\frac{(3+\tilde{\nu}\hat{\nu})^2}{(3+\hat{\nu}^2)}-\hat{A}_{(2)}^2=0 \qquad (\hat{A}_{(2)}=\text{constant}). \qquad (8.4.18)$$

When equation (8.4.18) is compared with equation (8.4.8) it is taken to imply

$$\hat{N}=\frac{(3+\tilde{\nu}_n\hat{\nu})^2}{(3+\hat{\nu}^2)}\left(\frac{(3+\hat{\nu}^2)}{(3+\tilde{\nu}_n\hat{\nu})^2}\right)_{\hat{\Omega}=0} \qquad (n=0,1,2) \qquad (8.4.19)$$

where

$$|\tilde{\nu}_n|=|\tilde{\nu}_0|+\frac{(3-2|\tilde{\nu}_0|-\tilde{\nu}_0^2)}{(1+|\tilde{\nu}_0|)}n(2-n)+\frac{(3+\tilde{\nu}_0^2)}{(1-|\tilde{\nu}_0|)}\frac{n(n-1)}{2} \qquad (n=0,1,2). \qquad (8.4.20)$$

Also, for $n=0$, equations (8.4.17) gives

$$\tilde{\nu}_0=-\lim_{\hat{\varepsilon}_1'\to 0}\frac{\mathrm{d}}{\mathrm{d}\hat{\varepsilon}_1'}(\hat{\varepsilon}_3'-\hat{\varepsilon}_2') \qquad (8.4.21)$$

having noted that

$$n=\begin{cases}0\\1\\2\end{cases} \text{for} \begin{cases}(0\leqslant\hat{\nu}^2\leqslant1), & (0\leqslant\hat{\Omega}\leqslant1)\\(1\leqslant\hat{\nu}^2\leqslant9), & (1\geqslant\hat{\Omega}\geqslant0).\\(9\leqslant\hat{\nu}^2\leqslant\infty), & (0\leqslant\hat{\Omega}\leqslant1)\end{cases} \qquad (8.4.22)$$

(iii) ***Composite strain intensity function.*** A composite strain intensity function can be obtained by taking $\hat{N}$ to be of the form

$$\hat{N}=\hat{\phi}\tilde{N}_{(1)}+(\tilde{N}_{(2)}-\hat{\phi}\tilde{N}_{(1)})U. \qquad (8.4.23)$$

U is some appropriate form of step function, $\tilde{N}_{(1)}$ is restricted to the form of equation (8.4.19) and $\tilde{N}_{(2)}$ is restricted to the form of equation (8.4.14). In the context of equation (8.4.22), for each value of $n=0,1,2$, there is a particular value of $\hat{v}=\hat{v}^*$ at which the strain intensity function changes from that obtained using equation (8.4.8) with $\hat{N}=\tilde{N}_{(2)}$, for which $U=1$, to that obtained using equation (8.4.8) with $\hat{N}=\hat{\phi}\tilde{N}_{(1)}$, for which $U=0$. Thus, at the discontinuous change from $U=0$ to $U=1$,

$$[\hat{N}]_{\hat{v}=\hat{v}^*}=[\tilde{N}_{(2)}]_{\hat{v}=\hat{v}^*}=[\hat{\phi}\tilde{N}_{(1)}]_{\hat{v}=\hat{v}^*} \tag{8.4.24}$$

having noted that $\hat{\phi}$ is solely a function of $\hat{v}^*$.

Since there is a direct analogy between the formulation of the referential strain intensity function $\hat{g}$ and the spatial stress intensity function h, which itself depends upon the formulation of the initial yield function, further details regarding $\hat{g}$ can be obtained through Chapter 3.

8.5 RESTRICTIONS ON THE FORM OF THE STRESS AND STRAIN INTENSITY FUNCTIONS

In the context of the spectral representation

$$\hat{\mathbf{E}}'=\sum_{r=1}^{3}\hat{\varepsilon}'_r \boldsymbol{p}_r\otimes\boldsymbol{p}_r \tag{8.5.1}$$

let the principal axes of $\hat{\mathbf{E}}'$ be oriented so that they are equally inclined to a plane Π_r, through the origin O of the system (O, p). Each of the deviatoric strain space axes makes an angle $\cos^{-1}\sqrt{\frac{2}{3}}$ with the Π_r plane, and hence the projected deviatoric components are $\sqrt{\frac{2}{3}}\,\hat{\varepsilon}'_i$ $(i=1,2,3)$. The $(\bar{u},\bar{v})$ curve, where

$$\bar{u}=\sqrt{\tfrac{3}{2}}\,\hat{\varepsilon}'_1=\frac{1}{\sqrt{3}}\hat{v}\bar{v}\qquad \bar{v}=\frac{(\hat{\varepsilon}'_3-\hat{\varepsilon}'_2)}{\sqrt{2}}=\frac{\sqrt{3}}{(3+\hat{v}^2)^{1/2}\hat{N}^{1/2}}\sqrt{2\hat{b}} \tag{8.5.2}$$

constitutes a geometrical representation in the Π_r plane of any referential strain system defined by the referential strain intensity function $\hat{g}$. It is of particular interest to note that,

$$\tilde{v}=-\sqrt{3}\frac{d\bar{v}}{d\bar{u}}=-\frac{d}{d\hat{\varepsilon}_1}(\hat{\varepsilon}'_3-\hat{\varepsilon}'_2)\equiv-\frac{d}{d\hat{t}_1}(\hat{t}_3-\hat{t}_2)=\tilde{\mu}. \tag{8.5.3}$$

Thus, for $\hat{v}=\tilde{v}$ for all $\hat{v}$, the $(\bar{u},\bar{v})$ curve in the Π_r plane is a circle. Deviations from a circle in the Π_r plane are obtained by taking $\hat{N}$ to be given by equation (8.4.14). For a piecewise linear, continuous, referential strain intensity function, the $(\bar{u},\bar{v})$ curve in the Π_r plane consists of twelve linear sections, having noted that $\tilde{v}\equiv\tilde{v}_n=$constant for all $\hat{v}$ in each of the three ranges of $\hat{v}^2$ identified by the three values of n in equation (8.4.22).

Thus far the referential stress and strain parameters, $\hat{\mu}$ and $\hat{\nu}$, have been considered separately. However, in terms of the constitutive equation (8.3.25), the principal axes of $\hat{\mathbf{K}}'$ and $\hat{\mathbf{E}}'$ are also those of $\hat{\mathbf{G}}'$ and $\hat{\mathbf{T}}'$. Hence the π_r plane must be indistinguishable from the Π_r plane, this common plane being the deviatoric plane, having noted that the π_r plane is the counterpart in the referential description to the π plane in the spatial description (see §3.1.2). This conclusion leads to the interesting result that, in principle, the referential strain systems defined by the referential strain intensity function $\hat{g}$, can have geometrical representations in the deviatoric plane which differ from that for the corresponding referential stress system. Compatibility of the two different geometrical representations is ensured, at least in principle, by the identity of equation (8.3.34), realising that

$$\hat{\mu}^2 = \tilde{\mu}^2 \equiv \tilde{\nu}^2 = \hat{\nu}^2 = \begin{cases} 0, 9 \\ 1, \infty \end{cases} \qquad \text{for } \hat{\omega} = \hat{\Omega} = \begin{cases} 0 \\ 1 \end{cases} \tag{8.5.4}$$

where $\hat{\omega}$ is defined by equation (8.3.13).

The identity of equation (8.3.34) does, however, exclude the possible use of a continuously differentiable, referential, stress intensity function with a piecewise *linear*, continuous, referential strain intensity function. This is because in the case of a continuously differentiable, referential, stress intensity function, $\tilde{\mu}$ varies continuously in each of the three ranges of $n=0, 1, 2$, whereas $\tilde{\nu} \equiv \tilde{\nu}_n =$ constant for each value of n. Similarly, the identity of equation (8.3.34) excludes the use of a continuously differentiable, referential strain intensity function with a piecewise linear, continuous, referential stress intensity function.

8.6 COMBINED STRESSING OF A SHEET OF INCOMPRESSIBLE MATERIAL

8.6.1 Effective Stress-Reduced Strain Relation

Let the cartesian coordinates (X^1, X^2, X^3) be the referential coordinates of a body in the natural state and the cartesian coordinates (x^1, x^2, x^3) be the spatial coordinates in the deformed configuration. The discussion will be restricted to an incompressible sheet of thickness $2h_0$ aligned with the cartesian coordinate system (X^1, X^2, X^3). The material, regarded as incompressible with respect to volume, is assumed to be isotropic relative to its undeformed and unstressed, that is its natural, configuration.

Assume a homogeneous deformation such that

$$x^1 = \lambda_1 X^1 \qquad x^2 = \lambda_2 X^2 \qquad x^3 = \lambda_3 X^3 \tag{8.6.1}$$

the condition of incompressibility giving

$$\lambda_1\lambda_2\lambda_3 = \det \mathbf{F} = 1 \tag{8.6.2}$$

where the λ_i $(i=1,2,3)$ are the principal stretches.

Consider the deformation to be produced by means of edge tractions only, so that the major surfaces $X^1 = \pm h_0$ are free from applied stress. The only non-zero components of stress are the normal components, thus in terms of the principal referential stresses,

$$\hat{t}_1 = \hat{T}_{11} = 0 \qquad \hat{t}_2 = \hat{T}_{22} \qquad \hat{t}_3 = \hat{T}_{33}.$$

For $\hat{t}_1 = 0$, the referential form of Lode's (1926) stress parameter can be expressed in the form

$$\hat{\mu} = -1 - 2\frac{\hat{t}_2}{(\hat{t}_3 - \hat{t}_2)} \tag{8.6.3}$$

where use has been made of equation (8.3.9).

Equation (8.6.2) implies that only two of the principal stretches are independently assignable. Since the major surfaces $X^1 = \pm h_0$ are free from applied stress, the two independently assignable principal stretches will be taken to be λ_2 and λ_3.

From equations (2.1.35), (2.1.44) and (8.6.1)

$$[F_{i\alpha}] = \begin{bmatrix} (\lambda_2\lambda_3)^{-1} & 0 & 0 \\ 0 & \lambda_2 & 0 \\ 0 & 0 & \lambda_3 \end{bmatrix}$$

$$[C_{\alpha\beta}] = \begin{bmatrix} (\lambda_2\lambda_3)^2 & 0 & 0 \\ 0 & \lambda_2^2 & 0 \\ 0 & 0 & \lambda_3^2 \end{bmatrix} \qquad [(C^{-1})_{\alpha\beta}] = \begin{bmatrix} \lambda_2^2\lambda_3^2 & 0 & 0 \\ 0 & \lambda_2^{-2} & 0 \\ 0 & 0 & \lambda_3^{-2} \end{bmatrix}. \tag{8.6.4}$$

The principal invariants are:

$$\begin{gathered} I_C = (1 + \lambda_2^2\lambda_3^4 + \lambda_2^4\lambda_3^2)/(\lambda_2^2\lambda_3^2) \\ II_C = (\lambda_2^2 + \lambda_3^2 + \lambda_2^4\lambda_3^4)/(\lambda_2^2\lambda_3^2) \qquad III_C = 1. \end{gathered} \tag{8.6.5}$$

Substituting for the $c_i = \lambda_i^2$ in equation (8.3.31) gives

$$\begin{aligned} \tilde{\varepsilon}_1 &= \frac{(2 + \lambda_2^2 + \lambda_3^2 + 2\lambda_2^2\lambda_3^2)(1 - \lambda_2^2\lambda_3^2)}{12\lambda_2^2\lambda_3^2} \\ \tilde{\varepsilon}_2 &= \frac{(1 + 2\lambda_3^2 + 2\lambda_2^2\lambda_3^2 + \lambda_2^2\lambda_3^4)(\lambda_2^2 - 1)}{12\lambda_2^2\lambda_3^2} \\ \tilde{\varepsilon}_3 &= \frac{(1 + 2\lambda_2^2 + 2\lambda_2^2\lambda_3^2 + \lambda_2^4\lambda_3^2)(\lambda_3^2 - 1)}{12\lambda_2^2\lambda_3^2} \end{aligned} \tag{8.6.6}$$

where use has been made of equation (8.6.2). Entering the values of the $\hat{\varepsilon}_i'$ into equation (8.3.33) from equation (8.6.6) gives

$$\hat{v}=-\frac{(2+\lambda_2^2+\lambda_3^2+2\lambda_2^2\lambda_3^2)(\lambda_2^2\lambda_3^2-1)}{(\lambda_2^2\lambda_3^2+1)(\lambda_3^2-\lambda_2^2)}. \tag{8.6.7}$$

Using the constitutive equation (8.3.25) in the form

$$\hat{\mathbf{G}}'=2\hat{G}\hat{\mathbf{K}}' \tag{8.6.8}$$

the following effective stress–effective strain relation can be obtained

$$t''=3\frac{[\hat{N}^{1/2}]_{\hat{v}=-1}}{[\hat{H}^{1/2}]_{\hat{\mu}=-1}}G\hat{\varepsilon}'' \qquad G=\hat{G}\left(\frac{\hat{\Psi}(3+\hat{\mu}\tilde{\mu})}{\hat{\Lambda}(3+\hat{v}\tilde{v})}\right)^{1/2}. \tag{8.6.9}$$

Here use has been made of the referential form of the generalised Lode relation, i.e. equation (8.3.34), to eliminate the terms involving $\tilde{\mu}$ and $\tilde{v}$ and where the *effective stress* is

$$t''=\frac{(3\hat{J}_2'\hat{H})^{1/2}}{[\hat{H}^{1/2}]_{\hat{\mu}=-1}}=\frac{[\hat{H}(3+\hat{\mu}^2)]^{1/2}}{[\hat{H}^{1/2}]_{\hat{\mu}=-1}}\hat{\tau}_1 \qquad \hat{\tau}_1=\tfrac{1}{2}(\hat{t}_3-\hat{t}_2) \tag{8.6.10}$$

and the *effective strain*

$$\hat{\varepsilon}''=\frac{(\frac{4}{3}\hat{K}_2'\hat{N})^{1/2}}{[\hat{N}^{1/2}]_{\hat{v}=-1}}=\frac{2[\hat{N}(3+\hat{v}^2)]^{1/2}}{[\hat{N}^{1/2}]_{\hat{v}=-1}}\hat{\gamma}_1 \qquad \hat{\gamma}_1=\tfrac{1}{2}(\hat{\varepsilon}_3'-\hat{\varepsilon}_2'). \tag{8.6.11}$$

Use has been made of the relations

$$\hat{\Lambda}=\hat{H}\frac{(3+\hat{\mu}^2)}{(3+\hat{\mu}\tilde{\mu})}=\hat{\delta}_1-\hat{\delta}_2\hat{t}_1' \qquad \hat{\Psi}=\hat{N}\frac{(3+\hat{v}^2)}{(3+\hat{v}\tilde{v})}=\hat{\varphi}_1-\hat{\varphi}_2\hat{\varepsilon}_1'$$

which follow from equations (8.3.17), (8.3.32), (8.4.12) and (8.4.13).

The remaining discussion is restricted to those materials for which

$$\lim_{\hat{\varepsilon}''\to 0}\hat{G}=G.$$

where G_0 is the ground state, that is the classical shear modulus. With this as the basis, equation (8.6.9) can be rearranged into the linear relation,

$$\hat{t}''=3G_0\hat{e}'' \tag{8.6.12}$$

where the *reduced stress*

$$\hat{t}''=(3\hat{J}_2')^{1/2}=(3+\hat{\mu}^2)^{1/2}\hat{\tau}_1 \qquad \hat{\tau}_1=\tfrac{1}{2}(\hat{t}_3-\hat{t}_2) \tag{8.6.13}$$

and the *reduced strain*

$$\begin{aligned}\hat{e}''&=\frac{\hat{G}}{G_0}\left(\frac{\hat{\Psi}(3+\hat{\mu}\tilde{\mu})}{\hat{H}\hat{\Lambda}(3+\hat{v}\tilde{v})}\right)^{1/2}\hat{\varepsilon}''\\&=\frac{2}{3}\frac{\hat{G}}{G_0}\left(\frac{\hat{N}\hat{\Psi}(3+\hat{\mu}\tilde{\mu})}{\hat{H}\hat{\Lambda}(3+\hat{v}\tilde{v})}\right)^{1/2}(3+\hat{v}^2)^{1/2}\hat{\gamma}_1 \qquad \hat{\gamma}_1=\tfrac{1}{2}(\hat{\varepsilon}_3'-\hat{\varepsilon}_2').\end{aligned} \tag{8.6.14}$$

8.6.2 Biaxial Extension

It is evident from the discussion given in §§3.3, 8.3.1, 8.3.2 and 8.4.2 that for all forms of continuously differentiable stress intensity and strain intensity functions

$$\hat{\mu}^2 = \tilde{\mu}^2 \equiv \tilde{\nu}^2 = \hat{\nu}^2 = \begin{cases} 0, 9 \\ 1, \infty \end{cases} \qquad \text{for } \hat{\omega} = \hat{\Omega} = \begin{cases} 0 \\ 1 \end{cases} \tag{8.6.15}$$

where $\hat{\omega}$ and $\hat{\Omega}$ are defined by equations (8.3.13) and (8.4.9) respectively, and where use has been made of the referential form of the generalised Lode relation, that is equation (8.3.34). For any piecewise continuous referential stress intensity and referential strain intensity functions the associated vertices occur at

$$\hat{\mu}^2 = \hat{\nu}^2 = \begin{cases} 0, 9 \\ 1, \infty \end{cases} \qquad \text{for } \hat{\omega} = \hat{\Omega} = \begin{cases} 0 \\ 1 \end{cases}.$$

Hence, in the transition through a given vertex, both $\tilde{\mu}$ and $\tilde{\nu}$ must pass discontinuously through the above values of $\hat{\mu} \equiv \hat{\nu}$ appropriate to the given vertex. Thus, even for the piecewise continuous forms of the referential stress intensity and strain intensity functions, equation (8.6.15) has to be satisfied at some stage. It is evident from equation (8.6.15) that the values of $\hat{\mu}$ corresponding to $\hat{\omega} = 0, 1$ and the values of $\hat{\nu}$ for $\hat{\Omega} = 0, 1$, cover the entire range of $\hat{\mu}$ and $\hat{\nu}$. These considerations are taken to imply that particular significance is to be attached to the values of $\hat{\nu} = 0, -1, -3, -\infty$, and hence to the particular conditions

$$\hat{\mu} = \tilde{\mu} \equiv \tilde{\nu} = \hat{\nu} = \begin{cases} 0, -3 \\ -1, -\infty \end{cases} \qquad \text{for } \hat{\omega} = \hat{\Omega} = \begin{cases} 0 \\ 1 \end{cases} \tag{8.6.16}$$

where use has been made of equation (8.6.15).

(i) $\hat{\nu} = 0$. Solving equation (8.6.7) for this value of $\hat{\nu}$ gives

$$\lambda_1 = 1 \qquad \lambda_2 = \lambda_3^{-1} \tag{8.6.17}$$

where use has been made of equation (8.6.2). Entering the condition of equation (8.6.17) into equations (8.6.5), (8.6.6) and (8.3.37) gives

$$I_C = II_C = \lambda_3^2 + \lambda_3^{-2} + 1, \qquad \hat{\Omega} = 0,\ \hat{\varphi}_0 = 0,\ \hat{\varphi}_1 = 1,\ \hat{\varphi}_2 = 0,\ \hat{\Psi} = \hat{N} = 1 \tag{8.6.18}$$

$$\hat{\varepsilon}_1' = 0, \qquad \hat{\varepsilon}_3' = -\hat{\varepsilon}_2' = \hat{\gamma}_1, \qquad \hat{\gamma}_1 = (\lambda_3^4 - 1)/4\lambda_3^2 \tag{8.6.19}$$

where use has been made of equations (8.4.8), (8.4.9) and (8.4.11). The conditions of equation (8.6.18), when entered into equation (8.3.36), give $\tilde{\nu} = 0$ and hence $\tilde{\mu} \equiv \tilde{\nu} = \hat{\nu} = 0$, where use has been made of the referential form of the generalised Lode relation of equation (8.3.34). Thus, for equation (8.6.16) to be satisfied (i.e. for $\hat{\mu} = 0$), equation (8.6.3) gives

$$\hat{t}_3 = -\hat{t}_2 \qquad (\hat{\mu}=0, \hat{t}_1=0). \tag{8.6.20}$$

Consequently the simple mode of deformation for which $\hat{v}=0$ is an example of combined tension ($\hat{t}_3 \geqslant 0$) and compression ($\hat{t}_2 \leqslant 0$). The conditions $\hat{\mu}=0$, $\tilde{\mu}=0$ require that,

$$\hat{\delta}_0=0 \qquad \hat{\delta}_1=1 \qquad \hat{\delta}_2=0 \qquad \hat{H}=\hat{\Lambda}=1 \qquad \hat{\omega}=0. \tag{8.6.21}$$

The conditions of equations (8.6.18) to (8.6.21) reduce equations (8.6.10), (8.6.11), (8.6.13) and (8.6.14) to

$$t''=\frac{\sqrt{3}}{[\hat{H}^{1/2}]_{\hat{\mu}=-1}}\hat{t}_3 \qquad \hat{\varepsilon}''=\frac{2}{\sqrt{3}[\hat{N}^{1/2}]_{\hat{v}=-1}}\hat{\gamma}_1$$

$$\hat{t}''=\sqrt{3}\,\hat{\tau}_1=\sqrt{3}\,\hat{t}_3 \qquad \hat{e}''=\frac{2[\hat{G}]_{\hat{v}=0}}{\sqrt{3}G_0}\hat{\gamma}_1 \tag{8.6.22}$$

which can be entered into equations (8.6.9) and (8.6.12) to give

$$\hat{\tau}_1=2[\hat{G}]_{\hat{v}=0}\hat{\gamma}_1 \qquad (\hat{\tau}_1=\hat{t}_3). \tag{8.6.23}$$

For infinitely small strains, equation (8.6.23) gives

$$\lim_{\lambda_3\to 1}\frac{\hat{\tau}_1}{2\hat{\gamma}_1}=[\hat{G}]_{\lambda_3=1}=G_0. \tag{8.6.24}$$

Since the proper numbers c_i of **C** are identical to the proper numbers b_i of **B**, it follows from the identity $b_i=\lambda_i^2=c_i$ that for this simple mode of deformation **B** and **C** are indistinguishable. Therefore it is concluded that there is zero rotation and this simple mode of deformation is an example of *pure shear*.

(ii) $\hat{v}=-1$. Solving equation (8.6.7) for $\hat{v}=-1$ gives

$$\lambda_1=\lambda_2=\lambda_3^{-1/2} \tag{8.6.25}$$

where use has been made of equation (8.6.2). Entering the condition of equation (8.6.25) into equations (8.6.5), (8.6.6) and (8.3.7) gives

$$I_C=\frac{2}{\lambda_3}+\lambda_3^2 \qquad II_C=2\lambda_3+\lambda_3^{-2} \qquad \hat{\Omega}=1$$

$$\hat{\varphi}_1=\frac{[(2\lambda_3^3+3\lambda_3+1)\hat{\beta}_1-(\lambda_3^3+3\lambda_3^2+2)\hat{\beta}_{-1}}{3\hat{G}(\lambda_3+1)(\lambda_3^2+1)} \tag{8.6.26}$$

$$\hat{\varphi}_2=\frac{8(\hat{\beta}_1+\hat{\beta}_{-1})\lambda_3^2}{\hat{G}(\lambda_3+1)^2(\lambda_3^2+1)} \qquad \hat{\Psi}=\frac{2(\hat{\beta}_1\lambda_3-\hat{\beta}_{-1})}{\hat{G}(\lambda_3+1)}=\hat{N}$$

$$\hat{\varepsilon}_1'=\hat{\varepsilon}_2'=-\tfrac{1}{2}\hat{\varepsilon}_3' \qquad \hat{\varepsilon}_3'=\frac{(\lambda_3+1)(\lambda_3^3-1)}{6\lambda_3^2} \tag{8.6.27}$$

where use has been made of equations (8.4.8), (8.4.9) and (8.4.11). The conditions of equation (8.6.26), when entered into equation (8.3.36) give $\hat{v} = -1$ and hence $\tilde{\mu} \equiv \tilde{v} = \hat{v} = -1$, where use has been made of the referential form of the generalised Lode relation of equation (8.3.34). Thus for equation (8.6.16) to be satisfied (i.e. for $\hat{\mu} = -1$), equation (8.6.3) gives

$$\hat{t}_2 = 0 \qquad \hat{t}_3 \geqslant 0 \qquad (\hat{\mu} = -1, \hat{t}_1 = 0) \tag{8.6.28}$$

from which it follows that the simple mode of deformation for which equation (8.6.16) is satisfied with $\hat{\mu} = -1$ is an example of simple extension. The conditions $\hat{\mu} = -1$, $\tilde{\mu} = -1$, require that

$$\hat{\omega} = 1 \qquad \hat{H} = \hat{\delta}_1 + \tfrac{1}{3}\hat{\delta}_2\hat{t}_3 = \hat{\Lambda}. \tag{8.6.29}$$

The conditions of equations (8.6.26) to (8.6.29) reduce equations (8.6.10), (8.6.11), (8.6.13) and (8.6.14) to

$$\begin{aligned} t'' &= \hat{t}_3 \qquad & \hat{\varepsilon}'' &= \hat{\varepsilon}_3' = \frac{(\lambda_3 + 1)(\lambda_3^3 - 1)}{6\lambda_3^2} \\ \hat{t}'' &= \hat{t}_3 \qquad & \hat{e}'' &= \left[\frac{\hat{G}\hat{\Psi}}{G_0\hat{\Lambda}}\right]_{\hat{v}=-1} \hat{\varepsilon}_3' \end{aligned} \tag{8.6.30}$$

which can be entered into equations (8.6.9) and (8.6.12) to give

$$\hat{t}_3 = 3\left[\hat{G}\frac{\hat{\Psi}}{\hat{\Lambda}}\right]_{\hat{v}=-1} \hat{\varepsilon}_3' = \hat{E}\hat{\varepsilon}_3' \tag{8.6.31}$$

where

$$\hat{E} = 3\left[\hat{G}\frac{\hat{\Psi}}{\hat{\Lambda}}\right]_{\hat{\mu}=\hat{v}=-1}. \tag{8.6.32}$$

For infinitely small strains, equation (8.6.31) gives

$$\lim_{\lambda_3 \to 1} \frac{\hat{t}_3}{\hat{\varepsilon}_3'} = 3\left[\hat{G}\frac{\hat{\Psi}}{\hat{\Lambda}}\right]_{\lambda_3=1} = [\hat{E}]_{\lambda_3=1} = 3G_0\left[\frac{\hat{\Psi}}{\hat{\Lambda}}\right]_{\lambda_3=1} = E_0 \tag{8.6.33}$$

having noted that $\hat{\Psi}/\hat{\Lambda} = 1$ for the ground state. For this condition equation (8.6.33) is just the stress–strain relation of classical infinitesimal elasticity for simple tension. The associated condition $E_0 = 3G_0$ is taken to imply that the $\hat{E}$ defined by equation (8.6.32) can be regarded as a generalised form of Young's modulus in the referential description. Equation (8.6.31) implies that the universal stress–strain relation of equation (8.6.9) has been normalised to the stress–strain relation in simple tension (or simple compression). Since $\hat{\varepsilon}_3'(\lambda_3) = -\hat{\varepsilon}_3'(1/\lambda_3)$, the material response in simple tension and simple compression has equal but opposite effect.

(iii) $\hat{v} = -3$. Solving equation (8.6.7) for this value of $\hat{v}$ gives

$$\lambda_2 = 1 \qquad \lambda_3 = \lambda_1^{-1} \tag{8.6.34}$$

where use has been made of equation (8.6.2). Entering the condition of equation (8.6.34) into equations (8.6.5), (8.6.6) and (8.3.37) gives

$$I_C = II_C = \lambda_3^2 + \frac{1}{\lambda_3^2} + 1 \quad \hat{\Omega}=0,\ \hat{\varphi}=0,\ \hat{\varphi}_1=1,\ \hat{\varphi}_2=0,\ \hat{\Psi}=\hat{N}=1 \tag{8.6.35}$$

$$\hat{\varepsilon}_1' = -\hat{\varepsilon}_3' \qquad \hat{\varepsilon}_2'=0 \qquad \hat{\varepsilon}_3'=(\lambda_3^4-1)/4\lambda_3^2=2\hat{\gamma}_1 \tag{8.6.36}$$

where use has been made of equations (8.4.8), (8.4.9) and (8.4.11). The conditions of equation (8.6.35), when entered into equation (8.3.36), give $\tilde{v}=-3$ and hence $\tilde{\mu}\equiv\tilde{v}=\hat{v}=-3$, where use has been made of the referential form of the generalised Lode relation of equation (8.3.34). Thus, for equation (8.6.16) to be satisfied (i.e. for $\hat{\mu}=-3$), equation (8.6.3) gives (Billington 1985b)

$$\hat{t}_3 = 2\hat{t}_2 \qquad (\hat{\mu}=-3,\ \hat{t}_1=0) \tag{8.6.37}$$

for all λ_3. The condition of equation (8.6.37) is independent of both the loading coefficients $\hat{\delta}_i$ $(i=0, 1, 2)$ and the material response coefficients $\hat{\varphi}_i$ $(i=0, 1, 2)$ and is therefore a *universal relation*. It is evident that this simple mode of deformation is an example of pure shear characterised by $\hat{v}=-3$. The conditions $\hat{\mu}=-3$, $\tilde{\mu}=-3$ require that

$$\hat{\delta}_0=0, \qquad \hat{\delta}_1=1, \qquad \hat{\delta}_2=0, \qquad \hat{H}=1, \qquad \hat{\Lambda}=1, \qquad \hat{\omega}=0. \tag{8.6.38}$$

The conditions of equations (8.6.35) to (8.6.38) reduce equations (8.6.10), (8.6.11), (8.6.13) and (8.6.14) to

$$t'' = \frac{\sqrt{3}}{2[\hat{H}]_{\hat{\mu}=-1}}\hat{t}_3 \qquad \hat{\varepsilon}'' = \frac{4}{\sqrt{3}\,[\hat{N}^{1/2}]_{\hat{v}=-1}}\hat{\gamma}_1$$

$$\hat{t}'' = \tfrac{1}{2}\sqrt{3}\,\hat{t}_3 \qquad \hat{e}'' = \frac{4[\hat{G}]_{\hat{v}=-3}}{\sqrt{3}\,G_0}\hat{\gamma}_1. \tag{8.6.39}$$

These can be entered into equations (8.6.9) and (8.6.12) to give

$$\hat{\tau}_1 = 2[\hat{G}]_{\hat{v}=-3}\hat{\gamma}_1 \qquad (\hat{\tau}_1=\tfrac{1}{4}\hat{t}_3). \tag{8.6.40}$$

For infinitely small strains, equation (8.6.40) gives

$$\lim_{\lambda_3\to 1}\frac{\hat{\tau}_1}{2\hat{\gamma}_1} = [\hat{G}]_{\lambda_3=1} = G_0 \tag{8.6.41}$$

where G_0 is the ground state, or classical shear modulus

(iv) $\hat{v}=-\infty$. Solving equation (8.6.7) for $\hat{v}=-\infty$ gives

$$\lambda_2=\lambda_3=\lambda_1^{-1/2} \tag{8.6.42}$$

where use has been made of equation (8.6.2). Entering the condition of equation (8.6.42) into equations (8.6.5), (8.6.6) and (8.3.7) gives

$$I_C = 2\lambda_3^2 + \frac{1}{\lambda_3^4} \qquad II_C = \frac{2}{\lambda_3^2} + \lambda_3^4 \qquad \hat{\Omega} = 1$$

$$\hat{\varphi}_1 = \frac{2[(2+3\lambda_3^4+\lambda_3^6)\hat{\beta}_1 - (1+3\lambda_3^2+2\lambda_3^6)\hat{\beta}_{-1}]}{3\hat{G}(\lambda_3^2+1)(\lambda_3^4+1)} \tag{8.6.43}$$

$$\hat{\varphi}_2 = \frac{8(\hat{\beta}_1+\hat{\beta}_{-1})\lambda_3^4}{\hat{G}(\lambda_3^2+1)^2(\lambda_3^4+1)} \qquad \hat{\Psi} = \frac{2(\hat{\beta}_1\lambda_3^4 - \hat{\beta}_{-1})}{\hat{G}(\lambda_3^4+1)} = \hat{N}$$

$$\hat{\varepsilon}_2' = \hat{\varepsilon}_3' = -\tfrac{1}{2}\hat{\varepsilon}_1' \qquad \hat{\varepsilon}_1' = -(\lambda_3^2+1)(\lambda_3^6-1)/6\lambda_3^4 \tag{8.6.44}$$

where use has been made of equations (8.4.8), (8.4.9) and (8.4.11). The conditions of equation (8.6.43), when entered into equation (8.3.36), give $\tilde{\nu} = -\infty$ and hence $\tilde{\mu} \equiv \tilde{\nu} = \hat{\nu} = -\infty$, where use has been made of the referential form of the generalised Lode relation of equation (8.3.34). Thus for equation (8.6.16) to be satisfied (i.e. for $\hat{\mu} = -\infty$) equation (8.6.3) gives

$$\hat{t}_2 = \hat{t}_3 \qquad (\hat{\mu} = -\infty,\ \hat{t}_1 = 0) \tag{8.6.45}$$

This is just the condition for equibiaxial extension, provided that $\lambda_2 = \lambda_3$ for this simple mode of deformation always which is just the condition of equation (8.6.42). The conditions, $\hat{\mu} = -\infty$, $\tilde{\mu} = -\infty$, require that

$$\hat{\omega} = 1 \qquad \hat{\Lambda} = \hat{\delta}_1 + \tfrac{2}{3}\hat{\delta}_2\hat{t}_3 = \hat{H}. \tag{8.6.46}$$

For equibiaxial extension it is necessary to rearrange equations (8.6.10), (8.6.11), (8.6.13) and (8.6.14) into the form

$$t'' = \frac{[(\hat{t}_3-\hat{t}_2)^2 + \hat{t}_3\hat{t}_2]^{1/2}\hat{H}^{1/2}}{[\hat{H}^{1/2}]_{\hat{\mu}=-1}} \qquad \varepsilon'' = \frac{[(\hat{\varepsilon}_3'-\hat{\varepsilon}_2')^2 + 3(\hat{\varepsilon}_3'+\hat{\varepsilon}_2')^2]^{1/2}\hat{N}^{1/2}}{\sqrt{3}[\hat{N}^{1/2}]_{\hat{\nu}=-1}} \tag{8.6.47}$$

$$\hat{t}'' = [(\hat{t}_3-\hat{t}_2)^2 + \hat{t}_3\hat{t}_2]^{1/2} \qquad \hat{e}'' = \frac{\hat{G}}{G_0}\left(\frac{\hat{\Psi}(3+\hat{\mu}\tilde{\mu})}{\hat{H}\hat{\Lambda}(3+\hat{\nu}\tilde{\nu})}\right)^{1/2}\varepsilon''. \tag{8.6.48}$$

The conditions of equations (8.6.43) to (8.6.46) reduce equations (8.6.47) and (8.6.48) to

$$t'' = \frac{[\hat{H}^{1/2}]_{\hat{\mu}=-\infty}}{[\hat{H}^{1/2}]_{\hat{\mu}=-1}}\hat{t}_3 \qquad \varepsilon'' = 2\frac{[\hat{N}^{1/2}]_{\hat{\nu}=-\infty}}{[\hat{N}^{1/2}]_{\hat{\nu}=-1}}\hat{\varepsilon}_3'$$

$$\hat{t}'' = \hat{t}_3 \qquad \hat{e}'' = 2\left[\frac{\hat{G}\hat{\Psi}}{G_0\hat{\Lambda}}\right]_{\hat{\nu}=-\infty}\hat{\varepsilon}_3' \tag{8.6.49}$$

which can be entered into equations (8.6.9) and (8.6.12) to give

$$\hat{t}_3 = 6\left[\hat{G}\frac{\hat{\Psi}}{\hat{\Lambda}}\right]_{\hat{\nu}=-\infty}\hat{\varepsilon}_E \qquad \hat{\varepsilon}_E = \frac{(\lambda_3^2+1)(\lambda_3^6-1)}{12\lambda_3^4}\ (\equiv \hat{\varepsilon}_3'). \tag{8.6.50}$$

For the material response in equibiaxial compression and equibiaxial tension to have equal but opposite effect, it is necessary that $\hat{\varepsilon}_E(\lambda_3) = -\hat{\varepsilon}_E(1/\lambda_3)$ a condition which equation (8.6.50) satisfies, and hence $\hat{\varepsilon}_E$ is an appropriate measure of simple equibiaxial tensile and compressive strain.

Given a form for the stress intensity and strain intensity functions, the four separate stress–strain relations obtained using equations (8.6.9), (8.6.22), (8.6.30), (8.6.39) and (8.6.49) should be indistinguishable, thus giving a single, universal (σ'', ε'') relation for an incompressible isotropic material.

8.7 EXTENSION AND TORSION OF A SOLID ROD

8.7.1 Stress–Strain Relations

Consider a right-circular solid rod of incompressible material. Let (R, Θ, Z) be the cylindrical referential coordinates in the initial state of a particle that is located in the deformed configuration by the cylindrical spatial coordinates (r, θ, z).

For a twisted rod with its principal axis aligned with the Z-axis, the simple deformations to be considered are

$$r = R/\sqrt{F} \qquad \theta = \Theta + DZ \qquad z = FZ \tag{8.7.1}$$

where $F = l/L$ is the ratio of the current length l to the undeformed length L of a solid rod. Using equation (8.7.1)

$$[F^i{}_\alpha] = \begin{bmatrix} \dfrac{1}{\sqrt{F}} & 0 & 0 \\ 0 & 1 & D \\ 0 & 0 & F \end{bmatrix} \tag{8.7.2}$$

and hence the non-vanishing physical components $B\langle ij\rangle$, $(B^{-1})\langle ij\rangle$ of $\mathbf{B}$ and its inverse can be obtained in the form

$$[B\langle ij\rangle] = \begin{bmatrix} \dfrac{1}{F} & 0 & 0 \\ 0 & \dfrac{1}{F}(1+R^2D^2) & rDF \\ 0 & rDF & F^2 \end{bmatrix}$$

$$[(B^{-1})\langle ij\rangle] = \begin{bmatrix} F & 0 & 0 \\ 0 & F & -rD \\ 0 & -rD & \dfrac{(1+R^2D^2)}{F^2} \end{bmatrix} \tag{8.7.3}$$

Similarly, the physical components of $\mathbf{C}$ and its inverse can be obtained in the form:

$$[C\langle\alpha\beta\rangle]=\begin{bmatrix}\frac{1}{F} & 0 & 0\\ 0 & \frac{1}{F} & \frac{RD}{F}\\ 0 & \frac{RD}{F} & \frac{(F^3+R^2D^2)}{F}\end{bmatrix}$$

$$[(C^{-1})\langle\alpha\beta\rangle]=\begin{bmatrix}F & 0 & 0\\ 0 & \left(F+\frac{R^2D^2}{F^2}\right) & -\frac{RD}{F^2}\\ 0 & -\frac{RD}{F^2} & \frac{1}{F^2}\end{bmatrix}. \qquad (8.7.4)$$

The non-vanishing physical components $T\langle ij\rangle$ of the Cauchy stress tensor $\mathbf{T}$ and the non-vanishing physical components $\hat{T}\langle\alpha\beta\rangle$ of the second referential stress tensor $\hat{\mathbf{T}}$ are

$$[T\langle ij\rangle]=\begin{bmatrix}\sigma_{rr} & 0 & 0\\ 0 & \sigma_{\theta\theta} & \tau_{\theta z}\\ 0 & \tau_{z\theta} & \sigma_{zz}\end{bmatrix} \qquad [\hat{T}\langle\alpha\beta\rangle]=\begin{bmatrix}\hat{\sigma}_{RR} & 0 & 0\\ 0 & \hat{\sigma}_{\Theta\Theta} & \hat{\tau}_{\Theta Z}\\ 0 & \hat{\tau}_{Z\Theta} & \hat{\sigma}_{ZZ}\end{bmatrix}. \qquad (8.7.5)$$

For an incompressible material, $J=1$ for all $\mathbf{F}$, and hence, from equation (8.2.4)

$$\mathbf{T}=\mathbf{F}\hat{\mathbf{T}}\mathbf{R}^{\mathrm{T}}. \qquad (8.7.6)$$

Noting that $\mathbf{B}=\mathbf{R}\mathbf{C}\mathbf{R}^{\mathrm{T}}$, it follows from the first parts of equations (8.7.3) and (8.7.4) that $\mathbf{R}$ must be of the form

$$[\mathbf{R}]=\begin{bmatrix}1 & 0 & 0\\ 0 & \cos\psi & \sin\psi\\ 0 & -\sin\psi & \cos\psi\end{bmatrix} \qquad (8.7.7)$$

where

$$\tan\psi=\frac{RD}{(F^{3/2}+1)}. \qquad (8.7.8)$$

Entering the form for $\mathbf{R}$ given by equation (8.7.7) into equation (8.7.6) gives

$$\sigma_{rr}=\frac{1}{\sqrt{F}}\hat{\sigma}_{RR} \qquad (8.7.9)$$

$$\sigma_{\theta\theta}=\frac{\cos\psi}{\sqrt{F}}(\hat{\sigma}_{\Theta\Theta}+RD\hat{\tau}_{Z\Theta})+\frac{\sin\psi}{\sqrt{F}}(\hat{\tau}_{\Theta Z}+RD\hat{\sigma}_{ZZ}) \tag{8.7.10}$$

$$\sigma_{zz}=F\cos\psi\hat{\sigma}_{ZZ}-F\sin\psi\hat{\tau}_{Z\Theta} \tag{8.7.11}$$

$$\tau_{\theta z}=\frac{\cos\psi}{\sqrt{F}}(\hat{\tau}_{\Theta Z}+RD\hat{\sigma}_{ZZ})-\frac{\sin\psi}{\sqrt{F}}(\hat{\sigma}_{\Theta\Theta}+RD\hat{\tau}_{Z\Theta})=\tau_{z\theta} \tag{8.7.12}$$

$$\tau_{z\theta}=F\cos\psi\hat{\tau}_{Z\Theta}+F\sin\psi\hat{\sigma}_{ZZ}=\tau_{\theta z}. \tag{8.7.13}$$

The constitutive equation (8.3.25) will be used in the form

$$\hat{\mathbf{G}}'(\hat{\mathbf{T}}')=2\hat{G}\hat{\mathbf{K}}'(\hat{\mathbf{E}}') \tag{8.7.14}$$

where the physical components of the referential loading function are given by

$$[\hat{G}'\langle\alpha\beta\rangle]=\hat{\Lambda}\hat{\tau}_{\Theta Z}\begin{bmatrix}\frac{\tilde{\mu}}{\hat{\mu}}\zeta & 0 & 0\\ 0 & -\frac{1}{2}\left(\xi+\frac{\tilde{\mu}}{\hat{\mu}}\zeta\right) & 1\\ 0 & 1 & \frac{1}{2}\left(\xi-\frac{\tilde{\mu}}{\hat{\mu}}\zeta\right)\end{bmatrix}. \tag{8.7.15}$$

The referential stress ratios ξ and ζ are defined by the relations

$$\xi=\frac{\hat{\sigma}_{ZZ}-\hat{\sigma}_{\Theta\Theta}}{\hat{\tau}_{\Theta Z}}=\frac{\hat{G}'\langle ZZ\rangle-\hat{G}\langle\Theta\Theta\rangle}{\hat{G}'\langle\Theta Z\rangle} \tag{8.7.16}$$

$$\zeta=\frac{2\hat{\sigma}_{RR}-\hat{\sigma}_{\Theta\Theta}-\hat{\sigma}_{ZZ}}{3\hat{\tau}_{\Theta Z}}=\frac{\hat{\mu}}{\tilde{\mu}}\frac{\hat{G}'\langle RR\rangle}{\hat{G}'\langle\Theta Z\rangle} \tag{8.7.17}$$

having noted that

$$\hat{\mu}=\frac{3\hat{t}_1}{\hat{t}_3-\hat{t}_2}=\frac{3\zeta}{(4+\xi^2)^{1/2}} \tag{8.7.18}$$

and that

$$\hat{\Lambda}=\delta_1-\delta_2\hat{t}_1=\delta_1-\delta_2\hat{\sigma}'_{RR}. \tag{8.7.19}$$

Entering the forms for $\mathbf{C}$ and $\mathbf{C}^{-1}$ given by equation (8.7.4) into equation (8.3.31) gives

$$[\hat{E}\langle\alpha\beta\rangle]=\begin{bmatrix}\frac{(1-F^2)}{4F} & 0 & 0\\ 0 & \frac{(1-F^2)}{4F}-\frac{R^2D^2}{4F^2} & (F+1)\frac{RD}{4F^2}\\ 0 & (F+1)\frac{RD}{4F^2} & \frac{(F^4-1)}{4F^2}+\frac{R^2D^2}{4F}\end{bmatrix}. \tag{8.7.20}$$

The physical components of the referential strain response function are given by

$$[\hat{K}'\langle\alpha\beta\rangle]=\hat{\Psi}\hat{E}\langle\Theta Z\rangle\begin{bmatrix}\frac{\tilde{v}}{\hat{v}}\hat{\rho} & 0 & 0\\ 0 & -\frac{1}{2}\left(\hat{\chi}+\frac{\tilde{v}}{\hat{v}}\hat{\rho}\right) & 1\\ 0 & 1 & \frac{1}{2}\left(\hat{\chi}-\frac{\tilde{v}}{\hat{v}}\hat{\rho}\right)\end{bmatrix}. \quad (8.7.21)$$

Here the referential strain ratios $\hat{\chi}$ and $\hat{\rho}$ are defined by the relations

$$\hat{\chi}=\frac{\hat{E}\langle ZZ\rangle-\hat{E}\langle\Theta\Theta\rangle}{\hat{E}\langle\Theta Z\rangle}=\frac{\hat{K}'\langle ZZ\rangle-\hat{K}'\langle\Theta\Theta\rangle}{\hat{K}'\langle\Theta Z\rangle}\equiv\frac{\hat{G}'\langle ZZ\rangle-\hat{G}'\langle\Theta\Theta\rangle}{\hat{G}'\langle\Theta Z\rangle}=\hat{\xi}$$

$$=\frac{(F^3-1)}{RD}+RD\equiv\hat{\xi} \quad (8.7.22)$$

$$\hat{\rho}=\frac{\hat{E}'\langle RR\rangle}{\hat{E}\langle\Theta Z\rangle}=\frac{2(\hat{E}\langle RR\rangle-\hat{E}\langle\Theta\Theta\rangle)}{3\hat{E}\langle\Theta Z\rangle}-\tfrac{1}{3}\hat{\chi}=\frac{\hat{v}}{\hat{\mu}}\hat{\zeta}$$

$$=-\frac{1}{3}\left(\frac{(F^3-1)}{RD}+\frac{(F-1)}{(F+1)}RD\right)=\frac{\hat{v}}{\hat{\mu}}\hat{\zeta} \quad (8.7.23)$$

and where

$$\hat{v}=\frac{3\hat{\varepsilon}_1'}{\hat{\varepsilon}_3'-\hat{\varepsilon}_2'}=\frac{3\hat{\rho}}{(4+\hat{\chi}^2)^{1/2}} \quad (8.7.24)$$

$$\hat{\Psi}=\hat{\varphi}_1-\hat{\varphi}_2\hat{\varepsilon}_1'=\hat{\varphi}_1-\hat{\varphi}_2\hat{E}'\langle RR\rangle. \quad (8.7.25)$$

Entering the form for $\hat{\mathbf{G}}'$ and $\hat{\mathbf{K}}'$ into equation (8.7.14) from equations (8.7.15) and (8.7.21) and rearranging gives the referential stress relations

$$\hat{\sigma}_{\Theta\Theta}=\hat{\sigma}_{RR}-\left[\hat{G}\frac{\hat{\Psi}}{\hat{\Lambda}}\right]\left[\frac{3}{2}\left(1-\frac{\hat{\mu}}{\hat{v}}\right)\hat{\varepsilon}_{ZZ}+\left(1-\frac{\hat{\mu}}{\hat{v}}\frac{(F-1)}{(F+1)}\right)\frac{(F+1)}{4F^2}R^2D^2\right] \quad (8.7.26)$$

$$\hat{\sigma}_{ZZ}=\hat{\sigma}_{RR}+\left[\hat{G}\frac{\hat{\Psi}}{\hat{\Lambda}}\right]\left[\frac{3}{2}\left(1+\frac{\hat{\mu}}{\hat{v}}\right)\hat{\varepsilon}_{ZZ}+\left(1+\frac{\hat{\mu}}{\hat{v}}\frac{(F-1)}{(F+1)}\right)\frac{(F+1)}{4F^2}R^2D^2\right] \quad (8.7.27)$$

$$\hat{\tau}_{\Theta Z}=\left[\hat{G}\frac{\hat{\Psi}}{\hat{\Lambda}}\right]\hat{\gamma}_{\Theta Z}=\hat{\tau}_{Z\Theta}. \quad (8.7.28)$$

Here

$$\hat{\varepsilon}_{ZZ}=\frac{(F+1)(F^3-1)}{6F^2} \qquad \hat{\gamma}_{\Theta Z}=\frac{1}{2}\frac{(F+1)}{F^2}RD \quad (8.7.29)$$

and use has been made of the referential form of the generalised Lode relation, that is the identity of equation (8.3.34).

It has been shown in §6.2.1, equation (6.6.11), that for a solid rod

$$\sigma_{rr}+\sigma_{\theta\theta}=0 \tag{8.7.30}$$

which can be used to rearrange the sum of equations (8.7.9) and (8.7.10) into the form

$$\hat{\sigma}_{RR}=-\frac{[(F^{3/2}+2)RD\hat{\tau}_{\Theta Z}+(F^{3/2}+1)(\hat{\sigma}_{\Theta\Theta}-\hat{\sigma}_{RR})+R^2D^2(\hat{\sigma}_{ZZ}-\hat{\sigma}_{RR})]\cos\psi}{[F^{3/2}+1+(F^{3/2}+1+R^2D^2)\cos\psi]}. \tag{8.7.31}$$

Equation (8.7.11) can be rearranged to give

$$\sigma_{zz}=F[\hat{\sigma}_{ZZ}-(F^{3/2}+1)^{-1}\hat{\tau}_{Z\Theta}]\cos\psi. \tag{8.7.32}$$

8.7.2 Resultant Longitudinal Force on the Plane End of a Solid Rod

Substitution of the relation for $\hat{\tau}_{\Theta Z}$ and the relations for the stress difference $\hat{\sigma}_{\Theta\Theta}-\hat{\sigma}_{RR}$ and $\hat{\sigma}_{ZZ}-\hat{\sigma}_{RR}$ given by equations (8.7.28), (8.7.26) and (8.7.27) respectively into equation (8.7.31) gives a relation for $\hat{\sigma}_{RR}$ which depends only on R, D and F. This form for $\hat{\sigma}_{RR}$ can then be substituted into equation (8.7.27) to give $\hat{\sigma}_{ZZ}$ in a form which depends only on R, D and F. The form for $\hat{\sigma}_{ZZ}$ obtained in this way can be entered into equation (8.7.32) to give a form for σ_{zz} which depends only on R, D and F. Entering this form for σ_{zz} into equation (6.5.62) gives the resultant force N_z acting on the plane end of a solid rod in a form which depends only on R, D and F.

The remaining discussion will be restricted to those materials which satisfy a von Mises type of referential stress intensity function which is characterised by the conditions

$$\begin{aligned}&\hat{\delta}_0=0, \quad \hat{\delta}_1=1, \quad \hat{\delta}_2=0 \quad \hat{H}=1, \quad \hat{\Lambda}=1, \quad \hat{\omega}=1\\ &\hat{\varphi}_0=0, \quad \hat{\varphi}_1=1, \quad \hat{\varphi}_2=0 \quad \hat{N}=1, \quad \hat{\Psi}=1, \quad \hat{\Omega}=1.\end{aligned} \tag{8.7.33}$$

These limiting conditions reduce equations (8.7.26), (8.7.27) and (8.7.28) to the form

$$\hat{\sigma}_{\Theta\Theta}=\hat{\sigma}_{RR}-\tfrac{1}{2}\hat{G}(R^2D^2/F^2) \tag{8.7.34}$$

$$\hat{\sigma}_{ZZ}=\hat{\sigma}_{RR}+\tfrac{1}{2}\hat{G}(6\hat{\varepsilon}_{ZZ}+R^2D^2/F) \tag{8.7.35}$$

$$\hat{\tau}_{\Theta Z}=\tfrac{1}{2}\hat{G}RD(F+1)/F^2. \tag{8.7.36}$$

Entering the form for $\hat{\tau}_{\Theta Z}$, $\hat{\sigma}_{\Theta\Theta}-\hat{\sigma}_{RR}$, and $\hat{\sigma}_{ZZ}-\hat{\sigma}_{RR}$ given by equations (8.7.34), (8.7.35) and (8.7.36) into equations (8.7.31) and (8.7.32) gives

(Billington 1985c)

$$\hat{\sigma}_{RR} = -\tfrac{1}{2}\hat{G}\frac{(F^3+F^2+F^{3/2}+1+R^2D^2)R^2D^2}{F[F^{3/2}+1+(F^{3/2}+1+R^2D^2)\cos\psi]}\cos\psi \tag{8.7.37}$$

$$\sigma_{zz} = F\left[\hat{\sigma}_{RR}+\tfrac{1}{2}\hat{G}\left(6\hat{\varepsilon}_{ZZ}+\frac{(F^{5/2}-1)}{(F^{3/2}+1)}\frac{R^2D^2}{F^2}\right)\right]\cos\psi. \tag{8.7.38}$$

For infinitely small strains ($D \to 0$, $F \to 1$) equations (8.7.37) and (8.7.38) reduce to

$$\hat{\sigma}_{RR} = -\tfrac{1}{2}G_0 r^2 D^2 \qquad (r \simeq R) \tag{8.7.39}$$

$$\sigma_{zz} = 3G_0 e_{zz} - \tfrac{1}{2}G_0 r^2 D^2 \qquad (r \simeq R) \tag{8.7.40}$$

where

$$e_{zz} = F-1 \qquad G_0 = [\hat{G}]_{F=1} = \text{constant}. \tag{8.7.41}$$

Entering the form for σ_{zz} given by equation (8.7.40) into equation (6.5.62), with $r_2 = R_2 = 0$, gives

$$\frac{N_z}{\pi R_1^2} = 3G_0 e_{zz} - \tfrac{1}{4}G_0 R_1^2 D^2 \qquad (r_1 \simeq R_1) \tag{8.7.42}$$

which is to be compared with equation (6.6.17). The Poynting effect is observed by setting $N_z = 0$, and $e_{zz} = e_{(\mathrm{P})}$ in equation (8.7.42) to give

$$e_{(\mathrm{P})} = \tfrac{1}{12}R_1^2 D^2 \tag{8.7.43}$$

which can be compared with equation (7.1.45).

8.8 COMPATIBILITY CONDITIONS

In the context of the spatial description of stress, the constitutive equation of a simple elastic material can be expressed in the form

$$\mathbf{T} = \beta_0\mathbf{I} + \beta_1\mathbf{B} + \beta_{-1}\mathbf{B}^{-1}. \tag{8.8.1}$$

From equation (8.8.1)

$$\mathbf{T}^2 = \overset{(2)}{\beta_0}\mathbf{I} + \overset{(2)}{\beta_1}\mathbf{B} + \overset{(2)}{\beta_{-1}}\mathbf{B}^{-1} \tag{8.8.2}$$

where use has been made of the Cayley–Hamilton theorem to give

$$\overset{(2)}{\beta_0} = \beta_0^2 + 2\beta_1\beta_{-1} - \beta_1^2 II_{\mathbf{B}} - \beta_{-1}^2(I_{\mathbf{B}}/III_{\mathbf{B}})$$

$$\overset{(2)}{\beta_1} = 2\beta_0\beta_1 + \beta_1^2 I_{\mathbf{B}} + (\beta_{-1}^2/III_{\mathbf{B}}) \tag{8.8.3}$$

$$\overset{(2)}{\beta_{-1}} = 2\beta_0\beta_{-1} + \beta_1^2 III_{\mathbf{B}} + \beta_{-1}^2(II_{\mathbf{B}}/III_{\mathbf{B}}).$$

Alternatively, in the context of the referential description of stress the constitutive equation of a simple elastic material can be expressed in the form

$$\hat{\mathbf{T}}=\alpha_0\mathbf{I}+\alpha_1\mathbf{C}+\alpha_{-1}\mathbf{C}^{-1}. \tag{8.8.4}$$

From equation (8.8.4)

$$\hat{\mathbf{T}}^2=\overset{(2)}{\alpha_0}\mathbf{I}+\overset{(2)}{\alpha_1}\mathbf{C}+\overset{(2)}{\alpha_{-1}}\mathbf{C}^{-1} \tag{8.8.5}$$

where use has been made of the Cayley–Hamilton theorem to give

$$\begin{aligned}\overset{(2)}{\alpha_0}&=\alpha_0^2+2\alpha_1\alpha_{-1}-\alpha_1^2 II_C-\alpha_{-1}^2(I_C/III_C)\\ \overset{(2)}{\alpha_1}&=2\alpha_0\alpha_1+\alpha_1^2 I_C+(\alpha_{-1}^2/III_C)\\ \overset{(2)}{\alpha_{-1}}&=2\alpha_0\alpha_{-1}+\alpha_1^2 III_C+\alpha_{-1}^2(II_C/III_C).\end{aligned} \tag{8.8.6}$$

Setting $n=2$ in equation (8.1.15) gives

$$\hat{\mathbf{T}}^2=J^2\mathbf{F}^{-1}\mathbf{T}^2(\mathbf{F}^{-1})^{\mathrm{T}} \tag{8.8.7}$$

which when applied to equations (8.8.2) and (8.8.5) gives

$$\begin{aligned}\overset{(2)}{\alpha_0}&=J^2\overset{(2)}{\beta_1}-I_C\overset{(2)}{\beta_{-1}}\\ \overset{(2)}{\alpha_1}&=\overset{(2)}{\beta_{-1}}\\ \overset{(2)}{\alpha_{-1}}&=J^2\overset{(2)}{\beta_0}+II_C\overset{(2)}{\beta_{-1}}.\end{aligned} \tag{8.8.8}$$

Equations (8.8.7) and (8.8.8) would appear to imply compatibility between the two constitutive equations (8.8.1) and (8.8.4). However there are situations for which they are not compatible and this is evident from the following consideration.

For pure shear, characterised by $\nu=-3$, §6.3.3(iii) gives

$$\sigma_3=2\sigma_2 \qquad (\mu=-3,\ \sigma_1=0) \tag{8.8.9}$$

for the spatial description.

In the case of pure shear, characterised by $\hat{\nu}=-3$, §8.6.2(iii) gives

$$\hat{t}_3=2\hat{t}_2 \qquad (\hat{\mu}=-3,\ \hat{t}_1=0) \tag{8.8.10}$$

for the referential description.

Equations (8.8.9) and (8.8.10) imply that if a set of experimental results satisfy one of these relations, then the same set of results cannot satisfy the other relation, the associated constitutive equation being inadmissable.

Thus, if a material satisfies equation (8.8.9), then the constitutive equations must be of the form

$$\mathbf{T}=\beta_0\mathbf{I}+\beta_1\mathbf{B}+\beta_{-1}\mathbf{B}^{-1} \tag{8.8.11}$$

$$\hat{\mathbf{T}}=J(\beta_0\mathbf{I}+\beta_1\mathbf{C}+\beta_{-1}\mathbf{C}^{-1})\mathbf{U}^{-1} \tag{8.8.12}$$

whereas if the material satisfies equation (8.8.10), then the constitutive equations must be of the form

$$\hat{\mathbf{T}}=\alpha_0\mathbf{I}+\alpha_1\mathbf{C}+\alpha_{-1}\mathbf{C}^{-1} \tag{8.8.13}$$

$$\mathbf{T}=J^{-1}(\alpha_0\mathbf{I}+\alpha_1\mathbf{B}+\alpha_{-1}\mathbf{B}^{-1})\mathbf{V}. \tag{8.8.14}$$

For a material which satisfies equation (8.8.10), it is thus necessary to formulate the constitutive equation (8.8.13), independently of the spatial description, noting however, that the method of formulation parallels that for equation (8.8.11).

REFERENCES

Atluri S N 1984 *Computers and Structures* **18** 93–116
Billington E W 1985a *Acta Mechan.* **55** 263–6
—— 1985b *Acta Mechan.* **55**
—— 1985c *Acta Mechan.* **55**
Bufler H 1983 *Comp. Meths. Appl. Mech. Eng.* **36** 95–124
Koiter W T 1976 *On the Complementary Energy Theorem in Nonlinear Elasticity Theory* in *Trends in Application of Pure Mathematics to Mechanics* ed. G Fichera pp 20–232 (London: Pitman)
Lode W 1926 *Zeits. Phys.* **36** 913–39
Toupin R A 1964 *Arch. Ration. Mech. Anal.* **17** 85–112

GENERAL REFERENCES

Biot M A 1965 *Mechanics of Incremental Deformations* (New York: Wiley) chs I, II
Truesdell C and Noll W 1965 *The Nonlinear Field Theories of Mechanics* in *Handbuch der Physik* vol. III/3 ed. S Flügge (Berlin: Springer)

9

Material Response in the Referential Description

9.1 RUBBER-LIKE SOLIDS

9.1.1 Universal Relation from Pure Shear

Any physically defensible test of the constitutive equation for a simple elastic solid must be based on a quantitative prediction regarding the mechanical properties of the test material. Thus, insertion of a property of the material, such as incompressibility, into the kinematics of the system must lead to a specific prediction regarding the dynamics of the system. It is to be noted that simple tension, characterised by $\hat{\mu}=-1, \hat{\nu}=-1$ for all $\lambda_3 \geqslant 1$, does not provide a prediction regarding the dynamics of the system. In contrast, equations (8.6.20), (8.6.37) and (8.6.45) give the predictions

$$\left.\begin{aligned} \hat{t}_3 &= -\hat{t}_2 \\ \hat{t}_3 &= 2\hat{t}_2 \\ \hat{t}_3 &= \hat{t}_2 \end{aligned}\right\} \quad \text{for} \quad \hat{\nu} = \begin{cases} 0 \\ -3 \\ -\infty \end{cases} \tag{9.1.1}$$

for all $\lambda_3 \geqslant 1$. The simple mode of deformation for which $\hat{\nu}=0$ cannot be obtained in practice using a thin sheet of material in the way discussed in §8.6.2. The prediction of most interest is that for $\hat{\nu}=-3$: that is $\hat{t}_3 = 2\hat{t}_2$.

There would appear to be only two sets of measurements available in a form suitable for comparison with the proposed constitutive equation (8.3.25), used in the form of the reduced stress–reduced strain relation of equation (8.6.12). These are the experimental studies of Jones and Treloar (1975) and James *et al* (1975). In the case of the measurements of Jones and Treloar, the numerical values of $\hat{t}_2$, $\hat{t}_3$, λ_2 and λ_3 have been obtained from table 2 of Haines and Wilson (1979). It is to be noted that in all the studies of rubber-like solids

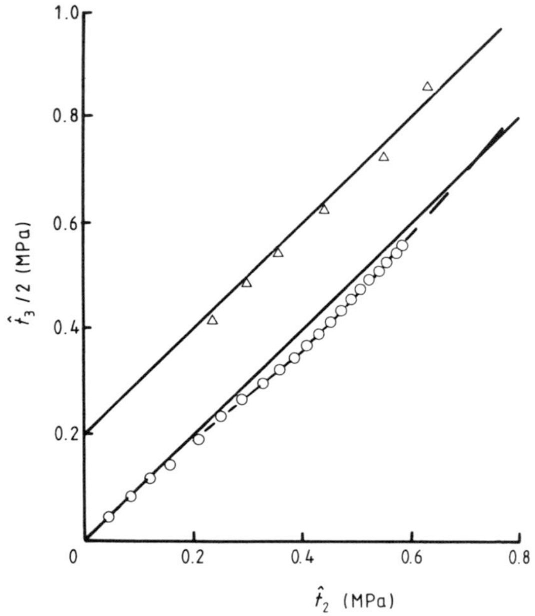

Figure 9.1 Variation of $\hat{t}_3/2$ with $\hat{t}_2$. ○, from values of Jones and Treloar (1975) in table 2 of Haines and Wilson (1979); △, table 2 of James *et al* (1975).

considered in this chapter, the present $\hat{t}_3$ obtainable from these studies is in all cases given in terms of the principal Cauchy stress σ_1.

The variation of $\hat{t}_3/2$ with $\hat{t}_2$, using the results given by Jones and Treloar (1975), is shown in figure 9.1 (from table 2 of Haines and Wilson (1979)). The three smallest values of $\hat{t}_2$ and the corresponding values of $\hat{t}_3/2$ are seen to be in good agreement with the theoretical prediction represented by the full line. The next three $(\hat{t}_3/2, \hat{t}_2)$ points also agree with the theoretical prediction to well within the level of experimental error estimated by Jones and Treloar (1975). The remaining $(\hat{t}_3/2, \hat{t}_2)$ measurements are seen to lie on a smooth curve for which $2\hat{t}_2/\hat{t}_3>1$. With regard to figure 9.1, it is of interest that Jones and Treloar (1975) remark that the accuracy with which the elastic properties of a rubber can be defined is limited in practice by the degree of reversibility which can be attained. The hysteresis effects for this material have been considered by Jones and Treloar (1975), and in the context of their discussion, the observed deviation from the condition $2\hat{t}_2/\hat{t}_3=1$ could have its origin in hysteresis effects.

James *et al* (1975) give values of $\hat{t}_2$ and $\hat{t}_3$ for $\lambda_2=1$ in their table 2. These values of $\hat{t}_3/2$ and $\hat{t}_2$ are shown in figure 9.1 (triangles) and are seen to be in general accord with the theoretical prediction represented by the full straight line. The origin for this $(\hat{t}_3/2, \hat{t}_2)$ relation has been shifted by 0.2 units parallel to the ordinate.

It can be concluded (Billington 1985a) that the results shown in figure 9.1 are in general accord with the theoretical prediction that if $\lambda_2=1, 2\hat{t}_2/\hat{t}_3=1$ for all $\lambda_3 \geqslant 1$. This conclusion implies the further conclusion that the mechanical response of these rubber-like solids can only be satisfactorily formulated within the proposed referential description.

9.1.2 Biaxial Extension

The generalised shear modulus of equation (8.3.26) is a function of the invariants $\hat{K}'_2$ and $\hat{K}'_3$, the $\hat{K}'_i$ $(i=2, 3)$ being defined by equation (8.3.27). The deviators $\hat{\varepsilon}'_i$ $(i=1, 2, 3)$ of the proper numbers $\hat{\varepsilon}_i$ of $\hat{\mathbf{E}}$ are the positive solutions for $\hat{\varepsilon}'$ of the equation

$$\hat{\varepsilon}'^3 - \hat{K}'_2\hat{\varepsilon}' - \hat{K}'_3 = 0. \tag{9.1.2}$$

For, $\hat{\varepsilon}_3 > \hat{\varepsilon}_1 > \hat{\varepsilon}_2$, equation (9.1.2) gives

$$\hat{\varepsilon}'_i = \frac{2}{\sqrt{3}}(\hat{K}'_2)^{1/2} \cos\left(\tfrac{1}{3}\cos^{-1}\hat{\Omega}^{1/2} + \frac{2(i-1)\pi}{3}\right) \qquad (i=1, 2, 3) \tag{9.1.3}$$

where $\hat{\Omega}$ is defined by equation (8.4.9). From equation (8.3.32)

$$\hat{k}'_i = \hat{\varphi}_0 + \hat{\varphi}_1\hat{\varepsilon}'_i + \hat{\varphi}_2\hat{\varepsilon}'^2_i \qquad (i=1, 2, 3) \tag{9.1.4}$$

having noted that

$$\hat{\varphi}_i = \hat{\varphi}_i(\hat{K}'_2, \hat{K}'_3) \qquad (i=0, 1, 2) \tag{9.1.5}$$

and that the effective strain

$$\hat{\varepsilon}'' = (\tfrac{2}{3}\operatorname{tr}\hat{\mathbf{E}}'^2)^{1/2} = \sqrt{\tfrac{2}{3}}\,(\hat{\varepsilon}'^2_1 + \hat{\varepsilon}'^2_2 + \hat{\varepsilon}'^2_3)^{1/2}. \tag{9.1.6}$$

In the context of equations (9.1.2) to (9.1.6), it is assumed that

$$\hat{G} = \hat{G}(\hat{\varepsilon}'', \hat{v}) \tag{9.1.7}$$

where, in regard to equation (8.4.9), a dependence upon $\hat{v}$ has also been assumed. There remains the problem of formulating the dependence of $\hat{G}$ upon $\hat{\varepsilon}''$ and $\hat{v}$. For present purposes it has been shown (Billington 1985a), that the results of biaxial tests can be correlated using the generalised shear modulus

$$\hat{G} = G_0 \ln(1+\phi\hat{\varepsilon}'')/\phi\hat{\varepsilon}'' \tag{9.1.8}$$

where

$$\phi = \phi_1/[1-(\phi_2/v)] \tag{9.1.9}$$

and where G_0, ϕ_1 and ϕ_2 are disposable constants characteristic of material properties.

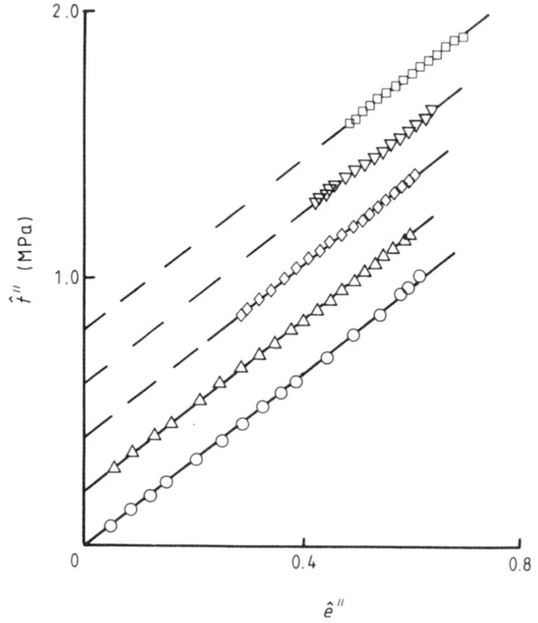

Figure 9.2 Variation of the effective stress $\hat{t}''$ with the reduced strain $\hat{e}''$, using the results given in table 2 of Haines and Wilson (1979). ○, simple extension; △, $\lambda_2 = 1$; λ_3: ◇, 1.502; ▽, 1.984; □, 2.295.

The pure shear mode of deformation characterised by $\hat{v} = -3$, i.e. case (iii) of §8.6.2, is of particular interest in so far as $\hat{H} = \hat{\Lambda} = 1$, $\hat{N} = \hat{\Psi} = 1$, and hence the proposed correlation for pure shear is independent of the form of both the stress intensity function and the strain intensity function. The variation of

$$\hat{t}'' = \tfrac{1}{2}\sqrt{3}\,\hat{t}_3 \qquad \text{with} \quad \hat{e}'' = \frac{\ln[1 + (4\phi\hat{\gamma}_1/\sqrt{3})]}{\phi} \tag{9.1.10}$$

is shown in figure 9.2 for the results given by Jones and Treloar (1975) (obtained from table 2 of Haynes and Wilson (1979)). The values of G_0 and of

$$\phi = 3\phi_1/[(3 + \phi_2)] \qquad (\hat{v} = -3) \tag{9.1.11}$$

obtained in this way are given in the present table 9.1. It is evident from figure 9.2 that $\hat{t}''$ varies with $\hat{e}''$ according to equation (8.6.12).

For simple extension, characterised by $\hat{v} = -1$, that is case (ii) of §8.6.2, equation (8.6.29) gives

$$\hat{\Lambda} = \hat{H} = \text{constant} \qquad \hat{\omega} = 1 \tag{9.1.12}$$

and similarly, it follows from equations (8.4.11) and (8.6.26) that

$$\hat{N} = \hat{\Psi} = \text{constant} \qquad \hat{\Omega} = 1 \tag{9.1.13}$$

Table 9.1 Values of the parameters appearing in the mechanical equation of state.

λ_3	G_0 (MPa)	ϕ_1	ϕ_2
$\lambda_2^{-2}=\lambda_1^{-2}$	0.539	$2.80(1+\phi_2)$	—
λ_1^{-1}, $(\lambda_2=1)$	0.538	$3.87[1+(\phi_2/3)]$	—
1.50_2	0.538	4.310	0.664
1.98_4	0.533	4.651	0.744
2.29_5	0.533	4.785	0.775

for all λ_3. The condition that $\hat{H}=$constant and $\hat{N}=$constant, irrespective of the form of either the stress intensity function or the strain intensity function, follows from the limiting conditions

$$\hat{\omega}=1, \qquad \hat{\Omega}=1, \qquad \hat{\mu}=\tilde{\mu}\equiv\tilde{v}=\hat{v}=-1 \quad \text{for all } \lambda_3.$$

These conditions, together with equations (9.1.8) and (9.1.9) give equation (8.6.30) in the form

$$\hat{t}''=\hat{t}_3 \qquad \hat{e}''=\ln[1+\phi(\hat{N}/\hat{H}^{1/2})\hat{\varepsilon}_3']/\phi \tag{9.1.14}$$

where use has been made of equation (8.6.12). Because $\hat{N}$ and $\hat{H}$ are both constants for all λ_3, it is possible to evaluate

$$\phi[\hat{N}/\hat{H}^{1/2}]_{\hat{\mu}=\hat{v}=-1}=\text{constant} \qquad G_0/[\hat{H}^{1/2}]_{\hat{\mu}=-1}=\text{constant} \tag{9.1.15}$$

assuming that $\hat{t}''$ varies with $\hat{e}''$ according to equation (8.6.12). In this way it has been shown that for the Jones and Treloar (1975) measurements, within the limits of experimental accuracy, $\hat{H}=1$, $\hat{N}=1$ for all λ_3. The variation of

$$\hat{t}''=\hat{t}_3 \qquad \text{with} \quad \hat{e}''=\ln(1+\phi\hat{\varepsilon}_3')/\phi \qquad (\hat{v}=-1) \tag{9.1.16}$$

is shown in figure 9.2 for the results given by Jones and Treloar (1975) (triangles) (from table 2 of Haynes and Wilson (1979)). The origin of the $(\hat{t}'', \hat{e}'')$ curve for $\hat{v}=-1$ has been shifted by 0.2 units parallel to the ordinate in figure 9.2. In figure 9.2 the gradient of the full line for $\hat{v}=-1$ is identical to that for $\hat{v}=-3$. The values of G_0 and

$$\phi=\phi_1/(1+\phi_2) \qquad (\hat{v}=-1) \tag{9.1.17}$$

are given in table 9.1 for $\hat{v}=-1$.

From figure 9.2, it is evident that the measurements in pure shear and simple tension have been correlated in such a way that they combine to give a single stress–strain relation, use being made of equations (8.6.12), (8.6.13) and (8.6.14), subject to the limiting conditions $\hat{H}=1$, $\hat{N}=1$ for all λ_3. Thus it can be concluded that for these simple modes of deformation, the mechanical response of these materials identify the stress intensity function and the strain intensity function with the von Mises type for which

$$\begin{array}{lllllll} \hat{\delta}_0=0, & \hat{\delta}_1=1, & \hat{\delta}_2=0, & \hat{H}=1, & \hat{\Lambda}=1, & \hat{\mathbf{G}}'=\hat{\mathbf{T}}', & \hat{\mu}=\tilde{\mu} \\ \hat{\varphi}_0=0, & \hat{\varphi}_1=1, & \hat{\varphi}_2=0, & \hat{N}=1, & \hat{\Psi}=1, & \hat{\mathbf{K}}'=\hat{\mathbf{E}}', & \hat{v}=\tilde{v} \end{array} \quad (9.1.18)$$

for all λ_3.

Assuming the conditions of equation (9.1.18), and using the values of G_0, ϕ_1 and ϕ_2 given in table 9.1, the variation of the reduced stress $\hat{t}''$ with the reduced strain $\hat{e}''$ is shown in figure 9.2 for the results of Jones and Treloar (1975) (as given by Haines and Wilson (1979)). Use is made of equations (8.6.13) and (8.6.14) for $\lambda_3=1.502$ (diamonds), 1.984 (inverted triangles) 2.295 (squares). The origin of each of the $(\hat{t}'', \hat{e}'')$ curves corresponding to $\lambda_3=1.502$, 1.984, 2.295 have been shifted by 0.2 units parallel to the ordinate, the shift being relative to the curve immediately below. It is evident that the measurements of Jones and Treloar (1975) are in good accord with the linear relation of equation (8.6.12). This can be seen because all the straight lines have identical slopes and represent the linear relation of equation (8.6.12) with $G_0=$ 0.536 MPa, having noted from table 9.1 that the present analysis gives $G_0=$ 0.536 ± 0.003 MPa.

The results of the experimental study by James *et al* (1975) have been analysed in the same way, subject to the conditions of equation (9.1.18). The variation of $\hat{t}''$ with $\hat{e}''$, using the values of ϕ_1 and ϕ_2 given in table 9.2 is shown in figure 9.3 for $\lambda_3=1.3$ (triangles), 1.5 (diamonds), 1.7 (inverted triangles), 2.0 (circles) from which it is concluded that these measurements are also in good accord with the linear relation of equation (8.6.12). This can be seen in figure 9.3 since the straight line represents the linear relation of equation (8.6.12) with $G_0=0.619$ MPa.

9.1.3 Assumption of Isotropy

The observation that the individual stress–strain relations, obtained from biaxial tests, compound together to give a single reduced stress–reduced strain

Table 9.2 Values of the parameters appearing in the mechanical equation of state.

λ_3	G_0 (MPa)	ϕ_1	ϕ_2
1.3	0.618	4.900	0.285
1.5	0.618	5.100	0.420
1.7	0.622	5.390	0.550
2.0	0.618	5.550	0.721

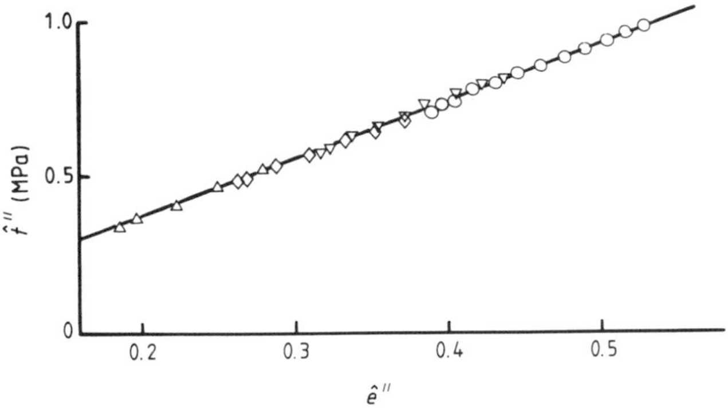

Figure 9.3 Variation of the effective stress $\hat{t}''$ with reduced strain $\hat{e}''$, using the results given in table 2 of James *et al* (1975). λ_3: △, 1.3; ◇, 1.5; ▽, 1.7; ○, 2.0.

curve over a wide range of strain is taken as evidence in support of the assumption that the materials can be regarded as being isotropic.

Deviations of the stress–strain relations from the linear, reduced stress–reduced strain relation, have been observed in the above reanalysis of biaxial tests. These deviations are attributed to the onset of anisotropy.

For the equibiaxial mode of deformation discussed as case (iv) in §8.6.2, and which is characterised by $\hat{v} = -\infty$, $\lambda_3 = \lambda_2$ for all λ_3, deviations from the prediction $\hat{t}_3 = \hat{t}_2$ for certain values of $\lambda_3 = \lambda_2$ occur for large values of the reduced stress. These observed deviations are tentatively attributed to the onset of anisotropy.

Correlation of the individual stress–strain relations to give a single reduced stress–reduced strain relation has been restricted to those results which satisfy the prediction associated with equibiaxial extension.

9.1.4 Poynting Effect

Equation (8.7.43)

$$b = e_{(P)}/\gamma_{\theta z}^2 = \tfrac{1}{12} \qquad (9.1.19)$$

can be compared with equation (7.1.45) and gives $b = \frac{1}{8}$.

The non-linear mechanical response of polyurethane rubbers with a high volume percentage of inorganic filler has been studied by Lenoe *et al* (1965). A value of $b = 0.095$ at a strain rate of $0.0013\ \mathrm{s}^{-1}$, using a solid rod specimen for which $(R_1/L)^2 = 0.04$, is in good accord with the value $b = \frac{1}{12}$ predicted by equation (9.1.19). For a strain rate of $0.000\,13\ \mathrm{s}^{-1}$, Lenoe *et al* (1965) give $b = 0.14$. There are two possible explanations for the significant difference between this value of b and the one predicted by equation (9.1.19). Examination of their results for this lower strain rate show what appears to be a significant tensile strain superimposed upon the $(e_{(P)}, \gamma_{\theta z})$ curve. Also, the scatter in the experimental results for this lower strain rate is much greater than for the measurements at the higher strain rate. At the much lower strain rate, the axial elongation could be affected by the onset of the phenomenon of creep. That this is the most likely explanation can be inferred from the discussion given by Lenoe *et al* (1965) of their torsion creep tests with a free end.

9.1.5 Simple Torsion of a Solid Rod

For a material which satisfies von Mises type stress intensity and strain intensity functions, equation (8.7.28) can be expressed in the form

$$\hat{\tau}_{\Theta Z} = \hat{G}\hat{\gamma}_{\Theta Z} = \tfrac{1}{2}\hat{G}\frac{(F+1)}{F^2}RD. \tag{9.1.20}$$

For simple torsion $(F \simeq 1, \hat{\gamma} \simeq 0)$, and equation (9.1.8) reduces to

$$\lim_{\hat{\gamma}\to 0} \hat{G} = G_0. \tag{9.1.21}$$

Entering this condition into equation (9.1.20) gives

$$\hat{\tau}_{\Theta Z} = G_0 \frac{(F+1)}{2F^2} RD. \tag{9.1.22}$$

From equation (9.1.22)

$$\hat{\Gamma} = 2\pi \int_0^{R_1} R^2 \hat{\tau}_{\Theta Z}\, \mathrm{d}R = \tfrac{1}{4}\pi G_0 R_1^4 \frac{(F+1)}{F^2} D \tag{9.1.23}$$

where R_1 is the outer radius of the solid rod.

The prediction of equation (9.1.23) is in good accord with the experimental studies of Rivlin and Saunders (1951).

With N_z adjusted to give $e_{zz} = 0$ for all D, the prediction of equation (8.7.42) in the form $N_z = -\frac{1}{4}\pi G_0 R_1^4 D^2$ is in good accord with the experimental studies of Rivlin and Saunders (1951) for small strains. However, for large strains (Lenoe *et al* 1965), the relation between N_z and D^2 (with $e_{zz} = 0$ for all D) is no longer linear, an observation in accord with equation (8.7.42) (Billington 1985b).

9.2 METALS

The method developed by Taylor and Quinney (1931) for determining the yield function is concerned only with the purely elastic deformation of the test materials. For most metals, the purely elastic strains are small compared to unity. The Green–St Venant strain can be written as

$$\mathbf{E}=\tfrac{1}{2}(\mathbf{C}-\mathbf{I}). \tag{9.2.1}$$

For classical infinitesimal linearised elasticity, the assumption that the displacements and displacement gradient components are small compared to unity implies the use of the same reference axes for both $\boldsymbol{x}$ and $\boldsymbol{X}$, such that the displacement vector

$$\boldsymbol{u}=\boldsymbol{x}-\boldsymbol{X}=\boldsymbol{u}(\boldsymbol{X}, t). \tag{9.2.2}$$

This approximation allows the introduction of the displacement gradient

$$\tilde{\mathbf{S}}=\mathbf{F}-\mathbf{I}. \tag{9.2.3}$$

Substituting for $\mathbf{F}$ in equation (2.1.44) from equation (9.2.3) gives

$$\mathbf{C}=(\mathbf{I}+\tilde{\mathbf{S}}^{\mathrm{T}})(\mathbf{I}+\tilde{\mathbf{S}}). \tag{9.2.4}$$

To a first order in $\tilde{\mathbf{S}}$, equation (9.2.4) gives

$$\mathbf{C}=\mathbf{I}+(\tilde{\mathbf{S}}+\tilde{\mathbf{S}}^{\mathrm{T}}) \tag{9.2.5}$$

which when entered into equation (9.2.1) gives

$$\tilde{\mathbf{E}}=\tfrac{1}{2}(\tilde{\mathbf{S}}+\tilde{\mathbf{S}}^{\mathrm{T}}). \tag{9.2.6}$$

Hence, for sufficiently small strains

$$2\tilde{\mathbf{E}}=\mathbf{C}-\mathbf{I}. \tag{9.2.7}$$

From the definition of $\mathbf{C}$, it follows that

$$\mathbf{C}^{-1}=\mathbf{I}-(\tilde{\mathbf{S}}+\tilde{\mathbf{S}}^{\mathrm{T}})=\mathbf{I}-2\tilde{\mathbf{E}}. \tag{9.2.8}$$

Substituting for $\mathbf{C}$ and $\mathbf{C}^{-1}$ in equation (8.3.20) gives

$$\hat{\mathbf{E}}=\tilde{\mathbf{E}}=\tfrac{1}{2}(\tilde{\mathbf{S}}+\tilde{\mathbf{S}}^{\mathrm{T}}) \tag{9.2.9}$$

for sufficiently small strains. With this as the basis, the constitutive equation (8.3.25) can be expressed in the form

$$\hat{\mathbf{G}}'(\hat{\mathbf{T}}')=2G_0\hat{\mathbf{E}}'. \tag{9.2.10}$$

Using the notation of §8.7, the simple deformations of a thin-walled tube are

$$r=(\eta/F)^{1/2}R \qquad \theta=\Theta+DZ \qquad z=FZ. \tag{9.2.11}$$

Here

$$\eta=(r/R)^2F=v/V \tag{9.2.12}$$

is the ratio of the deformed volume v to the undeformed volume V of a cylinder of radius R and where $F=l/L$ is the ratio of the current length l to the undeformed length L of a tube. For infinitely small strains, $D\rightarrow 0$, $F\rightarrow 1$,

$\eta \rightarrow 1$, and the components of $\tilde{\mathbf{E}}'$ are

$$[\tilde{\mathbf{E}}'\langle\alpha\beta\rangle] = \begin{bmatrix} -\frac{1}{2}e_{zz} - \frac{1}{2}(\eta - 1) & 0 & 0 \\ 0 & -\frac{1}{2}e_{zz} + \frac{1}{2}(\eta - 1) & \frac{1}{2}\gamma_{\theta z} \\ 0 & \frac{1}{2}\gamma_{\theta z} & e_{zz} \end{bmatrix}. \tag{9.2.13}$$

Using equation (9.2.13) in the context of the constitutive equation (9.2.10) gives

$$\lim_{D \rightarrow 0} \hat{\xi} = 3\frac{e_{zz}}{\gamma_{\theta z}} - \frac{(\eta - 1)}{\gamma_{\theta z}} = \frac{\hat{\sigma}_{ZZ} - \hat{\sigma}_{\Theta\Theta}}{\hat{\tau}_{\Theta Z}} \tag{9.2.14}$$

$$\lim_{D \rightarrow 0} \hat{\zeta} = -\frac{\hat{\mu}}{\tilde{\mu}}\left(\frac{e_{zz}}{\gamma_{\theta z}} + \frac{\eta - 1}{\gamma_{\theta z}}\right) = \frac{\hat{\sigma}'_{RR}}{\hat{\tau}_{\Theta Z}}. \tag{9.2.15}$$

These can be combined to give

$$\frac{e_{zz}}{\gamma_{\theta z}} = \frac{1}{12}\hat{\xi}\left[3 - \tilde{\mu}\left(1 + \frac{4}{\hat{\xi}^2}\right)^{1/2}\right] \tag{9.2.16}$$

where

$$e_{zz} = F - 1 \ (\equiv \tilde{E}_{ZZ}) \qquad \gamma_{\theta z} = r\frac{D}{F} \ (\equiv \hat{\gamma}_{\Theta Z}). \tag{9.2.17}$$

For infinitely small strains, $\mathbf{F} \rightarrow \mathbf{I}$, $\mathbf{R} \rightarrow \mathbf{I}$, from which it follows that

$$\hat{\mathbf{T}} \rightarrow \mathbf{T} \quad \text{for} \quad \mathbf{F} \rightarrow \mathbf{I} \qquad \mathbf{R} \rightarrow \mathbf{I} \tag{9.2.18}$$

and hence, $\hat{\xi} \rightarrow \xi$, $\hat{\zeta} \rightarrow \zeta$, $\tilde{\mu} \rightarrow \bar{\mu}$. These limiting conditions make equation (9.2.16) indistinguishable from equation (7.1.3) and it is therefore concluded that the Taylor and Quinney (1931) method for determining the yield condition does not distinguish between the referential and spatial descriptions.

From equation (8.7.43)

$$b = e_{(\mathrm{P})}/\gamma_{\theta z} = \frac{1}{12} \tag{9.2.19}$$

which when compared with equation (7.1.45) gives $b = \frac{1}{8}$. It is evident from figure 7.8 that the measurements of Foux (1964) do not resolve the question of whether the mechanical response of metals should be given in the spatial description or in the referential description.

REFERENCES

Billington E W 1985a *Acta Mechan.* **55**

—— 1985b *Acta Mechan.* **55**

Foux A 1964 *Second Order Effects* in *Elasticity, Plasticity and Fluid Dynamics* ed. M Reiner and D Abir (Oxford: Pergamon) pp 228–51

Haines D W and Wilson W D 1979 *J. Mech. Phys. Solids* **27** 345–60
James A G, Green A and Simpson G M 1975 *J. Appl. Polymer Sci.* **19** 2033–58
Jones D F and Treloar L R G 1975 *J. Phys. D: Appl. Phys.* **8** 1285–304
Lenoe E M, Heller R A and Freudenthal A M 1965 *Trans. Soc. Rheol.* **9** 77–102
Rivlin R S and Saunders D W 1951 *Phil. Trans. R. Soc.* A **243** 251–88
Taylor G I and Quinney H 1931 *Phil. Trans. R. Soc.* A **230** 323–62

Index